JOHN ROSS, Department of Chemistry, Massachusetts Institute of Technology, Cambridge, Massachusetts, U. S. A.

R. SCHECTER, Department of Chemical Engineering, University of Texas at Austin, Austin, Texas, U. S. A.

I. SHAVITT, Battelle Memorial Institute, Columbus, Ohio, U. S. A.

JAN STECKI, Institute of Physical Chemistry of the Polish Academy of Sciences, Warsaw, Poland

GEORGE SZASZ, General Electric Corporate R & D, Zurich, Switzerland

KAZUHISA TOMITA, Department of Physics, Faculty of Science, Kyoto University, Kyoto, Japan

M. V. VOLKENSTEIN, Institute of Molecular Biology, Academy of Science, Moscow, USSR

E. BRIGHT WILSON, Department of Chemistry, Harvard University, Cambridge, Massachusetts, U. S. A.

FOR ILYA PRIGOGINE

ADVANCES IN CHEMICAL PHYSICS

VOLUME XXXVIII

For Ilya Prigogine

Edited by STUART A. RICE

Department of Chemistry
and
The James Franck Institute
University of Chicago
Chicago, Illinois

VOLUME XXXVIII

AN INTERSCIENCE ® PUBLICATION

JOHN WILEY & SONS

NEW YORK · CHICHESTER · BRISBANE · TORONTO

AN INTERSCIENCE® PUBLICATION

Library of Congress Catalog Card Number: 58-9935
ISBN 0-471-03883-0

Printed in the United States of America

10 9 8 7 6 5 4 3 2 1

PREFACE

On January 25, 1977, Ilya Prigogine was 60 years old. This volume is a birthday present from some of his friends. It has not been my intent, nor that of the contributors, to assess or in any way comment on his work—that will stand on its own merits. It is our belief that the most rewarding form of recognition for a working scientist is to demonstrate his influence, both direct and indirect, on new science. That is what this volume attempts. Although Ilya Prigogine's broad interests overlap many of the subjects discussed, it is not for that reason alone that they were selected for inclusion. We hope that each article will be a significant contribution to the scientific literature and that, as this volume is used, the reader will be reminded, in a natural way, of the stimulation which Ilya Prigogine has provided to so many of us.

It is a delightful coincidence that this volume, conceived as a birthday present and form of public recognition of influence on science, should coincide with the award of the 1977 Nobel Prize for Chemistry, the most prestigious and most public form of acclaim within the scientific community, to Ilya Prigogine.

STUART A. RICE

Chicago, Illinois
March 1978

CONTRIBUTORS TO VOLUME XXXVIII

R. BALESCU, Faculté des Sciences, Université Libre de Bruxelles, Brussels, Belgium

MANFRED EIGEN, Max Planck Institut für Biophysikalische Chemie, Gottingen, Germany

T. ERNEUX, Faculté des Sciences, Université Libre de Bruxelles, Brussels, Belgium

A. GOLDBETER, Faculté des Sciences, Université Libre de Bruxelles, Brussels, Belgium

A. P. GRECOS, Faculté des Sciences, Université Libre de Bruxelles, Brussels, Belgium

P. HANUSSE, Department of Chemistry, Massachusetts Institute of Technology, Cambridge, Massachusetts

M. HERSCHKOWITZ-KAUFMAN, Faculté des Sciences, Université Libre de Bruxelles, Brussels, Belgium

B. HESS, Max Planck Institut für Ernährungsphysiologie, Dortmund, Germany

YU. L. KLIMONTOVICH, Moscow State University, Moscow, USSR

JOEL L. LEBOWITZ, Service de Physique Theorique, Gif-sur-Yvette, France

R. LEFEVER, Faculté des Sciences, Université Libre de Bruxelles, Brussels, Belgium

U. MÜLLER-HEROLD, Laboratory of Physical Chemistry, Swiss Federal Institute of Technology, Zurich, Switzerland

A. D. NAZAREA, University of Texas, Center for Statistical Mechanics and Thermodynamics, Austin, Texas

G. NICOLIS, Faculté des Sciences, Université Libre de Bruxelles, Brussels, Belgium

P. ORTOLEVA, Department of Chemistry, Indiana University, Bloomington, Indiana

H. PRIMAS, Laboratory of Physical Chemistry, Swiss Federal Institute of Technology, Zurich, Switzerland

P. RÉSIBOIS, Faculté des Sciences, Université Libre de Bruxelles, Brussels, Belgium

JOHN ROSS, Department of Chemistry, Massachusetts Institute of Technology, Cambridge, Massachusetts

HERBERT SPOHN, Belfer Graduate School of Science, Yeshiva University, New York

INTRODUCTION

Few of us can any longer keep up with the flood of scientific literature, even in specialized subfields. Any attempt to do more, and be broadly educated with respect to a large domain of science, has the appearance of tilting at windmills. Yet the synthesis of ideas drawn from different subjects into new, powerful, general concepts is as valuable as ever, thus the desire to remain educated persists in all scientists. This series, *Advances in Chemical Physics*, is devoted to helping the reader obtain general information about a wide variety of topics in chemical physics, which field we interpret very broadly. Our intent is to have experts present comprehensive analyses of subjects of interest, and to encourage the expression of individual points of view. We hope that this approach to the presentation of an overview of a subject will both stimulate new research and serve as a personalized learning text for beginners in a field.

ILYA PRIGOGINE
STUART A. RICE

CONTENTS

FOR ILYA PRIGOGINE

ADVANCES IN CHEMICAL PHYSICS

VOLUME XXXVIII

QUANTUM-MECHANICAL SYSTEM THEORY: A UNIFYING FRAMEWORK FOR OBSERVATIONS AND STOCHASTIC PROCESSES IN QUANTUM MECHANICS

H. PRIMAS and U. MÜLLER-HEROLD

Laboratory of Physical Chemistry
Swiss Federal Institute of Technology
Zurich, Switzerland

CONTENTS

I. QUANTUM-MECHANICAL SYSTEM THEORY VERSUS TRADITIONAL QUANTUM MECHANICS

A. Theories as Guiding Principles for the Experimenter

It is strange, but a fact, that we know how to do very sophisticated experiments and that we are able to explore very successfully the structure of matter even though we do not have a satisfactory theory of measurements in quantum mechanics. During the heroic period of the late 1920s the measuring problem of quantum mechanics was discussed in great depth and the probabilistic interpretation of quantum mechanics was the result of a careful analysis of the measuring process. However, in spite of this great effort expended during the early stage of quantum mechanics, the measuring problem was not solved but only clearly posed by Johann von Neumann in his famous text of 1932.[1] The problems of the measuring process are related to the difficulties in the interpretation of nonrelativistic quantum mechanics and have been discussed extensively in a great number of

1

papers, which can at least partially be traced from two excellent books by Jammer.[2,3] During recent years the research on the foundations of quantum mechanics has made great progress; in particular the measuring problem of quantum mechanics has been essentially solved in a mathematically rigorous manner by Grib[4] and Hepp.[5] Characteristically, the experimenters are not at all impressed by these developments—a good indication that the measuring theory of quantum mechanics is not the crucial link between theory and experience.

In spite of its operationalistic jargon ("observables," "measurement of the first kind") and notwithstanding the immense literature on the so-called measurement problem of quantum mechanics, the genuine questions of the relation between quantum theory and experiment are only scantily discussed, and traditional quantum mechanics does not give many theoretically well-founded guidelines for the experimenter. In fact, there is still no fundamental theory of actual measuring instruments. In spite of many assertions to the contrary, we would like to stress that no experimenter is performing a measurement of the first kind—every actual experiment (including the Stern–Gerlach experiment!) is tremendously more complicated. On the other hand, no experimenter cares about the specific nature of the initial state of an experiment, presupposed to be a product state by the measuring theory of traditional quantum mechanics.

In many respects the foundations of traditional quantum mechanics are superseded by the de facto applications of quantum mechanics in experimental science. It is therefore no longer possible to base the statistical interpretation of quantum mechanics on the projection postulate, which presupposes the feasibility of measurements of the first kind. On the other hand, theoreticians interested in the foundations have to face the modern experimental techniques in which the analysis of stochastic processes plays the crucial role.

Quantum-mechanical system theory deals with (essentially) nonrelativistic systems of any kind, microscopic or macroscopic. It replaces the controversial theory of the measuring process by a much more general theory of directly perceptible stochastic processes. Because no decomposition into a microsystem and an observing system is presupposed, the system concept includes always at least the essential parts of the observational tools.

For traditional quantum mechanics the Stern–Gerlach experiment has been paradigmatic. We prefer Felix Bloch's[6] system-theoretical description of nuclear magnetic resonance as a paradigm for quantum-mechanical system theory. Although the theory of measurements of the first kind gives no hint at all how to design and to improve the Stern–Gerlach experiment, the Bloch equations have a most useful intermediate position. On the one hand, they actually allow the experimenter to design and to refine his experiment; on the other hand, they are theoretically meaningful and can be rigorously related to the first principles of the theory of matter.

B. Limitations of Traditional Quantum Mechanics

Notwithstanding its immense success in describing atomic and molecular phenomena, traditional quantum mechanics (as formulated by Dirac[7] in 1930 and by Neumann[1] in 1932) suffers from several limitations. The most serious deficiency of traditional quantum mechanics is its inability to deal in a candid way with classical systems. The notorious measurement problem of quantum mechanics is a particularly well-known example of this disease. Because this measuring problem has historically been given excessive weight, it has pushed into the background many related problems of a greater practical importance like that of a simultaneous and consistent theoretical description of both the purely quantal and the purely classical properties in the very same object. Most objects of chemistry and molecular biology are of this type, involving both microphysical events and classical properties in an essential way. As a paradigmatic example one could take a biomolecule with its molecular stability, its photochemical properties, its primary, secondary, and tertiary structure.

The limitations of traditional quantum mechanics have been overcome by some ad hoc assumptions. One of these is the claim that the unitary time evolution (as given by the time-dependent Schrödinger equation) is valid only if the system is not under observation, and that a measurement induces a stochastic time evolution whose result is summarized by the so-called reduction of the wave packet, or, more precisely, by the projection postulate. Nowadays the limited validity of Neumann's projection postulate is generally recognized. The conditional expectations based on the projection postulate only exist for observables with a discrete spectrum,[8,9] so that traditional quantum mechanics whose statistical interpretation is based on the projection postulate is not qualified for a theory of continuous quantal stochastic processes.[10]

The need for a generalization of traditional quantum mechanics has been stressed not so much by the philosophers of science as (not surprisingly for the experts) by the engineers. Quantum electronics has one of the best-developed measurement theories, created in the context of detection and estimation problems in radar systems and optical communication systems with lasers.[11-14] For example, the simultaneous measurement of noncommuting observables is not a metaphysical but a practical problem in engineering quantum electronics, and it is known that the technique of optical heterodyning is equivalent to an optimal measurement of the Schrödinger pair (P, Q) or, equivalently, to an optimal measurement of the photon annihilation operator $Q + iP$.[15-18] These practical examples show perhaps better than anything else that the traditional concept of a measurement of the first kind is not only unrealistic but also too restrictive.

There is another, less fundamental but nevertheless severe limitation of traditional quantum mechanics: It deals only with systems having finitely many degrees of freedom. This limitation is severe because every molecular system interacts with its own electromagnetic radiation field. This field is a system with infinitely many degrees of freedom whose correct mathematical description goes beyond the formalism of traditional quantum mechanics. This is only a technical problem whose solution is well known today. However, the correct inclusion of the radiation field (in an essentially nonrelativistic approximation) can exhibit quite dramatic effects, such as molecular phase transitions and emergence of classical observables, effects that cannot be included in a coherent manner in the formalism of traditional quantum mechanics.

C. Generalized Quantum Mechanics

By generalized quantum mechanics we mean a mathematically rigorous and conceptually well-founded theory of molecular matter that contains the formalism of traditional quantum mechanics as a special case and whose framework is broad enough to apply to *all* forms of molecular matter, including its interaction with the radiation field. In particular, in the range of their validity the classical physical theories and quantum theories should be treated on an equal footing. Such a unified description contains the theories of microsystems and the theory of macrosystems as special cases, so that it becomes possible to include the so-called measuring system into the theoretical formalism and to treat the entire observed system in accordance with a unitary time evolution.

Generalized quantum mechanics is not a radical departure of traditional quantum mechanics but a genuine generalization that can be obtained by eliminating some of the postulates of traditional quantum mechanics that are neither well founded nor necessary. In particular, the correspondence principle, the projection postulate, and the requirements of irreducibility and atomicity of the algebras of observables can be dropped. The following examples may illustrate the necessity for and the feasibility of such generalizations:

1. There exist systems having discrete or continuous superselection rules.[19-24]
2. There exist systems whose classical behavior is not due to some macroscopic properties.[25]

3. There exist quantum stochastic processes.[10]
4. There exist quantal systems whose algebra of observables is of type III, hence nonatomic (e.g., thermodynamic systems).[26]

During the last decade considerable effort has been directed toward a unified, general, and mathematically rigorous description of both quantal and classical systems. The introduction of highly elaborate functional-analytic methods into quantum mechanics, especially in its B^*-algebraic version, has given us most powerful tools to discuss finite and infinite quantal systems. But tools are not enough; we also need clear conceptual definitions of the most fundamental terms, such as *system*, *state*, and *observable*. Conceptual definitions depend on the interpretation adopted for the mathematical formalism of the theory. The statistical interpretation of traditional quantum mechanics[1] and of algebraic quantum mechanics[5] is based on the projection postulate, which we do not want to accept because of its limited validity. Therefore we adopt a more general individual interpretation where the referent of the theory is one single system in contrast to the fictitious ensemble of the statistical interpretations. Such an approach has become possible because of the modern developments in quantum logics.

Quantum logics goes back to a suggestion by Johann von Neu-mann[1,sec.III.5] to consider the projection operators in the Hilbert space of state vectors as propositions, interpreting their eigenvalues 1 and 0 as the truth values *true* and *false*. This idea was developed into a lattice-theoretical formulation of quantum mechanics by Birkhoff and Neumann.[27] The modern development of quantum logics is well summarized in the texts by Jauch,[28] Varadarajan,[29] Gudder,[30] and Piron.[31] An attractive feature of the quantum logics approach is that it provides us with a deep insight into the conceptual structure of the theories of matter.

In contrast to the mentioned pioneering work in quantum logics, we use the more recent scheme initiated by Jauch and Piron[32] in which only propositions are used as primitive entities. This approach allows a state concept that applies to individual systems. It is notable that probabilities in the usual sense do not enter into such a foundation of quantum mechanics. The traditional statistical interpretation can, however, be derived as a rather special case of this more general approach.

D. Pitfalls We Have To Avoid

To develop a system-theoretical framework for a large class of molecular phenomena, we have to proceed in a formal and general way on the basis of the most fundamental theory available. It is not easy to avoid historically well-established preconceptions and empirically well-founded ad hoc rules.

For these reasons we collect some concepts that—in contrast to other approaches—we will *not* admit as *first principles*.

1. Semigroup Time Evolution. It is assumed that the fundamental laws of physics are all without exception symmetrical with respect to the time direction. The nonanticipatory nature of many engineering systems is a fact but the question of the impossibility of a backward causation cannot be settled by armchair philosophers.[33] Although the description of the time evolution of open systems by semigroups is a powerful tool having a mathematically sound basis,[34] we prefer not to suppress the physical problem of the "arrow of time" and to work always with a time evolution *group*.

2. Oversimplified Operationalism. The explications of the fundamental concepts *observables* and *states* in traditional quantum mechanics[1,7] are neither conceptually clear beyond any doubt nor sufficient for modern experimental science. The repeatability hypothesis should be eliminated from the first principles.[8] Accordingly, a measurement or observation theory must not be founded on the projection postulate. A realistic theory of measurements should not be the basis but the final result of theory.

3. Equivalence of the Schrödinger and the Heisenberg Picture. The equivalence between the Schrödinger picture (time evolution as an automorphism on the set of all states) and the Heisenberg picture (time evolution as an automorphism on the set of all observables) is valid only under very restrictive assumptions that we do not intend to make in general. We base the theory on a state description, so that we work always in the Schrödinger picture.

4. Statistical Interpretation. Quantum mechanics is intrinsically a probabilistic theory. Nevertheless it is possible to derive the statistical frequency interpretation from a formulation of quantum mechanics that avoids the concept of probability altogether. In an a priori statistical interpretation it is difficult to avoid circular arguments (e.g., the derivation of the projection postulate in an interpretation based just on this postulate) and to achieve a conceptually transparent mathematical construction of quantal stochastic processes.

5. Thermodynamics. On the most fundamental level, such concepts as *temperature, canonical ensemble, heat bath, entropy*, and *irreversibility* are not explained. They have to be derived in a conceptually irreproachable and mathematically rigorous way from the first principles. Subjective concepts (such as lack of knowledge, information loss) and approximations or ad hoc averaging methods (such as coarse graining, phase averaging) have to be strictly avoided.

E. Preview

To avoid the pitfalls mentioned and to overcome the limitations of older approaches in quantum theory, we have to pay careful attention to the foundations of generalized quantum mechanics. For this reason we formulate in Section II the basic assumptions as clearly as possible and without undue restrictions. We would like to stress that this section does *not* give a full axiomatization of generalized quantum mechanics—this can be done but it is not our intention. The aim is only to give conceptually clear and mathematically sound definitions of the crucial concepts *time, system, state,* and *observation*. This discussion requires knowledge of some elementary facts of lattice theory, which are summarized in an appendix to Section II.

Although a lattice-theoretical formulation of quantum mechanics provides conceptual simplicity and generality, it is not very convenient for the mathematical development of the theory and its practical applications. We feel forced to use a convenient Hilbert-space embedding of quantum logics. The main postulates of the Hilbert-space model of generalized quantum mechanics are stated in Section III. This formulation encompasses algebraic quantum mechanics, systems with discrete or continuous, commutative or noncommutative superselection rules and admits a very general state concept for a single system.

The statistical interpretation of quantum mechanics is derived from the individual interpretation in Section IV. As a rather special example, measurements of the first kind are discussed and it is shown that they imply the usual statistical interpretation of traditional quantum mechanics. A straightforward generalization leads to a more useful measurement concept, based on probability operator measures.

The dynamics of directly perceptible output processes of an arbitrary quantal system is discussed in Section V. The mathematical characterization of these stochastic output processes is the heart of the intrinsic observation theory, which makes no reference to measurement instruments and therefrom distinct measured objects. It is shown that these processes can quite generally be characterized by probability operator measures.

Section VI has a more programmatic character. We present a classification of directly observable stochastic processes induced by the dynamics of quantal systems. This classification is of interest for the discussion of such concepts as *determinism, irreversibility, evolution,* and *finality*. The close relations of the theory of directly perceptible stochastic processes in quantal systems to modern ergodic theory and topological dynamics[35] are not discussed but should be in evidence for the experts.

II. THE LOGICAL AND CONCEPTUAL STRUCTURE OF GENERALIZED QUANTUM MECHANICS OF SINGLE SYSTEMS

A. Primitive Concepts

To avoid vicious circles, every logically consistent theory must use undefined terms (so-called primitive concepts) and unproved premises (the axioms of the theory).[36] All we can require is that we have as few of such primitive concepts and axioms as possible, that the primitive concepts be elucidated in a metalanguage, and that the axioms be intuitively appealing. It would be mistaken to expect that the axioms can be derived from experience or that primitive concepts have a direct operational meaning. We do not accept the operationalist thesis that every scientifically meaningful concept must be capable of full definition in terms of performable physical operations, but we insist that it is essential to distinguish between meaning and method of testing. What a measuring instrument must look like cannot be incorporated into the postulates of a fundamental theory. The crucial contact between theory and experiment can only be made when the theory is well developed.

We use as primitive concepts the notion of a *universal time*, the abstraction of a *universe of discourse* and its *qualities*. The notions of systems, states, and observables are not taken as primitive. Furthermore, the existence of a three-dimensional physical space is not included as a fundamental axiom so that all our concepts will be independent of it. If necessary, it can be introduced at a later stage via an appropriate projective representation of the Galilei group.

B. Universal Time

Time is represented by the linearly ordered real axis $(\mathbb{R}, \leq)$ where every $t \in \mathbb{R}$ represents an instant of time and the ordering $\leq$ means the relation of "being prior to or simultaneous with." The direction of time is assumed to be universal in the sense that the preferred direction of time must not be tied to special molecular systems or molecular processes.

We do not base the anisotropy of time on thermodynamic arguments because we do not presuppose the phenomenological entropy law of classical thermodynamics. The arrow of time is assumed to be a large-scale feature of our world. It may be of electromagnetic origin because the universe is in fact an almost perfect sink for radiation, or it may be of cosmological origin because the universe is expanding.[37,38] There exist cosmological models that can explain the nonexistence of contracting electromagnetic waves collapsing inward from infinity,[39] hence justifying the traditional use of retarded potentials in electrodynamics. Because every molecular system interacts with its radiation field, the predominantly dissipative behavior of molecular matter may be traced back to the electromagnetic arrow of time, which in turn may be generated by the cosmological time asymmetry. However, in discussing molecular systems, we have also to take into account the choice of initial conditions. It may be difficult to prepare initial states for which molecular systems show an easily observable antidissipative behavior; nevertheless such a possibility cannot be excluded.

The empirical fact that most spontaneous processes in well-isolated molecular systems increase the entropy in the direction of positive universal time neither conflicts with the fundamental laws of physics (which are symmetrical to both time directions) nor reflects a law of nature. The time anisotropy of molecular systems is due to the de facto conditions, such as boundary conditions and initial conditions. The problem of understanding the boundary conditions is essentially one of understanding how molecular systems are related to the rest of the world. Since the radiation field of a molecular system can never be completely isolated from the rest of the world, the electromagnetic arrow of time has always to be used as a boundary condition. However, from this boundary condition alone the thermodynamic arrow of time does not yet follow; the factlike initial conditions are of equal importance.

The temporally asymmetric character of the evolution of some molecular systems has to be compared with the arrow of universal time. The factlike character of the initial conditions leaves it open whether a particular molecular process is associated with an increase or decrease of entropy, and this has a number of important consequences. A priori it is by no means plain that every molecular stochastic process is necessarily past-determined and nonanticipatory. Only a detailed mathematical discussion can elucidate the conditions under which a statistical retrodiction of the past or a statistical prediction of the future is possible. *The asymmetry of retrodictability and predictability in a particular molecular system is not entirely fixed by the arrow of universal time*; it also depends on the initial conditions and on the chosen point of view (i.e., on the chosen algebra of observables).

C. Systems and State of a System

The notion of a system is a rather difficult one. The traditional meaning as a part of the physical universe that we have isolated for separate study is a useful guideline but has to be used with great care because there exist entangled systems (in the sense of Schrödinger[40,41]) that can have nonclassical correlations (in the sense of Einstein, Podolsky, and Rosen[42]) even for spatially well-separated subsystems. It would be mistaken only to *imagine* that we have isolated a part of the world; we have to *perform* this isolation. Isolation is a process that can destroy holistic properties.

The primitive concept of the *universe of discourse* is an inclusive class containing all those entities with which the theories under consideration are concerned. The proper choice of the universe of discourse is a difficult empirical question; a bad choice will lead to useless or empirically wrong theories. In any case, the universe of discourse has to be chosen to be large enough so as to encompass all known phenomena concerning the object under investigation. Furthermore, the universe has to be large enough to allow the bidirectionally deterministic time evolution required by Axiom 4a, discussed in Section II.G. The inherent traits of the universe of discourse are here called *qualities*. In elaborating the theory this concept is operationalized and some of the qualities are identified with directly observable properties. On the other hand, all feasible observable properties of an object under investigation are represented by qualities.

Instead of using qualities it is technically more convenient to operate with statements concerning the qualities. Such a statement is called a proposition; it may be *true*, or *false*, or *neuter*.

Example. "Beethoven's fifth symphony is in C minor" is a true proposition; "Beethoven's fifth symphony has five movements" is a false statement; "Beethoven's fifth symphony is green" is a meaningless proposition, it is neither true nor false and has to be taken as neuter.

To every quality f we associate a *proposition F*, and if the universe of discourse has the quality f we say that the proposition F is true. If a quality f implies the quality g, we write $F \le G$,

$$F \le G \quad \text{iff "F is true"} \quad \text{implies "G is true"} \tag{1}$$

Two qualities that imply each other are *equal*, and we write

$$F = G \quad \text{iff } F \le G \quad \text{and} \quad G \le F \tag{2}$$

A quality that is not implied by another quality is called an *atomic quality* and the corresponding proposition an *atomic proposition*. To every proposition F we associate another proposition $F^{\perp}$ such that whenever F is true then $F^{\perp}$ is false, and if F is false then $F^{\perp}$ is true,

$$\text{"F is true"} \quad \text{implies "$F^{\perp}$ is false"} \tag{3}$$

It is characteristic for quantal systems that there exist incompatible propositions; thus we do not assume that all propositions one can set up are either true or false. Saying that a particular proposition is true does not imply that the corresponding quality can be verified by an appropriate single experiment because one cannot assume a priori that a corroboration is reproducible or due to mere chance. However, after a full elaboration and operationalization of the theory, one can show that it is possible to refute a proposition by a single experiment and to corroborate statistically a proposition by a large number of experiments.

In the set of all feasible propositions of the universe of discourse there are two quite distinct kinds of propositions: temporal propositions and timeless propositions. The *temporal propositions* refer to a certain time; they can be true now and false or neuter at another time. The *timeless propositions* are true or false without any reference to time. The timeless propositions characterize the *system*, whereas the temporal propositions characterize the *state* of the system. Accordingly, we define the notions of a *system* and of a *state of a system* in terms of the primitive concepts *time*, *universe of discourse*, and *quality* as follows:

Definition: System (4)
A system is the set of all timeless propositions of the universe of discourse that are true.

Note that this definition neither supposes nor implies that a system is in some way an aggregation of objects or that component units exist.

Definition: State (5)
A state of a system at time t is the set of all temporal propositions of the universe of discourse that are true at time t.

Essentially the same state concept* has been proposed independently by Scheibe,[43,44] Hartle,[45] and Jauch and Piron.[32] It applies to quantal as well as to classical systems. It is important that the definition (5) does not involve the notion of probability so that it defines a state of one individual system. In contrast, a statistical state refers to a fictitious infinite ensemble of identically prepared but uncorrelated individual systems. Note that we do *not* assume that a state can necessarily be prepared by means of physical operations. However, by definition, any system is at any time in a definite state. Such a description using an instantaneous state is often referred to as the *Schrödinger picture*.

D. The Multilevel Aspect of Natural Phenomena

In every scientific description of nature we have to disregard certain features. Every formal treatment depends in a crucial way on abstractions

---

* There are some inexactitudes in the important papers by Hartle[45] and Jauch and Piron[32] because they do not explicitly distinguish between temporal and timeless propositions. Jauch and Piron[32] sometimes say "system" when they mean "system in a specified state." In contradistinction to Jauch[46] we accept definition (5) to be valid under any circumstances and accept as an empirical fact that there are spatially well-separated systems that are correlated in the sense of Einstein, Podolsky, and Rosen.[42]

that are not yet given by choosing the universe of discourse but that are determined by our interest and the experimental method chosen. For example, for some objects it is possible to give either a quasither-modynamic description, or a hydrodynamic description, or a more detailed kinetic description, or an even more detailed molecular description. Which description we have to use depends on the questions we ask. Every such logically consistent description will be called a *theory* of the object under discussion. The various theories of an object allow one to work at very different levels of description and with very different purposes in mind. A particular theory is characterized by a restriction of the admitted theoretical propositions. The set of all temporal propositions admitted for a particular theoretical description is called the domain $\mathscr{L}$ of the theory. If the domain $\mathscr{L}$ equals the set of *all* temporal propositions of the universe of discourse, we call the corresponding theory the *universal theory* and the corresponding domain $\mathscr{U}$. A theory with domain $\mathscr{L} \subset \mathscr{U}$ is called a *subtheory* of the universal theory.

The special nature of a subtheory is determined by the algebraic structure of its propositional system $\mathscr{L}$. From this point of view, generalized quantum mechanics is a *temporal logic* of the propositions about the qualities of a physical system of any kind, rather than a generalization of Newtonian mechanics.

E.   The Structure of Propositional Systems

In all prevailing scientific theories the timeless propositions are assumed to fulfill the rules of the usual two-valued logic. Therefore we postulate

*Axiom 1: **Timeless Propositions*** (6)
The set of all timeless propositions of the universe of discourse form a Boolean lattice that is interpreted in the usual way as a classical propositional logic where the lattice operations join, meet, and orthocomplement correspond to the logical operations disjunction, conjunction, and negation.

Much more delicate is the question of the appropriate algebraic structure of the set $\mathscr{L}$ of all temporal propositions. Since the pioneering work of Birkhoff and Neumann[27] in 1936 it has been known that the logic of quantum mechanics is not Boolean. That is, a temporal proposition is not in general either true or false so that the classical two-valued logic does not apply but must be replaced by another less restricted logic. The propositional system of traditional quantum mechanics (as formalized by Neumann[1] in 1932) is a complete, irreducible, atomic, semimodular, and orthomodular lattice. Generalized quantum mechanics admits classical observables by dropping the postulate of irreducibility. A further generalization admits infinite systems and thermodynamic equilibrium systems with their associated type II and type III algebras of observables by dropping the postulate of atomicity (compare, for example, the text by Emch[26]). The semimodularity is related to the projection postulate and could be dropped if necessary. Accordingly, the minimal algebraic structure for a workable theory seems to be a complete othomodular lattice. The modern work on the foundation of quantum logic has indeed good and intuitively appealing arguments for such a choice.[28-31,47-49] Moreover, a reasonable interpretation of $\mathscr{L}$ as a logic requires that $\mathscr{L}$ admits an implication connective so that a *modus ponens* inference schema is possible. Every orthomodular lattice allows such a Stalnaker conditional and, on the other hand, every quantum implication algebra is an orthomodular lattice.[50-54] For these reasons we postulate

Axiom 2: **Logic $\mathscr{L}$**   (7)
The set of all temporal propositions of a theory is a complete orthomodular lattice where the lattice-theoretical partial ordering is interpreted as logical implication and the orthocomplementation as logical negation.

Orthomodular lattices are nondistributive generalizations of Boolean lattices. Examples of orthomodular lattices include all Boolean lattices, the lattice of all closed subspaces of a Hilbert space, and the lattice of the self-adjoint idempotents in a W^*-algebra. However, all these well-known examples have an additional structure; they are all also semimodular lattices.[55]

F. The Interpretation of Propositional Systems

Two propositions F, G of an orthomodular lattice $\mathscr{L}$ that satisfy the relation

$$(F \wedge G^{\perp}) \vee G = (G \wedge F^{\perp}) \vee F \tag{8}$$

are called *compatible*, written as $F \leftrightarrow G$. In an orthomodular lattice the binary relation $\leftrightarrow$ is reflexive and symmetric. The lattice-theoretical notion of compatibility is conceptually meaningful because in every orthomodular lattice we have

$$F \leq G \quad \text{implies } F \leftrightarrow G \tag{9}$$

That is, if the proposition F logically entails the proposition G, then F and G are two compatible propositions. Moreover, for compatible propositions it is possible to extend the interpretation in the sense of the classical propositional logic because an orthomodular lattice is *Boolean* if and only if all its elements are compatible,

$$\mathscr{L} \text{ is Boolean} \quad \text{iff } F \leftrightarrow G \quad \forall F, G \in \mathscr{L} \tag{10}$$

The simplest Boolean lattice is the *Boolean algebra of truth values* with the two elements 0 and 1 and the algebraic rules

$$0+0=0 \qquad 0+1=1 \qquad 1+1=0$$

$$0 \cdot 0 = 0 \qquad 0 \cdot 1 = 0 \qquad 1 \cdot 1 = 1$$

$$0^{\perp} = 1 \qquad 1^{\perp} = 0$$

Every Boolean lattice $\mathscr{B}$ can be interpreted as a two-valued Boole–Whitehead propositional calculus where the truth values τ,

$$\tau(F) = 1 \qquad \text{iff } F \text{ is true}$$

$$\tau(F) = 0 \qquad \text{iff } F \text{ is false}$$

induce a homomorphism τ of $\mathscr{B}$ into the Boolean algebra of truth values:

$$\tau(E) = 1$$

$$\tau(0) = 0$$

$$\tau(F \vee G) = \tau(F) + \tau(G)$$

$$\tau(F \wedge G) = \tau(F) \cdot \tau(G)$$

$$\tau(F^{\perp}) = \{\tau(F)\}^{\perp}$$

where E is the largest element of $\mathscr{B}$ and 0 is the smallest element of $\mathscr{B}$. Accordingly, in a Boolean lattice $\mathscr{B}$ each proposition is either true or false (law of the excluded middle), and we have the following interpretation of the lattice operation $\vee$ as the logical connective *or* and of the lattice operation $\wedge$ as the logical connective *and*. That is, if $F, G \in \mathscr{B}$, then

$$F \vee G \text{ is true} \quad \text{iff } F \text{ is true } or \text{ } G \text{ is true} \tag{11a}$$

$$F \wedge G \text{ is true} \quad \text{iff } F \text{ is true } and \text{ } G \text{ is true} \tag{11b}$$

Because of the existence of incompatible propositions in a general orthomodular lattice, these interpretations are possible *only* in a Boolean lattice.

The fact that every Boolean lattice can be interpreted as a classical propositional logic gives us the possibility of a full interpretation of quantum mechanics. We call a theory whose domain is a Boolean lattice a *classical subtheory* of the universal theory. To every classical subtheory we associate the traditional interpretation of the classical physical theories:

Axiom 3: Interpretation (12)

Let $\mathscr{U}$ be the set of all temporal propositions of the universe of discourse, and let $\mathscr{B} \subset \mathscr{U}$ be an arbitrary Boolean sublattice of the orthomodular lattice $\mathcal{U}$. Then in the classical subtheory with domain $\mathcal{B}$ the propositions of $\mathcal{B}$ have to be interpreted by a two-valued truth function in the sense of classical propositional logic.

The lattice $\mathcal{U}$ of all temporal propositions of the universe of discourse is not Boolean, but it is generated by the family of all its Boolean sublattices. Therefore the orthomodular lattice $\mathcal{U}$ can also be interpreted as a partial Boolean algebra.

G. States and Their Time Evolution

According to Axiom 2, the domain $\mathcal{U}$ of the universal theory is a complete lattice so that every nonvoid subset of $\mathcal{U}$ has an infimum in $\mathcal{U}$. Let $\mathcal{F}_t \subset \mathcal{U}$ be the set of all temporal propositions of the universe of discourse that are true at time t; then the proposition S_t

$$S_t \stackrel{\text{def}}{=} \inf \mathcal{F}_t = \bigwedge_{F \in \mathcal{F}_t} F \tag{13a}$$

implies every proposition F that is true at time t, that is,

$$S_t \leq F \qquad \forall F \in \mathcal{F}_t \tag{13b}$$

Accordingly, we can represent a state in the sense of definition (5) by the single proposition S_t, so that we have the following theorem:

States as Propositions (14)
A state at time t is realized by the unique maximal proposition that implies all the temporal propositions that are true at time t.

The states are always related to the propositions of a theory. It would be unreasonable to admit states for which the propositions in the domain of a theory are undecidable in principle. If a proposition F in the domain $\mathcal{L}$ is compatible with the state S, it is sensible to require that the proposition S implies either that F is true or that F is false. Such states are here called $\mathcal{L}$-admissible; with respect to them all compatible propositions of the domain $\mathcal{L}$ are decidable.

Definition: $\mathcal{L}$-Admissible States (15)

$$\text{If} \begin{Bmatrix} F \in \mathcal{L} \\ \text{and } F \leftrightarrow S \end{Bmatrix} \quad \text{implies} \quad \begin{Bmatrix} \text{either } S \leq F \\ \text{or } S \leq F^{\perp} \end{Bmatrix}$$

then the state $S \in \mathcal{U}$ is called $\mathcal{L}$-admissible.

In the following, all states of a theory with domain $\mathcal{L}$ are tacitly assumed to be $\mathcal{L}$-admissible.

A proposition is called *atomic* if it is not implied by a nontrivial proposition different from itself. An atomic proposition can be interpreted as representing an ultimate quality. Correspondingly, a state S is called atomic if S is an atom, that is,

$$S \text{ is an atomic state} \quad \text{iff} \quad 0 \le R \le S \quad \text{implies } R = 0 \text{ or } R = S \quad (16)$$

A lattice is said to be *continuous* if it contains no atoms. A lattice is called *atomic* if every nonzero element of the lattice contains an atom. Note that there are lattices that are neither continuous nor atomic, and that the atoms of a sublattice $\mathscr{L} \subset \mathscr{U}$ are not necessarily atoms in $\mathscr{U}$. If the domain $\mathscr{U}$ of the universal theory is atomic, the $\mathscr{U}$-admissible states are the atoms of $\mathscr{U}$.

The universe of discourse is assumed to represent a closed system. A system is called *closed* if all phenomena or variables that can influence the system have been taken into account in the initial specification.[56,57] More precisely, we assume that the state S_t of the universe of discourse is only a function of the initial state S_{t_0} and of the time difference $t - t_0$ for every time $t > t_0$ *and* $t < t_0$. The change of the state with the passage of the universal time is called the *time evolution*.

The fundamental assumption in any mechanical theory is that the time evolution is defined by symmetries. A *symmetry transformation* is an automorphism of the propositional system $\mathscr{L}$ of a theory that preserves its logical structure. For this it is necessary and sufficient that a symmetry is an automorphism of the orthomodular lattice $\mathscr{L}$, that is, a bijective mapping $\alpha : \mathscr{L} \to \mathscr{L}$ preserving the partial order $\le$ (i.e., the logical implication) and the orthocomplementation $\perp$ (i.e., the logical negation):[58,59]

$$\alpha : \mathscr{L} \to \mathscr{L} \tag{17a}$$

$$F \le G \quad \text{iff } \alpha(F) \le \alpha(G), \qquad \forall F, G \in \mathscr{L} \tag{17b}$$

$$\alpha(F^{\perp}) = \{\alpha(F)\}^{\perp}, \qquad \forall F \in \mathscr{L} \tag{17c}$$

In the usual elementary manner it follows that the set of all automorphisms of $\mathscr{L}$ is a group, called the *group Aut* $(\mathscr{L})$ *of automorphisms of* $\mathscr{L}$.

The time evolution is assumed to be given by a one-parameter group $\{\alpha_t : t \in \mathbb{R}\}$ defined as a subgroup of Aut $(\mathscr{U})$ and as a representation of the additive group $\mathbb{R}$

$$\alpha_{t_1} \cdot \alpha_{t_2} = \alpha_{t_1 + t_2}, \qquad t_1, t_2 \in \mathbb{R}, \qquad \alpha_t \in \text{Aut }(\mathscr{U}) \tag{18}$$

As the time evolution of a closed system is assumed to be bidirectionally deterministic, the state S_t at time t is given by

$$S_t = \alpha_{t - t_0}\{S_{t_0}\}, \qquad \forall t \in \mathbb{R} \tag{19}$$

where S_{t_0} is the initial state at an arbitrary time t_0. The time evolution is a property of the universe of discourse, hence not dependent on a particular point of view that determines the propositional system $\mathscr{L}$ of a subtheory. Accordingly, it is neither required nor generally true that the time evolution is an automorphism of $\mathscr{L} \neq \mathscr{U}$. To summarize, we have

Axiom 4: Time Evolution $\qquad\qquad\qquad\qquad\qquad\qquad\qquad$ (20)
a. The time evolution $S \to S_t$ of a state S of the universe of discourse is given by

$$S_t = \alpha_t(S), \qquad t \in \mathbb{R}$$

where $\{\alpha_t \in \text{Aut}(\mathscr{U}) : t \in \mathbb{R}\}$ is an Abelian one-parameter group $t \to \alpha_t$ of automorphisms of the universal propositional system $\mathscr{U}$.
b. The time evolution of a subtheory with the propositional system $\mathscr{L} \subset \mathscr{U}$ is given by the time evolution of the universal theory with the propositional system $\mathscr{U}$.

H. Observations of Quantal Systems

Empirically we know that there exist mutually exclusive experimental procedures so that there exist incompatible empirical propositions. Therefore the totality of all empirical propositions is not embedable into a Boolean lattice. Nevertheless science as we know it requires domains with a language based on the classical two-valued predicate logic. In particular, Niels Bohr has stressed not only the necessity to describe all experimental outcomes within the classical propositional calculus, but also the fact that the domain of validity of such a classical description is confined by the relevant conditions defining the experimental situation. We accept Bohr's[60] definition of the word *phenomenon* "to refer exclusively to observations obtained under specified circumstances, including an account of the whole experiment." For that reason we never consider an isolated microsystem without the relevant observational tools.

To every theory with the propositional system $\mathscr{L} \subset \mathscr{U}$ we can associate a unique classical subtheory that contains all mutually compatible propositions of $\mathscr{L}$. The domain of this classical subtheory is the *center* $\mathscr{Z}(\mathscr{L})$ of the orthomodular lattice $\mathscr{L}$

$$\mathscr{Z}(\mathscr{L}) \overset{\text{def}}{=} \{Z \in \mathscr{L} : Z \leftrightarrow F, \quad \forall F \in \mathscr{L}\} \qquad (21)$$

By (10), the center $\mathscr{Z}(\mathscr{L})$ of every orthomodular lattice is a Boolean lattice, and by Axiom 3 $\mathscr{Z}(\mathscr{L})$ has to be interpreted in the sense of a two-valued classical propositional logic. The propositions in $\mathscr{Z}(\mathscr{L})$ are called the *classical propositions* of the theory with the domain $\mathscr{L}$. Our final Axiom 5 claims that the classical propositions correspond to empirically decidable alternatives. Hence the link between theory and experiment is described by a mapping of an operationally specified empirical domain into the center of the classical propositions of the theory.

The classical propositions of a theory do not necessarily have a definite value. However, as we discuss in detail in Section III.G, every state can be decomposed uniquely into so-called primary states that are compatible with all classical propositions. Such states are truth-definite in the sense that every classical proposition is either true or false with respect to this state. Therefore we adopt:

Definition: **Primary States**   (22)

A state $S \in \mathcal{U}$ of the universe of discourse is called *primary* relative to a theory with domain $\mathcal{L}$ if it is truth-definite with respect to all classical propositions of $\mathcal{L}$, that is, if

$$S \le Z \quad \text{or} \quad S \le Z^{\perp}, \qquad \forall Z \in \mathcal{Z}(\mathcal{L})$$

We assume that the truth value of a classical proposition is empirically decidable. By taking appropriate precautions, we assume that it is possible to reduce the disturbances necessarily accompanying every such decision procedure to any desired level, so that we can state our final axiom as follows:

Axiom 5: **Observations**   (23)

For every classical proposition $Z \in \mathcal{Z}(\mathcal{L})$ of a theory with domain $\mathcal{L}$ there exist experimental methods such that for every primary state (with respect to $\mathcal{L}$) the truth value of Z can be determined with an arbitrarily small error of decision and with an arbitrarily small disturbance of the state.

Axiom 5 replaces the controversial concept of a measurement by the innocuous notion of an observation. In the framework of Neumann's[1] theory of measurements, a *measurement* of the first kind is a hypothetical procedure that

1. changes the state of the system such that
2. the final state is an eigenstate of the measured observable and
3. hence brings into existence the quality to be measured.

In contrast to this drastic intervention, we call a procedure that (*1*) only determines what is already the case and (*2*) can be performed without significant disturbance, an *observation*. That is, an observation is understood as the perception of something that objectively exists. That may be a difficult problem of applied science but does not present any deep problem like the measuring process of traditional quantum mechanics.

Appendix: Definitions and Some Elementary Results
from Lattice Theory

In this appendix we collect all the results concerning the theory of orthomodular and Boolean lattices we need; for a full account of lattice theory compare Birkhoff,[61] Holland,[62] and Maeda and Maeda.[63]

A *partially ordered set* (abbreviation: *poset*) is a pair $(\mathcal{L}, \leq)$, where $\mathcal{L}$ is a set equipped with a binary relation that fulfills for every $a, b, c \in \mathcal{L}$ the relations:

  (i) $a \leq a$ (reflexive)
  (ii) $a \leq b, b \leq a$ implies $a = b$ (antisymmetric)
  (iii) $a \leq b, b \leq c$ implies $a \leq c$ (transitive)

The relation $\leq$ is called a *partial ordering* on $\mathcal{L}$, it has to be defined only for *some* pairs of elements of a poset.

In every poset one can define the supremum (least upper bound) and the infimum (greatest lower bound). If the supremum of two elements $a, b \in \mathcal{L}$ exists, it is denoted by $a \vee b$

$$a \vee b = \sup \{a, b\}$$

and defined to be the unique element $c \in \mathcal{L}$ for which $a \leq c$, $b \leq c$, and if $a \leq d$ and $b \leq d$ then $c \leq d$. Dually, the infimum of two elements $a, b \in \mathcal{L}$, if it exists, is denoted by $a \wedge b$,

$$a \wedge b = \inf \{a, b\}$$

and defined to be the unique element $c \in \mathcal{L}$ such that $c \leq a$, $c \leq b$, and if $d \leq a$ and $d \leq b$ then $d \leq c$.

More generally, one writes

$$\bigvee_{i \in \mathfrak{I}} a_i \overset{\text{def}}{=} \sup_{i \in \mathfrak{I}} \{a_i\}$$

$$\bigwedge_{i \in \mathfrak{I}} a_i \overset{\text{def}}{=} \inf_{i \in \mathfrak{I}} \{a_i\}$$

A *lattice* $(\mathcal{L}, \wedge, \vee)$ is a partially ordered set $(\mathcal{L}, \leq)$ in which every pair of elements has a supremum, called join $\vee$, and an infimum, called meet $\wedge$. A lattice is called *complete* if every nonvoid subset has a supremum and an infimum. In a complete lattice there exists a largest element e and a smallest element o. An *orthomodular lattice* is a lattice with universal bounds o, e and an orthocomplementation $a \rightarrow a^{\perp}$ such that for every $a, b \in \mathcal{L}$

  (i) $(a^{\perp})^{\perp} = a$
  (ii) $a \wedge a^{\perp} = o, \qquad a \vee a^{\perp} = e$

(iii) $a \le b$ implies $b^\perp \le a^\perp$

(iv) $a \le b$ implies $b = a \vee (b \wedge a^\perp)$

If the infimum of a family $\{a_i\}$ in $\mathscr{L}$ exists, then so does the supremum, and $\vee a_i = (\wedge a_i^\perp)^\perp$. A subset of an orthomodular lattice $\mathscr{L}$ is called an *orthomodular sublattice* of $\mathscr{L}$ if it is a sublattice of $\mathscr{L}$ containing o and e and is closed under the orthocomplementation of $\mathscr{L}$. Two elements a, b from a lattice $(\mathscr{L}, \le)$ form a *modular pair* (a, b) if

$$(c \vee a) \wedge b = c \vee (a \wedge b) \qquad \text{for every } c \le b, c \in \mathscr{L}$$

A lattice $(\mathscr{L}, \le)$ is called *semimodular* if

$$(a, b) = \text{modular pair implies } (b, a) = \text{modular pair}$$

A nonzero element $a \in \mathscr{L}$ such that

$$o \le b \le a, \, b \in \mathscr{L} \text{ implies either } b = o \text{ or } b = a$$

is called an *atom*. A lattice $\mathscr{L}$ is called *atomic* if *every* nonzero element $b \in \mathscr{L}$ contains an atom a, that is, if $a \le b$. A lattice $\mathscr{L}$ is called *continuous* if it contains no atoms. Two elements a, b of an orthomodular lattice $\mathscr{L}$ are called compatible if the sublattice generated by $a, a^\perp, b, b^\perp$ is distributive, that is,

$$a \leftrightarrow b \quad \text{iff } a = (a \wedge b) \vee (a \wedge b^\perp)$$

An orthocomplemented lattice is orthomodular if and only if the compatibility relation $\leftrightarrow$ is symmetric. If $a \leftrightarrow b$, then $b \leftrightarrow a$, $a \leftrightarrow b^\perp$, $a^\perp \leftrightarrow b$, and $a^\perp \leftrightarrow b^\perp$. If $a \le b$, then $a \leftrightarrow b$. An orthomodular lattice is called Boolean if all its elements are mutually compatible,

$$\mathscr{L} \text{ is Boolean} \quad \text{iff } a \leftrightarrow b \qquad \text{for all } a, b \in \mathscr{L}.$$

III.   THE HILBERT-SPACE MODEL OF GENERALIZED QUANTUM MECHANICS

A.   Why a Hilbert-Space Model?

Although the lattice-theoretical formulation provides a good approach to the conceptual and logical problems of quantum theory, it is not so convenient for further development as formulations using more familiar tools from functional analysis. In particular, a general and well-developed theory of group representations in orthomodular lattices is not yet available. This is the main reason why a development of a workable quantum theory in the abstract lattice-theoretical setting is difficult so that one embeds the lattice of propositions into a richer structure where more analytical tools are available. We prefer this point of view (which resembles the use of group algebras in group theory) over the more popular attempts to find isomorphic realizations of the lattice of propositions by postulating additional axioms, because such new axioms (e.g., atomicity and semi-modularity) seem to be of questionable conceptual significance.

As a concrete model for the orthomodular lattice $\mathscr{L}$ of quantum logic we use the Hilbert-space model, where the lattice $\mathscr{L}$ is realized by a particular orthomodular sublattice of the lattice of all subspaces of a separable Hilbert space over the complex numbers. Equivalently but more conveniently, we use the projection lattice of a Neumann algebra of operators acting on this Hilbert space. This model is quite general and does *not* exclude the study of infinite systems such as fields.*

B. Realization of the Propositional System $\mathscr{L}$ as a Projection Lattice

To begin with, we introduce some notations and definitions. We denote by $\mathfrak{H}$ a separable complex Hilbert space with the inner product of vectors Φ, $\Psi \in \mathfrak{H}$ written as $\langle \Phi | \Psi \rangle$, linear in Ψ. The norm $\| \cdot \|$ induced by the inner product is given by $\|\Psi\|^2 = \langle \Psi | \Psi \rangle$. A linear operator A acting on $\mathfrak{H}$ is called *bounded* if the set $\{\|A\Psi\| : \Psi \in \mathfrak{H}, \|\Psi\| = 1\}$ is bounded. The supremum of this set is called the norm of A and will be denoted by $\|A\|$.

The algebra of all bounded linear operators acting on $\mathfrak{H}$ is denoted by $\mathfrak{B}(\mathfrak{H})$ or simply by $\mathfrak{B}$. If $A \in \mathfrak{B}(\mathfrak{H})$, then A^* denotes the adjoint of A, that is, the unique operator $A^* \in \mathfrak{B}$ such that $\langle \Phi | A\Psi \rangle = \langle A^*\Phi | \Psi \rangle$ for all vectors Φ, $\Psi \in \mathfrak{H}$.

If $A = A^*$, then A is called self-adjoint. The projection F onto a closed subspace $\mathfrak{J} \subset \mathfrak{H}$ is an idempotent and self-adjoint operator, and we write

$$\mathfrak{F} = F\mathfrak{H}, \qquad F = F^2 = F^* \in \mathfrak{B}(\mathfrak{H}) \tag{1}$$

Clearly, the norm of every projection equals one. The orthogonal complement of a closed subspace $\mathfrak{F}$ is denoted by $\mathfrak{F}^\perp$, and the projection

---

* There are other realizations of generalized quantum mechanics, the best known being the C^*-algebraic approach.[64–68] In general, a C^*-algebra does not contain enough projections to form a lattice so that it is not an appropriate structure to embed a quantum logic. Nevertheless, the Hilbert-space model we consider does include the C^*-algebraic approach in the following sense. Algebraic quantum mechanics starts with an abstract B^*-algebra $\mathfrak{A}$ of observables. From this algebraic realization of quantum mechanics we can get the corresponding Hilbert-space model by taking as Hilbert space the representing Hilbert-space $\mathfrak{H}$ of the universal representation $(\pi, \mathfrak{H})$ of the B^*-algebra $\mathfrak{A}$. The enveloping Neumann algebra $\{\pi,(\mathfrak{A})\}''$ is then isomorphic (as a Banach space) to the second dual of $\mathfrak{A}$ (compare Dixmier[69,par. 12]). Using this relationship, we can take advantage of the mathematically highly developed techniques of algebraic quantum mechanics such as the GNS-construction, which is a powerful tool for the discussion of systems having infinitely many degrees of freedom.

onto $\mathfrak{F}^{\perp}$ by $F^{\perp}$

$$\mathfrak{F}^{\perp} = F^{\perp}\mathfrak{H}, \qquad F^{\perp} = 1 - F \in \mathfrak{B}(\mathfrak{H}) \tag{2}$$

where 1 denotes the unit operator. Let F, G be two projections onto the subspaces $\mathfrak{F}$ and $\mathfrak{G}$, respectively. Then we write $F \leq G$ to mean that $\mathfrak{F}$ is contained in $\mathfrak{G}$,

$$F \leq G \quad \text{iff } \mathfrak{F} \subset \mathfrak{G} \tag{3a}$$

This partial ordering of the projections can be expressed in the equivalent purely algebraic form

$$F \leq G \quad \text{iff } F = FG \quad \text{iff } F = GF \tag{3b}$$

(note that $F \leq G$ implies $FG = GF$). Furthermore $G \wedge F$ stands for the projection onto the set-theoretical intersection of $\mathfrak{F}$ and $\mathfrak{G}$, whereas $F \vee G$ denotes the projection onto the closure of the sum space $\mathfrak{F} + \mathfrak{G}$ (i.e., the set of all $\Phi + \Psi$ with $\Phi \in \mathfrak{F}$ and $\Psi \in \mathfrak{G}$):

$$(F \wedge G)\mathfrak{H} = \mathfrak{F} \cap \mathfrak{G} \tag{4a}$$

$$(F \vee G)\mathfrak{H} = \text{closure}(\mathfrak{F} + \mathfrak{G}) \tag{5a}$$

An equivalent algebraic expression is given by

$$F \wedge G = s\text{-}\lim_{n \to \infty} (FG)^n \tag{4b}$$

$$F \vee G = 1 - s\text{-}\lim_{n \to \infty} (F^{\perp}G^{\perp})^n \tag{5b}$$

Let $\mathfrak{P}(\mathfrak{B})$ denote the set of all projections in $\mathfrak{B}(\mathfrak{H})$,

$$\mathfrak{P}(\mathfrak{B}) \overset{\text{def}}{=} \{P : P \in \mathfrak{B}(\mathfrak{H}), \quad P = P^2 = P^*\} \tag{6}$$

It is easy to check that all projections in $\mathfrak{P}(\mathfrak{B})$ fulfill the orthomodular identity[62]

$$F \leq G \quad \text{implies } G = F \vee (G \wedge F^{\perp}) \tag{7}$$

With the partial ordering (3) and the lattice operations (4) and (5), and the orthocomplementation map

$$F \to F^{\perp} \overset{\text{def}}{=} 1 - F \tag{8}$$

$\mathfrak{P}(\mathfrak{B})$ becomes a *complete, irreducible atomic orthomodular lattice*. The largest projection of $\mathfrak{P}(\mathfrak{B})$ equals the identity operator 1, and the smallest projection equals the zero operator 0. To say that $\mathfrak{P}(\mathfrak{B})$ is atomic means that every nonzero projection has an atom under it, that is, for every $P \in \mathfrak{P}(\mathfrak{B})$ with $P \neq 0$ there exists an atom A with $A \leq P$. An element

$A \in \mathfrak{P}(\mathfrak{B})$ is an atom if and only if the projection A has rank 1, or

$$A \in \mathfrak{P}(\mathfrak{B}) \text{ is an atom} \quad \text{iff tr}(A) = 1 \qquad (9)$$

The compatibility relation $\leftrightarrow$ as defined by (II.8) has the following simple algebraic expression for F, $G \in \mathfrak{P}(\mathfrak{B})$:

$$F \leftrightarrow G \quad \text{iff } FG = GF \qquad (10)$$

That is, *two projections are compatible if and only if they commute*. The center of $\mathfrak{P}(\mathfrak{B})$ [defined as the set of all elements of $\mathfrak{P}(\mathfrak{B})$ that commute with all elements of $\mathfrak{P}(\mathfrak{B})$] is trivial (i.e., it contains only one and zero), so that $\mathfrak{P}(\mathfrak{B})$ is irreducible.

The projection lattice $\mathfrak{P}(\mathfrak{B})$ is the propositional system of traditional quantum mechanics as formulated in a mathematically rigorous way by Neumann.[1] From the modern point of view it corresponds to the "pure quantum case," and we take $\mathfrak{P}(\mathfrak{B})$ as the domain $\mathscr{U}$ of the universal theory of the universe of discourse. The domain $\mathscr{L}$ of a particular subtheory of the universal theory corresponds to an orthomodular sublattice of $\mathscr{U} = \mathfrak{P}(\mathfrak{B})$. With this we have the following realization of Axiom 2:

Realization Postulate (11)

a. The domain $\mathscr{U}$ of the universal theory is taken to be the projection lattice $\mathfrak{P}(\mathfrak{B})$ consisting of all projections in $\mathfrak{B}(\mathfrak{H})$, where $\mathfrak{H}$ is a separable complex Hilbert space.

b. The domain $\mathscr{L}$ of a subtheory is realized by a complete orthomodular sublattice of $\mathfrak{P}(\mathfrak{B})$.

Remark: The Hilbert space $\mathfrak{H}$ is always thought of as an abstract one; hence it is fully characterized by the cardinality of its dimension. We do not exclude finite-dimensional Hilbert spaces, but many of our discussions become trivial for this special case. To exclude unessential complications, we assume tacitly that the dimension of $\mathfrak{H}$ is larger than 2.

There are very many and very different kinds of orthomodular sublattices of $\mathfrak{P}(\mathfrak{N})$; their structural classification is of key importance to us. The domain $\mathfrak{P}(\mathfrak{B})$ is only needed as the most general framework to fix the universe of discourse. All directly perceptible phenomena are related to particular subtheories, hence to sublattices of $\mathfrak{P}(\mathfrak{B})$. Fortunately, we have a very powerful tool to investigate the orthomodular sublattices of $\mathfrak{P}(\mathfrak{B})$: Every complete orthomodular sublattice of $\mathfrak{P}(\mathfrak{B})$ is the projection lattice $\mathfrak{P}(\mathfrak{N})$ of a Neumann algebra $\mathfrak{N}$, and every projection lattice of a Neumann algebra is a complete orthomodular lattice.[62] As the theory of Neumann algebras is a highly developed field and a rich source of nontrivial results, it is worthwhile to investigate the Neumann algebras associated with quantum theories.

C. Algebras of Observables

The books by Dixmier[69,70] and by Sakai[71] are the basic references on the algebras of operators acting on a Hilbert space. For a fascinating account of the relation between orthomodular lattices and Neumann algebras one may consult Holland,[62] and the book by Berberian[72] is a source of a great deal of information on the interplay between operator algebras and lattice theory.

Neumann algebras are blessed with a rich structure so that they can be viewed from either a predominantly geometric, algebraic, or topological frame of reference. We avoid geometrical and topological considerations as far as possible and start with a purely algebraic definition of a Neumann algebra of surprising simplicity. The set $\mathfrak{B}(\mathfrak{H})$ of all bounded linear operators acting on the separable complex Hilbert space $\mathfrak{H}$ is a *-algebra; that is, it is closed under the algebraic operations of addition, of multiplication, of multiplication by complex scalars, and of taking the adjoint.

For an arbitrary subset $\mathfrak{M} \subset \mathfrak{B}(\mathfrak{H})$ we denote by $\mathfrak{M}'$ the set of all elements of $\mathfrak{B}(\mathfrak{H})$ that commute with all elements of $\mathfrak{M}$,

$$\mathfrak{M}' \stackrel{\text{def}}{=} \{B : B \in \mathfrak{B}(\mathfrak{H}) \text{ with } BA = AB \quad \forall A \in \mathfrak{M}\} \tag{12}$$

$\mathfrak{M}'$ is called the *commutant* of $\mathfrak{M}$; it is always an algebra that contains the identity operator. Moreover, if $\mathfrak{M}$ is self-adjoint, $\mathfrak{M} = \mathfrak{M}^*$ (i.e., $M \in \mathfrak{M}$ implies $M^* \in \mathfrak{M}$), then $\mathfrak{M}'$ is a *-algebra. The commutant of $\mathfrak{M}'$ is written as $\mathfrak{M}''$ and called the *double commutant* of $\mathfrak{M}$. Neumann's celebrated double commutant theorem allows us to give an algebraic definition of a Neumann algebra: *Every self-adjoint subset $\mathfrak{N} \subset \mathfrak{B}(\mathfrak{H})$ that equals its double commutant, $\mathfrak{N} = \mathfrak{N}''$, is called a Neumann algebra.* It is trivial to check that every Neumann algebra contains the identity operator and is closed under the algebraic operations and the *-operation. If $\mathfrak{M} \subset \mathfrak{B}(\mathfrak{H})$ is an arbitrary self-adjoint set of operators, then $\mathfrak{M}''$ is the smallest Neumann algebra that contains $\mathfrak{M}$. It is plain that the intersection of two Neumann algebras is again a Neumann algebra.

The outstanding importance of Neumann algebras for quantum mechanics is due to the following property: *Every Neumann algebra is generated by the projections it contains.* We denote by $\mathfrak{P}(\mathfrak{N})$ the set of all projection operators in the Neumann algebra $\mathfrak{N}$

$$\mathfrak{P}(\mathfrak{N}) \stackrel{\text{def}}{=} \{P : P \in \mathfrak{N}, \quad P = P^2 = P^*\} \tag{13}$$

so that

$$\mathfrak{N} = \{\mathfrak{P}(\mathfrak{N})\}'' \tag{14}$$

Moreover, $\mathfrak{P}(\mathfrak{N})$ is a complete orthomodular lattice with respect to the previously defined lattice operations for $\mathfrak{P}(\mathfrak{B})$. With this and the realization postulate (11b) we can associate to every domain $\mathscr{L}$ of a subtheory a particular Neumann algebra $\mathfrak{N}$ such that

$$\mathscr{L} = \mathfrak{P}(\mathfrak{N}) \tag{15}$$

and

$$\mathfrak{N} = \mathscr{L}'' \tag{16}$$

Because of this one-to-one correspondence between $\mathscr{L}$ and $\mathfrak{N}$ we can pass from the lattice-theoretical description to the more convenient operator-algebraic description. For merely historical reasons we call the Neumann algebra $\mathfrak{N}$ generated by $\mathscr{L}$ the *algebra of observables* of this theory.

Neumann algebras are algebras of *bounded* operators. In the context of group representations it is convenient to discuss generators that are in general unbounded self-adjoint operators. An unbounded self-adjoint operator is said to be *associated* with a Neumann algebra $\mathfrak{N}$ if all its bounded functions belong to $\mathfrak{N}$, or equivalently, if all its spectral projections belong to $\mathfrak{N}$. Accordingly, we choose

Definition: **Observables** (17)
The algebra of observables of a theory with the propositional system $\mathscr{L}$ is defined as the Neumann algebra $\mathfrak{N}$ generated by $\mathscr{L}$, $\mathfrak{N} = \mathscr{L}''$. The self-adjoint operators associated with $\mathfrak{N}$ are called *observables*, the self-adjoint elements of $\mathfrak{N}$ *bounded observables*.

Adopting the traditional terminology does not imply that we necessarily adopt the traditional interpretation of observables. We certainly do not presuppose that observables correspond to "physical quantities that can be measured," but we use them primarily as a convenient mathematical tool. Often it is said that the requirement of the self-adjointness of observables is due to the fact that measured values correspond to *real* numbers. This remark is grossly misleading because such a requirement is neither necessary nor sufficient. It is absolutely no problem to construct an apparatus that gives a complex number as the result of a measurement. On the other hand, the restriction to real-valued expectation values leads us only to operators whose numerical range consists of real numbers. Any Hermitean operator fulfills this requirement, but, of course, not every Hermitean operator is self-adjoint. The crucial point is *not* that a self-adjoint operator has a real-valued spectrum but that it generates an *Abelian* algebra.

An Abelian algebra is by definition commutative, so that for an Abelian Neumann algebra $\mathfrak{A}$ we have

$$\mathfrak{A} \text{ is Abelian} \quad \text{iff} \quad \mathfrak{A} \subset \mathfrak{A}' \tag{18}$$

The projection lattice of an Abelian Neumann algebra is Boolean, and we have

$$\mathfrak{P}(\mathfrak{A}) \text{ is Boolean} \quad \text{iff} \quad \mathfrak{A} \text{ is Abelian} \tag{19}$$

According to a famous result by Neumann,[73] every Abelian Neumann algebra $\mathfrak{A}$ is generated by a single self-adjoint operator A

$$\mathfrak{A} = \{A\}'' \quad \text{with} \quad A = A^* \in \mathfrak{B}(\mathfrak{H}) \tag{20}$$

and conversely, every self-adjoint operator generates an Abelian Neumann algebra. The connection between a self-adjoint operator A and the Boolean projection lattice $\mathfrak{P}(\mathfrak{A})$ is given by the spectral theorem. Every bounded or unbounded self-adjoint operator A acting on $\mathfrak{H}$ can be represented as

$$A = \int_\Lambda \lambda E(d\lambda) \tag{21}$$

where $\Lambda \subset \mathbb{R}$ is the spectrum of A, and $E : \Sigma \to \mathfrak{B}(\mathfrak{H})$ is a normalized spectral measure on the σ-field Σ of Borel subsets of Λ. Recall that a normalized *spectral measure* $E : \Sigma \to \mathfrak{B}(\mathfrak{H})$ on an arbitrary measurable space (Λ, Σ) is an operator-valued set function having the following properties:

E is *nonnegative-definite*, that is

$$E(\mathscr{A}) \geq 0 \text{ for each } \mathscr{A} \in \Sigma \tag{22a}$$

E is *additive*, that is,

$$\text{if } \mathscr{A} \cap \mathscr{B} = \varnothing, \quad \text{then } E(\mathscr{A} \cup \mathscr{B}) = E(\mathscr{A}) + E(\mathscr{B}) \tag{22b}$$

E is *multiplicative*, that is,

$$E(\mathscr{A} \cap \mathscr{B}) = E(\mathscr{A})E(\mathscr{B}), \tag{22c}$$

E is *continuous*, that is,

$$\sup E(\mathscr{A}_n) = E(\mathscr{A}) \text{ whenever } \{\mathscr{A}_n\} \text{ is an} \\ \text{increasing sequence of sets in } \Sigma \text{ whose} \\ \text{union } \mathscr{A} \text{ is also in } \Sigma \tag{22d}$$

E is *normalized*, that is,

$$E(\Lambda) = 1 \tag{22e}$$

With the spectral resolution (21), the Boolean projection lattice $\mathfrak{P}(\mathfrak{A})$ is given by

$$\mathfrak{P}(\mathfrak{A}) = \{E(\mathscr{A}) : \mathscr{A} \in \Sigma\} \tag{23}$$

and the Abelian algebra $\mathfrak{A}$ by

$$\mathfrak{A} = \{E(\mathscr{A}) : \mathscr{A} \in \Sigma\}'' \tag{24}$$

Even when A is unbounded, we can write this relation in the simple form

$$\mathfrak{A} = \{A\}''\tag{25}$$

provided we adopt the usual convention that an unbounded self-adjoint operator A is said to commute with $B \in \mathfrak{B}(\mathfrak{H})$ if and only if all spectral projections of A commute with B.

These relations show that there is a one-to-one correspondence between spectral measures, Abelian Neumann algebras, and Boolean lattices. According to Axiom 3, Boolean lattices are prone to an operational interpretation under appropriate experimental circumstances. Specific observables can be constructed from the spectral projections (which represent propositions) by attaching to them some spectral values. A proper labeling of the feasible propositions of an experiment may be important for the experimentalist but can hardly be considered to be fundamental. From a modern point of view the essential characteristic of an observable is the Abelian Neumann algebra, or equivalently, the Boolean projection lattice it generates.

D. States and Generalized States

According to the result (II.15), a state is realized by a proposition $S \in \mathfrak{U}$, hence in the Hilbert-space model by a nonzero projection,

$$S \text{ is a state iff } S \in \mathfrak{P}(\mathfrak{B}), \quad S \neq 0 \tag{26}$$

Of particular interest are the atomic states, which are realized by projections of rank 1,

$$S \text{ is an atomic state } \quad \text{iff tr}(S) = 1, \quad S \in \mathfrak{P}(\mathfrak{B}) \tag{27}$$

An atomic state corresponds to a *ray* in the Hilbert space $\mathfrak{H}$ (i.e., to a one-dimensional subspace of $\mathfrak{H}$). A ray can be generated by a single vector $\Psi \in \mathfrak{H}$; a unit vector representing an atomic state is called a *state vector*:

$$\left.\begin{array}{l} \Psi \text{ is a state vector associated} \\ \text{with an atomic state } S \end{array}\right\} \quad \text{iff} \begin{cases} \Psi \in \mathfrak{H}, \|\Psi\| = 1 \\ S\Psi = \Psi \end{cases} \tag{28}$$

In traditional quantum mechanics it is assumed that all physical states correspond to rays, that is, to atomic states. However, as also stressed by Bogolubov et al.,[74, p. 127] such an additional postulate is not obligatory, and moreover not appropriate for theories with noncommutative superselection rules.

The basic definitions (II.5) and (II.15) imply that a proposition F is true with respect to a state S if and only if S implies F, $S \leq F$, or using (3b), if and only if $FS = S$. On this account we say

a system in the state S has the quality f iff $FS = S$ (29)

If F is true, then $F^{\perp} = 1 - F$ is false, so that the eigenvalue 1 of F corresponds to the truth value "true," $\tau(F) = 1$, and the eigenvalue 0 of F corresponds to the truth value "false," $\tau(F) = 0$. Therefore we call a proposition F *truth-definite* with respect to the state S if S implies either F or $F^{\perp}$,

F is truth-definite   iff $FS = S$ *or* $F^{\perp}S = S$ (30)

and we write

$$FS = \tau_S(F)S, \qquad \tau_S(F) = 1 \text{ or } 0$$ (31)

Correspondingly, we say that an observable A is truth-definite with respect to a state S if all its spectral projections are truth-definite. In this case one can write

$$AS = \tau_S(A)S$$ (32)

where $\tau_S(A)$ is given by the spectral resolution (21) and (31b) as

$$\tau_S(A) = \int_{\Lambda} \lambda \tau_S\{E(d\lambda)\}$$ (33)

and its absolute value can be expressed with (32) as

$$|\tau_S(A)| = \|AS\|$$ (34)

Consequently, *if a system is in the state S and if the observable A is truth-definite with respect to S, we can say that A has the value $\tau_S(A)$.* If the state S is finite-dimensional one also can express τ_S as a trace

$$\tau_S(A) = \frac{\text{tr}(AS)}{\text{tr}(S)}$$ (35)

and if S is an atomic state, we can rewrite (32) in terms of a state vector $\Psi = S\Psi$ and an eigenvalue $\tau_S(A) = a$ in the more familiar form

$$A\Psi = a\Psi$$ (36)

Clearly, the value $\tau_S(A)$ in (36) corresponds always to an eigenvalue of A, so that the only values that can be realized by truth-definite atomic states are elements from the discrete spectrum of an observable. A similar discussion would be useful for the continuous part of the spectrum but in this case an eigenvector in the sense of (36) would have necessarily an infinite norm. That is, eigenvectors belonging to the continuous spectrum of an observable are not contained in the Hilbert space $\mathfrak{H}$ but only exist in some larger space, for example, in a rigged Hilbert space[75] or in a nonstandard Hilbert space.[76] Nevertheless, it is not really necessary to leave the Hilbert space $\mathfrak{H}$; there exist eigenpackets that play the role of approximate eigenvectors.

Recall that a number a is said to be an *approximate eigenvalue* of an operator A on $\mathfrak{H}$ if there exists a sequence of unit vectors $\Psi_n \in \mathfrak{H}$ such that

$$\lim_{n \to \infty} \|(A - a)\Psi_n\| = 0, \qquad \|\Psi_n\| = 1 \tag{37}$$

The set of all approximate eigenvalues of A is called the approximate point spectrum of A; if A is self-adjoint then the approximate point spectrum equals the spectrum of A (cf. Halmos[77, p.51]). Therefore if A is an arbitrary observable with the spectrum Λ, then for every $a \in \Lambda$ and every $\varepsilon > 0$ there exists a unit vector $\Psi \in \mathfrak{H}$ such that

$$\|(A - a)\Psi\| < \varepsilon \tag{38}$$

Equivalently, for every $a \in \Lambda$ and every $\varepsilon > 0$ there exists a projection $P \in \mathfrak{P}(\mathfrak{B})$ such that

$$\|(A - a)P\| < \varepsilon \tag{39}$$

Such an approximate eigenprojector can easily be constructed with the spectral measure $E : \Sigma \to \mathfrak{P}(\mathfrak{B})$ of the observable A [cf. (21)]. Let $\mathfrak{B}$ be a Borel subset of the real axis and an ε-neighborhood of a point in the spectrum Λ of the observable A,

$$\varepsilon = \sup\{|a - \lambda| : \lambda \in \mathfrak{B}\}, \qquad a \in \Lambda \tag{40}$$

Then (22c) implies

$$E(\mathfrak{B})A = \int_{\mathfrak{B}} \lambda E(d\lambda) \tag{41}$$

so that

$$(a - \varepsilon)E(\mathfrak{B}) \le AE(\mathfrak{B}) \le (a + \varepsilon)E(\mathfrak{B}) \tag{42}$$

$$\|(A - a)E(\mathfrak{B})\| \le \varepsilon \tag{43}$$

Let $\Phi \in \mathfrak{H}$ be any vector for which $E(\mathfrak{B})\Phi \neq 0$. Then the normalized vector

$$\varphi(\mathfrak{B}) \stackrel{\text{def}}{=} \frac{E(\mathfrak{B})\Phi}{\|E(\mathfrak{B})\Phi\|} \tag{44}$$

fulfills the defining relation for an eigenpacket of A,[78]

$$A\varphi(\mathfrak{B}) = \int_{\mathfrak{B}} \lambda \varphi(d\lambda) \tag{45}$$

and can be considered as an approximate eigenvector in the sense of (38).

Consider now a sequence $\{\mathscr{B}_n\}$, $\mathscr{B}_n \subset \mathscr{B}$, that contracts to the point a. If a does not belong to the spectrum, then $\varphi(\mathscr{B}_n)$ converges to zero. If a belongs to the point spectrum, then $\varphi(\mathscr{B}_n)$ converges to a normalized eigenvector of A. If a belongs to the continuous spectrum, the sequence $\varphi(\mathscr{B}_n)$ does *not* converge. Nevertheless, the mean values and the variances of A do converge for *every* $a \in \Lambda$,

$$\langle \varphi(\mathscr{B}_n)|A\varphi(\mathscr{B}_n)\rangle \to a \qquad (46a)$$

$$\langle \varphi(\mathscr{B}_n)|(A-a)^2\varphi(\mathscr{B}_n)\rangle \to 0 \qquad (46b)$$

so that with respect to the sequence $\{\varphi(\mathscr{B}_n)\}$ the observable A has a dispersion-free value a even if a belongs to the continuous part of the spectrum. The crucial idea is now to regard any sequence $\{\varphi(\mathscr{B}_n)\}$ that fulfills (46b) as representing a *generalized eigenvector* corresponding to a *generalized eigenvalue a*. This point of view corresponds to the sequential approach to distributions.[79,80] Similarly, a *generalized eigenprojection P(a)* belonging to the generalized eigenvalue a is defined as the equivalence class of all sequences $\{E(\mathscr{B}_n)\}$ with

$$\|(A-a)E(\mathscr{B}_n)\| \to 0, \qquad E(\mathscr{B}_n) \in \mathfrak{P}(\mathfrak{B}) \qquad (47)$$

for $\mathscr{B}_n \to a$, and we write

$$AP(a) = aP(a) \qquad (48)$$

If one considers the sequential approach as awkward, one can extend the basic Hilbert space $\mathfrak{H}$ and define generalized eigenvectors as elements of a larger space via (45) by the relation $\varphi(d\lambda) = \Phi(\lambda)\,d\lambda$. Put $\varphi_n = \varphi(\mathscr{B}_n)$ and consider two sequences $\{\varphi_n\}$ and $\{\varphi_n'\}$ belonging to the generalized eigenvalues a and a', respectively. Then (46) implies that

$$|(a-a')\langle\varphi_n|\varphi_n'\rangle| \to 0, \qquad n \to \infty$$

so that for $a \neq a'$ we have $\langle\varphi_n|\varphi_n'\rangle \to 0$. That is, *the generalized eigenvectors belonging to distinct generalized eigenvalues of an observable are orthogonal.* Berberian[81] has shown that this simple fact can be used to define a nonseparable Hilbert space $\mathfrak{K}$ of sequences $\{\varphi_n\}$ with the inner product defined by $\langle\varphi|\varphi'\rangle = \lim \langle\varphi_n|\varphi_n'\rangle$ such that every observable $A \in \mathfrak{B}(\mathfrak{H})$ determines an extension $\tilde{A}$ acting on $\mathfrak{K}$ and that the spectrum of A equals the point spectrum of $\tilde{A}$. The generalized eigenvectors of A are then ordinary eigenvectors of $\tilde{A}$ and elements of the Hilbert space $\mathfrak{K}$.

Example. Let P be the momentum operator in the Hilbert space $\mathscr{L}_2(\mathbb{R}, dx)$ of all Lebesgue square integrable complex-valued functions on the real axis $\mathbb{R}$. P is self-adjoint and has a simple, purely continuous spectrum that equals the real axis.

$$V(q) \stackrel{\text{def}}{=} \exp\{-iqP\}, \qquad q \in \mathbb{R} \qquad (49a)$$

is the unitary operator generated by P. It acts as a shift operator

$$\{V(q)\Psi\}(x) = \Psi(x - q), \qquad \Psi \in \mathscr{L}_2(\mathbb{R}, dx) \tag{49b}$$

The normalized functions $\Psi_k^{(n)} \in \mathscr{L}_2(\mathbb{R}, dx)$

$$\Psi_k^{(n)}(x) \overset{\text{def}}{=} \begin{cases} (2n)^{-1/2} \exp(ikx) & \text{if } |x| < n \\ 0 & \text{if } |x| \geq n \end{cases} \tag{49c}$$

$$n = 1, 2, \ldots, \qquad k \in \mathbb{R}$$

fulfill for every k the relation

$$\|(V(q) - e^{-ikq})\Psi_k^{(n)}\|^2 = \left|\frac{q}{n}\right|, \qquad |q| \leq n \tag{49d}$$

so that the sequence $\Psi_k^{(1)}, \Psi_k^{(2)}, \ldots$ represents a generalized eigenvector to the eigenvalue e^{-ikq}. The inner product in $\mathscr{L}_2(\mathbb{R}, dx)$ is given by

$$\langle \Psi_k^{(n)} | \Psi_{k'}^{(n)} \rangle = \{n(k' - k)\}^{-1} \sin\{n(k' - k)\} \tag{49e}$$

so that for $n \to \infty$

$$\langle \Psi_k^{(n)} | \Psi_{k'}^{(n)} \rangle \to \delta_{k,k'} \overset{\text{def}}{=} \begin{cases} 1 & \text{for } k = k' \\ 0 & \text{for } k \neq k' \end{cases} \tag{49f}$$

The limiting functions

$$(2n)^{1/2} \Psi_k^{(n)}(x) \to \exp(ikx) \overset{\text{def}}{=} e_k(x), \qquad k \in \mathbb{R} \tag{49g}$$

are not elements of $\mathscr{L}_2(\mathbb{R}, dx)$, but they span the nonseparable Hilbert space $\mathfrak{K}$ of Harald Bohr's[82] almost periodic functions with inner product

$$\langle f | g \rangle_{\mathfrak{K}} \overset{\text{def}}{=} \lim_{T \to \infty} \frac{1}{2T} \int_{-T}^{T} f^*(x) g(x)\, dx, \qquad g, f \in \mathfrak{K} \tag{49h}$$

They are normalized with respect to $\mathfrak{K}$

$$\langle e_k | e_{k'} \rangle = \delta_{k,k'}, \qquad k, k' \in \mathbb{R} \tag{49i}$$

and constitute a basis for $\mathfrak{K}$. Moreover, they are eigenvectors of $V(q)$ (now considered as an operator acting on $\mathfrak{K}$) to the eigenvalue $\exp(-ikq)$

$$V(q) e_k = e^{-ikq} e_k, \qquad e_k \in \mathfrak{K}, \quad k \in \mathbb{R} \tag{49j}$$

A more popular extension uses a rigged Hilbert space, that is, a triplet of spaces $\mathfrak{H}_+ \subset \mathfrak{H} \subset \mathfrak{H}_-$ where $\mathfrak{H}_+$ is a dense subspace of $\mathfrak{H}$ endowed with a topology finer than the one induced on $\mathfrak{H}_+$ by the inner product of $\mathfrak{H}$; $\mathfrak{H}_-$ is the space of continuous linear functionals on $\mathfrak{H}_+$. In this realization the generalized eigenvector $\Phi(a) = \varphi(da)/da$ exists as an element of the space $\mathfrak{H}_-$, $\Phi \in \mathfrak{H}_-$. The generalized eigenprojection $P(a)$ is then defined as the projection onto the subspace of $\mathfrak{H}_-$ spanned by the generalized eigenvectors $\Phi(a)$ corresponding to the generalized eigenvalue a (cf. Berezanskii[75, p. 336]).

There are qualities that can be well characterized by a sequence of propositions but whose limiting element is not an admissible proposition of the theory. In such cases it may be useful to extend the theory by adjoining new ideal elements, represented by *generalized propositions*. Note that such ideal objects are not needed for the formulation of the most fundamental axioms; they are generated in a natural way by the theory itself. Their introduction is a matter of convenience that does not imply any change of the basic axioms. In this sense we characterize generalized propositions by generalized projection operators. In accordance with the definition (II.5), we characterize a *generalized state* by a sequence of states and represent it by a generalized projection operator.

D. Duals of Algebras of Observables

The dual $\mathfrak{N}^*$ of a Neumann algebra $\mathfrak{N}$ is defined as the set of all bounded linear functionals on $\mathfrak{N}$. A *functional* $\rho : \mathfrak{N} \to \mathbb{C}$ is a map from the Neumann algebra $\mathfrak{N}$ into the complex numbers $\mathbb{C}$ assigning a complex number $\rho(N) \in \mathbb{C}$ to every element $N \in \mathfrak{N}$. A functional $\rho : \mathfrak{N} \to \mathbb{C}$ is called:

linear

$$\rho(c_1 N_1 + c_2 N_2) = c_1 \rho(N_1) + c_2 \rho(N_2) \qquad (50)$$
$$c_1, c_2 \in \mathbb{C};\ N_1, N_2 \in \mathfrak{N}$$

bounded

$$\text{if for some } c \geq 0 : |\rho(N)| \leq c\|N\| \qquad (51)$$

positive

$$\text{if } \rho(N^*N) \geq 0 \quad \forall N \in \mathfrak{N} \qquad (52)$$

normalized

$$\text{if } \rho(1) = 1 \qquad (53)$$

In the mathematical investigation of the structure of Neumann algebras the duals play an important role and are therefore studied extensively by the mathematicians. Unfortunately, the mathematicians have borrowed the term *state* from the statistical interpretation of quantum mechanics as a synonym for a "normalized positive linear functional." However, in our context a state is a physical rather than a mathematical term; moreover, a state is represented by a projection and not by a positive linear functional. To avoid a conceptual confusion we do not use the name *state* for a normalized positive linear functional but abbreviate it to *NPL-functional*. We use NPL-functionals as important but purely mathematical tools. A priori, they have no physical interpretation for the quantum theory of single systems but they acquire a physical meaning for the description of macrostates (see Section III.H) and in the context of the statistical ensemble interpretation of quantum mechanics (see Section IV). The set of all NPL-functionals of a Neumann algebra is a convex set whose extreme points generate the set. These extreme points are called *pure NPL-functionals*. An NPL-functional ρ is called *mixed* if there are different

NPL-functionals ρ_1 and ρ_2 such that $\rho = c\rho_1 + (1-c)\rho_2$ with $0 < c < 1$. If $\rho = c\rho_1 + (1-c)\rho_2$ with $0 < c < 1$ implies $\rho_1 = \rho_2 = \rho$, then ρ is pure.

An important class of NPL-functionals are generated by the well-known density operators. We use the notion of a *density operator* as a synonym for a normalized, nonnegative definite, self-adjoint, nuclear operator acting on the Hilbert space $\mathfrak{H}$ of state vectors.

$$D \in \mathfrak{B}_1(\mathfrak{H}) \tag{54a}$$

$$D \text{ is a density operator iff} \quad D = D^* \geq 0 \tag{54b}$$

$$\text{tr}(D) = 1 \tag{54c}$$

Recall that an operator is said to be nuclear if it belongs to the trace class, denoted here by $\mathfrak{B}_1(\mathfrak{H})$, consisting of all operators in $\mathfrak{B}(\mathfrak{H})$ with finite trace.[83] A simple characterization of nuclear operators is given by the following result:[84, sec. I-2.3]

$$D \in \mathfrak{B}_1(\mathfrak{H}) \quad \text{iff} \begin{cases} \text{there is at least one orthonormal} \\ \text{basis } \{\varphi_n\} \text{ of } \mathfrak{H} \text{ such that} \\ \sum_n \|D\varphi_n\| < \infty \end{cases} \tag{55}$$

Every density operator has a purely discrete spectrum consisting of non-negative eigenvalues λ_n so that its spectral resolution can be written as

$$D = \sum_{n=1}^{\infty} \lambda_n P_n, \qquad P_n \in \mathfrak{P}(\mathfrak{B}) \tag{56a}$$

with

$$\text{tr}(P_n) = 1, \qquad \sum_{n=1}^{\infty} \lambda_n = 1, \qquad \lambda_n \geq 0 \tag{56b}$$

Accordingly, every density operator equals a convex linear combination of atomic projections. To every density operator $D \in \mathfrak{B}_1(\mathfrak{H})$ we can associate a unique NPL-functional $\rho_D : \mathfrak{B}(\mathfrak{H}) \to \mathbb{C}$ by the definition

$$\rho_D(A) \overset{\text{def}}{=} \text{tr}(DA), \qquad A \in \mathfrak{B}(\mathfrak{H}) \tag{57}$$

In particular, every finite-dimensional state S defines a density operator and an NPL-functional on $\mathfrak{B}(\mathfrak{H})$ by

$$\rho_S(A) \overset{\text{def}}{=} \frac{\text{tr}(SA)}{\text{tr}(S)}, \qquad A \in \mathfrak{B}(\mathfrak{H}) \tag{58}$$

If S is an atomic state, the corresponding NPL-functional is called a *vector*

NPL-functional ρ_Ψ,

$$\rho_\Psi(A) \overset{\text{def}}{=} \cdot \langle \Psi | A\Psi \rangle, \qquad A \in \mathfrak{B}(\mathfrak{H}) \tag{59}$$

where Ψ is a state vector corresponding to S, $S\Psi = \Psi$.

An NPL-functional that can be represented by a density operator is called *normal*.* The set of all normal linear functionals on a Neumann algebra $\mathfrak{N}$, denoted by $\mathfrak{N}_*$, is generated by the normal NPL-functionals on $\mathfrak{N}$. It is known that normality is equivalent to the lattice-theoretical condition of complete additivity of mutually orthogonal projections (cf. Sakai[71, p.30]).

Normal NPL-functionals on Abelian Neumann algebras give the crucial mathematical link between the formalism of quantum mechanics and mathematical probability theory. According to (24), every Abelian Neumann algebra is generated by some spectral measure $E : \Sigma \to \mathfrak{B}(\mathfrak{H})$, defined by (22), so that every NPL-functional $\rho : \mathfrak{B}(\mathfrak{H}) \to \mathbb{C}$ defines a normalized positive measure $\mu : \Sigma \to \mathbb{C}$ by

$$\mu(\mathcal{A}) \overset{\text{def}}{=} \rho\{E(\mathcal{A})\}, \qquad \mathcal{A} \in \Sigma \tag{60}$$

An important theorem says that μ is a *σ-additive probability measure* if and only if ρ is normal.[87]

The set of all *normal* NPL-functionals does not exhaust the set of *all* NPL-functionals on a Neumann algebra acting on an infinite-dimensional Hilbert space. Nevertheless, the set of all normal NPL-functionals is an extremal convex subset of the set of all NPL-functionals. That is, an extreme point of the set of all normal NPL-functionals on a Neumann algebra is also a pure NPL-functional; it is called a *normal pure NPL-functional*. The nonnormal NPL-functionals are of interest because they are necessary for the construction of ergodic states[88] and because they give rise to superselection rules. In a natural way, nonnormal states arise in the thermodynamic limit of statistical mechanics.[26]

---

* For the convenience of those readers who look for theorems about normal NPL-functionals in the mathematical literature, we mention (without providing the definitions) that for an NPL-functional ρ on a Neumann algebra $\mathfrak{N}$ the following is equivalent (cf. Refs. 85, 86):

(i) ρ is normal on $\mathfrak{N}$
(ii) ρ is completely additive on $\mathfrak{P}(\mathfrak{N})$
(iii) ρ is a normalized element in the positive cone $\mathfrak{N}_*^+$ of the predual $\mathfrak{N}_*$ of $\mathfrak{N}$
(iv) ρ is σ-continuous on $\mathfrak{N}$
(v) ρ is ultraweakly continuous on $\mathfrak{N}$
(vi) ρ is ultrastrongly continuous on $\mathfrak{N}$
(vii) ρ is weakly continuous on the unit ball of $\mathfrak{N}$
(viii) ρ is strongly continuous on the unit ball of $\mathfrak{N}$

Every NPL-functional ρ on a Neumann algebra $\mathfrak{N}$ has a unique decomposition into a *normal* NPL-functional ρ_n and a *singular* NPL-functional ρ_s,[89–91]

$$\rho = c_n \rho_n + c_s \rho_s \tag{61a}$$

with

$$c_n + c_s = 1, \qquad c_n \geq 0, \quad c_s \geq 0 \tag{61b}$$

The singular part ρ_s is characterized by the fact that it vanishes at all atoms of $\mathfrak{P}(\mathfrak{N})$, so that ρ_s annihilates all compact operators. By construction, ρ_s cannot be represented by a density operator. If $\mathfrak{N}_*$ is the space of all linear combinations of normal NPL-functionals on $\mathfrak{N}$, and $\mathfrak{N}_*^\perp$ the space of all linear combinations of singular NPL-functionals on $\mathfrak{N}$, then (61) implies that the dual $\mathfrak{N}^*$ can be decomposed as the direct sum of $\mathfrak{N}_*$ and $\mathfrak{N}_*^\perp$,

$$\mathfrak{N}^* = \mathfrak{N}_* \oplus \mathfrak{N}_*^\perp \tag{61c}$$

For every point λ in the spectrum of an arbitrary observable $N \in \mathfrak{N}$ there exists at least one pure NPL-functional ρ on $\mathfrak{N}$ such that $\rho(N) = \lambda$ and $\rho\{(N - \lambda)\}^2 = 0$ (cf. Emch[26. theorem 11]). If λ belongs to the continuous part of the spectrum, ρ is a pure singular NPL-functional.

Example. Let Q be the multiplication operator in the Hilbert space $\mathfrak{H} = \mathscr{L}_2(\mathbb{R}, dx)$ of all complex-valued functions whose absolute squares are Lebesgue integrable on the real axis $\mathbb{R}$,

$$\{Q\Psi\}(x) = x\Psi(x), \qquad \Psi \in \mathscr{L}_2(\mathbb{R}, dx) \tag{62a}$$

Let $\mathfrak{N}$ be the maximal Abelian Neumann algebra generated by Q,

$$\mathfrak{N} \overset{\text{def}}{=} \{Q\}'' \tag{62b}$$

This algebra $\mathfrak{N}$ is of the continuous type; that is, it does not contain atomic projections. Every operator $N \in \mathfrak{N}$ is represented by a bounded function $n \in \mathscr{L}_\infty(\mathbb{R}, dx)$,

$$\{N\Psi\}(x) = n(x)\Psi(x), \qquad \Psi \in \mathscr{L}_2(\mathbb{R}, dx) \tag{62c}$$

In example (49) we introduced the sequence $\Psi_k^{(1)}, \Psi_k^{(2)}, \ldots$, representing the generalized eigenvector (corresponding to the generalized eigenvalue e^{-ikq}) of the shift operator $V(q)$. This sequence permits us to define the NPL-functional

$$\rho_k(A) \overset{\text{def}}{=} \lim_{n \to \infty} \langle \Psi_k^{(n)} | A \Psi_k^{(n)} \rangle \tag{62d}$$

for all $A \in \mathfrak{B}(\mathfrak{H})$ such that the required limit exists. One easily verifies that

$$\rho_k\{n(Q)\} = \lim_{T \to \infty} \frac{1}{2T} \int_{-T}^{T} n(x)\,dx \qquad \text{for all almost periodic functions } n : \mathbb{R} \to \mathbb{C} \tag{62e}$$

(note the independence of k). Hence ρ_k is well defined on a C^*-subalgebra of $\mathfrak{N}$ and may thus be extended in a canonical fashion to an NPL-functional on the whole of $\mathfrak{N}$ (cf. Emch[26, p.121]).

$$\rho_k\{n(P)\} = n(k) \qquad \text{for all bounded continuous } n : \mathbb{R} \to \mathbb{C} \tag{62f}$$

$$\rho_k\{F\} = 0 \qquad \text{for all atomic } F \in \mathfrak{P}(\mathfrak{B}) \tag{62g}$$

so that ρ_k is singular. Accordingly, *the sequence* $\Psi_k^{(1)}, \Psi_k^{(2)}, \ldots$, *representing a generalized eigenvector of the shift operator generates a singular NPL-functional on the maximal Abelian algebra* $\mathfrak{N} = \{Q\}''$ *of the position operator Q.*

The $\mathcal{L}$-admissible states of a quantal system with $\mathcal{L} = \mathfrak{P}(\mathfrak{N})$ generate a (not necessarily normal) NPL-functional on the algebra $\mathfrak{N}$ of observables. If a projection $F \in \mathfrak{P}(\mathfrak{N})$ commutes with an $\mathcal{L}$-admissible state S, then condition (II.15) implies that FS equals either S or zero, that is,

$$FS = SF \quad \text{implies } FS = \tau_S(F)S$$

with $\tau_S(F) = 1$ or $\tau_S(F) = 0$. If an observable $A \in \mathfrak{N}$ with the spectral resolution (21) commutes with a $\mathfrak{P}(\mathfrak{N})$-admissible state, all its spectral projections $E(\mathscr{A})$ commute with S so that

$$E(\mathscr{A})S = \tau_S\{E(\mathscr{A})\}S$$

hence

$$AS = SA \quad \text{implies } AS = \tau_S(A)S$$

with

$$\tau_S(A) \stackrel{\text{def}}{=} \int_\Lambda \lambda \tau_S\{E(d\lambda)\}$$

Let now N be an arbitrary observable from $\mathfrak{N}$ and put $A \stackrel{\text{def}}{=} SNS$. Then A commutes with S so that for every $N \in \mathfrak{N}$ and every $\mathfrak{P}(\mathfrak{N})$-admissible state S we have

$$SNS = \tau_S(SNS)S$$

Putting

$$\rho_S(N) \stackrel{\text{def}}{=} \tau_S(SNS), \qquad N \in \mathfrak{N}$$

it follows trivially that ρ_S is an NPL-functional on $\mathfrak{N}$. In the following we consider only $\mathfrak{P}(\mathfrak{N})$-admissible states and call the NPL-functional ρ_S defined by

$$SNS = \rho_S(N)S, \qquad N \in \mathfrak{N} \tag{63}$$

the *NPL-functional on* $\mathfrak{N}$ *generated by the state S.* In particular, a vector state $S = |\Psi\rangle\langle\Psi|$, $\Psi \in \mathfrak{H}$, is $\mathfrak{P}(\mathfrak{N})$-admissible for every Neumann algebra

$\mathfrak{N} \subset \mathfrak{B}(\mathfrak{H})$ so that the corresponding NPL-functional ρ_Ψ is normal and given by

$$\rho_\Psi(N) = \langle \Psi | N\Psi \rangle, \qquad N \in \mathfrak{N}$$

F.   Representations of Algebras of Observables

Every unit vector $\Psi \in \mathfrak{H}$ defines by

$$\rho_\Psi(N) \overset{\text{def}}{=} \langle \Psi | N\Psi \rangle, \qquad \forall N \in \mathfrak{N}$$

an NPL-functional ρ_Ψ on a Neumann algebra $\mathfrak{N} \subset \mathfrak{B}(\mathfrak{H})$. It is of considerable technical importance that the converse is also true in the sense that *every* NPL-functional on an algebra of observables can be realized in a new Hilbert space by a state vector. For this it is convenient to think of an algebra of observables as a mere algebraic structure. In every Neumann algebra the purely algebraic operations of taking linear combinations with complex coefficients, products, adjoints, together with a norm $\| \cdot \|$ fulfilling the condition $\|A^*A\| = \|A\|^2$, comply with the defining relations of an abstract B^*-algebra, so that every Neumann algebra can also be considered as a B^*-algebra. (The converse is not true; those B^*-algebras that are *-isomorphic to a Neumann algebra are called W^*-algebras.[71]) Taking this point of view, we may look at an algebra $\mathfrak{N}$ of observables as an abstract B^*-algebra that can in turn be represented on an appropriate Hilbert space.

A representation $(\pi, \mathfrak{K})$ of a B^*-algebra $\mathfrak{A}$ on a Hilbert space $\mathfrak{K}$ is a map $\pi : \mathfrak{A} \to \mathfrak{B}(\mathfrak{K})$ that assigns to each operator $A \in \mathfrak{A}$ a bounded operator $\pi(A)$ acting in $\mathfrak{K}$ such that $(a_i \in \mathbb{C}, A_i \in \mathfrak{A})$

(i) $\pi(a_1 A_1 + a_2 A_2) = a_1 \pi(A_1) + a_2 \pi(A_2)$

(ii) $\pi(A_1 A_2) = \pi(A_1)\pi(A_2)$

(iii) $\pi(A^*) = \{\pi(A)\}^*$

The various possible representations of an algebra of observables lead to physically meaningful classifications. We recall the most important concepts. A representation is said to be *faithful* if it preserves the zero, that is,

$$(\pi, \mathfrak{K}) \text{ is faithful } \quad \text{iff } \pi(A) = 0 \text{ implies } A = 0 \qquad (64)$$

A representation $(\pi, \mathfrak{K})$ is called *cyclic* if there exists a vector $\Phi \in \mathfrak{K}$ such that the set $\pi(\mathfrak{A})\Phi$ is everywhere dense in $\mathfrak{K}$, that is,

$$(\pi, \mathfrak{K}) \text{ is cyclic } \quad \text{iff closure } \{\pi(A)\Phi : A \in \mathfrak{A}\} = \mathfrak{K} \qquad (65)$$

In this case Φ is called the *cyclic vector* for π. A representation $(\pi, \mathfrak{K})$ is called *irreducible* if the only bounded operators on $\mathfrak{K}$ commuting with all

$\pi(A)$ are the scalar multiples of the identity 1

$$(\pi, \mathfrak{K}) \text{ is called irreducible} \quad \text{iff } \{\pi(\mathfrak{A})\}' = 1 \cdot \mathbb{C} \tag{66a}$$

$$\text{iff } \{\pi(\mathfrak{A})\} = \mathfrak{B}(\mathfrak{K}) \tag{66b}$$

A representation $(\pi, \mathfrak{K})$ is called a *factor representation* if $\{\pi(\mathfrak{A})\}''$ is a factor; that is, if $\{\pi(\mathfrak{A})\}''$ has a trivial center,

$$(\pi, \mathfrak{K}) \text{ is a factor representation} \quad \text{iff } \{\pi(\mathfrak{A})\}'' \cap \{\pi(\mathfrak{A})\}' = \mathbb{C} \cdot 1 \tag{67}$$

Two representations $(\pi_1, \mathfrak{K}_1)$ and $(\pi_2, \mathfrak{K}_2)$ are called *unitarily equivalent* (or simply, equivalent) if there exists a unitary transformation U from $\mathfrak{K}_1$ onto $\mathfrak{K}_2$ such that

$$\pi_2(A) = U\pi_1(A)U^{-1} \qquad \text{for each } A \in \mathfrak{A} \tag{68}$$

Note that for an irreducible representation π, the representation $\pi \oplus \pi$ is *not* equivalent to π. This relationship can be described by the notion of a *subrepresentation*. If $(\pi_1, \mathfrak{K}_1)$ and $(\pi_2, \mathfrak{K}_2)$ are representations of $\mathfrak{A}$, we say π_1 is a subrepresentation of π_2 if

$$\mathfrak{K}_1 \subset \mathfrak{K}_2 \tag{69a}$$

$$\pi_1(A)\Psi = \pi_2(A)\Psi \qquad \text{for all } A \in \mathfrak{A} \text{ and all } \Psi \in \mathfrak{K}_1 \tag{69b}$$

Two representations are called *disjoint* if they contain no subrepresentations that are unitarily equivalent. Two representations are called *quasiequivalent* if no subrepresentation of the one is disjoint from the other. For example, π and $\pi \oplus \pi$ are quasiequivalent. Unitary equivalence implies quasiequivalence, whereas inequivalent irreducible representations are always disjoint. Disjoint and quasiequivalent representations are two extreme cases; however, in factor representations there are no others: *two-factor representations are either disjoint or equivalent.*[92]

All these concepts can be transferred to NPL-functionals because there is a one-to-one correspondence between representations and NPL-functionals. A procedure known as the *GNS construction* (due to Gel'fand, Naĭmark, and Segal) allows for every NPL-functional ρ the construction of a representation $(\pi_\rho, \mathfrak{K}_\rho)$ in a complex (not necessarily separable) Hilbert space $\mathfrak{K}_\rho$ such that

$$\rho(A) = \langle \Phi_\rho | \pi_\rho(A)\Phi_\rho \rangle, \qquad A \in \mathfrak{A} \tag{70}$$

where $\langle \cdot | \cdot \rangle$ denotes the inner product in $\mathfrak{K}_\rho$ and Φ_ρ is a cyclic vector $\Phi_\rho \in \mathfrak{K}_\rho$, with $\|\Phi_\rho\| = 1$. This cyclic representation $(\pi_\rho, \mathfrak{K}_\rho)$ of the B^*-algebra $\mathfrak{A}$ is called the *GNS representation* generated by the NPL-functional ρ. The triple $(\pi_\rho, \mathfrak{K}_\rho, \Phi_\rho)$ is unique up to unitary equivalence; therefore it is also called the *canonical cyclic representation* associated with ρ. The GNS theorem is a very powerful tool and plays a central role in the algebraic formulation of quantum mechanics. An easy but important result is that there is a relation between indecomposable linear functionals and irreducible representations:

$$\pi_\rho \text{ is irreducible} \quad \text{iff } \rho \text{ is a pure NPL-functional} \tag{71}$$

Example. Let the algebra of observables be $\mathfrak{B}(\mathfrak{H})$ and let ρ be an NPL-functional given by a strictly positive definite density operator D on $\mathfrak{H}$, with the eigenvalues $\lambda_n > 0$ and the orthonormal eigenvectors φ_n,

$$D\varphi_n = \lambda_n \varphi_n; \quad \langle \varphi_n | \varphi_m \rangle = \delta_{nm}; \quad n, m = 1, 2, 3, \ldots \tag{72a}$$

Define a separable Hilbert space $\mathfrak{K}$ by

$$\mathfrak{K} \stackrel{\text{def}}{=} \mathfrak{H} \otimes \mathfrak{H} \tag{72b}$$

and a normalized vector $\Phi \in \mathfrak{K}$ by

$$\Phi \stackrel{\text{def}}{=} \sum_{n=1}^{\infty} \sqrt{\lambda_n}\, \varphi_n \otimes \varphi_n \tag{72c}$$

It is straightforward to verify that the Neumann algebra $\mathfrak{N} = \mathfrak{B} \otimes 1 \cdot \mathbb{C}$ on $\mathfrak{K}$ defines a faithful representation $(\rho, \mathfrak{K})$ of the algebra $\mathfrak{B}(\mathfrak{H})$ with the cyclic vector Φ:

$$\pi(\mathfrak{B}) \stackrel{\text{def}}{=} \mathfrak{B} \otimes 1 \cdot \mathbb{C} \stackrel{\text{def}}{=} \mathfrak{N} \tag{73a}$$

$$\rho(A) = \langle \Phi | \pi(A) \Phi \rangle_{\mathfrak{K}}, \quad A \in \mathfrak{B}(\mathfrak{H}) \tag{73b}$$

Clearly, $\mathfrak{N}' = 1 \cdot \mathbb{C} \otimes \mathfrak{B}$, so that according to (66a) π is a reducible representation. Because $\mathfrak{N} \cap \mathfrak{N}' = 1 \cdot \mathbb{C}$, the definition (67) says that π is a factor representation.

Based on the one-to-one correspondence between NPL-functionals and representations, we can shift the terminology used for representations to the linear functionals. If ρ is an NPL-functional on an algebra of observables and $(\pi_\rho, \mathfrak{K}_\rho)$ the GNS representation induced by ρ, then we say

ρ is a factor NPL-functional	if π is a factor representation	(74)
ρ_1 and ρ_2 are equivalent	if π_1 and π_2 are quasiequivalent	(75)
ρ_1 and ρ_2 are quasiequivalent	if π_1 and π_2 are quasiequivalent	(76)
ρ_1 and ρ_2 are disjoint	if π_1 and π_2 are disjoint	(77)

G. Classical Observables

In Section II.8 we have associated to every propositional system $\mathcal{L}$ a unique classical subtheory whose propositional system is Boolean and given by the center $\mathfrak{Z}(\mathcal{L})$, defined by (II.21). In the Hilbert-space model, $\mathcal{L}$ is given by the projection lattice (15) generated by some Neumann algebra $\mathfrak{N}$ so that $\mathcal{Z}$ is given by

$$\mathcal{Z}(\mathfrak{N}) \stackrel{\text{def}}{=} \{G : G \in \mathfrak{P}(\mathfrak{N}), \quad GF = FG \quad \forall F \in \mathfrak{P}(\mathfrak{N})\} \tag{78}$$

By (20) the Neumann algebra generated by the Boolean projection lattice

$\mathscr{Z}(\mathfrak{N})$ is Abelian; it equals the center $\mathfrak{Z}(\mathfrak{N})$ of the Neumann algebra $\mathfrak{Z}$,

$$\{\mathscr{Z}(\mathfrak{N})\}'' = \mathfrak{Z}(\mathfrak{N}) \tag{79}$$

where $\mathfrak{Z}(\mathfrak{N})$ is defined as the set of those operators of $\mathfrak{N}$ that commute with *all* operators of $\mathfrak{N}$,

$$\mathfrak{Z}(\mathfrak{N}) \overset{\text{def}}{=} \{Z : Z \in \mathfrak{N}, \quad ZN = NZ \quad \forall N \in \mathfrak{Z}\} \tag{80a}$$

$$= \mathfrak{N} \cap \mathfrak{N}' \tag{80b}$$

Clearly, the center $\mathfrak{Z}(\mathfrak{N})$ is an Abelian Neumann algebra. With this we adopt the following terminology:

*Definition: **Classical Observables*** (81)

The elements of the Boolean projection lattice $\mathscr{Z}(\mathfrak{N})$ of a theory with the algebra $\mathfrak{N}$ of observables represent the *classical propositions*. The self-adjoint operators associated with the center $\mathfrak{Z}(\mathfrak{N}) = \{\mathscr{Z}(\mathfrak{N})\}''$ are called the *classical observables* of the theory.

The classical observables of a theory play a crucial role for the observation theory of quantal systems. Within the class of primary states, introduced in the definition (II.22), the classical propositions are truth definite. If S is a primary state, then (32) implies that for every classical observable $X \in \mathfrak{Z}$ there exists a real number x such that

$$XS = xS$$

Therefore every classical observable has a definite dispersion-free value x. With Axiom 5 (II.23) we can interpret the classical observables of a quantal system in a primary state as the *directly perceptible properties of the system*.

Of course, not every state is primary. Nevertheless, the so-called *central decomposition* allows the decomposition of an arbitrary state into primary states. More generally, every NPL-functional ρ on the algebra $\mathfrak{N}$ of observables has an integral representation of the form

$$\rho = \int_{\Lambda} \rho_{\lambda} \nu(d\lambda) \tag{82}$$

where the probability measure ν on Λ is carried by the factor NPL-functionals ρ_{λ} (cf. also Sakai[71, Chap. 3]). The measure ν is called the *central measure* and is defined by the integral decomposition of the Neumann algebra $\mathfrak{N}$ into factors.

A Neumann algebra $\mathfrak{F}$ is called a *factor* if its center contains only scalar multiples of the identity,

$$\mathfrak{F} \text{ is a factor} \quad \text{iff } \mathfrak{F} \cap \mathfrak{F}' = 1 \cdot \mathbb{C} \tag{83}$$

It is no serious limitation to restrict the mathematical discussion of the structure of Neumann algebras to factors because—roughly speaking—every Neumann algebra is a direct sum of factors. More precisely, the indexing family for such a "direct sum" may also be a measure space so that in general one has to integrate over the constituent factors. Every Neumann algebra on a separable Hilbert space has an essentially unique representation as a direct integral of factors. For a lucid presentation of the general theory of direct integral decompositions of Neumann algebras one may consult the text of Schwartz;[93] a short introduction is given in Gel'fand and Vilenkin.[84]

The central decomposition can be constructed as follows. The center $\mathfrak{Z}(\mathfrak{N}) = \mathfrak{N} \cap \mathfrak{N}'$ is an Abelian Neumann algebra and can therefore be generated by a single self-adjoint operator $Z \in \mathfrak{Z}$.[94] The spectral decomposition of Z is given by (21), or

$$Z = \int_\Lambda \lambda E(d\lambda) \qquad (84)$$

where Λ is the spectrum of Z and $E : \Sigma \to \mathfrak{Z}$ is a normalized spectral measure on the σ-field Σ of Borel subsets of Λ.

Every Abelian Neumann algebra acting on a separable Hilbert space has a separating vector; that is, there exists a normalized vector $\Phi \in \mathfrak{H}$ with $E(\mathscr{A})\Phi \neq 0$ for all $\mathscr{A} \in \Sigma$. For such a vector Φ we define a normalized probability measure ν on (Λ, Σ) by

$$\nu(\mathscr{A}) \overset{\text{def}}{=} \langle \Phi | E(\mathscr{A})\Phi \rangle, \qquad \mathscr{A} \in \Sigma \qquad (85)$$

It then follows (cf. Ref. 93) that there exists a direct integral decomposition called the central decomposition of the Hilbert space $\mathfrak{H}$ and of the Neumann algebra $\mathfrak{N}$,

$$\mathfrak{H} = \int_\Lambda^\oplus \mathfrak{H}(\lambda)\nu(d\lambda) \qquad (86)$$

$$\mathfrak{N} = \int_\Lambda^\oplus \mathfrak{N}(\lambda)\nu(d\lambda) \qquad (87)$$

with the following properties:

1. For each $\lambda \in \Lambda$, $\mathfrak{H}(\lambda)$ is either a one-dimensional, a finite-dimensional, or a countably infinite-dimensional Hilbert space with inner product $\langle \cdot | \cdot \rangle_\lambda$. For each $\lambda \in \Lambda$, $\mathfrak{N}(\lambda)$ is a factor on $\mathfrak{H}(\lambda)$; that is, $\mathfrak{N}(\lambda) = \mathfrak{N}(\lambda)'' \subset \mathfrak{B}(\mathfrak{H}(\lambda))$ with $\mathfrak{N}(\lambda) \cap \mathfrak{N}(\lambda)' = 1_\lambda \cdot \mathbb{C}$, where 1_λ denotes the unit operator on $\mathfrak{H}(\lambda)$.
2. Each vector $\Psi \in \mathfrak{H}$ may be identified with a Borel function $\lambda \to \Psi(\lambda) \in$

$\mathfrak{H}(\lambda)$, $\lambda \in \Lambda$, such that for any two vectors Ψ, $\Phi \in \mathfrak{H}$ we have

$$\langle \Psi | \Phi \rangle = \int_\Lambda \langle \Psi(\lambda) | \Phi(\lambda) \rangle_\lambda \nu(d\lambda) \tag{88}$$

One also writes symbolically

$$\Psi = \int_\Lambda^\oplus \Psi(\lambda) \nu(d\lambda) \tag{89}$$

where $\Psi(\lambda)$ is called the *component* of Ψ in $\mathfrak{H}(\lambda)$.

3. Each operator $N \in \mathfrak{N}$ may be identified with a Borel function $\lambda \to N(\lambda) \in \mathfrak{N}(\lambda)$, $\lambda \in \Lambda$, such that for any $\Psi = \int_\Lambda^\oplus \Psi(\lambda) \nu(d\lambda)$ we have

$$N\Psi = \int_\Lambda^\oplus N(\lambda)\Psi(\lambda) \nu(d\lambda) \tag{90}$$

and we say that the operator $N \in \mathfrak{N}$ is reduced by the direct integral decomposition; $\mathfrak{N}(\lambda)$ is called the *component* of $\mathfrak{N}$ on $\mathfrak{H}(\lambda)$. In particular, every $X \in \mathfrak{Z}$ is represented by a multiplication operator,

$$X\Psi = \int_\Lambda^\oplus x(\lambda)\Psi(\lambda) \nu(d\lambda), \qquad \Psi \in \mathfrak{H} \tag{91}$$

where the Borel function $x : \Lambda \to \mathbb{C}$ is bounded.

4. The central decomposition is unique up to unitary equivalence.

The decomposition (86) of a Hilbert space $\mathfrak{H}$ into a direct integral of Hilbert spaces $\mathfrak{H}(\lambda)$ is a generalization of a decomposition of a Hilbert space into a direct sum of Hilbert spaces. It accounts for the possibility that the spectrum Λ of the generator Z of the center has a continuous part. If the generator Z has a pure point spectrum

$$Z = \sum_n \lambda_n P_n, \qquad P_n P_m = \delta_{nm} P_n \tag{92}$$

then the measure ν defined in (85) is atomic so that the decomposition (86) is simply a direct sum

$$\mathfrak{H} = \oplus_n \mathfrak{H}_n \tag{93}$$

where the subspace $\mathfrak{H}_n \subset \mathfrak{H}$ equals the range of the projection P_n

$$\mathfrak{H}_n = P_n \mathfrak{H} \tag{94}$$

In contrast, the Hilbert spaces $\mathfrak{H}(\lambda)$ of the integral decomposition (86) will in general *not* be subspaces of $\mathfrak{H}$: $\mathfrak{H}(\lambda)$ is a subspace of $\mathfrak{H}$ if and only if $\nu(\{\lambda\}) \neq 0$. This fact causes the usual (minor) trouble, well known from eigenvalue problems of operators having continuous spectra. The remedy is the same as the one already discussed in Section III.D. One can always rig the Hilbert space $\mathfrak{H}$ and introduce a Gel'fand triple $\mathfrak{H}_+ \subset \mathfrak{H} \subset \mathfrak{H}_-$ such that for every $\lambda \in \Lambda$ the Hilbert space $\mathfrak{H}(\lambda)$ is a subspace[95] of $\mathfrak{H}_-$ and that

$$\mathfrak{H}(\lambda) = P(\lambda)\mathfrak{H}_- \tag{95}$$

where $P(\lambda)$ is the generalized eigenprojection corresponding to the generalized eigenvalue λ of the generator Z. The decomposition (86) is then to be interpreted as a direct integral of the pair $(\mathfrak{H}_+, \mathfrak{H}_-)$ of dual spaces.[96]

The importance of the central decomposition is due to the fact that it allows the introduction of dispersion-free states. An NPL-functional ρ is said to be *dispersion-free* on a Neumann algebra $\mathfrak{A}$ if all self-adjoint elements of $\mathfrak{A}$ have a sharp value; that is,

$$\rho \text{ is dispersion-free on } \mathfrak{A} \quad \text{iff } \rho(A^2) = \{\rho(A)\}^2, \qquad \forall A = A^* \in \mathfrak{A} \tag{96}$$

If ρ is dispersion-free on $\mathfrak{A}$, it easily follows that

$$\rho(AB) = \rho(A)\rho(B) \qquad \text{for all } A, B \in \mathfrak{A} \tag{97}$$

Every *Abelian* Neumann algebra admits dispersion-free NPL-functionals. An *atomic* Abelian Neumann algebra admits dispersion-free *normal* NPL-functionals, whereas on a *continuous* Abelian Neumann algebra every dispersion-free NPL-functional is necessarily *singular*.[97] On the other hand, Misra[98] has shown that a nontrivial factor does not admit any dispersion-free NPL-functional. Therefore *the only observables for which it is possible to assign dispersion-free values are the classical observables.*

According to (63), every $\mathfrak{P}(\mathfrak{N})$-admissible state generates an NPL-functional ρ on $\mathfrak{N}$, whose central decomposition (82) is carried by the factor NPL-functionals ρ_λ, $\lambda \in \Lambda$, which are dispersion-free on the center $\mathfrak{Z}$,

$$\rho_\lambda(X^2) = \{\rho_\lambda(X)\}^2, \qquad X \in \mathfrak{Z} \tag{98}$$

hence primary states in the sense of definition (II.22). They even fulfill the relation

$$\rho_\lambda(NX) = \rho_\lambda(N)\rho_\lambda(X), \qquad N \in \mathfrak{N}, \quad X \in \mathfrak{Z} \tag{99}$$

The uniqueness of the central decomposition allows us to interpret non-primary states as a *classical mixture*, where the central measure ν in (82) describes our ignorance of the system. That is, we can take a nonprimary state to mean that the system is actually in one and only one primary state and that it is always possible in principle to determine which one of the primary states represents the system.This viewpoint is in accordance with Axiom 5 (II.23) and is a logically permissible but not necessary way of talking. We adopt it by the following.

Postulate: **Physically Pure States** (100)

The primary states represent the physically pure states and are factor NPL-functionals. The unique central decomposition of an arbitrary state into primary states is to be interpreted as a physical mixture in the classical sense.

Remarks:

1. The statistical interpretation of the mixtures described by nonprimary states is discussed in Section IV.

2. If the algebra of observables is Abelian, postulate (100) implies that all physically pure states are dispersion-free. With this, we recover the usual interpretation adopted in Newtonian mechanics.

3. As there are no dispersion-free states on factors, the central decomposition is the finest possible physically sensible decomposition of a state into a physical mixture of states. Note that the central decomposition is into factors and not into irreducible algebras. In general, a decomposition into irreducible algebras is highly nonunique and has no physical interpretation.

With postulate (100), the central decomposition acquires a fundamental physical meaning. In this context the Hilbert spaces $\mathfrak{H}(\lambda)$, $\lambda \in \Lambda$, of the direct integral decomposition (86) are called *sectors*. A sector that is a subspace of $\mathfrak{H}$ can also be characterized as a maximal subset of $\mathfrak{H}$ in which the superposition principle of traditional quantum mechanics[7] holds without restrictions. We did not postulate (nor even formulate!) the superposition principle because it is in fact a consequence of the non-Boolean structure of the propositional system.[29] All physically pure states are projections whose range lies in a particular sector. Physically pure states lying in different sectors are said to be separated by a superselection rule. If there is only one sector, namely, $\mathfrak{H}$, then there are no superselection rules, the corresponding algebra of observables is irreducible, and the system is called *purely quantal*. If all sectors are one-dimensional, then the algebra of observables is maximal Abelian, and the system is called *purely classical*. In a purely classical theory the propositional system is Boolean and every state is a mixture of physically pure states.

H. Macrostates

The commutant $\mathfrak{N}'$ of the algebra $\mathfrak{N}$ of observables of a quantal system has interesting properties. We call two states S_1 and S_2 $\mathfrak{N}$-*equivalent* if they are unitarily equivalent and if this unitary equivalence is implemented by a unitary operator from the commutant $\mathfrak{N}'$. This equivalence relation decomposes the set of all states into disjoint equivalence classes $[\cdot]_{\mathfrak{N}}$ defined by

$$[S]_{\mathfrak{N}} \overset{\text{def}}{=} \{T : T = USU^*, \quad U^* = U^{-1}, \quad U \in \mathfrak{N}'\} \tag{101}$$

If an observable $N \in \mathfrak{N}$ in the state S has a sharp value $\tau(N)$,

$$NS = \tau(N)S$$

then N has the same value with respect to every $\mathfrak{N}$-equivalent state, that is,

$$NS = \tau(N)S \quad \text{implies} \quad NT = \tau(N)T \qquad \text{for every } T \in [S]_{\mathfrak{N}} \tag{102}$$

That is, $\mathfrak{N}$-equivalent states cannot be distinguished by means of the observables available in $\mathfrak{N}$. It is reasonable to identify indistinguishable elements and to call the equivalence class $[S]_{\mathfrak{N}}$ a macrostate with respect to $\mathfrak{N}$. If (102) holds for any (hence for all) elements of $[S]_{\mathfrak{N}}$, we say that the observable N has the value $\tau(N)$ with respect to the macrostate $[S]_{\mathfrak{N}}$. More generally, we can use the NPL-functional ρ_S generated by a state S,

$$SNS = \rho_S(N)S, \qquad N \in \mathfrak{N}$$

Clearly, we have for $N \in \mathfrak{N}$

$$TNT = \rho_S(N)T \qquad \text{for every } T \in [S]_{\mathfrak{N}} \tag{103}$$

so that the NPL-functional ρ_S can also be transferred to the macrostate $[S]_{\mathfrak{N}}$.

If S is a primary state, all elements of the macrostate $[S]_{\mathfrak{N}}$ are primary, hence physically pure. Therefore a macrostate generated by a primary state is called a *physically pure macrostate*. The postulate (100) implies that every macrostate can be decomposed into a mixture of physically pure macrostates. A physically pure macrostate is said to be trivial if it contains one element only. Since every primary state commutes with all classical observables,

$$\text{if } S \text{ is a primary state, then } S \in \mathfrak{Z}', \tag{104}$$

every primary state that is an observable generates only a trivial macrostate,

$$\left. \begin{array}{c} \text{if } S \text{ is primary} \\ \text{and if } S \in \mathfrak{N} \end{array} \right\}, \quad \text{then } [S]_{\mathfrak{N}} = S \tag{105}$$

On the other hand, if in a quantal system the center equals the commutant of the algebra of observables, there exist only trivial physically pure macrostates,

$$\text{if } \mathfrak{N}' = \mathfrak{Z}, \quad \text{then every physically pure macrostate is trivial} \tag{106}$$

I. Examples for Quantal Systems

Generalized quantum mechanics allows the description of a wide variety of physical and engineering systems from a unified point of view. The kinematical structure of these systems is characterized by the specification of the Neumann algebra $\mathfrak{N}$ of the so-called observables of the system. Because every classification of Neumann algebras implies a classification of quantal systems, we first collect some elementary facts from the theory of Neumann algebras (for details see Refs. 70, 71).

A Neumann algebra $\mathfrak{N}$ is called *atomic* if the lattice $\mathfrak{P}(\mathfrak{N})$ of its projections is atomic. A Neumann algebra $\mathfrak{N}$ is called *continuous* if its projection lattice $\mathfrak{P}(\mathfrak{N})$ is continuous. Recall that a Neumann algebra $\mathfrak{N}$ is called a *factor* if $\mathfrak{N} \cap \mathfrak{N}'$ consists only of scalar operators and that every Neumann algebra can be realized canonically as a direct sum or a direct integral of factors. Factors possessing atoms are called *factors of type I*; factors without atoms are divided into types II and III. Factors possessing finite projections but no atoms are called *factors of type II*; factors possessing neither atoms nor finite projections are called *factors of type III*. A Neumann algebra is said to be of *type X* ($X = $ I, II, or III) if all the factors in its central decomposition are of type X. If a Neumann algebra $\mathfrak{N}$ is of type X, its commutant $\mathfrak{N}'$ is also of type X. Every Abelian Neumann algebra is of type I.

The following list presents a few important classes of quantal systems. Here $\mathfrak{H}$ denotes the Hilbert space of the universe of discourse, $\mathfrak{N}$ is the algebra of observables, $\mathfrak{N}'$ is its commutant, which determines the macrostates, and $\mathfrak{Z}$ is the center $\mathfrak{N} \cap \mathfrak{N}'$, which determines the directly perceptible events.

1. Traditional quantum mechanics
Basic assumption $\mathfrak{N}$ is irreducible.

Consequences
- $\mathfrak{N} = \mathfrak{B}(\mathfrak{H})$ is atomic and of type I.
- $\mathfrak{N}' = 1 \cdot \mathbb{C}$, that is, there are no nontrivial macrostates.
- $\mathfrak{Z} = 1 \cdot \mathbb{C}$, that is, there are no nontrivial directly perceptible events.
- All states are primary.

Remark   This choice was made by von Neumann[1] in 1932.

2. Reduced description in traditional quantum mechanics
Basic assumption   In a combined system $\mathfrak{H} = \mathfrak{H}_1 \otimes \mathfrak{H}_2$, the second system is not observed, and $\mathfrak{N}$ is chosen as $\mathfrak{N} = \mathfrak{B}(\mathfrak{H}_1) \otimes 1 \cdot \mathbb{C}$.

Consequences
- $\mathfrak{N}$ is atomic and of type I.
- $\mathfrak{N}' = 1 \cdot \mathbb{C} \otimes \mathfrak{B}(\mathfrak{H}_2)$, that is, there exist nontrivial macrostates.

- $\mathfrak{Z} = 1 \cdot \mathbb{C} \otimes 1 \cdot \mathbb{C}$, that is, there are no nontrivial directly perceptible events.
- All states are primary.

Remarks   This choice represents an incomplete description; the macrostates arise only because of a lack of knowledge, which is in principle avoidable. An example is a molecule where only the electronic motions ($\mathfrak{H}_1$) but not the nuclear motions ($\mathfrak{H}_2$) are observed.

3. Abelian superselection rules

Basic assumption $\mathfrak{N}'$ is Abelian and nontrivial.

Consequences
- $\mathfrak{N}$ is of type I.
- $\mathfrak{N}' \subset \mathfrak{N}'' = \mathfrak{N}$.
- $\mathfrak{Z} = \mathfrak{N}'$; that is, there exist directly perceptible events but no nontrivial physically pure macrostates.

Remark   The hypothesis of commuting superselection rules[20] has been inspired by the empirical fact that the well-known superselection operators of particle physics (electric charge, baryon number, electron lepton number, muon lepton number) do not only commute with all observables but also among themselves.

4. Dirac system

Basic assumption   Existence of a complete set of commuting observables.

Consequences
- $\mathfrak{N} = \mathfrak{Z}'$, so that $\mathfrak{N}$ is of type I.
- $\mathfrak{Z} = \mathfrak{N}'$.
- There exist no nontrivial physically pure macrostates.
- If $\mathfrak{N}'$ is nontrivial, there are directly perceptible events; all superselection rules are Abelian.

Remark   Dirac's[7] requirement of the existence of a complete set of commuting observables has been interpreted in a mathematically rigorous form by Jauch[21] to mean that $\mathfrak{N}$ contains at least one maximal Abelian algebra.

5. Classical systems

Basic assumption   Nonexistence of incompatible propositions.

Consequences
- $\mathfrak{N} = \mathfrak{Z}$.
- $\mathfrak{N}$ is Abelian; that is, $\mathfrak{N} \subset \mathfrak{N}'$, hence of type I.

6. Deterministic classical systems

Basic assumption   All motions are bidirectionally deterministic and classical.

Consequences
- $\mathfrak{N}$ is Abelian and atomic.

Remark   Compare Kronfli.[99]

7. Stochastic classical system

Basic assumption   All motions are purely nondeterministic and classical.

Consequences   • $\mathfrak{N}$ is Abelian and continuous.

Remark   Compare also the discussion in Section VI.

8. Perfectly classical systems

Basic assumption   Absence of quantal correlations in the sense of Schrödinger[40,41] and Einstein, Podolsky, and Rosen.[42]

Consequences
- $\mathfrak{N}$ is maximal Abelian; that is, $\mathfrak{N} = \mathfrak{N}'$.
- $\mathfrak{Z} = \mathfrak{N} = \mathfrak{N}'$.
- There exists no superposition of states.
- There exist no nontrivial physically pure macrostates.

Remark   Note that there exist maximal Abelian algebras that are atomic and some that are continuous. Only perfectly classical systems can never be entangled with their surroundings.

9. Purely quantal systems in interaction with perfectly classical systems

Basic assumption   Consistent description of the interactions between a purely quantal system with $\mathfrak{N}_1 = \mathfrak{B}(\mathfrak{H}_1)$ and a perfectly classical system with $\mathfrak{N}_2 = \mathfrak{A} = \mathfrak{A}' \subset \mathfrak{B}(\mathfrak{H}_2)$.

Consequences
- $\mathfrak{H} = \mathfrak{H}_1 \otimes \mathfrak{H}_2$.
- $\mathfrak{N} = \mathfrak{B}(\mathfrak{H}_1) \otimes \mathfrak{A}$.
- $\mathfrak{N}' = 1 \cdot \mathbb{C} \otimes \mathfrak{A}$.
- $\mathfrak{Z} = 1 \cdot \mathbb{C} \otimes \mathfrak{A}$.
- There exist no nontrivial physically pure macrostates.
- All primary states are product states with respect to $\mathfrak{H}_1 \otimes \mathfrak{H}_2$.

Remark   The best-known example is the first-order Born–Oppenheimer approximation,[100] describing the motions of the electrons ($\mathfrak{H}_1$) interacting with a classical nuclear framework.

10. General type I quantal systems

Basic assumption $\mathfrak{N}$ is a homogeneous Neumann algebra of type I.

Consequences
- $\mathfrak{H} = \mathfrak{H}_1 \otimes \mathfrak{H}_2 \otimes \mathfrak{H}_3$.
- $\mathfrak{N} = \mathfrak{B}(\mathfrak{H}_1) \otimes 1 \cdot \mathbb{C} \otimes \mathfrak{A}$ with $\mathfrak{A} = \mathfrak{A}'$.
- $\mathfrak{N}' = 1 \cdot \mathbb{C} \otimes \mathfrak{B}(\mathfrak{H}_2) \otimes \mathfrak{A}$.
- $\mathfrak{Z} = 1 \cdot \mathbb{C} \otimes 1 \cdot \mathbb{C} \otimes \mathfrak{A}$.

Remark   Compare Dixmier.[70, sec. ed., p.241] Accordingly, every homogeneous type I quantal system is a combination of cases 2 and 4.

11. *Standard type III quantal systems*

Basic assumption $\mathfrak{N}$ is standard (i.e., there exists a vector that is cyclic for $\mathfrak{N}$ and $\mathfrak{N}'$) and of type III.

Consequences   • There exists a conjugation J in $\mathfrak{H}$ with

$\mathfrak{N} = J\mathfrak{N}'J,$

$\mathfrak{N}' = J\mathfrak{N}J,$

$JXJ = X^*$ for every $X \in \mathfrak{Z}$.

• There exists a unique modular group of *outer* automorphisms of $\mathfrak{N}$ generating thermodynamic equilibrium states characterized by an inverse temperature β that is a generalized eigenvalue of a directly perceptible classical observable.

• As type III algebras have no irreducible representations, there exist no pure states, and a decomposition of a physically pure macrostate makes no physical sense.

• In contrast to example 2 the existence of macrostates is not due to an avoidable lack of knowledge.

• It is sound to associate to a *single* system with a type III algebra a temperature and an entropy.

Remarks   For the mathematical details, see Takesaki;[101] for applications in statistical mechanics, see Emch.[26] Note that there are no classical systems of type III.

J. Dynamical Quantal Systems

According to our Axiom 4 [compare (II.20b)], the time evolution of every subtheory is given by an Abelian one-parameter group $t \to \alpha_t$ of automorphisms of the universal propositional system $\mathcal{U} = \mathfrak{P}(\mathfrak{B})$. A theorem by Uhlhorn[58] implies that this time evolution group is implemented by a unitary or antiunitary projective representation in the Hilbert space $\mathfrak{H}$ (a projective representation is a representation up to a factor of unit modulus),

$$\alpha_t(F) = T_t F T_{-t}, \qquad \forall F \in \mathfrak{P}(\mathfrak{B}), \quad \forall t \in \mathbb{R}$$

$$T_t T_s = c(t, s) T_{t+s}, \qquad |c(t, s)| = 1, \quad t, s \in \mathbb{R}$$

Since the square of an antiunitary operator is unitary, the case of an antiunitary operator T_t can be ruled out. Furthermore, for an Abelian group the factors $c(t, s)$ can always be chosen to be identically equal to one so that the projective representations reduce to a vector representation[102] with

$$T_t T_s = T_{t+s}, \qquad t, s \in \mathbb{R} \tag{107}$$

For technical reasons some continuity assumptions are needed. Fortunately, an exceedingly weak assumption is sufficient, because, in a separable Hilbert space, every weakly measurable one-parameter unitary group is strongly continuous.[103] In this case we can apply Stone's theorem and write the unitary operator T_t in the form

$$T_t = \exp(-itH), \qquad t \in \mathbb{R} \tag{108}$$

where the infinitesimal generator H is an (in general, unbounded) self-adjoint operator, called the *Hamiltonian* of the system. To summarize:

Dynamical Group (109)

The time evolution of the universe of discourse is uniquely determined by a weakly measurable Abelian one-parameter group $t \to T_t$ of unitary operators acting on the Hilbert space $\mathfrak{H}$, called the dynamical group $\mathfrak{T}$

$$\mathfrak{T} = \{T_t : t \in \mathbb{R}\}$$

If S is a state of the universe of discourse at time $t = 0$, the state S_t at time t is given by

$$S_t = T_t S T_{-t}, \qquad \forall S \in \mathfrak{P}(\mathfrak{B}), \quad \forall t \in \mathbb{R}$$

Remarks:

1. If S is an atom in $\mathfrak{P}(\mathfrak{B})$, and $\Psi = S\Psi$ a corresponding state vector, the state vector Ψ_t at time t is given by

$$\Psi_t = T_t \Psi, \qquad t \in \mathbb{R} \tag{110}$$

In traditional quantum mechanics this is called the *Schrödinger picture* and is often expressed via the time-dependent Schrödinger equation

$$i\frac{d\Psi_t}{dt} = H\Psi_t \tag{111}$$

Note that (110) is more general than (111), because (110) applies to every vector $\Psi \in \mathfrak{H}$, whereas (111) applies only to vectors from the domain of the Hamiltonian H.

2. The dynamical group $\mathfrak{T}$ induces an automorphism on the algebra $\mathfrak{B}(\mathfrak{H})$ of all bounded operators acting on $\mathfrak{H}$,

$$A \to A(t) \stackrel{\text{def}}{=} T_{-t} A T_t \in \mathfrak{B}(\mathfrak{H}) \qquad \forall A \in \mathfrak{B}(\mathfrak{H})$$

However, this automorphism of $\mathfrak{B}(\mathfrak{H})$ does not always map the algebra $\mathfrak{N}$ of observables into itself,

$$T_{-t} N T_t \notin \mathfrak{N} \quad \text{in general}, \qquad N \in \mathfrak{N}$$

so that in our context the Heisenberg picture has, in general, no physical meaning. For the very special case when the Hamiltonian is an observable, the Heisenberg picture is equivalent to the Schrödinger picture. The Hamiltonian is not a bona fide observable, but just the infinitesimal generator of the dynamical group. In general, the Hamiltonian does not represent the energy so that we do not assume that the Hamiltonian is semibounded. In particular, for infinite systems the Hamiltonian is never associated with the algebra of observables so the dynamics must be expected to be algebraically nonautomorphic.

3. In general, the dynamical group transforms primary states into nonprimary states. This fact is the root for the occurrence of irreversible phenomena in quantum mechanics and is discussed in detail in Section V.

Therewith, the key concept in the Hilbert space model of generalized quantum mechanics is a *dynamical quantal system* $(\mathfrak{H}, \mathfrak{T}, \mathfrak{N})$ consisting of

1. a separable complex Hilbert space $\mathfrak{H}$, representing the universe of discourse,
2. a dynamical group $\mathfrak{T}$, describing the time evolution of every state of the universe of discourse,
3. a Neumann algebra $\mathfrak{N}$ of observables that specifies a particular point of view.

The universe of discourse is fully characterized by the pair $(\mathfrak{H}, \mathfrak{T})$. Every universe $(\mathfrak{H}, \mathfrak{T})$ allows myriads of subtheories $(\mathfrak{H}, \mathfrak{T}, \mathfrak{N}_\alpha)$ of the universal theory $(\mathfrak{H}, \mathfrak{T}, \mathfrak{B})$, $\mathfrak{N}_\alpha \subset \mathfrak{B}(\mathfrak{H})$. Moreover, all these subtheories can be partially ordered, and the set of all subtheories forms a lattice that is induced by the following lattice operations:[104]

$$\mathfrak{N}_\alpha \wedge \mathfrak{N}_\beta \overset{\text{def}}{=} \mathfrak{N}_\alpha \cap \mathfrak{N}_\beta$$

$$\mathfrak{N}_\alpha \vee \mathfrak{N}_\beta \overset{\text{def}}{=} (\mathfrak{N}_\alpha \cup \mathfrak{N}_\beta)''$$

The relevant algebra $\mathfrak{N}_\alpha$ has to be determined by pattern recognition devices used by the investigator. Of course, not every choice of $\mathfrak{N}_\alpha$ has to make sense. The usefulness of the particular subtheory depends on the integrity[105] and the dynamical stability of the selected subsystem and on the degree of dynamical autonomy of the description.

IV. THEORY OF OBSERVATIONS IN QUANTUM MECHANICS

A. Orientation

For the reasons given in Section I, a statistical interpretation of quantum mechanics that is based on the projection postulate is too restrictive. The purpose of the present section is to derive a much stronger statistical interpretation based on a general theory of observations and containing the traditional statistical interpretation of quantum mechanics as a special case.

By an observation theory we mean a theory that refers to directly perceptible phenomena in the sense of Section II.H but that makes no explicit reference to a measured system and a distinct measurement instrument. Neumann's[1] caricature of a laboratory measurement is included in such a scheme but it represents an oversimplification that, in general, is not tenable.

The classical nature of the observational tools is allowed for by the existence of a nontrivial center in the algebra of observables of the observed system. This classical part has an unproblematic observation theory since the interactions between the classical object system and external observational tools can be made as small as desired. In a factor state the classical observables always *have* definite values that can be determined without significant disturbance of the state. In this case we are allowed to refer to the values of the classical observables without reference to the observational tools so that the center is in a natural way the *output* of a quantal system.

If there are no external influences, a quantal system $(\mathfrak{H}, \mathfrak{T}, \mathfrak{N})$ is a *free dynamical system* whose motions are governed entirely by the unitary time evolution group $\mathfrak{T}$. If we restrict our attention to the directly perceptible phenomena only, the motions in the output can be represented by the classical dynamical system $(\mathfrak{H}, \mathfrak{T}, \mathfrak{Z})$ with $\mathfrak{Z} = \mathfrak{N} \cap \mathfrak{N}'$. In this section we show that $(\mathfrak{H}, \mathfrak{T}, \mathfrak{Z})$ is a *classical stochastic system* and discuss its relation to the "measurement process" of traditional quantum mechanics. This analysis is nontrivial because, as a rule, the time evolution does not map factor states into factor states. We prove here that the values of the classical observables are sample values of the random variables of the classical stochastic system $(\mathfrak{H}, \mathfrak{T}, \mathfrak{Z})$. In Section V we then discuss the dynamics of this classical stochastic system, that is, the directly perceptible stochastic processes induced in the classical part of a quantal system by the unitary and reversible evolution group $\mathfrak{T}$.

B. The Completeness of Quantum Mechanics

To derive the statistical interpretation of quantum mechanics from the individual one, we have to add a postulate that asserts the completeness of quantum mechanics. Such a postulate must be compatible with the axioms of quantum mechanics of individual systems, has the function to exclude the possibility of a still more detailed description, and is necessary to complete the theory. It has to be noted that the claims[45,106-111] that the statistical interpretation can be derived from an individual Everett-type interpretation are incorrect.[112-114] Heuristically, the postulate needed in addition has to warrant that in the long run it is impossible to win money with a gambling system governed by a quantal random generator. It is important that such a statement *cannot* be derived from the postulates formulated in Sections II and III. On the other hand, it is an empirically meaningful assertion because it is a statement about the directly perceptible phenomena of a quantal system. It is based on the empirically well-established existence of an irreducible randomness as a basic trait of nature.

More precisely, we assume that the outcome sequence obtained by a repetition of observations in the classical part of identically prepared systems must be either always identical or random. The randomness of a finite sequence of events is a purely logical concept and can be measured by its complexity. Note that no concepts of traditional probability theory (like probabilities or a probability measure) are involved in the definition of randomness. According to the pioneering ideas of Kolmogorov[115,116] and Chaitin,[117] the randomness of a finite binary sequence is measured by its complexity, which is defined as the length of the shortest binary program that produces this sequence in a given Turing machine (a Turing machine is essentially an automatic computer having an infinitely expandable memory; it can be taken as a precise definition of the concept of an algorithm). In the limit of very long sequences this definition of randomness becomes asymptotically independent of the particular Turing machine chosen. Therefore we can state the postulate of the excluded quantal gambling systems in the following precise way:

Postulate: **Completeness** (1)
Consider an ensemble of n ($n = 1, 2, \ldots$) structurally identical, mutually noninteracting and uncorrelated closed quantal systems $(\mathfrak{H}, \mathfrak{T}, \mathfrak{N})$, each of them having the same prehistory. Let p_k be the value of a projector $P \in \mathfrak{Z} = \mathfrak{N} \cap \mathfrak{N}'$ actualized in the kth member of the ensemble ($p_k = 0$ or $p_k = 1$). Then for every fixed value of n, *either* all the values of p_k are identical

$$p_1 = p_2 = \cdots = p_n$$

or there exists a constant $\bar{p}$, $0 < \bar{p} < 1$, such that the numbers $x_k = p_k - \bar{p}$ form a random sequence $\{x_1, x_2, \ldots, x_n\}$ in the sense of algorithmic probability theory so that every computable real-valued function C of bounded difference fulfills the relation

$$\lim_{n \to \infty} \frac{1}{n} C(x_1, x_2, \ldots, x_n, 0, 0, \ldots) = 0$$

Remarks:
1. Note that the completeness postulate does not presuppose any probability measure.

2.   There exist different definitions of randomness. We have chosen the weak concept of a "random sequence of exponential type," as defined by Schnorr.[118,sec. 17] Roughly, this characterization is equivalent to the validity of the strong law of large numbers. For a discussion of stronger concepts of randomness and their relation to nonisomorphic models of Zermelo–Fraenkel set theory and quantum mechanics, compare Benioff.[119,120]

C.   Statistical Interpretation of the Classical Part of a Quantal System

A statistical interpretation always refers to a fictitious *ensemble* of infinitely many mutually noninteracting and uncorrelated systems of the same kind. If all members of the ensemble are in the same state, the ensemble is said to be homogeneous; if not, it is said to be mixed. The statistical interpretation can be derived by applying the individual interpretation to the infinite ensemble, considering it as a single system.

We begin with the discussion of a homogeneous *finite pre-ensemble* consisting of n mutually uncorrelated quantal systems $(\mathfrak{H}, \mathfrak{T}, \mathfrak{N})$, all being in the state $S \in \mathfrak{P}(\mathfrak{B})$, which generates an NPL-functional ρ on $\mathfrak{N}$ by

$$SNS = \rho(N)S, \qquad N \in \mathfrak{N}$$

According to the results of Section III.G, the central decomposition of ρ

$$\rho = \int_{\Lambda} \rho_{\lambda} \nu(d\lambda) \tag{2}$$

is carried by the factor NPL-functionals ρ_{λ}, which are dispersion-free on $\mathfrak{Z} = \mathfrak{N} \cap \mathfrak{N}'$. With this, for every classical observable $X \in \mathfrak{Z}$ we have

$$\rho_{\lambda}(X) = x(\lambda) \tag{3}$$

where $x(\lambda)$ is the value of X in the sector $\mathfrak{H}(\lambda)$.

The individual members of the ensemble are by definition uncorrelated so that the state $\mathbf{S}$ of a homogeneous pre-ensemble with n members is given by

$$\mathbf{S} = \otimes^n S \tag{4}$$

and the corresponding NPL-functional $\boldsymbol{\rho}$ by

$$\boldsymbol{\rho}(X_1 \otimes X_2 \otimes \cdots \otimes X_n) = \rho(X_1)\rho(X_2) \cdots \rho(X_n) \tag{5}$$

Inserting the central decomposition (2) into (5), we get the central decomposition of $\boldsymbol{\rho}$ as

$$\boldsymbol{\rho} = \int_{\Lambda} \boldsymbol{\rho}_{\lambda} \nu[d\lambda] \tag{6}$$

where we have introduced the following multi-index notation:

$$\boldsymbol{\lambda} \overset{\text{def}}{=} (\lambda_1, \lambda_2, \ldots, \lambda_n)$$

$$\rho_{\boldsymbol{\lambda}}(X_1 \otimes X_2 \otimes \cdots X_n) = \rho_{\lambda_1}(X_1)\rho_{\lambda_2}(X_2) \cdots \rho_{\lambda_n}(X_n)$$

$$\boldsymbol{\nu}[d\boldsymbol{\lambda}] \overset{\text{def}}{=} \nu(d\lambda_1)\nu(d\lambda_2) \ldots \nu(d\lambda_n)$$

$$\boldsymbol{\Lambda} \overset{\text{def}}{=} \Lambda_1 \times \Lambda_2 \times \cdots \times \Lambda_n$$

Clearly, $\boldsymbol{\rho_\lambda}$ is a factor NPL-functional with respect to the algebra $\mathfrak{N}$

$$\mathfrak{N} \overset{\text{def}}{=} \otimes^n \mathfrak{N} \tag{7}$$

whose center $\mathfrak{Z} \overset{\text{def}}{=} \mathfrak{N} \cap \mathfrak{N}'$ is given by

$$\mathfrak{Z} \overset{\text{def}}{=} \otimes^n \mathfrak{Z} \tag{8}$$

Let $\mathbf{X}$ be the symmetric and normalized extension of the classical observable $X \in \mathfrak{Z}$ to the pre-ensemble

$$\mathbf{X} \overset{\text{def}}{=} \frac{1}{n} \sum_{j=1}^{n} \otimes^{j-1} 1 \otimes X \otimes^{n-j} 1 \tag{9}$$

so that

$$\boldsymbol{\rho_\lambda}(\mathbf{X}) = \left\{ \frac{1}{n} \sum_{j=1}^{n} x(\lambda_j) \right\} \tag{10}$$

According to postulate (III.100) the factor NPL-functional $\boldsymbol{\rho_\lambda}$ represents a physically pure state of the pre-ensemble. Thus $\{x(\lambda_1), x(\lambda_2), \ldots, x(\lambda_n)\}$ is a sequence of independently actualized values of the classical observable X, so that the conditions of the completeness postulate (1) are satisfied and we can apply the strong law of large numbers to the arithmetic mean and define a mean value $\bar{x}$ by

$$\lim_{n \to \infty} \frac{1}{n} \sum_{j=1}^{n} x(\lambda_j) \overset{\text{def}}{=} \bar{x} \tag{11}$$

Note that the mean value $\bar{x}$ defined by (11) for a particular sample sequence $\{\lambda_1, \lambda_2, \ldots\}$ is in fact independent of the particularities of this sequence. That is, $\bar{x}$ is independent of $\boldsymbol{\lambda}$, so that for every $\boldsymbol{\lambda}$ we have

$$\boldsymbol{\rho_\lambda}(\mathbf{X}) \to \bar{x} \qquad \text{for } n \to \infty$$

hence with (6)

$$\lim_{n \to \infty} \boldsymbol{\rho}(\mathbf{X}) = \bar{x}$$

and with (5) and (10)

$$\bar{x} = \rho(X) \tag{12}$$

We have now derived the following frequency interpretation of the NPL-functional ρ induced by an arbitrary state S:

Frequency Interpretation (13)

To every classical observable X and every state S there corresponds a real number $\rho(X)$ defined by

$$SXS = \rho(X)S$$

This number equals the mean value of the classical observable X in an ensemble whose individual members are in dispersion-free factor states whose frequency distribution is determined via (2) by the central measure ν.

Because every bounded function φ of a classical observable X is again a classical observable, the mean value $\bar{\varphi}$ of the classical observable $\varphi(X)$ is given by

$$\bar{\varphi} = \rho\{\varphi(X)\} \tag{14}$$

In the case of an atomic state S with the state vector $\Psi = S\Psi$ the mean values are given by the familiar expressions

$$\rho(X) = \langle \Psi | X\Psi \rangle \tag{15a}$$

$$\bar{\varphi} = \langle \Psi | \varphi(X)\Psi \rangle \tag{15b}$$

Remark: Sometimes it is claimed that the expectation value postulate of traditional quantum mechanics is a simple consequence of the fact that whatever the state vector $\Psi \in \mathfrak{H}$ and the observable $X \in \mathfrak{B}(\mathfrak{H})$ may be, the vector $\Phi = \otimes^n \Psi$ approaches an eigenvector of the operator $\mathbf{X}$, defined by (9), in the sense that

$$\|(\mathbf{X} - \bar{x})\Phi\| \to 0 \quad \text{as } n \to \infty$$

(see, for example, Finkelstein[121, p.59]). That is, even if X is *not* a classical observable, the extended operator $\mathbf{X}$ has always a definite value in every state of the ensemble. But this is not to say that the probabilities corresponding to the mean value $\bar{x}$ can be interpreted as a frequency of some event in the ensemble because the observable X has, in general, no definite value in a member of the ensemble. This is the case for all states if and only if X is a classical observable. For arbitrary observables a statistical interpretation can only be derived under an additional conditioning of the system by an external intervention such as a measurement process (cf. Sections IV.E and IV.F).

D. The Classical Part of a Quantal System Is a Probability Space

To prove that classical observables represent real-valued random variables in the sense of mathematical probability theory, we have to show two things:

1. The formalism of quantum mechanics generates the formalism of mathematical probability theory.
2. The interpretation adopted for quantum mechanics generates the usual frequency interpretation of probability theory.

We set the stage for the subsequent discussion by recalling the terminology of Kolmogorov's mathematical probability theory,[121] nowadays almost universally accepted in mathematical probability theory. For all omitted details, compare any modern text on measure-theoretical probability (e.g., Loève[122] or Breiman[123]). A *probability space* is a triple (Ω, Σ, μ). In it, Ω is a nonempty set, called the *sample space*. Its elements $\omega \in \Omega$ are called the *elementary events*. A probability cannot, in general, be defined on every subset of Ω but only on sufficiently nice subsets, which are obtainable from countable intersections and unions. These nice subsets are collected into a *σ-field* Σ, defined as a nonempty collection of subsets of Ω closed under complements, countable unions, and countable intersections. Every σ-field contains the empty set $\varnothing$ (interpreted as the impossible event) and the whole space Ω (interpreted as the sure event) as members. A subset $\mathscr{A}$ belonging to Σ is called a *measurable set* and interpreted as an *event*. A probability measure μ on Σ is a countably additive function defined on all events $\mathscr{A} \in \Sigma$ and possessing the properties $0 \le \mu(\mathscr{A}) \le 1$, $\mu(\varnothing) = 0$, and $\mu(\Omega) = 1$. The number $\mu(\mathscr{A})$ is called the *probability* of the event $\mathscr{A}$.

This setup of probability theory is much too general; in all physical applications the sample space Ω is always a topological space, and in our case it is even a metrizable compact space so that we can work with Borel σ-fields. A *Borel σ-field* is the smallest σ-field containing all open sets; the elements of a Borel σ-field are called *Borel sets*. A real-valued point function $x : \Omega \to \mathbb{R}$ is called a *Borel-measurable function* (or simply a *Borel function*) on (Ω, Σ) if for every Borel set $\mathscr{B}$ of the real line $\mathbb{R}$ the set $\{\omega : x(\omega) \in \mathscr{B}\}$ is in Σ. In the language of probability theory a Borel-measurable function is called a *random variable*. For some fixed $\omega \in \Omega$, $x(\omega)$ is called a *sample value* or a *realization* of the random variable $x : \Omega \to \mathbb{R}$.

The integral

$$\int_{\Omega} x(\omega) \mu(d\omega)$$

(if it exists) is called the expected value $\mathscr{E}(x)$ of the random variable x, and

$$\int_\Omega \{x(\omega) - \mathscr{E}(x)\}^2 \mu(d\omega)$$

(if it exists) is called the *variance*. A real-valued Borel function $\varphi : \mathbb{R} \to \mathbb{R}$ of random variable $x : \Omega \to \mathbb{R}$ is also a random variable; its expectation is defined to be

$$\mathscr{E}\{\varphi(x)\} = \int_\Omega \varphi\{x(\omega)\}\mu(d\omega) \tag{16}$$

For every probability space (Ω, Σ, μ) the space $\mathscr{L}_p(\Omega, \Sigma, \mu)$, $1 \le p < \infty$, consists of those μ-measurable complex-valued functions on Ω for which the norm

$$\|f\|_p \overset{\text{def}}{=} \left\{ \int_\Omega |f(\omega)|^p \mu(d\omega) \right\}^{1/p} \tag{17}$$

is finite. The space $\mathscr{L}_\infty(\Omega, \Sigma, \mu)$ consists of all μ-essentially bounded complex-valued functions with the norm

$$\|f\|_\infty = \inf_{\substack{\mathscr{A} \in \Sigma \\ \mu(\mathscr{A})=0}} \sup\{|x(\omega)| : \omega \in \Omega, \omega \notin \mathscr{A}\} \tag{18}$$

All these $\mathscr{L}_p$-spaces are Banach spaces; $\mathscr{L}_2$ is even a Hilbert space with the inner product

$$\langle \Phi | \Psi \rangle \overset{\text{def}}{=} \int_\Omega \Phi(\omega)^* \Psi(\omega) \mu(d\omega), \qquad \Phi, \Psi \in \mathscr{L}_2(\Omega, \Sigma, \mu) \tag{19}$$

A random variable $x : \Omega \to \mathbb{R}$ belonging to the Banach space $\mathscr{L}_\infty(\Omega, \Sigma, \mu)$ is called a *bounded random variable*. If x and x' are elements of $\mathscr{L}_\infty(\Omega, \Sigma, \mu)$, the pointwise product is well defined and in $\mathscr{L}_\infty(\Omega, \Sigma, \mu)$. Moreover, $\|xx'\|_\infty \le \|x\|_\infty \|x'\|_\infty$, so that $\mathscr{L}_\infty(\Omega, \Sigma, \mu)$ is a commutative Banach algebra. It is well known that $\mathscr{L}_\infty(\Omega, \Sigma, \mu)$ is even a maximal Abelian Neumann algebra acting on the Hilbert space $\mathscr{L}_2(\Omega, \Sigma, \mu)$. On the other hand, Segal[125] has shown that every Abelian Neumann algebra is unitarily equivalent to a subalgebra of an algebra $\mathscr{L}_\infty$ acting on a Hilbert space $\mathscr{L}_2$ over some probability space. This result gives the basic connection between classical observables and bounded random variables on some probability space.

The isomorphism between classical observables and random variables has already been constructed in Section III.G by realizing the center $\mathfrak{Z}$ as a multiplication algebra. Accordingly, there exists an isometric isomorphism $\pi : \mathfrak{Z} \to \mathscr{L}_\infty(\Lambda, \Sigma, \nu)$ from the center $\mathfrak{Z}$ onto the multiplication algebra

$\mathscr{L}_\infty(\Lambda, \Sigma, \nu)$ of all bounded, complex-valued Borel-measurable functions,

$$\{\pi(X)\Psi\}(\lambda) = x(\lambda)\Psi(\lambda) \quad \text{for all } \Psi \in \mathscr{L}_2(\Lambda, \Sigma, \nu), \tag{20}$$

where $X \in \mathfrak{Z}$, $x \in \mathscr{L}_\infty(\Lambda, \Sigma, \nu)$. The sample space Λ is given by the spectrum of a generating classical observable of the center, Σ is the σ-field of Borel subsets of Λ, and the probability measure ν is given by (III.85).

A point λ of the sample space Λ characterizes a pure NPL-functional on $\mathscr{L}_\infty(\Lambda, \Sigma, \nu)$, whereas in the central decomposition (III.89a) it specifies a particular sector $\mathfrak{H}(\lambda)$. A Borel set $\mathscr{A} \in \Sigma$ is uniquely characterized by its indicator function $\chi_{\mathscr{A}}$, defined by

$$\chi_{\mathscr{A}}(\lambda) \stackrel{\text{def}}{=} \begin{cases} 1 & \text{if } \lambda \in \mathscr{A} \\ 0 & \text{otherwise} \end{cases} \tag{21}$$

In the Hilbert-space model of quantum mechanics, the indicator function $\chi_{\mathscr{A}}$ is represented by a central projection $P(\mathscr{A})$, defined by

$$\{\pi[P(\mathscr{A})]\Psi\}(\lambda) = \chi_{\mathscr{A}}(\lambda)\Psi(\lambda), \qquad \Psi \in \mathfrak{H}, \quad \Psi(\lambda) \in \mathfrak{H}(\lambda) \tag{22}$$

and corresponds therefore to a classical proposition. Accordingly, we can call the Borel field Σ the σ-field of *directly perceptible events*. A random variable is by definition a real-valued Borel function $x : \Lambda \to \mathbb{R}$; it defines via (20) a self-adjoint operator in $\mathfrak{Z}$, that is, a classical observable.

According to the frequency interpretation (13), the mean value of any classical observable is given by the NPL-functional ρ, defined on $\mathfrak{Z}$ by

$$SXS = \rho(X)S, \qquad X \in \mathfrak{Z} \tag{23}$$

where S is the state of the universe of discourse. By the representation theorem of Riesz, this NPL-functional can be represented by a Borel measure μ on (Λ, Σ) so that

$$\rho(X) = \int_\Lambda x(\lambda)\mu(d\lambda) \tag{24}$$

Accordingly, the mean value $\rho(X)$ in the sense of the frequency interpretation (13) of quantum mechanics is exactly the expectation $\mathscr{E}(x)$ of the random variable x with respect to the probability space (Λ, Σ, μ). The probability measure μ can also be expressed with the aid of the spectral measure $E : \Sigma \to \mathfrak{Z}$ of (III.84) as

$$\mu(\mathscr{A}) = \rho\{E(\mathscr{A})\}, \qquad \mathscr{A} \in \Sigma \tag{25}$$

The central measure ν has been defined with respect to a separating vector for $\mathfrak{Z}$ [cf. (III.85)], so that the probability measure μ is absolutely continuous with respect to ν,

$$\mu \ll \nu \tag{26}$$

For the special case of an atomic state with a state vector Ψ, the Radon–Nikodym derivative is given by

$$\frac{\mu(d\lambda)}{\nu(d\lambda)} = \|\Psi(\lambda)\|_\lambda^2 \tag{27}$$

where $\Psi(\lambda)$ is the component of Ψ in the sector $\mathfrak{H}(\lambda)$, and $\|\cdot\|_\lambda$ is the norm in $\mathfrak{H}(\lambda)$. In this case we can write

$$\mathscr{E}(x) = \int_\Lambda x(\lambda)\|\Psi(\lambda)\|_\lambda^2 \nu(d\lambda) \tag{28}$$

In the foregoing explicit construction of the isomorphism between $\mathfrak{Z}$ and $\mathscr{L}_\infty(\Lambda, \Sigma, \nu)$ we have assumed that the spectrum Λ is a subset of the real axis. However, the isomorphism in question depends only on the topological properties of the space Λ and not on the specific nature of its points, so that two homeomorphic sample spaces can be considered as being essentially identical. Two topological spaces are called homeomorphic if their points can be put into a one-to-one correspondence in such a way that their open sets also correspond to one another. We call any space homeomorphic to the spectrum Λ of a generating classical observable of the center $\mathfrak{Z}$ a *spectrum space* of the center and denote it again by Λ. The spectrum space of the center is always a metrizable compact space. There are various standard realizations of the spectrum space of an Abelian algebra $\mathfrak{Z}$; the mathematicians favor its realization as the character space of $\mathfrak{Z}$, or as the space of regular maximal ideals of $\mathfrak{Z}$. Physically more relevant is the realization of the spectrum space as the set of all pure NPL-functionals of $\mathfrak{Z}$,

$$\Lambda = \{\rho : \rho \in \mathfrak{Z}^*, \rho = \text{pure NPL-functional}\}$$

$$= \{\rho : \rho \in \mathfrak{Z}^*, \rho(X^*X) \geq 0, \rho(X)\rho(Y) = \rho(XY), \forall X, Y \in \mathfrak{Z}\} \tag{29}$$

where the topology of Λ is the weak topology of $\mathfrak{Z}^*$, the dual of the center $\mathfrak{Z}$.

In this slightly more general setting the connection between classical observables and the random variables of probability theory is given by the extended functional calculus (cf., for example, Douglas[126, p. 112]). As a consequence, there exists a *-isometrical isomorphism of an Abelian Neumann algebra $\mathfrak{Z}$ onto a multiplication algebra $\mathscr{L}_\infty(\Lambda, \Sigma, \nu)$ where Λ is the spectrum space of $\mathfrak{Z}$, Σ the σ-field of the Borel sets of Λ, and ν a regular probability measure on Σ. Thereby the space $\mathscr{L}_\infty(\Lambda, \Sigma, \nu)$ is uniquely determined, and the measure ν is unique up to absolute continuity. With this we can summarize the main results as follows:

Classical Observables as Random Variables   (30)

The classical observables of every quantal system $(\mathfrak{H}, \mathfrak{T}, \mathfrak{N})$ in a particular state S are real-valued random variables on a probability space (Λ, Σ, μ) in the sense of Kolmogorov's probability theory in its statistical frequency interpretation.

The sample space Λ equals the spectrum space of the center $\mathfrak{Z} = \mathfrak{N} \cap \mathfrak{N}'$; the directly perceptible events are the elements of the σ-field Σ of the Borel subsets of Λ. The probability measure $\mu : \Lambda \to \mathbb{R}$ is determined by the generating spectral measure $E : \Sigma \to \mathfrak{Z}$ of the center and the NPL-functional ρ by

$$\mu(\mathscr{A}) = \rho\{E(\mathscr{A})\}, \qquad \mathscr{A} \in \Sigma$$

where ρ represents the state S and is defined by $SXS = \rho(X)S$, $X \in \mathfrak{Z}$. Every classical observable X is uniquely represented by a random variable $x : \Sigma \to \mathbb{R}$ on the probability space (Λ, Σ, μ); its characteristic function $\hat{\mu}$ is given by

$$\hat{\mu}(y) \overset{\text{def}}{=} \int_{\Lambda} e^{iyx(\lambda)} \mu(d\lambda) = \rho\{e^{iyX}\}$$

E.   First Illustration: Measurements of the First Kind

The traditional interpretation of quantum mechanics is founded on an extremely simplified model of actual measurement procedures, invented by Johann von Neumann[1] and called *measurements of the first kind* by Wolfgang Pauli.[127] In this model an object system with a Hilbert space $\mathfrak{H}_1$ and a measuring system with a Hilbert space $\mathfrak{H}_2$ are discussed as a joint system with a Hilbert space $\mathfrak{H} = \mathfrak{H}_1 \otimes \mathfrak{H}_2$. Furthermore, the following assumptions are made:

1. At some initial time the system is assumed to be in an   (31a)
    uncorrelated state, represented by a product state vector
    $\Psi \otimes \Phi$ of the joint Hilbert space $\mathfrak{H}_1 \otimes \mathfrak{H}_2$.
2. Then the object and the apparatus interact for some time,   (31b)
    the effect of this interaction being represented by a unitary
    transformation of the state vector.
3. The final state of the apparatus state can be determined by   (31c)
    reading the scale of the measuring instrument. The cor-
    relation established in the interacting period allows an
    inference about the object observables.
4. The remaining arbitrariness is fixed by the following   (31d)
    repeatability hypothesis: After an ideal measurement an
    immediately subsequent measurement is constrained to give
    a result agreeing with that of the first.

The traditional discussion of a measurement of the first kind is restricted to object observables having a pure point spectrum, so that the eigenvalue problem of $A = A^* \in \mathfrak{B}(\mathfrak{H}_1)$ can be written as

$$A\alpha_n = a_n\alpha_n, \qquad \alpha_n \in \mathfrak{H}_1, \qquad \langle \alpha_n | \alpha_m \rangle = \delta_{nm} \tag{32}$$

The measuring instrument is characterized by an observable $M = M^* \in \mathfrak{B}(\mathfrak{H}_2)$ with a corresponding nondegenerate discrete spectrum,

$$M\varphi_n = \lambda_n\varphi_n, \qquad \varphi_n \in \mathfrak{H}_2, \qquad \langle \varphi_n | \varphi_m \rangle = \delta_{nm} \tag{33}$$

with $\lambda_n \neq \lambda_m$ for $n \neq m$.

In order to fix the ideas, we assume that both Hilbert spaces $\mathfrak{H}_1$ and $\mathfrak{H}_2$ are countably infinite-dimensional; for convenience we may assume that the indices in (32) and (33) are given by $n = 0, \pm 1, \pm 2, \ldots$ and that φ_0 corresponds to the neutral state of the measuring instrument. A measuring operation can then be characterized by a unitary operator T (which may be understood as the time evolution operator) defined by

$$T\{\alpha_n \otimes \varphi_m\} = \alpha_n \otimes \varphi_{m+n}, \qquad n, m = 0, \pm 1, \pm 2, \ldots \tag{34}$$

An arbitrary initial state $\Psi \in \mathfrak{H}_1$ of the object system can be expanded as

$$\Psi = \sum_{n=-\infty}^{\infty} \langle \alpha_n | \Psi \rangle \alpha_n \tag{35}$$

so that we get for the transformation of the initial state $\Psi \otimes \varphi_0$

$$T\{\Psi \otimes \varphi_0\} = \sum_{n=-\infty}^{\infty} \langle \alpha_n | \Psi \rangle \alpha_n \otimes \varphi_n \tag{36}$$

So far all assumptions stem from von Neumann.[1] To avoid the difficulties of the traditional discussion of measurements of the first kind, we have in addition to characterize the measuring instrument as a classical device. That is, we have to assume that the instrument observable M is a classical observable. The simplest, although extreme, characterization is to choose the algebra $\mathfrak{N}_1$ of object observables as irreducible and the algebra $\mathfrak{N}_2$ of measuring observables as maximal Abelian

$$\mathfrak{N}_1 = \mathfrak{B}(\mathfrak{H}_1) \qquad \text{with} \quad \mathfrak{Z}_1 \overset{\text{def}}{=} \mathfrak{N}_1 \cap \mathfrak{N}_1' = 1 \cdot \mathbb{C} \tag{37a}$$

$$\mathfrak{N}_2 = \mathfrak{N}_2' \qquad \text{with} \quad \mathfrak{Z}_2 \overset{\text{def}}{=} \mathfrak{N}_2 \cap \mathfrak{N}_2' = \mathfrak{N}_2 \tag{37b}$$

so that the algebra $\mathfrak{N}$ of observables of the joint system is given by

$$\mathfrak{N} \overset{\text{def}}{=} \mathfrak{N}_1 \otimes \mathfrak{N}_2 = \mathfrak{B}(\mathfrak{H}_1) \otimes \mathfrak{Z}_2 \tag{38a}$$

$$\mathfrak{Z} \overset{\text{def}}{=} \mathfrak{N} \cap \mathfrak{N}' = 1 \cdot \mathbb{C} \otimes \mathfrak{Z}_2 \tag{38b}$$

With this the central decomposition of an arbitrary normalized vector $\Xi \in \mathfrak{H}$ is given by

$$\Xi = \sum_{n=-\infty}^{\infty} \Psi_n \otimes \varphi_n \tag{39a}$$

where

$$\Psi_n = \langle \varphi_n' | \Xi \rangle_{\mathfrak{H}_2} \tag{39b}$$

and $\Psi_n \otimes \varphi_n$ is a factor state vector, representing a physically pure state. According to the frequency interpretation (13), the probability that the system is in the factor state $\Psi_n \otimes \varphi_n$ is given by

$$p_n = \|\Psi_n \otimes \varphi_n\|^2 \tag{40}$$

Under a measuring operation T, the final state vector is given by (36), so that the factor state $\Psi_n \otimes \varphi_n$ equals $\langle \alpha_n | \Psi \rangle \alpha_n \otimes \varphi_n$ and the probability p_n can be written as

$$p_n = |\langle \alpha_n | \Psi \rangle|^2 \tag{41}$$

In this factor state the measuring observable M *has* the value λ_n and the object observable A *has* the value a_n. Accordingly, in an ensemble of systems with initial state vector Ψ, the expected value of A after an intervention by the unitary operator T is given by

$$\sum_{n=-\infty}^{\infty} p_n a_n = \sum_{n=-\infty}^{\infty} a_n \langle \Psi | \alpha_n \rangle \langle \alpha_n | \Psi \rangle = \langle \Psi | A \Psi \rangle \tag{42}$$

which is exactly the result claimed by the projection postulate of traditional quantum mechanics.

In this example it is important that the measuring system is characterized by a *maximal* Abelian algebra of observables. Only in this case the factor states are automatically product states, so that not only is requirement (31a) fulfilled, but the object system is again disentangled from the measuring system after the measuring intervention. Accordingly, after the performed measurement we can abstract from the measuring system and say a measured value a_n implies that the state of the object system has been transformed into a state with the state vector α_n.

F. Second Illustration: Generalized Measurements

The orthogonality of the eigenvector of the eigenvalue problem (32) of the preceding section warrants that the measuring operation (34) leads to a measurement of the first kind, characterized by the repeatability requirement (31d). As we have discussed in Section I, this "projection postulate" of traditional quantum mechanics can and should be relaxed. For example, we can consider a unitary measuring operation T defined by

$$T\{\chi_n \otimes \varphi_m\} = \chi_n \otimes \varphi_{m+n}, \qquad n, m = 0, \pm 1, \pm 2, \ldots \qquad (43a)$$

with

$$\langle \varphi_n | \varphi_m \rangle = \delta_{nm}, \qquad |\varphi_n \rangle \langle \varphi_n| \in \mathfrak{B}_2 \qquad (43b)$$

$$\|\chi_n\| = 1, \qquad \chi_n \in \mathfrak{H}_1 \qquad (43c)$$

but with no orthogonality requirement for χ_n. The central decomposition (39) and the probability interpretation (40) are then still valid but the factor state vector $\Psi_n \otimes \varphi_n$ is now given by $\langle \tilde{\chi}_n | \Psi \rangle \chi_n \otimes \varphi_n$, where $\{\tilde{\chi}_n\}$ is a system of vectors biorthogonal to $\{\chi_m\}$,

$$\langle \tilde{\chi}_n | \chi_m \rangle = \delta_{nm} \qquad (43d)$$

so that

$$T\{\Psi \otimes \varphi_0\} = \sum_{n=-\infty}^{\infty} \langle \tilde{\chi}_n | \Psi \rangle \chi_n \otimes \varphi_n \qquad (44)$$

and

$$p_n = |\langle \tilde{\chi}_n | \Psi \rangle|^2 \qquad (45)$$

The factor states of the combined systems are again product states, now given by $\chi_n \otimes \varphi_n$, so that before and after the measurement the object system is disentangled from the measuring system. If the measured value of the classical observable M equals λ_n, the object system is in the state χ_n. In the statistical frequency interpretation the probability distribution of the final states is governed by the probability of (45), that is, by

$$p_n = \langle \Psi | P_n | \Psi \rangle, \qquad n = 0, \pm 1, \pm 2, \ldots \qquad (46)$$

where P_n is the projection on the vector $\tilde{\chi}_n$,

$$P_n \stackrel{\text{def}}{=} |\tilde{\chi}_n \rangle \langle \tilde{\chi}_n| \qquad (47)$$

It is convenient to express the probability distribution (46) as a probability measure. For any semiclosed interval $\Delta = (\lambda', \lambda'']$ on the real line we set

$$\mu(\Delta) \stackrel{\text{def}}{=} \sum_{n=-\infty}^{[\lambda'']} p_n - \sum_{n=-\infty}^{[\lambda']} p_n \qquad (48)$$

where $[\lambda]$ is the greatest integer not exceeding λ. If $\mathscr{A}$ is the union of a finite number of disjoint intervals Δ, we define $\mu(\mathscr{A})$ as the sum of the corresponding $\mu(\Delta)$. By a standard measure-theoretical construction this measure μ can then be extended to the σ-field Σ of all Borel sets $\mathscr{A}$ of the real line. Defining an operator $F(\Delta)$ by

$$F(\Delta) \stackrel{\text{def}}{=} \sum_{n=-\infty}^{[\lambda'']} P_n - \sum_{n=-\infty}^{[\lambda']} P_n \qquad (49)$$

we can apply the very same construction and extend $F(\Delta)$ to all Borel sets of the real axis, so that we can write (46) as

$$\mu(\mathscr{A}) = \langle \Psi | F(\mathscr{A}) \Psi \rangle, \qquad \mathscr{A} \in \Sigma \qquad (50)$$

Such generalized measurements are well defined but do not fulfill the repeatability hypothesis (31d). This fact is an advantage because it is now easily possible to define generalized measurements that involve physical quantities associated with continuous spectra. Quite generally, families of operators $F(\mathscr{A})$, $\mathscr{A} \in \Sigma$ that define by (50) for every state vector Ψ a probability measure are called *probability operator measures*. Because of this relation Davies and Lewis[8] and Davies[34] have proposed to call a probability operator measure an observable. To avoid a confusion with the traditional terminology, we prefer Holevo's[128-130] terminology and define a *generalized measurement* as a map

$$\mathscr{A} \to F(\mathscr{A}), \qquad \mathscr{A} \in \Sigma \qquad (51)$$

where F is a probability operator measure and Σ is the σ-field of Borel sets of the real line.

We conclude this section by supplementing an explicit and general definition of a probability operator measure that will play an important role in the discussion of stochastic processes. A *positive operator-valued measure* (or *POV measure* for short) on a measurable space (Λ, Σ) is a map $F : \Sigma \to \mathscr{B}(\mathfrak{H})$ having the following properties (see, for example, Berberian[131, sec. 3]):

F is nonnegative definite, that is,
$F(\mathscr{A}) \geq 0$ for each $\mathscr{A} \in \Sigma$, $\qquad\qquad\qquad$ (52a)

F is additive, that is,
if $\mathscr{A} \cap \mathscr{B} = \varnothing$, then $F(\mathscr{A} \cup \mathscr{B}) = F(\mathscr{A}) + F(\mathscr{B})$, $\qquad$ (52b)

F is continuous, that is,
$\sup F(\mathscr{A}_n) = F(\mathscr{A})$ whenever $\{\mathscr{A}_n\}$ is an increasing
sequence of sets in Σ whose union $\mathscr{A}$ is also in Σ. $\qquad$ (52c)

If a POV-measure is in addition normalized,
$F : \Sigma \to \mathscr{B}(\mathfrak{H})$ is normalized iff $F(\Lambda) = 1$ $\qquad\qquad$ (52d)

then F is called a *probability operator measure*. The POV measures on a fixed measurable space form a convex set and every projection-valued measure is an extremal point (however, not all extremal points are spectral measures!). For every $\mathscr{A} \in \Sigma$, $F(\mathscr{A})$ is a nonnegative-definite self-adjoint operator but in contrast to the spectral measures $F(\mathscr{A})$ is not required to be idempotent. For a POV measure to be a spectral measure it is necessary and sufficient that it be multiplicative in the sense of (III.22c), that is, that

$$F(\mathscr{A} \cap \mathscr{B}) = F(\mathscr{A})F(\mathscr{B})$$

Every pair (F, ρ) of a probability operator measure $F : \Sigma \to \mathfrak{B}(\mathfrak{H})$ and a normal NPL-functional $\rho : \mathfrak{B}(\mathfrak{H}) \to \mathbb{R}$ generates a probability measure $\mu : \Sigma \to \mathbb{R}$ by

$$\mu(\mathscr{A}) = \rho\{F(\mathscr{A})\}, \qquad \mathscr{A} \in \Sigma \tag{53}$$

If $\mathscr{A} \to F(\mathscr{A})$ is a generalized measurement, and if ρ represents the state of the object system before the measurement, then the probability measure $\rho\{F(\mathscr{A})\}$ is called the probability distribution of the generalized measurement. If F is an atomic spectral measure, then $\mathscr{A} \to F(\mathscr{A})$ defines a measurement of the first kind. If F is not a spectral measure, then the repeatability hypothesis (31d) is not fulfilled, and in contrast to a measurement of the first kind the state after such a generalized measurement cannot be reconstructed from its results. The systematic use of probability operator measures allows a unified and mathematically rigorous treatment of measurements of observables having continuous spectra,[8,9,34] of generalized position observables for photons,[132-135] of optimal hypothesis testing in quantum estimation and detection theory,[14,128-130,136-145] of joint measurement of noncommuting observables,[9,129,130,136] and of sequential quantal measurements.[146,147]

V. DYNAMICS OF THE CLASSICAL PART OF A QUANTUM SYSTEM

A. Trajectories in the Classical Part

The directly perceptible events in a quantal system are represented by the classical observables. At every instant each classical observable of a single system has a definite and sharp numerical value determined by the sector realized in the system under consideration. As time goes on, the realized sector in one and the same physical system can change. As a consequence, the numerical value of a classical observable can change as a function of time. In this chapter we describe the dynamics of such directly perceptible events.

Instead of discussing arbitrary classical observables, it is more convenient to select one single generating classical observable $Z \in \mathfrak{Z}$ with

$$\{Z\}'' = \mathfrak{Z} \tag{1}$$

For each classical observable X there exists a real-valued Borel function φ_X with

$$X = \varphi_X(Z) \tag{2}$$

so that the discussion of the dynamics associated with an arbitrary classical observable X can easily be deduced from the dynamics of the generating observable Z. Therefore we call such a classical observable also *generic*.

Let Λ be the spectrum of a generic classical observable Z, and denote by $\zeta(t) \in \Lambda$ the value of Z in a particular physical system at time t. If we keep this physical system fixed, the map $t \to \zeta(t)$ describes the time evolution of the classical part of this particular system; it is called its *trajectory*. As the classical part of a quantal system represents a stochastic system, the trajectories of two identically prepared quantal systems will, in general, be different. In this situation the experimenter relies on the law of large numbers, roughly saying that if a well-designed experiment is repeated a large number of times, it is very likely that the average of the results represents a theoretically meaningful quantity, to be described in a theoretical model by an expectation value in the sense of mathematical probability theory. That is, the experimenter assumes that he has a large number of systems having the same structure and working simultaneously under identical conditions and starting from identical initial states. Let $\zeta_k(t)$ be the trajectory of the kth member of this finite pre-ensemble ($k = 1, 2, \ldots, N$), and define for every n ($n = 1, 2, \ldots$) and arbitrary times $t_1, t_2, \ldots, t_n$ the mean values

$$c_n^{(N)}(t_1, t_2, \ldots, t_n) \overset{\text{def}}{=} \frac{1}{N} \sum_{k=1}^{N} \zeta_k(t_1)\zeta_k(t_2), \ldots, \zeta_k(t_n) \tag{3}$$

For $N \to \infty$, the experimenter expects convergence in probability

$$c_n^{(N)}(t_1, t_2, \ldots, t_n) \to c_n(t_1, t_2, \ldots, t_n) \qquad \text{for } N \to \infty \tag{4}$$

and for a sufficiently large N he will consider $c_n^{(N)}$ as an estimate for c_n, called the *n-point correlation function* generated by the classical observable Z. From an experimental point of view, all regularities present in the trajectories can be represented by the family of all finite-order correlation functions.

To get a rigorous mathematical description of the properties of the trajectories we need some information about the nature of the space of all trajectories. If we assumed that the trajectories belong to a Lebesgue space

$\mathcal{L}_p(\mathbb{R}, dx)$, it would be impossible to describe many phenomena of great theoretical and engineering interest. For example, the realizations of stochastic processes with independent values at every point (like white noise) are not functions in the ordinary sense (for a detailed discussion, compare, for example, Gel'fand and Vilenkin[84] and Hida[148]). To eliminate such restrictions it is appropriate to use as trajectory space not an ordinary function space but a space of generalized functions on the real axis. For our purpose it is convenient to choose the space $\mathcal{S}^*$ of tempered distributions on the real axis, so that

$$\zeta \in \mathcal{S}^* \tag{5}$$

Recall that $\mathcal{S}^*$ is defined as the topological dual of the Schwartz space $\mathcal{S}$ of infinitely often differentiable and rapidly decreasing real-valued functions on the real axis (for a concise exposition of Schwartz distributions, compare, for example, the text by Bremermann[149]). A very elementary but useful characterization of the corresponding Gel'fand triple $\mathcal{S} \subset \mathcal{L}_2 \subset \mathcal{S}^*$ can be given in terms of the normalized Hermite orthogonal functions

$$h_n(t) \stackrel{\text{def}}{=} (2^n n! \sqrt{\pi})^{-1/2} \exp(\tfrac{1}{2}t^2) \frac{d^n \exp(-t^2)}{dt^n} \tag{6a}$$

and the expansion coefficients g_n of a (generalized) function g

$$g_n \stackrel{\text{def}}{=} \int_{-\infty}^{\infty} g(t) h_n(t) \, dt, \qquad g(t) = \sum_{n=0}^{\infty} g_n h_n(t) \tag{6b}$$

The following characterizations are well known:[150,151]

$$\mathcal{S} = \left\{ g : \sum_{n=0}^{\infty} |g_n|^2 (n+1)^m < \infty \quad \text{for every } m = 0, 1, 2, \ldots \right\} \tag{6c}$$

$$\mathcal{L}_2 = \left\{ g : \sum_{n=0}^{\infty} |g_n|^2 < \infty \right\} \tag{6d}$$

$$\mathcal{S}^* = \{ g : |g_n| < C \cdot (1+n)^m \quad \text{for some } C \text{ and } m \} \tag{6e}$$

With this it is clear that the integral between a test function $f \in \mathcal{S}$ and a tempered distribution $g \in \mathcal{S}^*$ is well defined, and we denote the pairing between $\mathcal{S}$ and $\mathcal{S}^*$ by $(\cdot | \cdot)$,

$$(f|g) \stackrel{\text{def}}{=} \int_{-\infty}^{\infty} f(t) g(t) \, dt, \qquad f \in \mathcal{S}, \quad g \in \mathcal{S}^* \tag{6f}$$

Though most results remain valid if we replace $\mathcal{S}$ by another nuclear space that is dense in $\mathcal{L}_2(\mathbb{R}, dx)$, we limit our study to trajectories in $\mathcal{S}^*$ in order to avoid a heavier mathematical terminology.

To characterize the statistical properties of the trajectories, we introduce an *expectation functional* $\mathscr{E}$ on the trajectory space and define a characteristic functional $\hat{m}$ by

$$\hat{m}[f] \overset{\text{def}}{=} \mathscr{E}\{e^{i(f|\zeta)}\}, \qquad f \in \mathscr{S} \tag{7}$$

This characteristic functional can be considered as the generating functional of all n-point correlation functions c_n of the trajectories,

$$c_m(t_1, \ldots, t_n) = (-i)^n \left. \frac{\delta^n \hat{m}[f]}{\delta f(t_1), \ldots, \delta f(t_n)} \right|_{f=0} \tag{8}$$

The characteristic functional $\hat{m}$ is a continuous positive definite functional on $\mathscr{S}$ with $\hat{m}[0] = 1$. According to a generalization of Bochner's theorem due to Minlos,[152] there exists a unique probability measure m on the σ-field Σ^* of Borel sets of $\mathscr{S}^*$ such that

$$\hat{m}[f] = \int_{\mathscr{S}^*} e^{i(f|\zeta)} m[d\zeta] \tag{9}$$

With this, $(\mathscr{S}^*, \Sigma^*, m)$ becomes a probability space, whereby a trajectory $\zeta \in \mathscr{S}^*$ plays the role of an elementary event, so that the statistical description of the trajectories reduces to the study of probability measures on the space $\mathscr{S}^*$.

B. Trajectories as Realizations of Stochastic Processes

To construct the characteristic functional $\hat{m}$ or the associated probability measure m on the trajectory space from the first principles of quantum mechanics, it is convenient to parametrize the trajectories ζ by the points ω of a sample space Ω, so that for the trajectory ζ_k of the kth member of a statistical ensemble we can write

$$\zeta_k(t) = z(t, \omega_k), \qquad \omega_k \in \Omega \tag{10}$$

It is sensible to choose the sample space Ω in such a way that two different points of Ω correspond to different trajectories, so that Ω is homeomorphic to the set of all possible trajectories. With this, z becomes a stochastic process whose realizations are given by the trajectories ζ. Recall that a *stochastic process* is an indexed family of random variables defined on a *common* probability space (Ω, Σ, μ). In our applications the index t represents the physical time, and the index set will be chosen to be real axis $\mathbb{R}$. In this case the stochastic process is a *continuous parameter process*. Thus, from a mathematical point of view, a stochastic process is a function $z : \mathbb{R} \times \Omega \to \mathbb{R}$ of two variables, $t \in \mathbb{R}$ and $\omega \in \Omega$, which is Σ-measurable on Ω for each $t \in \mathbb{R}$. When t is fixed, $z(t, \cdot)$ is merely a random variable on the sample space Ω. For each fixed $\omega \in \Omega$, $z(\cdot, \omega)$ defines a nonrandom function on the parameter set $\mathbb{R}$ and is called a realization of the stochastic process corresponding to the event ω.

The stochastic process $z: \mathbb{R} \times \Omega \to \mathbb{R}$ induces in a natural way a probability measure m on the σ-field Σ^* of the Borel sets of the trajectory space $\mathscr{S}^*$ by

$$m[\mathscr{B}] \overset{\text{def}}{=} \mu\{\omega : \omega \in \Omega, z(\cdot, \omega) \in \mathscr{B}\}, \qquad \mathscr{B} \in \Sigma^* \tag{11}$$

With this, the characteristic functional (7) can be expressed in terms of the fundamental probability space (Ω, Σ, μ) as

$$\hat{m}[f] = \int_\Omega \exp\left\{i \int_{-\infty}^\infty f(t) z(t, \omega)\, dt\right\} \mu(d\omega) \tag{12}$$

so that the experimentally accessible correlation functions c_n are given by

$$c_n(t_1, t_2, \ldots, t_n) = \int_\Omega z(t_1, \omega) z(t_2, \omega) \ldots z(t_n, \omega) \mu(d\omega) \tag{13}$$

C.   A Special Case: The Time Evolution Is an Automorphism of the Center

The stochastic process z induced by a generic classical observable Z of a quantal system can be characterized by the underlying probability space (Ω, Σ, μ) together with a trajectory space. The main purpose of this section is to construct the probability space (Ω, Σ, μ) from the first principles of quantum mechanics. Because this problem is not quite trivial, we proceed in two steps. First we consider in the present section the very special and simple case where the time evolution of the quantal system induces a motion of the center. The much more difficult general case can then be reduced to this special situation by an observationally equivalent extension of the quantal system.

According to the basic result (IV.30), every bounded classical observable $Z \in \mathfrak{Z} = \mathfrak{N} \cap \mathfrak{N}'$ of a quantal system $(\mathfrak{H}, \mathfrak{T}, \mathfrak{N})$ can be considered as a real-valued bounded random variable $z: \Lambda \to \mathbb{R}$ on the probability space (Λ, Σ, ν), where Λ is the spectrum space of the center $\mathfrak{Z}$, Σ is the σ-field of the Borel sets of Λ, and ν is the central measure. The relationship between Z and z is given by (III.91) as

$$\{\pi[Z]\Phi\}(\lambda) = z(\lambda)\Phi(\lambda), \qquad \forall \Phi \in \mathfrak{H}, \quad \Phi(\lambda) \in \mathfrak{H}(\lambda) \tag{14}$$

where $\mathfrak{H}(\lambda)$ is a sector arising from the central decomposition (III.86). If the quantal system $(\mathfrak{H}, \mathfrak{T}, \mathfrak{N})$ is in the state S with the corresponding NPL-functional ρ, the expectation of any Borel function φ of the random variable z is given by

$$\mathcal{E}\{\varphi(z)\} = \int_\Lambda \varphi\{z(\lambda)\}\mu(d\lambda) \tag{15}$$

where according to (IV.25) the probability measure μ is given by the spectral measure $E:\Sigma \to \mathfrak{Z}$ of the generating classical observable Z of the center $\mathfrak{Z}$,

$$\mu(\mathscr{A}) = \rho\{E(\mathscr{A})\}, \qquad \mathscr{A} \in \Sigma \tag{16}$$

$$Z = \int_\Lambda \lambda E(d\lambda) \tag{17}$$

If S is the state of the quantal system $(\mathfrak{H}, \mathfrak{T}, \mathfrak{N})$ at time $t = 0$, the state S_t at time t is given by

$$S_t = T_r S T_{-t}, \qquad T_t \in \mathfrak{T}, \quad t \in \mathbb{R} \tag{18}$$

so that the NPL-functional ρ_t associated with S_t is defined by

$$S_t N S_t = \rho_t(N) S_t, \qquad \forall N \in \mathfrak{N} \tag{19}$$

For the generator Z of the center we get

$$S_t Z S_t = \rho_t(Z) S_t \tag{20}$$

This relation can be transformed into

$$SZ(t)S = \rho_t(Z)S \tag{21}$$

where we introduced the abbreviation

$$Z(t) \stackrel{\text{def}}{=} T_{-t} Z T_t, \qquad t \in \mathbb{R} \tag{22}$$

In this section we consider only the very special case where for every $r \in \mathbb{R}$ the operator $Z(t)$ again belongs to the center,

$$Z(t) \in \mathfrak{Z}, \qquad \forall t \in \mathbb{R} \tag{23}$$

In this case the time evolution group $\mathfrak{T}$ induces an automorphism of $\mathfrak{Z}$ onto $\mathfrak{Z}$, so that we can define for each $t \in \mathbb{R}$ a random variable $z(t, \cdot)$ on the probability space (Λ, Σ, ν) by

$$\{\pi[Z(t)]\Phi\}(\lambda) = z(t, \lambda)\Phi(\lambda), \qquad \forall \Phi \in \mathfrak{H} \tag{24}$$

Accordingly, the family $\{z(t, \cdot): t \in \mathbb{R}\}$ of random variables $z(t, \cdot): \Lambda \to \mathbb{R}$ represents a stochastic process $z: \mathbb{R} \times \Lambda \to \mathbb{R}$ on the common probability space (Λ, Σ, ν). If we replace the central measure ν by the measure $\mu \ll \nu$ induced by the state S of the quantal system, the characteristic functional

$\hat{m}$ of the stochastic process z on the probability space (Λ, Σ, μ) is given by

$$\hat{m}[f] = \int_\Lambda \exp\left(i \int_{-\infty}^\infty f(t)z(t, \lambda) \, dt\right)\mu(d\lambda), \qquad f \in \mathcal{S} \tag{25}$$

or with (16)

$$\hat{m}[f] = \int_\Lambda \exp\left(i \int_{-\infty}^\infty f(t)z(t, \lambda) \, dt\right)\rho\{E(d\lambda)\} \tag{26}$$

Hence for a normal NPL-functional ρ

$$\hat{m}[f] = \rho\left\{\int_\Lambda \exp\left(i \int_{-\infty}^\infty f(t)z(t, \lambda) \, dt\right)E(d\lambda)\right\} \tag{27a}$$

$$= \rho\left\{\exp\left(i \int_{-\infty}^\infty f(t)Z(t) \, dt\right)\right\} \tag{27b}$$

Summary: *Directly Perceptible Processes in a Quantal System with Time-Invariant Center* (28)

If the time evolution group $\mathfrak{T}$ induces an automorphism on the center $\mathfrak{Z} = \mathfrak{N} \cap \mathfrak{N}'$ of a quantal system $(\mathfrak{H}, \mathfrak{T}, \mathfrak{N})$, the directly perceptible stochastic processes in the classical part have the underlying probability space (Λ, Σ, μ), where the sample space Λ equals the spectrum space of the center. The probability measure μ is given by

$$\mu(\mathcal{A}) = \rho\{E(\mathcal{A})\}, \qquad \mathcal{A} \in \Sigma$$

where $E : \Sigma \to \mathfrak{Z}$ is the spectral measure of the center and ρ is the NPL-functional generated by the state S of the system at time $t = 0$,

$$SNS = \rho(N)S, \qquad N \in \mathfrak{N}$$

If ρ is normal, the characteristic functional $\hat{m}$ of the directly perceptible stochastic processes is given by

$$\hat{m}[f] = \rho\left\{\exp\left(i \int_\infty^\infty f(t)Z(t) \, dt\right)\right\}, \qquad f \in \mathcal{S}$$

where

$$Z(t) \stackrel{\text{def}}{=} T_{-t}ZT_t, \qquad Z = Z^* \in \mathfrak{Z}$$

and

$$\{Z\}'' = \mathfrak{Z}$$

D. Extension of Quantal Systems

In general, the unitary evolution of a quantal system $(\mathfrak{H}, \mathfrak{T}, \mathfrak{N})$ does not induce an automorphism of the center $\mathfrak{Z} = \mathfrak{N} \cap \mathfrak{N}'$; that is, in general the operator $Z(t)$,

$$Z(t) \overset{\text{def}}{=} T_{-t} Z T_t, \qquad Z \in \mathfrak{Z}, \quad T_t \in \mathfrak{T} \tag{29}$$

does not belong to the center. In this case it may be very difficult to find directly an explicit expression for the characteristic functional of the stochastic process in the classical part of a quantal system. It is mathematically convenient to view these processes as taking part in a larger but fictitious quantal system $(\tilde{\mathfrak{H}}, \tilde{\mathfrak{T}}, \tilde{\mathfrak{N}})$ in which $\tilde{\mathfrak{T}}$ induces an automorphism of the center $\tilde{\mathfrak{Z}} \overset{\text{def}}{=} \tilde{\mathfrak{N}} \cap \tilde{\mathfrak{N}}'$. For such an extension of the original quantal system to be useful, the directly perceptible phenomena in $\mathfrak{Z}$ have to be related to those of $\tilde{\mathfrak{Z}}$. However, it is not required that the algebras $\mathfrak{N}$ or $\mathfrak{Z}$ be related by a homomorphism to $\tilde{\mathfrak{N}}$ or $\tilde{\mathfrak{Z}}$, respectively, so that it is possible to get a much simpler description of the dynamics in the classical part of the extended system. Note that it is not assumed that the extended quantal system $(\tilde{\mathfrak{H}}, \tilde{\mathfrak{T}}, \tilde{\mathfrak{N}})$ has a physical significance; the system is used merely as a mathematical tool. This procedure is reminiscent of the introduction of a state space description in mathematical system theory. It is not necessary but convenient to do so.

We relate an extension to the original quantal system by requiring that the original system be a subsystem of its extension and for all initial states the dynamics in the classical parts of the two systems be indistinguishable. More precisely:

*Definition: **Observationally Equivalent Extensions*** (30)
A quantal system $(\tilde{\mathfrak{H}}, \tilde{\mathfrak{T}}, \tilde{\mathfrak{N}})$ is said to be an observationally equivalent extension of the quantal system $(\mathfrak{H}, \mathfrak{T}, \mathfrak{N})$ if

1. $\mathfrak{H} \subset \tilde{\mathfrak{H}}$.
2. For every state S of $(\mathfrak{H}, \mathfrak{T}, \mathfrak{N})$ there exists a state $\tilde{S}$ of $(\tilde{\mathfrak{H}}, \tilde{\mathfrak{T}}, \tilde{\mathfrak{N}})$, and for every classical observable $X \in \mathfrak{Z} = \mathfrak{N} \cap \mathfrak{N}'$ there exists a classical observable $\tilde{X} \in \tilde{\mathfrak{Z}} = \tilde{\mathfrak{N}} \cap \tilde{\mathfrak{N}}'$ such that

$$\rho_t(X) = \tilde{\rho}_t(\tilde{X})$$

where

$$S_t X S_t = \rho_t(X) S_t$$

$$\tilde{S}_t \tilde{X} \tilde{S}_t = \tilde{\rho}_t(\tilde{X}) \tilde{S}_t$$

and

$$S_t \overset{\text{def}}{=} T_t S T_{-t}, \qquad T_t \in \mathfrak{T}$$

$$\tilde{S}_t \overset{\text{def}}{=} \tilde{T}_t \tilde{S} \tilde{T}_{-t}, \qquad \tilde{T}_t \in \tilde{\mathfrak{T}}$$

Since an automorphism of the center is very easy to discuss, we are looking for an observationally equivalent extension $(\tilde{\mathfrak{H}}, \tilde{\mathfrak{T}}, \tilde{\mathfrak{N}})$ of the original system $(\mathfrak{H}, \mathfrak{T}, \mathfrak{N})$ such that

$$\tilde{X}(t) \overset{\text{def}}{=} \tilde{T}_{-t}\tilde{X}\tilde{T}_t \in \tilde{\mathfrak{Z}} \qquad \text{for all } \tilde{X} \in \tilde{\mathfrak{Z}}, \quad t \in \mathbb{R}$$

For this construction we use Naĭmark's extension theorem in a version due to Sz.-Nagy.[153] According to this theorem, every finite or infinite sequence $\{\check{Z}_n\}$ of bounded self-adjoint operators acting on a complex Hilbert space $\mathfrak{H}$ can be represented by means of a sequence $\{\tilde{Z}_n\}$ of pairwise commuting self-adjoint bounded operators acting on the extension space $\tilde{\mathfrak{H}}$ in the form

$$Z_n\Psi = \tilde{P}\tilde{Z}_n\Psi \qquad \text{for each } \Psi \in \mathfrak{H} \tag{31a}$$

where

$$\tilde{Z}_n\tilde{Z}_m = \tilde{Z}_m\tilde{Z}_n \qquad \text{for every pair } (n, m) \tag{31b}$$

and $\tilde{P}$ is the projection from $\tilde{\mathfrak{H}}$ onto $\mathfrak{H}$

$$\mathfrak{H} = \tilde{P}\tilde{\mathfrak{H}} \tag{31c}$$

To apply Naĭmark's extension theorem we choose an arbitrary orthonormal basis $\{h_n : n = 0, 1, 2, \ldots\}$ in the Schwartz space $\mathscr{S}$, for example, the functions (6a), and define

$$Z_n \overset{\text{def}}{=} \int_{-\infty}^{\infty} Z(t)h_n(t)\, dt, \qquad n = 0, 1, 2, \ldots \tag{32}$$

$$Z(t) = T_{-t}ZT_t, \qquad t \in \mathbb{R} \tag{33}$$

where $Z \in \mathfrak{Z}$ is a generating classical observable for the center $\mathfrak{Z}$,

$$\mathfrak{Z} = \{Z\}'' \tag{34}$$

According to Naĭmark's extension theorem there exists an extension space $\tilde{\mathfrak{H}}$ such that the sequence $\{Z_n\}$ can be related to a sequence $\{\tilde{Z}_n\}$ of commuting self-adjoint bounded operators $\tilde{Z}_n$ fulfilling (31). These operators generate an Abelian Neumann algebra $\tilde{\mathfrak{Z}}$

$$\tilde{\mathfrak{Z}} \overset{\text{def}}{=} \{\tilde{Z}_n : n = 0, 1, 2, \ldots\}'' \tag{35}$$

where the double commutant has to be taken with respect to the bounded operators of $\tilde{\mathfrak{H}}$. Provided there exist two constants C and m such that for every n

$$\|\tilde{Z}_n\| < C(1 + n)^m \tag{36}$$

then the expression

$$\tilde{Z}(t) \overset{\text{def}}{=} \sum_{n=0}^{\infty} \tilde{Z}_n h_n(t) \tag{37}$$

is well defined and represents an Abelian field, that is, an operator-valued tempered distribution. In this case we define a time translation by

$$\alpha_t\{\tilde{Z}[f]\} \overset{\text{def}}{=} \tilde{Z}[f_t] \tag{38}$$

where

$$\tilde{Z}[f] \overset{\text{def}}{=} \int_{-\infty}^{\infty} \tilde{Z}(s)f(s)\,ds, \qquad f \in \mathscr{S} \tag{39}$$

and

$$f_t(s) \overset{\text{def}}{=} f(s-t) \tag{40}$$

The map $\alpha_t : \tilde{\mathfrak{Z}} \to \tilde{\mathfrak{Z}}$ is a continuous one-parameter automorphism of the Abelian Neumann algebra $\tilde{\mathfrak{Z}}$ that can therefore be unitarily implemented on the Hilbert space $\tilde{\mathfrak{H}}$.[154] That is, there exists a strongly continuous one-parameter group $\tilde{\mathfrak{T}}$ of unitary operators $\tilde{T}_t$ acting on the Hilbert space $\tilde{\mathfrak{H}}$ such that

$$\alpha_t\{\tilde{Z}[f]\} = \tilde{T}_{-t}\tilde{Z}[f]\tilde{T}_t, \qquad t \in \mathbb{R} \tag{41a}$$

or equivalently,

$$\tilde{Z}(t) = \tilde{T}_{-t}\tilde{Z}\tilde{T}_t, \qquad t \in \mathbb{R} \tag{41b}$$

The compression of the Abelian field (37) to the original Hilbert space $\mathfrak{H}$ gives with (31a), (32), and (37)

$$Z(t) = \tilde{P}\tilde{Z}(t)|_{\mathfrak{H}}, \quad t \in \mathbb{R} \tag{42}$$

where $|_{\mathfrak{H}}$ denotes the restriction to the Hilbert space $\mathfrak{H}$. As Z generates the center $\mathfrak{Z}$ and $\tilde{Z}(t)$ generates the center $\tilde{\mathfrak{Z}}$ we have constructed the correspondence of all classical observables required in definition (30). To construct the correspondence between the states, we select an arbitrary state S of the original quantal system $(\mathfrak{H}, \mathfrak{T}, \mathfrak{R})$ at time $t = 0$ and construct the corresponding time-evolved states $S_t \in \mathfrak{B}(\mathfrak{H})$ and $\tilde{S}_t \in \mathfrak{B}(\tilde{\mathfrak{H}})$ by

$$S_t \overset{\text{def}}{=} T_t S T_{-t} \tag{43}$$

$$\tilde{S}_t \overset{\text{def}}{=} \tilde{T}_t \tilde{S} \tilde{T}_{-t} \tag{44}$$

where $\tilde{S}$ denotes the trivial extension of S to $\tilde{\mathfrak{H}}$

$$\tilde{S} \overset{\text{def}}{=} S\tilde{P} = \tilde{P}S\tilde{P} \tag{45}$$

and the NPL-functionals associated therewith

$$S_t X S_t = \rho_t(X)S_t, \qquad X \in \mathfrak{Z} \tag{46a}$$

$$\tilde{S}_t \tilde{X} \tilde{S}_t = \tilde{\rho}_t(\tilde{X})\tilde{S}_t, \qquad \tilde{X} \in \tilde{\mathfrak{Z}} \tag{46b}$$

With (33), and (41b), (43) we can also write

$$SZ(t)S = \rho_t(Z)S \tag{47}$$

$$\tilde{S}\tilde{Z}(t)\tilde{S} = \tilde{\rho}_t(\tilde{Z})\tilde{S} \tag{48}$$

Inserting (42) into (47) and using the fact that $\tilde{P}$ and $\tilde{S}$ commute, we get

$$\rho_t(Z)\tilde{S} = \tilde{S}\tilde{P}\tilde{Z}(t)\tilde{P}\tilde{S} = \tilde{P}\tilde{S}\tilde{Z}(t)\tilde{S}\tilde{P}$$

$$= \tilde{P}\tilde{T}_{-t}\tilde{S}_t\tilde{Z}\tilde{S}_t\tilde{T}_t\tilde{P}$$

$$= \tilde{\rho}_t(\tilde{Z})\tilde{P}\tilde{T}_{-t}\tilde{S}_t\tilde{T}_t\tilde{P} \qquad \text{[by (46)]}$$

$$= \tilde{\rho}_t(\tilde{Z})\tilde{P}\tilde{S}\tilde{P} \qquad \text{[by (44)]}$$

$$= \tilde{\rho}_t(\tilde{Z})\tilde{S}$$

so that for every initial state we have

$$\rho_t(Z) = \tilde{\rho}_t(\tilde{Z}) \tag{49}$$

as required in definition (30). Thus we have fulfilled all essential requirements for an observationally equivalent quantal system. For a full description we have still to make a consistent choice for the algebra $\tilde{\mathfrak{N}}$ of the observables of the extended system. The simple possibilities $\tilde{\mathfrak{N}} = \tilde{\mathfrak{Z}}$ or $\tilde{\mathfrak{N}} = \tilde{\mathfrak{Z}}'$ are certainly consistent.

Summary: **Observationally Equivalent Naĭmark Extension** (50)

To every quantal system $(\mathfrak{H}, \mathfrak{T}, \mathfrak{N})$ there exists an observationally equivalent extension $(\tilde{\mathfrak{H}}, \tilde{\mathfrak{T}}, \tilde{\mathfrak{N}})$ such that the time evolution group $\tilde{\mathfrak{T}}$ induces an automorphism on the center $\tilde{\mathfrak{Z}} = \tilde{\mathfrak{N}} \cap \tilde{\mathfrak{N}}'$. The operator $Z(t)$

$$Z(t) \stackrel{\text{def}}{=} T_{-t}ZT_t, \qquad T_t \in \mathfrak{T}$$

where Z is a generating classical observable of the center $\mathfrak{Z}$, is the compression of the Abelian field

$$\tilde{Z}(t) \stackrel{\text{def}}{=} \tilde{T}_{-t}\tilde{Z}\tilde{T}_t, \qquad \tilde{T}_t \in \tilde{\mathfrak{T}}$$

to the Hilbert space $\mathfrak{H} = \tilde{P}\tilde{\mathfrak{H}}$,

$$Z(t) = \tilde{P}\tilde{Z}(t)|_{\mathfrak{H}}$$

The minimal dilation $\tilde{Z}(t)$ is unique up to unitary equivalence and given by the constructive Naĭmark extension. The center $\tilde{\mathfrak{Z}}$ of the minimal extension is (up to unitary equivalence) uniquely determined by the (in general, non-Abelian) Neumann algebra $\{Z(t) : t \in \mathbb{R}\}''$.

E. Characterization of the Directly Perceptible Stochastic Processes of a Quantal System

By combining the results of Sections V.C and V.D, we can easily derive a complete statistical description of the trajectories of the directly perceptible events in the classical part of an arbitrary quantal system. Let $(\mathfrak{H}, \mathfrak{T}, \mathfrak{M})$ be the quantal system under consideration by an experimenter, and let $(\tilde{\mathfrak{H}}, \tilde{\mathfrak{T}}, \tilde{\mathfrak{M}})$ be a minimal observationally equivalent Naïmark extension of it. Let $\tilde{Z} \in \tilde{\mathfrak{Z}}$ be a self-adjoint generator of the center $\tilde{\mathfrak{Z}} = \tilde{\mathfrak{M}} \cap \tilde{\mathfrak{M}}'$

$$\{\tilde{Z}\}'' = \tilde{\mathfrak{Z}} \tag{51}$$

and write its spectral resolution in the form

$$\tilde{Z} = \int_{\Omega} \omega \tilde{E}(d\omega) \tag{52}$$

As discussed in detail in Sections III.G and IV.D, the spectral resolution generates a probability space (Ω, Σ, ν) where the spectrum space Ω is a metrizable compact space, Σ is the σ-field of Borel sets of Ω, and ν is the central measure generated by the spectral measure $\tilde{E} : \Sigma \to \tilde{\mathfrak{Z}}$ and a normalized separating vector $\tilde{\Phi} \in \tilde{\mathfrak{H}}$ of $\tilde{\mathfrak{Z}}$,

$$\nu(\mathscr{A}) \overset{\text{def}}{=} \langle \tilde{\Phi} | \tilde{E}(\mathscr{A}) \tilde{\Phi} \rangle, \qquad \mathscr{A} \in \Sigma \tag{53}$$

The central decomposition of the extension space $\tilde{\mathfrak{H}}$ with respect to $\tilde{\mathfrak{Z}}$ is given by

$$\tilde{\mathfrak{H}} = \int_{\Omega}^{\oplus} \tilde{\mathfrak{H}}(\omega) \nu(d\omega) \tag{54}$$

and we denote the component of an arbitrary vector $\tilde{\Psi} \in \tilde{\mathfrak{H}}$ in the sector $\tilde{\mathfrak{H}}(\omega)$ by $\tilde{\Psi}(\omega)$. With this, we can associate to the Abelian field $\tilde{Z}(t) \in \tilde{\mathfrak{Z}}$, defined by (37), a bounded stochastic process $z : \mathbb{R} \times \Omega \to \mathbb{R}$ on the probability space (Ω, Σ, ν) by

$$\{\tilde{\pi}[\tilde{Z}(t)]\tilde{\Psi}\}(\omega) = z(t, \omega)\tilde{\Psi}(\omega), \qquad \tilde{\Psi} \in \tilde{\mathfrak{H}} \tag{55}$$

where $\tilde{\pi} : \tilde{\mathfrak{Z}} \to \mathscr{L}_{\infty}(\Omega, \Sigma, \nu)$ is an isometric isomorphism.

With respect to an arbitrary initial state S of the original system $(\mathfrak{H}, \mathfrak{T}, \mathfrak{M})$ at time $t = 0$, we define an NPL-functional $\tilde{\rho}$ as in (46)

$$\tilde{S}\tilde{X}\tilde{S} = \tilde{\rho}(\tilde{X})\tilde{S} \qquad \text{for all } \tilde{X} \in \tilde{\mathfrak{Z}} \tag{56}$$

where $\tilde{S}$ is given by (44b)

$$\tilde{S} = S\tilde{P} \tag{57}$$

so that the state S generates a probability measure μ on (Ω, Σ) by

$$\mu(\mathscr{A}) \overset{\text{def}}{=} \tilde{\rho}\{\tilde{E}(\mathscr{A})\}, \qquad \mathscr{A} \in \Sigma \tag{58}$$

With respect to the probability space (Ω, Σ, μ) the characteristic functional $\hat{m}$ of the stochastic process $z : \mathbb{R} \times \Omega \to \mathbb{R}$ is given by (25), that is, by

$$\hat{m}[f] = \int_{\Omega} \exp\left(i \int_{-\infty}^{\infty} f(t) z(t, \omega) \, dt\right) \mu(d\omega), \qquad f \in \mathscr{S} \tag{59}$$

By construction, a stochastic process z generated by the classical observable Z of the original system is directly perceptible in the system $(\tilde{\mathfrak{H}}, \tilde{\mathfrak{T}}, \tilde{\mathfrak{N}})$, and describes the possible trajectories of the events associated with the classical observable Z in the original system. As the Naïmark extension is only a convenient tool, it is appropriate to eliminate in the final formulation any reference to the fictitious extension system. Since the state $\tilde{S}$ commutes with the projection $\tilde{P}$ from $\tilde{\mathfrak{H}}$ onto $\mathfrak{H}$, we have

$$\tilde{\rho}\{\tilde{E}(\mathscr{A})\} = \tilde{\rho}\{\tilde{P}\tilde{E}(\mathscr{A})\tilde{P}\}$$

so that it is convenient to introduce the following operator probability measure $F : \Sigma \to \mathfrak{B}(\mathfrak{H})$ by

$$F(\mathscr{A}) \overset{\text{def}}{=} \tilde{P}\tilde{E}(\mathscr{A})|_{\mathfrak{H}} \tag{60}$$

so that

$$\tilde{\rho}\{\tilde{P}\tilde{E}(\mathscr{A})\tilde{P}\} = \rho\{F(\mathscr{A})\} \tag{61}$$

with

$$SNS = \rho(N)S, \qquad N \in \mathfrak{N}$$

The defining properties (III.22) of a normalized spectral measure imply that the compression of the normalized spectral measure $\tilde{E} : \Sigma \to \tilde{\mathfrak{Z}}$ on $\mathfrak{H}$ is a probability operator measure $F : \Sigma \to \mathfrak{B}(\mathfrak{H})$ having the properties (IV.52). The relation $\tilde{E}(\mathscr{A} \cap \mathscr{B}) = \tilde{E}(\mathscr{A})\tilde{E}(\mathscr{B})$ valid for the spectral measure $\tilde{E}$ does, however, not imply a corresponding relation for F, so that the probability operator F is, in general, not a spectral measure. With this, we have the following final result:

Result: *Structure of the Directly Perceptible*
Stochastic Processes of a Quantal System (62)

For every quantal system $(\mathfrak{H}, \mathfrak{T}, \mathfrak{N})$ there exists an (up to equivalence) unique measurable space (Ω, Σ) whereby Ω acts as sample space for the directly perceptible stochastic processes in the classical part of the quantal system $(\mathfrak{H}, \mathfrak{T}, \mathfrak{N})$ and Σ is the σ-field of Borel sets of Ω. The sample space Ω is metrizable and compact; it is determined by the center $\mathfrak{Z} \overset{\text{def}}{=} \mathfrak{N} \cap \mathfrak{N}'$

and its relation to the time evolution group $\mathfrak{T}$. Using the Naĭmark extension theorem, one can generate constructively a probability operator measure $F : \Sigma \to \mathfrak{B}(\mathfrak{H})$ that gives rise to a Neumann algebra having the property

$$\{F(\mathscr{A}) : \mathscr{A} \in \Sigma\}'' = \{T_{-t} Z T_t : Z \in \mathfrak{Z}, \quad T_t \in \mathfrak{T}\}''$$

The directly perceptible stochastic process $z : \mathbb{R} \times \Omega \to \mathbb{R}$ in the classical part of the system has the characteristic functional $\hat{m}$ given by

$$\hat{m}[f] = \int_{\Omega} \exp\left(i \int_{-\infty}^{\infty} f(t) z(t, \omega) \, dt \right) \mu(d\omega), \qquad f \in \mathscr{S}$$

where the probability measure μ is given by

$$\mu(\mathscr{A}) = \rho\{F(\mathscr{A})\}, \qquad \mathscr{A} \in \Sigma$$

and ρ is the NPL-functional determined by the initial state S at time $t = 0$ by

$$SNS = \rho(N)S, \qquad N \in \mathfrak{M}$$

The experimentally directly accessible n-point correlation function (4) of the trajectories of the events in the classical part is now given by

$$c_n(t_1, t_2, \ldots, t_n) = \int_{\Omega} z(t_1, \omega) z(t_2, \omega), \ldots, z(t_n, \omega) \mu(d\omega) \qquad (63)$$

If the NPL-functional ρ is normal, we can interchange the order of integration[129] and introduce an n-point correlation operator C_n by

$$C_n(t_1, t_2, \ldots, t_n) \overset{\text{def}}{=} \int_{\Omega} z(t_1, \omega) z(t_2, \omega), \ldots, z(t_n, \omega) F(d\omega) \qquad (64)$$

so that

$$c_n(t_1, t_2, \ldots, t_n) = \rho\{C_n(t_1, t_2, \ldots, t_n)\} \qquad (65)$$

In particular, we have

$$C_1(t) = \tilde{P}\tilde{Z}(t)|_{\mathfrak{H}} = Z(t) \qquad (66)$$

All n-point correlation operators are symmetric in the time arguments; for example,

$$C_2(t_1, t_2) = C_2(t_2, t_1) \qquad (67)$$

but only if F is a spectral measure do we have

$$C_2(t_1, t_2) = C_1(t_1) C_1(t_2) = Z(t_1) Z(t_2)$$

Accordingly, in the general case the operator variance is nonvanishing

$$C_2(t_1, t_2) \neq C_1(t_1) C_1(t_2) \quad \text{in general} \qquad (68)$$

F. First Example: Holevo's Extension

The explicit construction of the probability operator measure used in the result (62) for the description of the directly perceptible stochastic processes of a quantal system is as a rule a highly nontrivial matter. Often it is convenient to employ a reformulation of Naĭmark's extension theorem due to Holevo[129] and to use as an extension space a tensor product of the original Hilbert space $\mathfrak{H}$ and an auxiliary Hilbert space $\mathfrak{K}$. Holevo's extension theorem says that for every probability operator measure $F: \Sigma \to \mathfrak{B}(\mathfrak{H})$ on a measurable space (Ω, Σ) there exists a Hilbert space $\mathfrak{K}$ with an inner product $(\cdot | \cdot)_\mathfrak{K}$, and a fixed normalized vector $\Phi \in \mathfrak{K}$ such that

$$F(\mathscr{A}) = \langle \Phi | E(\mathscr{A})\Phi \rangle_\mathfrak{K}, \qquad \mathscr{A} \in \Sigma \tag{69}$$

where $E: \Sigma \to \mathfrak{B}(\mathfrak{H} \otimes \mathfrak{K})$ is a normalized spectral measure acting in the extension space $\mathfrak{H} \otimes \mathfrak{K}$.

It may be worthwhile to illustrate Holevo's extension theorem and the results of this chapter using one of the simplest possible examples: the harmonic oscillator. First we have to specify the dynamical quantal system $(\mathfrak{H}, \mathfrak{T}, \mathfrak{R})$. The complex and separable Hilbert space $\mathfrak{H}$ is chosen as the representation space of an irreducible representation of Weyl's commutation relations for one degree of freedom,[155] that is, for

$$U(p)V(q) = V(q)U(p)\, e^{ipq}, \qquad p, q \in \mathbb{R} \tag{70}$$

where both $\{U(p): p \in \mathbb{R}\}$ and $\{V(q): q \in \mathbb{R}\}$ are strongly continuous one-parameter groups of unitary operators acting irreducibly on the Hilbert space $\mathfrak{H}$. By Stone's theorem we can write

$$U(p) = e^{ipQ}, \qquad p \in \mathbb{R} \tag{71}$$

$$V(q) = e^{-iqP}, \qquad q \in \mathbb{R} \tag{72}$$

where the unbounded self-adjoint operators Q and P fulfill Heisenberg's commutation relation

$$QP - PQ \subset i$$

In the usual Schrödinger representation the operator Q is represented as the multiplication operator so that we can identify the Hilbert space $\mathfrak{H}$ with the space $\mathscr{L}_2(\mathbb{R}, dx)$ of the measurable complex-valued functions on the real axis $\mathbb{R}$ whose absolute square is integrable in the sense of Lebesgue. With this we can write

$$\mathfrak{H} = \mathscr{L}_2(\mathbb{R}, dx) \tag{73a}$$

$$\{U(p)\Psi\}(x) = e^{ipx}\Psi(x) \qquad \text{for every } \Psi \in \mathscr{L}_2(\mathbb{R}, dx) \tag{73b}$$

$$\{V(q)\Psi\}(x) = \Psi(x-q) \qquad \text{for every } \Psi \in \mathscr{L}_2(\mathbb{R}, dx) \tag{73c}$$

The dynamical group $\mathfrak{T}$ is assumed to be given by the Hamiltonian H of a harmonic oscillator of radial frequency ω; that is,

$$T_t \overset{\text{def}}{=} \exp(-itH), \qquad T_t \in \mathfrak{T}, \quad t \in \mathbb{R} \tag{74a}$$

$$H \overset{\text{def}}{=} \tfrac{1}{2}\omega\{P^2 + Q^2\}, \qquad \omega \in \mathbb{R} \tag{74b}$$

For the system to have directly perceptible phenomena, we must choose an algebra $\mathfrak{N}$ of observables having a nontrivial center. To keep the calculations as simple as possible, we choose directly a perfectly classical system characterized by a maximal Abelian algebra of observables. We decide for the algebra generated by Q

$$\mathfrak{N} \overset{\text{def}}{=} \{Q\}'' = \{U(p): p \in \mathbb{R}\}'' \tag{75}$$

It is well known that Q has a simple spectrum so that $\mathfrak{N}$ is maximal Abelian, $\mathfrak{N} = \mathfrak{N}'$, and

$$\mathfrak{Z} = \mathfrak{N} = \mathfrak{N}' \tag{76a}$$

In the Schrödinger representation (73), $\mathfrak{N}$ is represented by the multiplication algebra $\mathscr{L}_\infty$ over $\mathbb{R}$,

$$\mathfrak{N} = \mathscr{L}_\infty(\mathbb{R}, dx) \tag{76b}$$

Therewith the dynamical system $(\mathfrak{H}, \mathfrak{T}, \mathfrak{N})$ is completely specified. As generic classical observable Z we choose the unbounded operator Q,

$$Z = Q \tag{77}$$

The dynamical group (74) and the classical observable Z generate a directly perceptible stochastic process z that can be characterized by either the characteristic functional $\hat{m}$ or the probability measure μ. We assume that the state S is normal and given by the density operator D

$$D = \frac{S}{\text{tr}(S)} \tag{78}$$

so that

$$\mu(\mathscr{A}) = \text{tr}\{DF(\mathscr{A})\} \tag{79}$$

The aim of the example is to show how $\hat{m}$, μ, and F can be evaluated explicitly.

First of all, the example is nontrivial insofar as the dynamical group (74) does not induce an automorphism on the center. In fact, by a straightforward integration we get the well-known result

$$Z(t) \overset{\text{def}}{=} T_{-t}ZT_t = Q\cos\omega t + P\sin\omega t \tag{80}$$

hence

$$\{Z(t) : t \in \mathbb{R}\}'' = \mathcal{B}(\mathfrak{H}) \neq \mathfrak{Z} \tag{81}$$

so that we first have to construct an observationally equivalent extension of the quantal system $(\mathfrak{H}, \mathfrak{T}, \mathfrak{N})$. This can be done in an elegant manner by a Holevo extension.

To simplify the notation, we first introduce a complex parameter b by

$$b \stackrel{\text{def}}{=} \frac{1}{\sqrt{2}}(q + ip), \qquad q, p \in \mathbb{R} \tag{82}$$

and a boson operator B by

$$B = \frac{1}{\sqrt{2}}(Q + iP) \tag{83}$$

Furthermore, we introduce the unitary Weyl operator W by

$$W(b) \stackrel{\text{def}}{=} \exp(bB^* - b^*B) \tag{84a}$$

$$= U(p)V(q) \exp\left(\frac{-ipq}{2}\right) \tag{84b}$$

We also need the so-called normalized vacuum vector Φ, defined by

$$B\Phi = 0, \qquad \|\Phi\| = 1 \tag{85a}$$

or more elegantly by

$$\langle \Phi | W(b) | \Phi \rangle = \exp\left(-\tfrac{1}{2}|b|^2\right) \tag{85b}$$

In this notation the Hamiltonian (74b) is given by

$$H = \omega B^* B + \tfrac{1}{2}\omega \tag{86}$$

and the operator (80) by

$$Z(t) = \sqrt{2} \operatorname{Re}\{B e^{-i\omega t}\} \tag{87}$$

where the real part of an arbitrary operator A is defined by

$$\operatorname{Re}\{A\} = \tfrac{1}{2}(A + A^*)$$

In accordance with (39) we introduce a tested field $Z[f]$ by

$$Z[f] \stackrel{\text{def}}{=} \int_{-\infty}^{\infty} Z(t)f(t)\, dt, \qquad f \in \mathscr{S} \tag{88}$$

so that we get with (87)

$$Z[f] = \sqrt{2} \operatorname{Re}\{B \cdot \hat{f}(\omega)^*\} \tag{89}$$

where we denote the Fourier transform of the real-valued test function f by $\hat{f}$

$$\hat{f}(\omega) \stackrel{\text{def}}{=} \int_{-\infty}^{\infty} e^{i\omega t} f(t)\, dt, \qquad f, \hat{f} \in \mathscr{S} \tag{90}$$

For the Holevo extension (69) we choose as the auxiliary system a twin system of the very same structure and define

$$\tilde{B} \stackrel{\text{def}}{=} B \otimes 1 + 1 \otimes B^* \tag{91}$$

$$\tilde{Z}[f] \stackrel{\text{def}}{=} \sqrt{2}\, \mathrm{Re}\, \{\tilde{B} \cdot \hat{f}(\omega)^*\} \tag{92}$$

The commutation relation $BB^* - B^*B = 1$ implies that $\tilde{B}$ and $\tilde{B}^*$ commute

$$\tilde{B}\tilde{B}^* = \tilde{B}^*\tilde{B} \tag{93}$$

so that $\tilde{Z}$ generates an Abelian Neumann algebra $\tilde{\mathfrak{Z}}$ acting on $\mathfrak{H} \otimes \mathfrak{R}$,

$$\tilde{\mathfrak{Z}} \stackrel{\text{def}}{=} \{\tilde{Z}[f] : f \in \mathscr{S}\}'' \tag{94}$$

The Hamiltonian $\tilde{H}$ that generates the motion of $\tilde{Z}$ in $\tilde{\mathfrak{Z}}$ is given by

$$\tilde{H} \stackrel{\text{def}}{=} \omega B^*B \otimes 1 - \omega 1 \otimes B^*B \tag{95a}$$

$$\tilde{T}_t \stackrel{\text{def}}{=} \exp\{-it\tilde{H}\} \tag{95b}$$

$$\tilde{\mathfrak{T}} \stackrel{\text{def}}{=} \{\tilde{T}_t : t \in \mathbb{R}\} \tag{95c}$$

With this we get

$$\tilde{B}(t) \stackrel{\text{def}}{=} \tilde{T}_{-t}\tilde{B}\tilde{T}_t = \tilde{B}\, e^{-i\omega t} \tag{96}$$

hence

$$\tilde{B}[f] \stackrel{\text{def}}{=} \int_{-\infty}^{\infty} \tilde{B}(t)f(t)\, dt = \tilde{B} \cdot \hat{f}(\omega)^* \tag{97}$$

so that indeed the dynamical group (95) induces the motion (92).

Using the vacuum vector (85) of the auxiliary system as vector Φ in the sense of Holevo's theorem (69), we get with (91) and (92)

$$\sqrt{2}\tilde{Z}[f] = \hat{f}(\omega)\{B^* \otimes 1 + 1 \otimes B\} + \hat{f}(\omega)^*\{B \otimes 1 + 1 \otimes B^*\} \tag{98}$$

hence with $\langle\Phi|B\Phi\rangle = \langle\Phi|B^*\Phi\rangle = 0$ and with (89)

$$Z[f] = \langle\Phi|\tilde{Z}[f]\Phi\rangle_{\mathfrak{R}}, \qquad f \in \mathscr{S} \tag{99}$$

so that indeed $\tilde{Z}[f]$ is a commutative Holevo extension of $Z[f]$, representing the directly perceptible stochastic processes generated by the classical observable $Z = Q$ of the original system.

Using (27b), we get for the characteristic functional $\hat{m}$ of the Abelian field $\tilde{Z}$ with respect to a state represented by a density operator $\tilde{D}$

$$\hat{m}[f] = \text{tr}_{\mathfrak{H} \otimes \mathfrak{N}} \{\tilde{D} e^{i\tilde{Z}[f]}\} \tag{100}$$

where in the Holevo extension $\tilde{D}$ is the tensor product of the original state (78) and the vacuum state (85)

$$\tilde{D} = D \otimes |\Phi\rangle\langle\Phi| \tag{101}$$

Using the abbreviation

$$b \overset{\text{def}}{=} \frac{i}{\sqrt{2}} \hat{f}(\omega) \tag{102}$$

and (98) and (84a), we get

$$e^{i\tilde{Z}[f]} = W(b) \otimes W(-b^*) \tag{103}$$

so that with (85b)

$$\langle\Phi| e^{i\tilde{Z}[f]}\Phi\rangle_{\mathfrak{N}} = W(b) e^{-1/2|b|^2} \tag{104}$$

hence

$$\hat{m}[f] = e^{-1/2|b|^2} \text{tr}_{\mathfrak{H}}\{DW(b)\} \tag{105}$$

This result (105) can be related to well-known phase-space distribution functions. The expression

$$\hat{w}(b) \overset{\text{def}}{=} \text{tr}\{DW(b)\} \tag{106}$$

is the so-called state-generating function; its symplectic Fourier transform is the famous Wigner distribution function[156]

$$w(\beta) \overset{\text{def}}{=} \int_{\mathbb{R}^2} \hat{w}(b) e^{\beta b^* - \beta^* b} \frac{d^2 b}{\pi} \tag{107}$$

$$d^2 b \overset{\text{def}}{=} d(\text{Re } b) d(\text{Im } b)$$

which is, in general, *not* positive, hence not a probability distribution function. However, the so-called Husimi function h, defined in Ref. 157

$$h(\beta) \overset{\text{def}}{=} \langle\beta|D|\beta\rangle \tag{108}$$

where $|\beta\rangle$ are coherent states defined via the Weyl operator (84) and the vacuum vector (85)

$$|\beta\rangle = W(\beta)\Phi \tag{109}$$

is a positive probability distribution function. The relation between the

Wigner and the Husimi function is very simple. Denoting the symplectic Fourier transform of the Husimi function by $\hat{h}$

$$\hat{h}(b) \stackrel{\text{def}}{=} \int_{\mathbb{R}^2} h(\beta)\, e^{\beta b^* - \beta^* b} \frac{d^2\beta}{\pi} \tag{110}$$

we get

$$\hat{h}(b) = e^{-1/2|b|^2} \hat{w}(b) \tag{111}$$

so that we can express (105) by

$$\hat{m}[f] = \hat{h}(b) \tag{112}$$

with

$$b = \frac{i\hat{f}(\omega)}{\sqrt{2}}$$

With the relation

$$W(b)\, e^{-1/2|b|^2} = \int_{\mathbb{R}^2} e^{b\beta^* - b^*\beta} |\beta\rangle\langle\beta| \frac{d^2\beta}{\pi} \tag{113}$$

(for proof, evaluate the expectation values between coherent states) we can write (105) also as

$$\hat{m}[f] = \mathrm{tr}_{\hat{\mathfrak{d}}}\left\{ D \int_{\mathbb{R}^2} e^{b\beta^* - b^*\beta} |\beta\rangle\langle\beta| \frac{d^2\beta}{\pi} \right\} \tag{114}$$

$$= \int_{\mathbb{R}^2} e^{b\beta^* - b^*\beta} \mu(d^2\beta) \tag{115}$$

Hence the probability operator measure F and the probability measure μ mentioned in the main result (62) are given by

$$F(\mathcal{A}) = \int_{\mathcal{A}} |\beta\rangle\langle\beta| \frac{d^2\beta}{\pi}, \qquad \mathcal{A} \in \Sigma \tag{116}$$

$$\mu(\mathcal{A}) \stackrel{\text{def}}{=} \mathrm{tr}\,\{DF(\mathcal{A})\}, \qquad \mathcal{A} \in \Sigma \tag{117}$$

where Σ is the σ-field of the Borel sets of $\mathbb{R}^2$. With this the probability space for the directly perceptible processes of this example is given by (Ω, Σ, μ) with

$$\Omega = \mathbb{R}^2 \tag{118}$$

Comparing (112) and (110) with (115), we see that the Radon-Nikodym derivative of the probability measure μ with respect to the Lebesgue measure equals the Husimi function

$$\pi^{-1}h(\beta) = \frac{\mu(d^2\beta)}{d^2\beta} \qquad (119)$$

The equivalent operator expression is given by

$$\pi^{-1}|\beta\rangle\langle\beta| = \frac{F(d^2\beta)}{d^2\beta} \qquad (120)$$

G. Second Example: White Noise

Roughly speaking, white noise is the derivative of Brownian motion. A Brownian motion is a Gaussian stochastic process with mean zero and covariance Min (t, s). The trajectories of a Brownian motion are continuous but nowhere differentiable so that its formal derivative $z(t)$ makes sense only as a generalized function. With a real-valued test function f from the Schwartz space $\mathscr{S} \subset \mathscr{L}_2(\mathbb{R}, dx)$ [cf. also (6c)] we define

$$z[f] \overset{\text{def}}{=} \int_{-\infty}^{\infty} z(t)f(t)\, dt, \qquad f \in \mathscr{S} \qquad (121)$$

With this, $z[f]$ becomes an ordinary, real-valued Gaussian random variable of zero mean and unit variance so that its characteristic function is given by

$$\mathscr{E}\{e^{iz[f]}\} = e^{-1/2\|f\|^2} \qquad (122)$$

where

$$\|f\|^2 \overset{\text{def}}{=} \int_{-\infty}^{\infty} f(t)^2\, dt < \infty \qquad (123)$$

These considerations lead to the following rigorous definition of a Gaussian white-noise process on the real axis.[84,148] Consider the Gel'fand triple

$$\mathscr{S} \subset \mathscr{L}_2(\mathbb{R}, dx) \subset \mathscr{S}^* \qquad (124)$$

where $\mathscr{S}$ is the Schwartz space of infinitely differentiable and rapidly decreasing real-valued functions on the real axis, and define a characteristic functional $\hat{\eta}$ by

$$\hat{\eta}[f] \overset{\text{def}}{=} e^{-1/2\|f\|^2}, \qquad f \in \mathscr{S} \qquad (125)$$

By Minlos' theorem (quoted in Section V.A), there exists a unique probability measure η on the σ-field Σ^* of all Borel sets of $\mathscr{S}^*$ such that

$$\int_{\mathscr{S}^*} e^{i(f|\zeta)}\eta[d\zeta] = e^{-1/2\|f\|^2}, \qquad f \in \mathscr{S} \qquad (126)$$

We call the stochastic process z that induces this probability measure on the trajectory space $\mathscr{S}^*$ *Gaussian white noise*. Each element $\zeta \in \mathscr{S}^*$ can be thought of as a trajectory of white noise; for each $f \in \mathscr{S}$, $(f|\zeta)$ is a Gaussian random variable with mean zero and variance $\|f\|^2$. The covariance of z is given by

$$\mathscr{E}\{z[f]z[g]\} = (f|g), \qquad f, g \in \mathscr{S} \tag{127}$$

or formally by

$$\mathscr{E}\{z(t)z(s)\} = \delta(t-s) \tag{128}$$

where δ is the Dirac delta function.

The aim of this second example is to construct a dynamical quantal system $(\mathfrak{H}, \mathfrak{T}, \mathfrak{M})$ such that a generic classical observable $Z \in \mathfrak{Z}$ generates a Gaussian white noise process z in the directly perceptible classical part of the system. For this construction we use the results of the first example, discussed in Section V.F, where we have shown how to construct a single directly perceptible harmonic mode. As we know from harmonic analysis, a nondeterministic stochastic process can be thought of as consisting of infinitely many modes. That is, a mechanical model generating a nondeterministic stochastic process has to be a system with infinitely many degrees of freedom. The generalization of the example of Section V.F to finitely many degrees of freedom is trivial; one has just to replace the Hilbert space $\mathscr{L}_2(\mathbb{R}, dx)$ by $\mathscr{L}_2(\mathbb{R}^n, d^n x)$ and use Weyl's commutation relations for n degrees of freedom. The generalization to $n = \infty$ is nontrivial, because in the space $\mathbb{R}^\infty$ there exists no longer a quasiinvariant measure needed for the construction of the corresponding Hilbert space. However, the solution to this problem is well known;[84] instead of the nonexistent Hilbert space $\mathscr{L}_2(\mathbb{R}^\infty, ?)$ we have to use the Hilbert space $\mathscr{L}_2(\mathscr{S}^*, \Sigma^*, \eta)$ of all measurable complex-valued functions on $\mathscr{S}^*$ whose absolute squares are η-integrable. The inner product in this Hilbert space is given by

$$\langle \Phi | \Psi \rangle \stackrel{\text{def}}{=} \int_{\mathscr{S}^*} \Phi[\zeta]^* \Psi[\zeta] \eta[d\zeta], \qquad \Phi, \Psi \in \mathscr{L}_2(\mathscr{S}^*, \Sigma^*, \eta) \tag{129}$$

We proceed in strict analogy to the discussion of the example in Section V.F. To specify the quantal system $(\mathfrak{H}, \mathfrak{T}, \mathfrak{M})$ we first choose a convenient representation of $\mathfrak{H}$ as an $\mathscr{L}_2$-space over $\mathscr{S}^*$,

$$\mathfrak{H} = \mathscr{L}_2(\mathscr{S}^*, \Sigma^*, \eta) \tag{130}$$

It turns out that a Fock representation is appropriate; thus we choose as probability measure η a Gaussian measure defined by Minlos' theorem via the characteristic functional $\hat{\eta}$

$$\hat{\eta}[f] \stackrel{\text{def}}{=} \int_{\mathscr{S}^*} e^{i(f|x)} \eta[dx] = e^{-1/4\|f\|^2}, \qquad f \in \mathscr{S} \tag{131}$$

Note the difference between (126) and (131). It is well known that the strongly continuous Fock representation for Weyl's canonical commutation relations for a field (i.e., a system with infinitely many degrees of freedom)

$$U[f]V[g] = V[g]U[f] e^{i(f|g)}, \qquad f, g \in \mathscr{S} \tag{132}$$

admits an $\mathscr{L}_2$-realization in analogy to (73)[84,158] (for a short account see Reed and Simon[159] in their appendix to Sec. X.7)

$$\{U[f]\Psi\}[x] = e^{i(f|x)}\Psi[x] \tag{133a}$$

$$\{V[g]\Psi\}[x] = e^{(g|x)-(1/2)(g|g)}\Psi[x-g] \tag{133b}$$

valid for every $\Psi \in \mathscr{L}_2(\mathscr{S}^*, \Sigma^*, \eta)$. In analogy to (75) we choose as algebra $\mathfrak{N}$ of observables the maximal Abelian algebra generated by the operator field U

$$\mathfrak{N} \overset{\text{def}}{=} \{U[f] : f \in \mathscr{S}\}'' \tag{134}$$

which equals the multiplication algebra over $\mathscr{S}^*$,

$$\mathfrak{N} = \mathscr{L}_\infty(\mathscr{S}^*, \Sigma^*, \eta) \tag{135}$$

With this the center of the quantal system $(\mathfrak{H}, \mathfrak{T}, \mathfrak{N})$ is given by

$$\mathfrak{Z} = \mathfrak{N} = \mathscr{L}_\infty(\mathscr{S}^*, \Sigma^*, \eta) \tag{136}$$

In analogy to (71) and (72) we use Stone's theorem to introduce the self-adjoint Abelian fields $Q[f]$ and $P[g]$ by

$$U[f] = e^{iQ[f]}, \qquad f \in \mathscr{S} \tag{137}$$

$$V[g] = e^{-iP[g]}, \qquad g \in \mathscr{S} \tag{138}$$

In the first example, we have specified the time evolution by the Hamiltonian (74b). As a rule, in an infinite system the Hamiltonian does not exist as a well-defined operator. Nevertheless, it may be helpful to give at least a formal expression from which the dynamics can be derived heuristically. The result of the heuristic derivation will then be used to specify the dynamics in a mathematically rigorous manner. In analogy to (83) we define a non-self-adjoint field $B[f]$ by

$$B[f] \overset{\text{def}}{=} \frac{1}{\sqrt{2}}\{Q[f] + iP[f]\}, \qquad f \in \mathscr{S} \tag{139}$$

and write

$$B[f] = \int_{-\infty}^{\infty} B(s)f(s) \, ds \tag{140a}$$

$$= \int_{-\infty}^{\infty} \hat{B}(\omega)\hat{f}(\omega) \, d\omega \tag{140b}$$

where for convenience we introduced the Fourier transforms in an unsymmetrical way

$$\hat{f}(\omega) \stackrel{\text{def}}{=} \int_{-\infty}^{\infty} e^{i\omega t} f(t)\, dt \tag{141}$$

$$\hat{B}(\omega) \stackrel{\text{def}}{=} \frac{1}{2\pi} \int_{-\infty}^{\infty} e^{-i\omega t} B(t)\, dt \tag{142}$$

As a formal Hamiltonian H we choose

$$H \stackrel{\text{def}}{=} 2\pi \int_{-\infty}^{\infty} \omega \hat{B}(\omega)^* \hat{B}(\omega)\, d\omega \tag{143}$$

so that

$$e^{itH} \hat{B}(\omega)\, e^{-itH} = e^{-i\omega t} \hat{B}(\omega) \tag{144}$$

hence with

$$e^{itH} B[f]\, e^{-itH} \stackrel{\text{def}}{=} B_t[f] = B[f_t] \tag{145}$$

where

$$f_t(s) \stackrel{\text{def}}{=} f(s+t) \tag{146}$$

We use this heuristic argument to define the dynamical group $\mathfrak{T}$ by

$$T_{-t} U[f] T_t \stackrel{\text{def}}{=} U[f_t], \qquad t \in \mathbb{R} \tag{147a}$$

where f_t is defined by (146). Because the characteristic functional (131) is time translation invariant, the group

$$\mathfrak{T} \stackrel{\text{def}}{=} \{T_t : t \in \mathbb{R}\} \tag{147b}$$

is a strongly continuous unitary group on $\mathfrak{H}$, and it will be taken as the dynamical group. With this the dynamical quantal system $(\mathfrak{H}, \mathfrak{T}, \mathfrak{N})$ is completely and rigorously defined. As generic classical observable we will take the Abelian field $Q[f]$,

$$Z[f] \stackrel{\text{def}}{=} Q[f] \tag{148a}$$

with

$$\{Z[f] : f \in \mathscr{S}\}'' = \mathfrak{Z} \tag{148b}$$

and discuss the directly perceptible stochastic process generated by $Z[f]$.

   With this specification the main work is done. What is left is easy and precisely analogous to the example of Section V.F. One uses a Holevo extension with the extension space $\mathscr{L}_2(\mathscr{S}^*, \Sigma^*, \eta) \otimes \mathscr{L}_2(\mathscr{S}^*, \Sigma^*, \eta)$, and as reference vector Φ in the auxiliary Hilbert space [cf. (101)] the Fock vacuum vector, which is given by the function identically equal 1 on $\mathscr{S}^*$

$$\Phi[x] \overset{\text{def}}{=} 1 \qquad \text{for all } x \in \mathscr{S}^* \tag{149}$$

With (131) we have

$$\langle \Phi | U[f]\Phi \rangle = e^{-(1/4)\|f\|^2} \tag{150}$$

so that with a density operator $\tilde{D}$

$$\tilde{D} \overset{\text{def}}{=} |\Phi\rangle\langle\Phi| \otimes |\Phi\rangle\langle\Phi| \tag{151}$$

the characteristic functional $\hat{m}$ of the directly perceptible output process of the quantal system $(\mathfrak{H}, \mathfrak{T}, \mathfrak{N})$ is given via (100) as

$$\hat{m}[f] = e^{-(1/2)\|f\|^2} \tag{152}$$

that is, Gaussian white noise.

VI. CLASSIFICATIONS OF THE DIRECTLY PERCEPTIBLE PROCESSES OF QUANTAL SYSTEMS

A. Posing the Problem

If we presuppose the universal validity of generalized quantum mechanics in the molecular and the engineering domain, then everything we can directly observe occurs in the classical part of some quantal system. In particular, all data a scientist collects for a subsequent data processing are sample values of stochastic processes generated by classical observables. It has to be admitted that the present approach is abstract and formal, but it is also general and free from some limitations of the pragmatic engineering approach.

It is important to call to mind the mathematical definition of stochastic processes we used: A stochastic process is fully characterized by defining a normalized measure on the space of all feasible trajectories. Time occurs only as a parameter; the time set is the real axis but *an arrow of time does not occur.* Of course, the reversibility of the time evolution of the underlying quantal states does not imply the reversibility of directly perceptible processes in the classical part of the system. So the existence of irreversible processes is no miracle (although usually posing difficult mathematical problems). The deep problem is to explain why so often the direction of entropy increase of an irreversible process equals the direction of universal time.

It has already been remarked by Gibbs[160] "that while the probabilities of subsequent events may often be determined from the probabilities of prior events, it is rarely the case that probabilities of prior events can be determined from those of subsequent events." More categorically, C. F. von Weizsäcker[161] claims that the concept of probability is applicable for future events only, arguing that the future is possible but the past is factual. Certainly, everyday experience teaches us that the future is qualitatively different from the past. Nevertheless, there are no logical reasons to exclude situations in which retrodiction is possible while prediction is impossible.[162-165] In particular, Watanabe[166] has pointed out that the feasibility of retrodiction is related to the existence of subsystems governed by an entropy-decrease law.

It is of prime importance that the first principles of mechanics together with a nonsubjectivistic interpretation of the theory do not prefer a direction of time. That is, if the theory does allow a subsystem associated with an entropy increase in the direction of positive universal time, it automatically also allows another subsystem associated with an entropy decrease in the direction of positive universal time. Any preferential selection between these two classes is only due to initial and boundary conditions, and there are absolutely no reasons for an exclusive selection.

Engineering wisdom stresses the nonanticipatory nature of man-made systems. So strong is our habit of excluding any kind of backward causation that customarily one identifies the concepts "dynamical system" with "nonanticipatory system."[167] However, such an identification does not follow from quantum mechanics; and it seems to us quite conceivable that there will be problems of natural science that are to be described as antidynamical systems. By an antidynamical system we mean a dynamical system in the sense of the mathematicians in which the physical time equals $-t$. A stochastic antidynamical system can be best described using a teleological terminology: If the goal (the "final cause") is achieved, then we can make statistical retrodictions from the present into the past and evaluate the probabilities of former events that have been directed to the goal. A teleological language seems to be indispensable in biology ("a bird's wing is made for flying") but teleological explanations have a bad reputation in modern science. However, the fact that a concept has been misused does not imply that the concept is bad. Antecedent causality (which is generally accepted as a most useful concept) has the same logical status as final causality. Moreover, the following suggestion of Watanabe[166] is certainly intriguing: "The reasons why this kind of teleological explanation is more direct and more successful in certain biological phenomena may then be attributed to the fact that biological systems involve subsystems which disobey the entropy-increase law."

Fortunately, these difficult problems can be discussed in a purely mathematical manner by examining the temporal behavior of stochastic processes in terms of predictions and retrodictions. Prediction is the inference of future observational data from the present ones, whereas retrodiction is the inference of past observational data from the present ones. Every temporal asymmetry of a stochastic process can be discussed in terms of prediction and retrodiction, and for this problem we have a well-established mathematical formalism.

B. Deterministic and Nondeterministic Systems

An unperturbing observation of a family $\{z_i : i \in \mathfrak{J}\}$ of random variables $z_i : \Omega \to \mathbb{R}$ on a measurable space (Ω, Σ) is called a *stochastic experiment.* The collection of imaginable outcomes of a stochastic experiment is characterized by the sub-σ-field $\sigma\{z_i : i \in \mathfrak{J}\}$ generated by the family $\{z_i : i \in \mathfrak{J}\}$ of random variables z_i, that is,

$$\sigma\{z_i : i \in \mathfrak{J}\} \overset{\text{def}}{=} \{z_i^{-1}(\beta) : i \in \mathfrak{J}, \quad \beta \in \Sigma_\mathbb{R}\} \qquad (1)$$

where $z_i^{-1}(\beta)$ is the inverse image of the set β with respect to the map z_i and the σ-field $\Sigma_\mathbb{R}$ of Borel sets of the real axis,

$$z_i^{-1}(\beta) \overset{\text{def}}{=} \{\omega : \omega \in \Omega, \quad z_i(\omega) \in \beta\}, \quad \beta \in \Sigma_\mathbb{R} \qquad (2)$$

The σ-field Σ' is the smallest σ-field on Ω such that every random variable $z_i,$ $i \in \mathfrak{J}$, is measurable. An experiment determined by the σ-field $\Sigma' \subset \Sigma$ is called *cruder* than an experiment determined by the σ-field $\Sigma'' \subset \Sigma$ if $\Sigma' \subset \Sigma''$. A set of random variables on (Ω, Σ) is called *full* if its random variables generate the whole σ-field Σ.

Consider now the directly perceptible generic stochastic process $z : \mathbb{R} \times \Omega \to \mathbb{R}$ in the classical part of a quantal system in the sense of the result (V.62). We assume that the measurable space (Ω, Σ) is generated by a *minimal* Naĭmark extension so that the family $\{z[f] : f \in \mathcal{S}\}$ of random variables $z[f] : \Omega \to \mathbb{R}$,

$$z[f] \overset{\text{def}}{=} \int_{-\infty}^{\infty} z(t, \cdot) f(t) \, dt, \qquad f \in \mathcal{S} \qquad (3)$$

is *full.* For a discussion of the dependence of the future on the past (or the other way round!) it is appropriate to introduce the sub-σ-field generated by the stochastic process z in the time interval $a \le t \le b$,

$$\Sigma(a, b) \overset{\text{def}}{=} \sigma\{z[f] : f \in \mathcal{S}, \text{ support } f \subset [a, b]\} \qquad (4a)$$

If the "sharp-time" process $z(t, \cdot)$ exists, we can also write

$$\Sigma(a, b) = \sigma\{z(t, \cdot) : a \le t \le b\} \qquad (4b)$$

If the stochastic process refers to a probability space (Ω, Σ, μ), we for convenience identify sub-σ-fields coinciding μ-almost everywhere. Since the stochastic process z represents a full family of random variables, we have

$$\Sigma(-\infty, \infty) = \Sigma \tag{5}$$

Furthermore, we introduce the following sub-σ-fields:

the past $\qquad\qquad \Sigma(-\infty, 0) \tag{6}$

the remote past $\qquad \Sigma(-\infty) \overset{\text{def}}{=} \bigcap_{t \geq 0} \Sigma(-\infty, -t) \tag{7}$

the future $\qquad\qquad \Sigma(0, \infty) \tag{8}$

the remote future $\qquad \Sigma(\infty) \overset{\text{def}}{=} \bigcap_{t \geq 0} \Sigma(t, \infty) \tag{9}$

Suppose the observation of the classical observable Z associated to the stochastic process z is carried out continuously so that $\Sigma(-\infty, t)$ represents the σ-field of all observed events up to the moment t inclusively. Clearly, $\Sigma(-\infty, t)$ is a monotonically increasing family of σ-fields,

$$\Sigma(-\infty, t) \subset \Sigma(-\infty, s) \qquad \text{if } t < s \tag{10}$$

These concepts are important for the following extrapolation problem: Given the whole past $\Sigma(-\infty, t)$, can we *predict* the future behavior of a stochastic process? A stochastic process is called *forward deterministic* if the best (possibly nonlinear) predictor in terms of the past $\Sigma(-\infty, 0)$ allows an error-free prediction. In this case the process is completely determined by the remote past $\Sigma(-\infty)$ [for details, see, for example, Rosenblatt[168, sec. VI.2]. If an error-free prediction is not possible, the process is called forward nondeterministic. Every bounded stochastic process can be uniquely represented as the sum of a forward-deterministic process and a so-called purely nondeterministic process (where, of course, one component may be absent).[169] A process is said to be *forward purely nondeterministic* if the remote past $\Sigma(-\infty)$ is the trivial Borel field $\{\varnothing, \Omega\}$ consisting only of the impossible event $\varnothing$ and the sure event Ω. A forward purely nondeterministic process contains no components that can be exactly predicted from an arbitrarily long past record.

There is another extrapolation problem: retrodiction. Given the trajectory of the stochastic process on the positive real axis, can we *retrodict* the behavior of the stochastic process on the negative real axis? Of course, the answer is analogous to the problem of prediction; formally, t has just to be replaced by $-t$. We obtain the following terminology, generally adopted in modern mathematical probability and ergodic theory:

z is forward deterministic
$$\text{if } \Sigma(-\infty) = \Sigma(-\infty, +\infty) \tag{11}$$

z is backward deterministic
$$\text{if } \Sigma(+\infty) = \Sigma(-\infty, +\infty) \tag{12}$$

z is forward purely nondeterministic
$$\text{if } \Sigma(-\infty) = \{\varnothing, \Omega\} \tag{13}$$

z is backward purely nondeterministic
$$\text{if } \Sigma(+\infty) = \{\varnothing, \Omega\} \tag{14}$$

For a forward purely nondeterministic process the remote past does not contain any information useful for predictions; for a backward purely nondeterministic process the remote future does not contain information useful for retrodictions. A process that is forward and backward deterministic is called *bidirectional deterministic* and corresponds to a deterministic motion in the sense of Newtonian mechanics. It is a most important fact that, in general, forward determinism does not imply backward determinism. There exist even *stationary* stochastic processes that are forward deterministic and backward purely nondeterministic, and stationary processes that are forward purely nondeterministic and backward deterministic.[170,171]

C. Deterministic and Nondeterministic Output Processes of Quantal Systems

According to (V.55), the motion $\tilde{Z}(t)$ in the center $\tilde{\mathfrak{Z}}$ of the observationally equivalent Naĭmark extension $(\tilde{\mathfrak{H}}, \tilde{\mathfrak{T}}, \tilde{\mathfrak{N}})$ of the quantal system $(\mathfrak{H}, \mathfrak{T}, \mathfrak{N})$ produces the directly perceptible stochastic process $z : \mathbb{R} \times \Omega \to \mathbb{R}$ generated by the generic classical observable $Z \in \mathfrak{Z}$. The trajectories of this stochastic process have been assumed in (V.36) to lie in the trajectory space $\mathscr{S}^*$. It is convenient to represent the center $\tilde{\mathfrak{Z}}$ by an $\mathscr{L}_\infty$-space. For a minimal Naĭmark extension, the family $\{z[f]: f \in \mathscr{S}\}$ of random variables $z[f]$ is full so that the multiplication algebra $\mathscr{L}_\infty(\mathscr{S}^*, \Sigma^*, m)$ is *-isometrically isomorphic to $\tilde{\mathfrak{Z}}$, and the random variable $z[f]$, $f \in \mathscr{S}$, can be represented as

$$\{\pi(\tilde{Z}[f])\Psi\}(\zeta) = z[f](\zeta) \cdot \Psi(\zeta), \qquad \Psi \in \mathscr{L}_2(\mathscr{S}^*, \Sigma^*, m) \tag{15}$$

Clearly, the family of all random variables $z[f]$, $f \in \mathscr{S}$, generates $\mathscr{L}_\infty(\mathscr{S}^*, \Sigma^*, m)$; and the operators $\tilde{Z}[f]$, $f \in \mathscr{S}$, generate the center $\tilde{\mathfrak{Z}}$

$$\tilde{\mathfrak{Z}} = \{\tilde{Z}[f]: f \in \mathscr{S}\}'' \tag{16}$$

Let $\Sigma^*(a, b)$ be the σ-field generated by the random variables $z[f]$ with test functions $f \in \mathcal{S}$ whose support is restricted to the interval $[a, b]$. A standard argument shows that $\mathcal{L}_\infty\{\mathcal{S}^*, \Sigma^*(a, b), m\}$ equals the Neumann algebra generated by such random variables, which therefore corresponds to the Neumann algebra

$$\tilde{\mathfrak{Z}}(a, b) \stackrel{\text{def}}{=} \{\tilde{Z}[f]: f \in \mathcal{S}, \text{ support } f \subset [a, b]\} \tag{17}$$

Similarly to (6) to (9) we introduce the following Abelian Neumann algebras:

the past $\qquad\qquad \tilde{\mathfrak{Z}}(-\infty, 0) \tag{18}$

the remote past $\qquad \tilde{\mathfrak{Z}}(-\infty) \stackrel{\text{def}}{=} \bigcap_{t \geq 0} \tilde{\mathfrak{Z}}(-\infty, -t) \tag{19}$

the future $\qquad\qquad \tilde{\mathfrak{Z}}(0, \infty) \tag{20}$

the remote future $\quad \tilde{\mathfrak{Z}}(\infty) \stackrel{\text{def}}{=} \bigcap_{t \geq 0} \tilde{\mathfrak{Z}}(t, \infty) \tag{21}$

A classification fully equivalent to the definitions (11) to (14) can now be given in terms of the center $\tilde{\mathfrak{Z}}$ of the extended system:

Z is forward deterministic $\qquad\qquad$ if $\tilde{\mathfrak{Z}}(-\infty) = \tilde{\mathfrak{Z}}$ $\qquad$ (22)

Z is backward deterministic $\qquad\qquad$ if $\tilde{\mathfrak{Z}}(+\infty) = \tilde{\mathfrak{Z}}$ $\qquad$ (23)

Z is forward purely nondeterministic $\quad$ if $\tilde{\mathfrak{Z}}(-\infty) = 1 \cdot \mathbb{C}$ $\qquad$ (24)

Z is backward purely nondeterministic $\;$ if $\tilde{\mathfrak{Z}}(+\infty) = 1 \cdot \mathbb{C}$ $\qquad$ (25)

D. Third Example: One-Sided Processes

The simple examples traditionally used to illustrate general dynamical theories often possess a bilateral symmetry. Our first example (Section V.F) is a bilaterally deterministic process; our second example (Section V.G) is a stationary process with independent values at every instant, hence bilaterally purely nondeterministic (cf. also Ref. 148, p. 84). The existence of forward purely nondeterministic but backward-deterministic processes, or of forward-deterministic but backward purely nondeterministic processes is less trivial, and indeed *stationary* processes of these types behave quite pathologically.[170] However, for the discussion of molecular evolutions, nonstationary processes are of greater interest so that we give a very simple example of a nonstationary one-sided process.

We specify two quantal systems $(\mathfrak{H}, \mathfrak{T}, \mathfrak{N}_+)$ and $(\mathfrak{H}, \mathfrak{T}, \mathfrak{N}_-)$ that are subsystems of the same universal theory $(\mathfrak{H}, \mathfrak{T}, \mathfrak{B})$, that is,

$$\mathfrak{N}_+ \subset \mathfrak{B}(\mathfrak{H})$$

$$\mathfrak{N}_- \subset \mathfrak{B}(\mathfrak{H})$$

but that behave diametrically opposite under the very same dynamical group $\mathfrak{T}$. The complementary behavior of these two systems is reflected by the relations

$$\mathfrak{N}_+ \vee \mathfrak{N}_- = \mathfrak{B}(\mathfrak{H}) \tag{26}$$

$$\mathfrak{N}_+ \wedge \mathfrak{N}_- \text{ is maximal Abelian} \tag{27}$$

The relation (26) implies that the universal theory $(\mathfrak{H}, \mathfrak{T}, \mathfrak{B})$ is the smallest theory that dominates both $(\mathfrak{H}, \mathfrak{T}, \mathfrak{N}_+)$ and $(\mathfrak{H}, \mathfrak{T}, \mathfrak{N}_-)$, whereas (27) says that the largest theory dominated by $(\mathfrak{H}, \mathfrak{T}, \mathfrak{N}_+)$ and $(\mathfrak{H}, \mathfrak{T}, \mathfrak{N}_-)$ is a perfectly classical theory. In other words, the only theory from which both theories $(\mathfrak{H}, \mathfrak{T}, \mathfrak{N}_+)$ and $(\mathfrak{H}, \mathfrak{T}, \mathfrak{N}_-)$ can be derived is the purely quantal theory $(\mathfrak{H}, \mathfrak{T}, \mathfrak{B})$; the most comprehensive theory that can be derived from both $(\mathfrak{H}, \mathfrak{T}, \mathfrak{N}_+)$ and $(\mathfrak{H}, \mathfrak{T}, \mathfrak{N}_-)$ is perfectly classical.

Let $\{U(p), V(q) : p, q \in \mathbb{R}\}$ be a canonical Weyl system,

$$U(p) V(q) = V(q) U(p) e^{ipq}, \qquad p, q \in \mathbb{R} \tag{28}$$

acting irreducibly in the complex Hilbert space $\mathfrak{H}$. The spectral resolution of the strongly continuous one-parameter group $\{U(p) : p \in \mathbb{R}\}$,

$$U(p) = \int_{\mathbb{R}} e^{ipx} E(dx), \qquad p \in \mathbb{R} \tag{29}$$

defines a spectral measure $E : \Sigma \to \mathfrak{B}(\mathfrak{H})$, where Σ is the σ-field of Borel sets of the real line $\mathbb{R}$. For notational convenience we define a spectral family $\{E_t : t \in \mathbb{R}\}$ by

$$E_t \stackrel{\text{def}}{=} \int_{-\infty}^{t} E\,(dx) \tag{30a}$$

so that

$$E_s \leq E_t \qquad \text{for } s < t \tag{30b}$$

$$s - \lim_{t \to -\infty} E_t = 0 \tag{30c}$$

$$s - \lim_{t \to +\infty} E_t = 1 \tag{30d}$$

$$E_s E_t = E_t E_s = E_{s \wedge t} \tag{30e}$$

where $s \wedge t$ denotes the minimum of s, t. With this we define the following families of Abelian Neumann algebras

$$\mathfrak{Z}(a, b) \stackrel{\text{def}}{=} \{E_t : a \leq t \leq b\}'' \tag{31}$$

The algebras of observables of our two models are chosen to be

$$\mathfrak{N}_+ \overset{\text{def}}{=} \{\mathfrak{Z}(0, \infty)\}'$$ (32a)

$$\mathfrak{N}_- \overset{\text{def}}{=} \{\mathfrak{Z}(-\infty, 0)\}'$$ (32b)

so that their centers are given by

$$\mathfrak{Z}_+ \overset{\text{def}}{=} \mathfrak{N}_+ \cap \mathfrak{N}_+' = \mathfrak{Z}(0, \infty)$$ (33a)

$$\mathfrak{Z}_- \overset{\text{def}}{=} \mathfrak{N}_- \cap \mathfrak{N}_-' = \mathfrak{Z}(-\infty, 0)$$ (33b)

The dynamical group $\mathfrak{T} = \{T_t : t \in \mathbb{R}\}$ is assumed to be given by the canonically conjugated Weyl group $\{V(q) : q \in \mathbb{R}\}$, we choose

$$T_t \overset{\text{def}}{=} V(-t), \qquad t \in \mathbb{R}$$ (34)

With this the two mentioned quantal systems $(\mathfrak{H}, \mathfrak{T}, \mathfrak{N}_+)$ and $(\mathfrak{H}, \mathfrak{T}, \mathfrak{N}_-)$ are fully specified, and it is easy to verify the validity of (26) and (27).

Every generic classical observable $Z_\pm$ can be represented as

$$Z_+ = \int_0^\infty \varphi(s)\, dE_s + c \int_{-\infty}^0 dE_s$$ (35a)

or

$$Z_- = \int_{-\infty}^0 \varphi(-s)\, dE_s + c \int_0^\infty dE_s$$ (35b)

where c is a constant and $\varphi : \mathbb{R}^+ \to \mathbb{R}$ is an invertible measurable function on the positive real axis $\mathbb{R}^+$ so that

$$\mathfrak{Z}_\pm = \{Z_\pm\}''$$ (36)

In order to discuss the stochastic process generated by a generic classical observable we introduce the operator

$$Z_\pm(t) \overset{\text{def}}{=} T_{-t} Z_\pm T_t, \qquad t \in \mathbb{R}$$ (37)

Weyl's commutation relation (28) gives with (29) and (34)

$$T_{-t} E_s T_t = E_{s+t}$$ (38)

so that with (35) we get

$$Z_+(t) = \int_t^\infty \varphi(s - t)\, dE_s + c \int_{-\infty}^t dE_s$$ (39a)

$$Z_-(t) = \int_{-\infty}^t \varphi(t - s)\, dE_s + c \int_t^\infty dE_s$$ (39b)

Clearly, we have

$$Z_+(t) \in \mathfrak{Z}_+ \qquad \text{for } t \geq 0 \qquad\qquad (40a)$$

$$Z_+(t) \notin \mathfrak{Z}_+ \qquad \text{for } t < 0 \qquad\qquad (40b)$$

and

$$Z_-(t) \notin \mathfrak{Z}_- \qquad \text{for } t > 0 \qquad\qquad (41a)$$

$$Z_-(t) \in \mathfrak{Z}_- \qquad \text{for } t \leq 0 \qquad\qquad (41b)$$

so that the dynamical group $\mathfrak{T}$ is neither an automorphism for the center $\mathfrak{Z}_+$ nor for the center $\mathfrak{Z}_-$. In this example the extension in the sense of Section V.D is quite trivial. An observationally equivalent extension of both $(\mathfrak{H}, \mathfrak{T}, \mathfrak{Z}_+)$ and $(\mathfrak{H}, \mathfrak{T}, \mathfrak{Z}_-)$ is given by the dynamical system $(\mathfrak{H}, \mathfrak{T}, \mathfrak{Z})$, where $\mathfrak{Z}$ is given by the Abelian algebra generated by the Weyl operator $U(p)$, or in the notation of (31),

$$\mathfrak{Z} \overset{\text{def}}{=} \mathfrak{Z}(-\infty, \infty) \qquad\qquad (42)$$

Since $\mathfrak{Z}_+ \subset \mathfrak{Z}$, and since $\mathfrak{T}$ generates an automorphism of $\mathfrak{Z}$, we can define the stochastic process $z_\pm$ in the $\mathscr{L}_2$-representation $\mathfrak{H} = \mathscr{L}_2(\mathbb{R}, d\omega)$, $\mathfrak{Z} = \mathscr{L}_\infty(\mathbb{R}, \omega)$ by

$$\{Z_\pm(t)\Psi\}(\omega) = z_\pm(t, \omega)\Psi(\omega), \qquad \forall \Psi \in \mathscr{L}_2(\mathbb{R}, d\omega) \qquad (43)$$

Accordingly, the classical observable $Z_\pm$ and the dynamical group $\mathfrak{T}$ generate a directly perceptible stochastic process $z_\pm : \mathbb{R} \times \mathbb{R} \to \mathbb{R}$ on the probability space $(\mathbb{R}, \Sigma, \mu)$, with

$$\mu(\mathscr{A}) = \rho\{E(\mathscr{A})\}, \qquad \mathscr{A} \in \Sigma \qquad\qquad (44)$$

where ρ is the NPL-functional generated by the state of the system. The evaluation of the characteristic functional $m_\pm$ of the stochastic process $z_\pm$ is straightforward; it is given by

$$m_\pm[f] \overset{\text{def}}{=} \mathscr{E}\left\{ \exp\left(i \int_{-\infty}^{\infty} f(t) z_\pm(t, \cdot) \, dt \right) \right\} \qquad (45a)$$

$$= \rho\left\{ \exp\left(i \int_{-\infty}^{\infty} f(t) Z_\pm(t) \, dt \right) \right\}, \qquad f \in \mathscr{S} \qquad (45b)$$

so that

$$m_+[f] = \int_{\mathbb{R}} \exp\left(i \int_{\omega}^{\infty} \varphi(\omega - t) f(t) \, dt + ic \int_{-\infty}^{\omega} f(t) \, dt \right) \mu(d\omega) \qquad (46a)$$

and

$$m_-[f] = \int_{\mathbb{R}} \exp\left(i \int_{-\infty}^{\omega} \varphi(t - \omega) f(t) \, dt + ic \int_{\omega}^{\infty} f(t) \, dt \right) \mu(d\omega) \qquad (46b)$$

We assume that the probability measure μ is equivalent to the central measure so that the discussion about the deterministic character of the stochastic process $z_\pm$ can be transferred to the more convenient discussion of the Abelian field $Z_\pm(t)$. Using the notation of (31) we get the following with (39):

$$\{Z_+(t)\}'' = \mathfrak{Z}(t, \infty) \tag{47a}$$

$$\{Z_-(t)\}'' = \mathfrak{Z}(-\infty, t) \tag{47b}$$

so that the evaluation of the remote past (19) and the remote future (21) is easy:

the remote past of Z_+ equals the full algebra $\mathfrak{Z}$ (48a)

the remote future of Z_+ equals the trivial algebra $1 \cdot \mathbb{C}$ (48b)

the remote past of Z_- equals the trivial algebra $1 \cdot \mathbb{C}$ (49a)

the remote future of Z_- equals the full algebra $\mathfrak{Z}$ (49b)

According to (22) to (25) we get the following classification:

Z_+ is forward deterministic
and backward purely nondeterministic (50)

Z_- is forward purely nondeterministic
and backward deterministic (51)

With this the quantal system $(\mathfrak{H}, \mathfrak{T}, \mathfrak{N}_-)$ can also be characterized as follows. If the state S_t at an arbitrary time t is primary, then necessarily the state S_τ was primary at every earlier instant, $\tau < t$. In general, a primary state S_t will evolve into a nonprimary state. This behavior implies that the associated stochastic process is entropic, that is, that all entropies defined via convex functions (including the popular Shannon and Rényi entropies) are increasing in the positive direction of universal time. For every state, in the distant past the classical observable Z_- had the value c. We can summarize this behavior by saying that the classical part of the system $(\mathfrak{H}, \mathfrak{T}, \mathfrak{N}_-)$ is past determined by the unique cause c and shows an entropic behavior and its future can be predicted probabilistically.

The classical part of the system $(\mathfrak{H}, \mathfrak{T}, \mathfrak{N}_+)$ shows a diametrically opposite behavior. If the state S_t at an arbitrary initial time t is primary, then the state S_τ at every later instant $\tau > t$ is also primary so that the entropy of the system is always nonincreasing in the direction of positive universal time. On the other hand, every initial state evolves into a primary state for $t \to \infty$ in which the classical observable Z_+ has the value c. Hence for a nonprimary initial state the entropy is decreasing. We can summarize this by saying that the classical part of the system $(\mathscr{H}, \mathfrak{T}, \mathfrak{N}_+)$ is antientropic, behaves like a final (or learning) system having the goal c, and has a past that can be retrodicted probabilistically.

E. Causal versus Final Descriptions

We show that a directly perceptible output process z of a quantal system can be decomposed into a deterministic and a purely nondeterministic part, not only in the forward direction but also in the backward direction of universal time.

If $\tilde{Z}[f]$, $f \in \mathscr{S}$, is a generic element of the center $\tilde{\mathfrak{Z}}$, then the random element $z[f]$ defined by (15) belongs not only to $\mathscr{L}_\infty(\mathscr{S}^*, \Sigma^*, m)$, but also to the Hilbert space $\mathscr{L}_2(\mathscr{S}^*, \Sigma^*, m)$. Let $\mathscr{L}_2(a, b)$ be the closure in the mean square of the algebra generated by the random elements $z[f]$ with test functions $f \in \mathscr{S}$ whose support is restricted to the interval $[a, b]$. In terms of these Hilbert spaces we can define

$$\text{the past} \qquad \mathscr{L}_2(-\infty, 0) \tag{52}$$

$$\text{the remote past} \qquad \mathscr{L}_2(-\infty) \stackrel{\text{def}}{=} \bigcap_{t \geq 0} \mathscr{L}_2(-\infty, -t) \tag{53}$$

$$\text{the future} \qquad \mathscr{L}_2(0, \infty) \tag{54}$$

$$\text{the remote future} \qquad \mathscr{L}_2(\infty) \stackrel{\text{def}}{=} \bigcap_{t \geq 0} \mathscr{L}_2(t, \infty) \tag{55}$$

Denote the projection from $\mathscr{L}_2(\mathscr{S}^*, \Sigma^*, m)$ onto $\mathscr{L}_2(a, b)$ by $E(a, b)$, and onto $\mathscr{L}_2(\pm\infty)$ by $E(\pm\infty)$.

There exist two possibilities to decompose the stochastic process z into a deterministic and a purely nondeterministic part, namely, either as

$$z[f] = z_{\text{fd}}[f] + z_{\text{fn}}[f], \qquad f \in \mathscr{S} \tag{56a}$$

$$z_{\text{fd}}[f] \stackrel{\text{def}}{=} E(-\infty)z[f] \tag{56b}$$

$$z_{\text{fn}}[f] \stackrel{\text{def}}{=} \{1 - E(-\infty)\}z[f] \tag{56c}$$

or as

$$z[f] = z_{\text{bd}}[f] + z_{\text{bn}}[f], \qquad f \in \mathscr{S} \tag{57a}$$

$$z_{\text{bd}}[f] \stackrel{\text{def}}{=} E(+\infty)z[f] \tag{57b}$$

$$z_{\text{bn}}[f] \stackrel{\text{def}}{=} \{1 - E(+\infty)\}z[f] \tag{57c}$$

It is straightforward to show[169] that the stochastic process z_{fd} is forward deterministic, z_{fn} is forward purely nondeterministic, z_{bd} is backward deterministic, and z_{bn} is backward purely nondeterministic. Moreover, the decomposition is orthogonal in the sense that with respect to the inner product of $\mathscr{L}_2(\mathscr{S}^*, \Sigma^*, m)$ we have

$$z_{fd}[f] \perp z_{fn}[g] \qquad \text{for all } f, g \in \mathscr{S} \tag{58a}$$

$$z_{bd}[f] \perp z_{bn}[g] \qquad \text{for all } f, g \in \mathscr{S} \tag{58b}$$

or equivalently,

$$\mathscr{L}_2(-\infty, +\infty) = \mathscr{L}_2(\text{fd}) \oplus \mathscr{L}_2(\text{fn}) \tag{59a}$$

$$= \mathscr{L}_2(\text{bd}) \oplus \mathscr{L}_2(\text{bn}) \tag{59b}$$

with

$$z_{fd}[f] \in \mathscr{L}_2(\text{fd}) \overset{\text{def}}{=} E(-\infty)\mathscr{L}_2(-\infty, \infty) \tag{60a}$$

$$z_{fn}[f] \in \mathscr{L}_2(\text{fn}) \overset{\text{def}}{=} \{1 - E(-\infty)\}\mathscr{L}_2(-\infty, \infty) \tag{60b}$$

$$z_{bd}[f] \in \mathscr{L}_2(\text{bd}) \overset{\text{def}}{=} E(+\infty)\mathscr{L}_2(-\infty, \infty) \tag{60c}$$

$$z_{bn}[f] \in \mathscr{L}_2(\text{bn}) \overset{\text{def}}{=} \{1 - E(+\infty)\}\mathscr{L}_2(-\infty, \infty) \tag{60d}$$

However, there is no general relation between the forward and the backward decomposition. As a rule, the projectors $E(-\infty)$ and $E(+\infty)$ do not commute, so that they represent two incompatible propositions. That is to say, *both the forward decomposition* (56) *and the backward decomposition* (57) *represent two rational but usually incompatible descriptions.*

If a process is *forward deterministic* it has a unique goal in the future. A *backward-deterministic* process has a unique cause in the past; every event in a trajectory has a unique predecessor so that the history is uniquely fixed. If a stochastic process is *forward purely nondeterministic*, its entropy is increasing with universal time, and the process can be predicted probabilistically. If a stochastic process is *backward purely non-deterministic*, it has no unique history, but its past can be retrodicted probabilistically. Moreover, its entropy decreases with increasing universal time.

Because modern exact science has a preference for predictive descriptions of natural phenomena, the decomposition (56) is usually favored. Note, however, that the decomposition (56) does not automatically guarantee backward causation, another predilection of the Western world. From a purely formal point of view, the decomposition (57) has the same logical status as the forward decomposition (56); it is appropriate for retrodictions and leads to final descriptions. The success of the predictive decomposition (56) is well documented in physics and engineering science. The use of final concepts is an old tradition in descriptive biology and, as far as we can see, is compatible with all empirically well-established fundamental laws of physics. One may therefore conjecture that there exists a certain level of complexity such that for systems of complexity lesser than this level it is simpler to use the forward decomposition (56), whereas for systems with a larger complexity it is simpler to use the backward decomposition (57) and to retrodict the evolution of the system. Both viewpoints are theoretically legitimate, but only in special cases mutually compatible.

ACKNOWLEDGMENT

We express our gratitude to Dr. Peter Brand, Werner Gans, Wolfgang Gasche, Peter Pfeifer, and Guido Raggio for many stimulating and clarifying discussions as well as for their critical and constructive remarks on a preliminary version of this paper.

References

1. J. von Neumann, *Mathematische Grundlagen der Quantenmechanik*, Springer, Berlin, 1932. (English translation: *Mathematical Foundations of Quantum Mechanics*, Princeton University Press, Princeton, 1955).
2. M. Jammer, *The Conceptual Development of Quantum Mechanics*, McGraw-Hill, New York, 1966.
3. M. Jammer, *The Philosophy of Quantum Mechanics*, Wiley, New York, 1974.
4. A. A. Grib, *Vestn. Leningrad Univ., Ser. Fis. Khim.*, No. 16, 16–23 (1968).
5. K. Hepp, *Helv. Phys. Acta*, **45**, 237–248 (1972).
6. F. Bloch, *Phys. Rev.*, **70**, 460–474 (1946).
7. P. A. M. Dirac, *The Principles of Quantum Mechanics*, Clarendon, Oxford, 1st ed., 1930; 4th ed., 1958.
8. E. B. Davies and J. T. Lewis, *Commun. Math. Phys.*, **17**, 239–260 (1970).
9. E. B. Davies, *J. Funct. Anal.*, **6**, 318–346 (1970).
10. E. B. Davies, *Commun. Math. Phys.*, **15**, 277–304 (1969); **19**, 83–105 (1970); **22**, 51–70 (1971).
11. C. W. Helstrom, *Information and Control*, **10**, 254–291 (1967); **13**, 156–171 (1968).
12. C. W. Helstrom, *J. Statist. Phys.*, **1**, 231–252 (1969).
13. C. W. Helstrom, J. W. S. Liu, and J. P. Gordon, *Proc. IEEE*, **58**, 1578–1598 (1970).
14. C. W. Helstrom, *Quantum Detection and Estimation Theory*, Academic, New York, 1976.
15. E. Arthurs and J. L. Kelley, *Bell Syst. Tech. J.*, **44**, 725–729 (1965).
16. J. P. Gordon and W. H. Louisell, in P. L. Kelley, B. Lax, and P. E. Tannenwald, Eds., *Physics of Quantum Electronics*, McGraw-Hill, New York, 1966, pp. 833–840.
17. C. Y. She and H. Heffner, *Phys. Rev.*, **152**, 1103–1110 (1966).
18. S. D. Personik, *Bell Syst. Tech. J.*, **50**, 213–216 (1971).
19. G. C. Wick, A. S. Wightman, and E. P. Wigner, *Phys. Rev.*, **88**, 101–105 (1952).
20. A. S. Wightman, *Nuovo Cim. Suppl.*, **14**, 81–94 (1959).

21. J. M. Jauch, *Helv. Phys. Acta*, **33**, 711–720 (1960).
22. J. M. Jauch and B. Misra, *Helv. Phys. Acta*, **34**, 699–709 (1961).
23. C. Piron, *Helv. Phys. Acta*, **42**, 330–338 (1969).
24. R. Cirelli, F. Gallone, and B. Gubbay, *J. Math. Phys.* (*N.Y.*), **16**, 201–213 (1975).
25. J. M. Jauch, *Helv. Phys. Acta*, **37**, 293–316 (1964).
26. G. G. Emch, *Algebraic Methods in Statistical Mechanics and Quantum Field Theory*, Wiley–Interscience, New York, 1972.
27. G. Birkhoff and J. von Neumann, *Ann. Math.*, **37**, 823–843 (1936).
28. J. M. Jauch, *Foundations of Quantum Mechanics*, Addison-Wesley, Reading, Mass., 1968.
29. V. S. Varadarajan, *Geometry of Quantum Theory*, Vol. 1, Van Nostrand, Princeton, 1968.
30. S. Gudder, in A. T. Bharucha-Reid, Ed., *Probabilistic Methods in Applied Mathematics*, Vol. 2, Academic, New York, 1970, pp. 53–129.
31. C. Piron, *Foundations of Quantum Physics*, Benjamin, Reading, Mass., 1976.
32. J. M. Jauch and C. Piron, *Helv. Phys. Acta*, **42**, 842–848 (1969).
33. J. Earman, *Phil. Sci.*, **41**, 15–47 (1974).
34. E. B. Davies, *Quantum Theory of Open Systems*, Academic, London, 1976.
35. J. R. Brown, *Ergodic Theory and Topological Dynamics*, Academic, New York, 1976.
36. O. Veblen and J. W. Young, *Projective Geometry*, Vol. 1, Ginn, Boston, 1910, p. 1.
37. T. Gold, *The Nature of Time*, Cornell University Press, Ithaca, N.Y., 1967.
38. D. Layzer, *Astrophys. J.*, **206**, 559–569 (1976).
39. F. Jegerlehner, *Fortschr. Phys.*, **16**, 137–193 (1968).
40. E. Schrödinger, *Naturwissenschaft*, **23**, 807–812, 823–828, 844–849 (1935).
41. E. Schrödinger, *Proc. Camb. Phil. Soc.*, **31**, 555–563 (1935); **32**, 446–452 (1936).
42. A. Einstein, B. Podolsky, and N. Rosen, *Phys. Rev.*, **47**, 777–780 (1935).
43. E. Scheibe, *Die kontingenten Aussagen in der Physik. Untersuchungen zur Ontologie der klassischen Physik und der Quantentheorie*, Athenäum, Frankfurt, 1964.
44. E. Scheibe, *The Logical Analysis of Quantum Mechanics*, Pergamon, Oxford, 1973.
45. J. B. Hartle, *Am. J. Phys.*, **36**, 704–712 (1968).
46. J. M. Jauch, in B. d'Espagnat, Ed., Proceedings of the International School of Physics "Enrico Fermi," Course 49: *Foundations of Quantum Mechanics*, Academic, New York, 1971, pp. 20–55.
47. G. W. Mackey, *The Mathematical Foundations of Quantum Mechanics*, Benjamin, New York, 1963.
48. C. Piron, *Helv. Phys. Acta*, **37**, 439–468 (1964).
49. G. Ludwig, *Deutung des Begriffs "physikalische Theorie" und axiomatische Grundlegung der Hilbertraumstruktur der Quantenmechanik durch Hauptsätze des Messens*, Lecture Notes in Physics, Vol. 4, Springer, Berlin, 1970.
50. P. Mittelstaedt, *Z. Naturforsch.*, **25a**, 1773–1778 (1970); **27a**, 1358–1362 (1972).
51. G. M. Hardegree, *Z. Naturforsch.*, **30a**, 1347–1360 (1975).
52. G. M. Hardegree, *J. Phil. Logic*, **4**, 399–421 (1975).
53. R. Piziak, *J. Phil. Logic*, **3**, 413–418 (1974).
54. L. Herman, E. L. Marsden, and R. Piziak, *Notre Dame J. Formal Logic*, **16**, 305–328 (1975).
55. D. M. Topping, *Pac. J. Math.*, **20**, 317–325 (1967).
56. P. Havas, in Y. Bar-Hillel, Ed., *Logic, Methodology, and Philosophy of Science*, North-Holland, Amsterdam, 1965, pp. 347–362.
57. P. Havas, *Synthese*, **18**, 75–102 (1968).
58. U. Uhlhorn, *Ark. Fys.*, **23**, 307–340 (1962).

59. G. Emch and C. Piron, *J. Math. Phys. (N.Y.)*, **4**, 469–473 (1963).
60. N. Bohr, *Dialectica*, **2**, 312–319 (1948) [reprinted in *Science*, **111**, 51–54 (1950)].
61. G. Birkhoff, *Lattice Theory*, American Mathematical Society, Providence, R.I., 3rd (new) ed., 1967.
62. S. S. Holland, in J. C. Abbott, Ed., *Trends in Lattice Theory*, Van Nostrand-Reinhold, New York, 1970, pp. 41–126.
63. F. Maeda and S. Maeda, *Theory of Symmetric Lattices*, Springer, Berlin, 1970.
64. I. E. Segal, *Ann. Math.*, **48**, 930–948 (1947).
65. I. E. Segal, *Mathematical Problems of Relativistic Physics*, American Mathematical Society, Providence, R.I., 1963.
66. R. Haag and D. Kastler, *J. Math. Phys. (N.Y.)*, **5**, 848–861 (1964).
67. R. V. Kadison, *Topology*, **3**, Suppl. 2, 177–198 (1965).
68. R. J. Plymen, *Commun. Math. Phys.*, **8**, 132–146 (1968).
69. J. Dixmier, *Les C*-algèbres et leurs représentations*, Gauthier-Villars, Paris, 1964, 2nd ed., 1969.
70. J. Dixmier, *Les algèbres d'opérateurs dans l'espace Hilbertien (Algèbres de von Neumann)*, Gauthier-Villars, Paris, 1957, 2nd ed. 1969.
71. S. Sakai, *C*-algebras and W*-algebras*, Springer, Berlin, 1971.
72. S. K. Berberian, *Baer *-rings*, Springer, Berlin, 1972.
73. J. von Neumann, *Math. Ann.*, **102**, 370–427 (1929).
74. N. N. Bobolubov, A. A. Logunov, and I. T. Todorov, *Introduction to Axiomatic Quantum Field Theory*, Benjamin, Reading, Mass., 1975.
75. Ju. M. Berezanskiĭ, *Expansions in Eigenfunctions of Selfadjoint Operators*, American Mathematical Society, Providence, R.I., 1968.
76. M. O. Farrūkh, *J. Math. Phys. (N.Y.)*, **16**, 177–200 (1975).
77. P. R. Halmos, *Introduction to Hilbert Space and the Theory of Spectral Multiplicity*, Chelsea, New York, 1951.
78. P. Masani, in B. Sz.-Nagy, Ed., *Hilbert Space Operators and Operator Algebras*, North-Holland, Amsterdam, 1972, pp. 415–441.
79. M. J. Lighthill, *Introduction to Fourier Analysis and Generalized Functions*, Cambridge University Press, New York, 1958.
80. P. Antosik, J. Mikusinsky, and R. Sikorski, *Theory of Distributions: The Sequential Approach*, Elsevier, Amsterdam, 1973.
81. S. K. Berberian, *Proc. Amer. Math. Soc.*, **13**, 111–114 (1962).
82. H. Bohr, *Fastperiodische Funktionen*, Springer, Berlin, 1932.
83. R. Schatten, *Norm Ideals of Completely Continuous Operators*, Springer, Berlin, 1960.
84. I. M. Gel'fand and N. Ya. Vilenkin, *Generalized Functions*, Vol. 4, *Applications of Harmonic Analysis*, Academic, New York, 1964.
85. R. V. Kadison, in D. Kastler, Ed., Proceedings of the International School of Physics "Enrico Fermi", Course 60: *C*-Algebras and Their Applications to Statistical Mechanics and Quantum Field Theory*, North-Holland, Amsterdam, 1976, pp. 1–18.
86. O. E. Lanford, in D. Kastler, Ed., *Statistical Mechanics and Quantum Field Theory*, North-Holland, Amsterdam, 1976, pp. 1–18.
87. J. E. Moyal, *Adv. Appl. Prob.*, **4**, 39–80 (1972).
88. G. G. Emch, *J. Math. Phys. (N.Y.)*, **7**, 1413–1420 (1966).
89. M. Takesaki, *Tôhoku Math. J.*, **10**, 194–203 (1958).
90. M. Takesaki, *Proc. Jap. Acad.*, **35**, 365–366 (1959).
91. G. K. Pedersen, *Pac. J. Math.*, **37**, 795–800 (1971).
92. G. W. Mackey, *Ann. Math.*, **58**, 193–221 (1953).

93. J. T. Schwartz, *W*-algebras*, Gordon and Breach, New York, 1967.
94. J. von Neumann, *Ann. Math.*, **32**, 191–226 (1931).
95. K. Maurin, *Math. Ann.*, **165**, 204–222 (1966).
96. M. A. Naïmark, *Dokl. Akad. Nauk SSSR*, **217**, 762–765 (1974) [English translation: *Sov. Math. Dokl.*, **15**, 1126–1130 (1975)].
97. R. J. Plymen, *Nuovo Cimento*, **54A**, 862–870 (1968).
98. B. Misra, *Nuovo Cimento*, **47A**, 841–859 (1967).
99. N. S. Kronfli, *Int. J. Theor. Phys.*, **4**, 141–143 (1971).
100. M. Born and R. Oppenheimer, *Ann. Phys.*, **84**, 457–484 (1927).
101. M. Takesaki, *Tomita's Theory of Modular Hilbert Algebras and Its Applications*, Lecture Notes in Mathematics, Vol. 128, Springer, Berlin, 1970.
102. V. Bargmann, *Ann. Math.*, **59**, 1–46 (1954).
103. J. von Neumann, *Ann. Math.*, **33**, 567–573 (1932).
104. H. Primas, *J. Math. Biol.*, **4**, 281–301 (1977).
105. H. Primas, *Theor. Chim. Acta*, **39**, 127–148 (1975).
106. H. Everett, *Rev. Mod. Phys.*, **29**, 454–462 (1957).
107. J. A. Wheeler, *Rev. Mod. Phys.*, **29**, 463–465 (1957).
108. N. Graham, *The Everett Interpretation of Quantum Mechanics*, Ph.D. thesis, University of North Carolina at Chapel Hill, 1970.
109. B. S. DeWitt, *Phys. Today*, **23**:9, 30–35 (1970); **24**:4, 41–44 (1971).
110. B. S. DeWitt, in B. d'Espagnat, Ed., Proceedings of the International School of Physics "Enrico Fermi," Course 49: *Foundations of Quantum Mechanics*, Academic, New York, 1971, pp. 211–262.
111. B. S. DeWitt and N. Graham, *The Many-Worlds Interpretation of Quantum Mechanics*, Princeton University Press, Princeton, N.J., 1973.
112. L. E. Ballentine, *Found. Phys.* **3**, 229–240 (1973).
113. M. Mugur-Schachter, in A. Hartkämper and H. Neumann, Eds., *Foundations of Quantum Mechanics and Ordered Linear Spaces*, Lecture Notes in Physics, Vol. 29, Springer, Berlin, 1974, pp. 288–308.
114. K. Baumann, *Acta Phys. Austriaca*, **41**, 223–227 (1975).
115. A. N. Kolmogorov, *Sankhya, Indian J. Statist.* Ser. A, **25**, 369–376 (1963).
116. A. N. Kolmogorov, *Prob. Peredachi Inform.*, **1**, 3–11 (1965) [English translation: *Int. J. Comput. Math.*, **2**, 157–168 (1968)].
117. G. Chaitin, *J. Assoc. Comput. Mach.*, **13**, 547–569 (1966).
118. C. P. Schnorr, *Zufälligkeit und Wahrscheinlichkeit. Eine algorithmische Begründung der Wahrscheinlichkeitstheorie*, Lecture Notes in Mathematics, Vol. 218, Springer, Berlin, 1971.
119. P. Benioff, *Phys. Rev.*, **D7**, 3603–3609 (1973).
120. P. Benioff, *J. Math. Phys. (N.Y.)*, **17**, 618–628, 629–640 (1976).
121. D. Finkelstein, in R. G. Colodny, Ed., *Paradigms and Paradoxes*, University of Pittsburgh Press, 1972, pp. 47–66.
122. A. Kolmogoroff, *Grundbegriffe der Wahrscheinlichkeitsrechnung*, Springer, Berlin, 1933 [English translation: A. N. Kolmogorov, *Foundations of the Theory of Probability*, Chelsea, New York, 1950].
123. M. Loève, *Probability Theory*, 3rd ed., Van Nostrand-Reinhold, New York, 1963.
124. L. Breiman, *Probability*, Addison-Wesley, Reading, Mass., 1968.
125. I. E. Segal, *Decompositions of operator algebras I/II Mem. Am. Math. Soc.*, No. 9, American Mathematical Society, Providence, R.I., 1951.
126. R. G. Douglas, *Banach Algebra Techniques in Operator Theory*, Academic, New York, 1972.

127. W. Pauli, *Die allgemeinen Prinzipien der Wellenmechanik, Handbuch der Physik*, H. Geiger and K. Scheel, Eds., Vol. 24/1, Springer, Berlin, 1933.
128. A. S. Holevo, *Trudy Moscov. Mat. Obšč.*, **26**, 133–149 (1972) [English translation: *Trans. Moscow Math. Soc.*, **26**, 133–149 (1972)].
129. A. S. Holevo, *J. Multivariate Anal.*, **3**, 337–394 (1973).
130. A. S. Holevo, in *Proceedings of the Second Japan–USSR Symposium on Probability Theory*, Lecture Notes in Mathematics, Vol. 330, Springer, Berlin, pp. 104–119.
131. S. K. Berberian, *Notes on Spectral Theory*, Van Nostrand, Princeton, 1966.
132. J. M. Jauch and C. Piron, *Helv. Phys. Acta*, **40**, 559–570 (1967).
133. W. O. Amrein, *Helv. Phys. Acta* **42**, 149–190 (1969).
134. H. Neumann, *Commun. Math. Phys.*, **45**, 811–819 (1972).
135. S. T. Ali and G. G. Emch, *J. Math. Phys. (N.Y.)*, **15**, 176–182 (1974)
136. A. S. Holevo, *Teor. Mat. Fiz.*, **17**, 319–326 (1973) [English translation: *Theor. Math. Phys.*, **17**, 1172–1177 (1974)].
137. A. S. Holevo, *Dokl. Akad. Nauk SSSR*, **218**, 54–57 (1974) [English translation: *Sov. Math. Dokl.*, **15**, 1276–1281 (1974)].
138. A. S. Holevo, *IEEE Trans. Inform. Theory*, **IT-21**, 533–543 (1975).
139. C. W. Helstrom, *Int. J. Theor. Phys.*, **8**, 361–376 (1973).
140. C. W. Helstrom, *IEEE Trans. Inform. Theory*, **IT-20**, 374–376 (1974).
141. C. W. Helstrom, *Found. Phys.*, **4**, 453–463 (1974).
142. C. W. Helstrom, *Int. J. Theor. Phys.*, **11**, 357–378 (1974).
143. C. W. Helstrom and R. S. Kennedy, *IEEE Trans. Inform. Theory*, **IT-20**, 16–24 (1974).
144. H. P. Yuen and M. Lax, *IEEE Trans. Inform. Theory*, **IT-19**, 740–750 (1973).
145. H. P. Yuen, R. S. Kennedy, and M. Lax, *IEEE Trans. Inform. Theory*, **IT-21**, 125–134 (1975).
146. P. A. Benioff, *J. Math. Phys. (N.Y.)*, **13**, 231–242, 908–915, 1347–1355 (1972).
147. P. A. Benioff, *Found. Phys.*, **3**, 359–379 (1973).
148. T. Hida, *Stationary Stochastic Processes*, Princeton University Press, Princeton, N.J., 1970.
149. H. Bremermann, *Distributions, Complex Variables, and Fourier Transforms*, Addison-Wesley, Reading, Mass., 1965.
150. W. Güttinger, *Fortsch. Phys.*, **14**, 483–602 (1966).
151. B. Simon, *J. Math. Phys. (N.Y.)*, **12**, 140–148 (1971).
152. R. A. Minlos, *Trudy Moskov. Mat. Obšč.*, **8**, 497–518 (1969) [English translation: *Sel. Transl. Math. Statist. and Probab.*, **3**, 291–313 (1963)].
153. B. Sz.-Nagy, "*Extensions of Linear Transformations in Hilbert Space Which Extend Beyond this Space*," in F. Riesz and B. Sz.-Nagy, *Functional Analysis*, Ungar, New York, 1960, appendix.
154. H. Halpern, *Commun. Math. Phys.*, **25**, 253–275 (1972).
155. C. R. Putnam, *Commutation Properties of Hilbert Space Operators and Related Topics*, Springer, Berlin, 1967.
156. E. Wigner, *Phys. Rev.*, **40**, 749–759 (1932).
157. K. Husimi, *Proc. Phys. Math. Soc. (Japan)*, **22**, 264–314 (1940).
158. Y. Umemura, *Publ. Res. Inst. Math. Sci. (Kyoto)*, **A1**, 1–47 (1965).
159. M. Reed and B. Simon, *Methods of Modern Mathematical Physics*, Vol. 2, Academic, New York, 1975.
160. J. W. Gibbs, *Elementary Principles in Statistical Mechanics*, Yale University Press, New Haven, Conn., 1902, p. 151.
161. C. F. von Weizsäcker, *Ann. Phys.* **36**, 275–283 (1939).
162. S. Watanabe, *Rev. Mod. Phys.*, **27**, 179–186 (1955).

163. A. Grünbaum, *Arch. Phil.*, **7**, 184–185 (1957).
164. A. Grünbaum, *Phil. Sci.*, **29**, 146–170 (1962).
165. S. Watanabe, *Knowing and Guessing*, Wiley, New York, 1969.
166. S. Watanabe, *Progr. Theor. Phys., Suppl.*, Extra Number, 495–497 (1968).
167. R. E. Kalman, P. L. Falb, and M. A. Arbib, *Topics in Mathematical System Theory*, McGraw-Hill, New York, 1969, fn., p. 5.
168. M. Rosenblatt, *Markov Processes: Structure and Asymptotic Behavior*, Springer, Berlin, 1971.
169. H. Cramér and M. R. Leadbetter, *Stationary and Related Stochastic Processes*, Wiley, 1967, sec. 5.7.
170. U. Krengel, *J. Math. Anal. Appl.*, **35**, 611–620 (1971).
171. V. M. Gurevic, *Dokl. Akad. Nauk SSSR*, **210**, 763–766 (1973). [English translation: *Sov. Math. Dokl.*, **14**, 804–808 (1973)].

IRREVERSIBLE THERMODYNAMICS FOR QUANTUM SYSTEMS WEAKLY COUPLED TO THERMAL RESERVOIRS

HERBERT SPOHN*

*Belfer Graduate School of Science
Yeshiva University
New York, N.Y.*

JOEL L. LEBOWITZ†

*Service de Physique Théorique
CEN Saclay
Gif-sur-Yvette, France*

CONTENTS

I. INTRODUCTION

Statistical mechanics is the link between the microscopic and macroscopic world. Put more poetically—since this article is dedicated to Ilya Prigogine, whose life is devoted to the discovery and exposition of the grand poetry of nature—statistical mechanics is the bridge between the

* On leave of absence from the Fachbereich Physik der Universität München. Work supported by a Max Kade Foundation Fellowship.
† John Simon Guggenheim Fellow. Permanent address: Belfer Graduate School of Science, Yeshiva University, New York, N.Y., where work is supported by AFOSR Grant.

world of the atom and the world of the object. Man lives in the latter world but wishes, for aesthetic as well as for practical reasons, to understand the relation between the two worlds. This is the task of statistical mechanics.

Traditionally, statistical mechanics is split into two parts: equilibrium and nonequilibrium. Our understanding of the former is, formally at least, essentially complete. The Gibbs formalism gives a well-specified prescription for computing all the equilibrium properties of a macroscopic system from a knowledge of its microscopic Hamiltonian. Compared to this, our understanding of nonequilibrium statistical mechanics is still very incomplete. There is no general prescription for computing from first principles the great variety of nonequilibrium behavior observed in macroscopic systems. Some progress has been made, however, in recent years, and Prigogine is one of the pioneers in this field. It is therefore appropriate that this article, which describes recent developments in one small corner of this vast field, should be dedicated to him on the occasion of his sixtieth birthday.

The nonequilibrium phenomena we shall deal with in this article are transport processes. We wish to find the "Gibbs ensemble" appropriate for the description of a system in which there are flows of energy or material. Unfortunately, no such prescription exists at the present time even if we restrict ourselves to the case of steady fluxes. It is only in the special case of the weak coupling or van Hove limit that a mathematically precise microscopic theory has been developed. This theory is applied here to the case of systems coupled weakly to several idealized heat reservoirs. Some new results concerning entropy production in such situations are also presented. This yields in some special cases a microscopic justification for the principle of minimal entropy production that Prigogine has used so effectively in the thermodynamics of irreversible processes.

The situation we consider, then, is that of a finite system (say, a metal bar) coupled to inexhaustible heat reservoirs (heating the metal bar at one end and cooling it at the other). The case, where the reservoirs are of a certain stochastic nature, has been studied in detail by Lebowitz et al.[1-4] To obtain a truly dynamical model we choose in the present paper as reservoirs (infinitely extended) ideal gases at equilibrium. Even with such a simplifying assumption very little can be said. Already the problem to show the existence of a steady heat transporting state can be solved only in very special cases (such as the harmonic chain); even worse, these examples all show anomalous transport properties (their heat conductivity is infinite). Only in the limit, where the coupling between the system and the reservoirs becomes weak (van Hove limit) does the situation improve. However, a price has to be paid: All quantities can be determined only to lowest nonvanishing order in the coupling constant λ. For example, the total amount of energy flowing through the system $\langle J \rangle_\lambda$ has an expansion of the form $\langle J \rangle_\lambda = \lambda^2 \langle J_2 \rangle + \cdots$. In the weak coupling limit $\langle J_2 \rangle$ can be determined.

In the limit of weak coupling we can prove the approach to a steady state as $t \to \infty$, the Onsager relations and the principle of minimal entropy production. Our proofs are based on two fundamental properties of the time evolution (which is a quantum generalization of a classical Markov process) obtained in this limit: (1) the detailed balance property, which is a direct consequence of the fact that the reservoirs are at thermal equilibrium, and (2) the convexity of the entropy production.

Many things are omitted from this article. In particular we do not discuss at all the Kubo relations for computing transport coefficients, nor do we include recent work by the authors[5] on the heat flow in harmonic crystals or recent results about the heat flow in a random harmonic chain.[6] The reason is that we tried to cover the weak coupling situation as fully as possible at the moment.

II. THE MICROSCOPIC MODEL

We wish to describe steady-state nonequilibrium processes such as heat flow in a spatially confined system $\mathscr{S}$. To sustain a truly steady heat flow, diffusion, or chemical reactions throughout the system, we have to couple $\mathscr{S}$ to inexhaustible energy and particle reservoirs $\mathscr{R}$. In fact, we treat here only heat conduction, since the essential features of the problem appear already in this simplest case. Indeed, in the model we use, diffusion and chemical reactions would be treated in an analogous fashion and their inclusion would lead only to a "multiplication of indices" without giving any more physical insight. $\mathscr{R}$ is therefore chosen to be a collection of heat reservoirs at various temperatures.

Microscopically, the system $\mathscr{S}$ is described in the usual way. We associate with $\mathscr{S}$ a Hilbert space $\mathscr{H}$. The states of $\mathscr{S}$ are described by statistical operators or density matrices, which are positive operators on $\mathscr{H}$ normalized to trace one. Observables are self-adjoint operators on $\mathscr{H}$. The dynamics of the isolated system is governed by the unitary group $\exp(-iHt)$ generated by the Hamiltonian H. Since we want $\mathscr{S}$ to be a thermodynamically stable system, we have to assume that $Z = \mathrm{tr}\,(\exp(-\beta H))$ exists for all positive β. H must therefore have a purely discrete spectrum, which is bounded below and has no finite limit points; that is, there are only a finite number of energy levels in any finite interval ΔE.

An inexhaustible heat reservoir will be modeled microscopically here by an infinitely extended quantum mechanical system with certain additional structures. The necessity of an infinitely extended system can be seen from simple energy arguments: A finite reservoir can supply heat to (or extract heat from) the system only over a certain period of time. On a more precise level, we will later see that, in the "Markovian approximation," the influence of the reservoir on the system manifests itself through certain time correlation functions of the reservoir. To have an approach to stationarity it is necessary that these correlation functions decay faster than $1/t$. Such a decay is impossible for finite quantum systems, where because of the discreteness of the energy spectrum all correlation functions are almost periodic functions of t. Of course, if $\mathcal{R}$ is a very large quantum system, the correlation functions can be expected to have in general the right decay properties over long time spans. This expresses simply the fact that a large bucket of water "acts" for some purposes as a good heat reservoir. But clearly a piece of metal placed between two isolated buckets of water (or champagne) at different temperatures will never approach a truly stationary state in which heat is flowing through the metal bar (our system $\mathcal{S}$). It is therefore sensible to go over to the idealization of making $\mathcal{R}$ an infinitely extended system for which the faster than $1/t$ decay is expected to be true in general (and can be proved for certain simple systems like ideal gases).

The reservoir $\mathcal{R}$ is composed of r independent parts $\mathcal{R}^1, \ldots, \mathcal{R}^r$, corresponding to heat baths with different temperatures. With each one of them we associate a Hilbert space $\mathcal{H}_k^R$, a time evolution $\exp(-iH_k^R t)$, and an equilibrium state (density matrix) ω^k for the temperature β_k^{-1}. Then the Hilbert space of the reservoir is the product $\mathcal{H}^R = \mathcal{H}_1^R \otimes \cdots \otimes \mathcal{H}_r^R$. Since the various parts of the reservoir are isolated from each other the Hamiltonian H^R of $\mathcal{R}$ is the sum

$$H^R = H_1^R + \cdots + H_r^R \qquad \text{(II.1)}$$

and the state of $\mathcal{R}$ with the kth part at temperature β_k^{-1} is

$$\omega^R = \omega^1 \otimes \cdots \otimes \omega^r \qquad \text{(II.2)}$$

[The thoughtful reader might wonder how we incorporated the fact that the reservoir is infinitely extended and how we can define a Hamiltonian for a system with infinite energy. The reservoir enters in the description of the open system $\mathcal{S}$ only through certain time correlation functions $\langle A(t)B \rangle$. They are defined as the limit of the finite volume correlation functions $\langle A(t)B \rangle_V \to \langle A(t)B \rangle$ as $V \to \infty$, where $\langle \ \rangle_V$ denotes the average over the grand canonical ensemble. The crucial point is that we can always find a Hamiltonian H and a density matrix ω such that $\langle A(t)B \rangle = \text{tr}(\exp(iHt)A \exp(-iHt)B\omega)$. Of course, since H and ω are defined indirectly only through the limiting process $V \to \infty$, they will not have the form familiar from the finite system (e.g., H usually has the whole real line as continuous spectrum and an eigenvalue zero). See Refs. 7, 8.]

We couple $\mathscr{S}$ and $\mathscr{R}$ by some bounded interaction operator λH^{SR}, where $\lambda \geq 0$ is the coupling strength. Then the self-adjoint Hamiltonian of the coupled system $\mathscr{S} + \mathscr{R}$ is given by

$$H_\lambda = H \otimes 1^R + 1 \otimes H^R + \lambda H^{SR} \tag{II.3}$$

acting on the Hilbert space $\mathscr{H} \otimes \mathscr{H}^R$. The time evolution of the statistical operator W is given by

$$U_t^\lambda W = \exp(-iH_\lambda t) W \exp(iH_\lambda t) \tag{II.4}$$

It is convenient to think of U_t^λ as acting in the space of all density matrices, that is, in the so-called trace class $T(\mathscr{H} \otimes \mathscr{H}^R)$. (This is the space of all operators on $\mathscr{H} \otimes \mathscr{H}^R$ such that $\mathrm{tr}\,|W| < \infty$. It is a Banach space under the norm $\|W\|_1 = \mathrm{tr}\,|W|$.)

The reservoir by itself is of no interest to us. It forms merely a convenient microscopic device to induce steady flows throughout the system. Therefore we derive, following well-known methods,[9-18] a generalized master equation describing the reduced dynamics of the system. We assume that the initial state of $\mathscr{S} + \mathscr{R}$ is uncorrelated and of the form $\rho \otimes \omega^R$, where ρ is an arbitrary density matrix of $\mathscr{S}$. Then, in the Schrödinger picture, the reduced dynamics Λ_t^λ of the system is defined by taking the partial trace over all degrees of freedom of the reservoir

$$\Lambda_t^\lambda \rho = \mathrm{tr}_R \left[U_t^\lambda (\rho \otimes \omega^R) \right] \tag{II.5}$$

Λ_t^λ describes the time evolution of $\mathscr{S}$ coupled to the reservoirs. To derive the master equation, one thinks of (II.5) as a projection on $T(\mathscr{H} \otimes \mathscr{H}^R)$. Let us define an amplification $\mathscr{A}: T(\mathscr{H}) \to T(\mathscr{H} \otimes \mathscr{H}^R)$ by

$$\mathscr{A}\rho = \rho \otimes \omega^R \tag{II.6}$$

Then

$$P = \mathscr{A}\,\mathrm{tr}_R \tag{II.7}$$

projects $T(\mathscr{H} \otimes \mathscr{H}^R)$ to the subspace $T(\mathscr{H}) \otimes \omega^R$. We can now write (II.5) as $(\Lambda_t^\lambda \rho) \otimes \omega^R = P U_t^\lambda (\rho \otimes \omega^R)$. Let V_t^λ be the time evolution on $T(\mathscr{H} \otimes \mathscr{H}^R)$ generated by $P[H_\lambda, \cdot]P + (1-P)[H_\lambda, \cdot](1-P)$ and let us assume $\mathrm{tr}_R \omega^R H^{SR} = 0$. We consider U_t^λ as a perturbation of V_t^λ. Then, by a well-known formula (which can be verified directly by differentiation),

$$U_t^\lambda = V_t^\lambda - i\lambda \int_0^t ds\, V_{t-s}^\lambda \left(P[H^{SR}, \cdot](1-P) + (1-P)[H^{SR}, \cdot]P \right) U_s^\lambda \tag{II.8}$$

Projecting on the right and on the left, we obtain

$$PU_t^\lambda P = V_t^\lambda P - i\lambda \int_0^t ds \, V_{t-s}^\lambda P[H^{SR}, \cdot](1-P)U_s^\lambda P \tag{II.9}$$

and

$$(1-P)U_t^\lambda P = -i\lambda \int_0^t ds \, V_{t-s}^\lambda (1-P)[H^{SR}, \cdot]PU_s^\lambda P \tag{II.10}$$

and by substituting (II.10) into (II.9)

$$PU_t^\lambda P = V_t^\lambda P + \lambda^2 \int_0^t ds \int_0^s du \, V_{t-s}^\lambda P[H^{SR}, \cdot](1-P)$$
$$\times V_{s-u}^\lambda (1-P)[H^{SR}, \cdot]PU_u^\lambda P \tag{II.11}$$

If we identify $PT(\mathcal{H} \otimes \mathcal{H}^R) = T(\mathcal{H}) \otimes \omega^R$ with $T(\mathcal{H})$, then (II.11) can be written as

$$\Lambda_t^\lambda \rho = \Lambda_t^0 \rho + \lambda^2 \int_0^t ds \int_0^s du \, \Lambda_{t-s}^0 K^\lambda(s-u)\Lambda_u^\lambda \rho \tag{II.12}$$

which is the desired integral equation for the reduced dynamics. On differentiating, we obtain the generalized master equation of Nakajima, Prigogine, Résibois, and Zwanzig:[10-12]

$$\frac{d}{dt}\rho(t) = -i[H, \rho(t)] + \lambda^2 \int_0^t ds \, K^\lambda(t-s)\rho(s) \tag{II.13}$$

The integral kernel $K^\lambda(s)$ is given by

$$K^\lambda(s) = -\text{tr}_R ([H^{SR}, \cdot](1-P)V_s^\lambda (1-P)[H^{SR}, \cdot]\mathscr{A}) \tag{II.14}$$

$K^\lambda(s)$ admits a power series expansion in the coupling constant λ given by

$$K^\lambda(s) = \Lambda_s^0 \left[K_0(s) + \sum_{n=1}^\infty (-i)^n \lambda^n \int_{0 \le t_1 \le \cdots \le t_n \le s} dt_n \ldots dt_1 K_n(s, t_n, \ldots, t_1) \right] \tag{II.15}$$

where

$$K_0(s) = -\text{tr}_R ([H^{SR}(s), \cdot](1-P)[H^{SR}, \cdot]\mathscr{A}) \tag{II.16}$$

$$K_n(s, t_n, \ldots, t_1) = \text{tr}_R ([H^{SR}(s), \cdot](1-P)[H^{SR}(t_n), \cdot](1-P)$$
$$\ldots (1-P)[H^{SR}(t_1), \cdot](1-P)[H^{SR}, \cdot]\mathscr{A}) \tag{II.17}$$

and

$$[H^{SR}(t), \cdot] = U_{-t}^0[H^{SR}, \cdot]U_t^0 = [\exp(iH_0 t)H^{SR}\exp(-iH_0 t), \cdot] \tag{II.18}$$

If we choose for H^{SR} a linear coupling of the form $V \otimes V^R$ (or a sum of such linear couplings), then $K_n(s, t_n, \ldots, t_1)$ depends on the (equilibrium) multitime correlation functions $\text{tr}_R(\omega^R V^R(t_n') \ldots V^R(t_1')V^R)$ with $V^R(t) = \exp(iH^R t)V^R \exp(-iH^R t)$.

The first step in studying nonequilibrium steady state properties of $\mathcal{S}$ is to obtain the stationary state ρ_{st}^λ. If the reservoir temperatures are not too far apart and if H^{SR} represents a "good coupling," it seems safe to assume that there exists a unique stationary state ρ_{st}^λ and that every initial state approaches this state as $t \to \infty$. Therefore we should define

$$\rho_{st}^\lambda = \lim_{t \to \infty} \Lambda_t^\lambda \rho \qquad (II.19)$$

Unfortunately, our knowledge about the existence of this limit is at the present rather poor, which should come as no surprise, since (II.19) is related to establishing mixing (which is a stronger notion than ergodicity) for the infinite system $\mathcal{S} + \mathcal{R}$. Only in the case where $\mathcal{S} + \mathcal{R}$ together form either an infinite ideal solid (harmonic crystal) or an infinite ideal gas (or some other quasifree system) the existence of the limit (II.19) has been proved.[19-28] Of course, we are happy to admit any simplification on parts of the reservoir. But the system $\mathcal{S}$ should be something like a real physical system and the border cases ideal solid and ideal gas are of very limited interest; even worse, Fourier's law is not obeyed in such systems. They have an infinite thermal conductivity.[29,30] (This should be contrasted with the fact that an ideal gas obeys the laws of equilibrium thermodynamics.)

In view of this situation we look for a simplification of (II.13). If λ becomes small we might neglect all higher-order terms in (II.15). A further simplification is obtained if the memory terms present in $K_0(s)$ become negligible. From the structure of the integral kernel $K_0(s)$ we expect such a situation if the time scale on which $\rho(t)$ varies is much longer than the decay time of the kernel $K_0(s)$, that is, than the decay time of the correlation functions of the reservoir appearing in (II.16). For small λ and large t we then obtain the Born approximation to the master equation (II.13) as

$$\frac{d}{dt}\rho = -i[H, \rho] + \left(\lambda^2 \int_0^\infty ds\, K(s)\right)\rho \qquad (II.20)$$

which is the generalized Pauli equation. The original Pauli equation considered only transitions between energy eigenstates. Equation II.20 is a Markovian evolution equation on the whole space $T(\mathcal{H})$ of statistical operators. A precise definition of this weak coupling limit involves two points. First, we have to go from the Schrödinger picture to the interaction picture in order to suppress oscillations of parts of the system that happen to be uncoupled. Second, to make up for the slowing of the decay as $\lambda \to 0$, we have to rescale the physical time t in such a way that $\lambda^2 t = \tau$ is kept fixed. This is the van Hove limit[31,32] that has been made rigorous under some conditions by Davies. The theorem of Davies, to be discussed in the next section, asserts the existence of

$$\lim_{\lambda \to 0} \Lambda^0_{-\lambda^{-2}\tau} \Lambda^\lambda_{\lambda^{-2}\tau} \rho = e^{L\tau} \rho \tag{II.21}$$

which is, roughly, the integrated form of (II.20). L is a linear operator on $T(\mathcal{H})$.

All our microscopic treatment of the thermodynamics of irreversible processes is in terms of the time evolution $\{e^{L\tau} | \tau \ge 0\}$ of $\mathcal{S}$ obtained in the weak coupling limit. It is therefore essential to discuss at this point precisely what information about the irreversible processes is preserved in the weak coupling limit. After all, we work in the interaction picture, we let $\lambda \to 0$ and the physical time $t \to \infty$, and a priori it is unclear whether these limits just destroy all that we are interested in.

The coupling of a real physical system to a heat reservoir is always at the surface via short-range forces. The interaction energy H^{SR} should therefore be roughly proportional to the surface, that is, should be small compared to the bulk energy H. This crude argument gives some hope that at least some relevant information is preserved in the weak coupling limit.

To get a somewhat more precise idea, let us assume that the limit (III.19) exists and is unique. Furthermore, let $\{e^{L\tau}\} | \tau \ge 0\}$ have a unique stationary state ρ_{st},

$$e^{L\tau} \rho_{st} = \rho_{st} \tag{II.22}$$

for all $\tau \ge 0$. Then we expect that

$$\lim_{\lambda \to 0} \rho^\lambda_{st} = \rho_{st} \tag{II.23}$$

The limit $t \to \infty$ does not enter, since we look at a sequence of stationary states. The interaction picture drops out, since, as we show later, the stationary state obtained in the weak coupling limit is invariant under the isolated time evolution, which means that ρ_{st} commutes with H. Thus in the weak coupling limit we should obtain the lowest (zeroth) order contribution in λ to the stationary state ρ^λ_{st}, that is, $\rho^\lambda_{st} = \rho_0 + \lambda \rho_1 + \cdots$ with $\rho_0 = \rho_{st}$. In the case of a single reservoir temperature β^{-1}, ρ_0 is just the canonical ensemble $Z^{-1} \exp(-\beta H)$.

Since ρ_0 commutes with H, ρ_0 does not describe the local heat flow inside the system. This phenomenon would be described by the first-order contribution ρ_1, which cannot be obtained by the methods of the weak coupling limit. However, we can still compute the total amount of energy flowing through the system to lowest (λ^2) order in λ. Let us assume that the reservoir consists of a left part $\mathcal{R}^l$ and a right part $\mathcal{R}^r$ at different temperatures. The steady-state heat flow $\langle J \rangle_\lambda$ can be defined as the change of energy of $\mathcal{S}$ due to the coupling to the left heat reservoir averaged over the stationary state

$$\langle J \rangle_\lambda = \left\langle \left(\frac{d}{dt} H(t) \right)_{\text{left}} \right\rangle_\lambda \tag{II.24}$$

where $\langle \cdot \rangle_\lambda$ denotes the average over the stationary state of $\mathcal{S} + \mathcal{R}$ and $(\cdot)_{\text{left}}$ denotes the time evolution coming from the coupling to the left reservoir. (The Hamiltonian generating this time evolution is given by $H \otimes 1 + 1 \otimes H_l^R + \lambda H_l^{SR}$.) As $\lambda \to 0$, but $\lambda^2 t = \tau$ fixed, we should have the convergence

$$\langle (H(t))_{\text{left}} \rangle_\lambda \to \text{tr } (\rho_{st}(H(\tau))_{\text{left}}) = \text{tr } (\rho_{st} \exp (L^*_{\text{left}} \tau)(H)) \tag{II.25}$$

The interaction picture has again been "absorbed" by the invariance of ρ_{st} under the isolated time evolution of $\mathcal{S}$. If we assume that also the derivatives converge

$$\frac{1}{\lambda^2} \langle J \rangle_\lambda = \frac{1}{\lambda^2} \left\langle \left(\frac{d}{dt} H(t) \right)_{\text{left}} \right\rangle_\lambda \to \text{tr } \left(\rho_{st} \left(\frac{d}{d\tau} H(\tau) \right)_{\text{left}} \right) = \text{tr } (\rho_{st} L^*_{\text{left}}(H)) = \langle J \rangle \tag{II.26}$$

then it follows that the lowest-order contribution to the average heat flow is of order λ^2 and that $\lambda^{-2} \langle J \rangle_\lambda$ converges to a quantity $\langle J \rangle$, which is defined in Section V as the heat flow in the weak coupling limit.

Of course, these are plausibility arguments only. Their proof would require as prerequisite the proof of the existence of the limit (II.19). We investigated[5] these questions at a model system, where $\mathcal{S}$ is a classical harmonic chain of N particles (arbitrary couplings and masses) and $\mathcal{R}$ consists of two semi-infinite harmonic chains of equal masses ($=1$) and nearest neighbor coupling of unit strength coupled to the first and the Nth particle of $\mathcal{S}$. The conditions on "good coupling" are that all eigenfrequencies ω of $\mathcal{S}$ are in the range $0 < \omega < 2$ and that the sum of the first and Nth component of every eigenmode does not vanish. If they are fulfilled, then one can prove the existence of the limits (II.19) to (II.26) and the interchange of the various limits involved.

Let us summarize: We consider a spatially confined quantum mechanical system $\mathcal{S}$ coupled to inexhaustible heat reservoirs at various temperatures. We perform the weak coupling limit and obtain a Markovian time evolution for the system $\mathcal{S}$. We study irreversible processes in $\mathcal{S}$ in terms of this time evolution. Then, as long as we consider a steady-state situation, we obtain results that are valid to the lowest nonvanishing order in the coupling constant λ.

III. THE WEAK COUPLING LIMIT

We investigate in more detail under what conditions the weak coupling limit (II.21) of the generalized master equation (II.12) exists and study some of its properties. We follow with minor extensions (see Ref. 33) the work of Davies.[18] References 34 and 35 contain an extension of the theory developed in Ref. 18. A study of the weak coupling limit for specific model systems can be found in Refs. 36 to 39 and in Ref. 5. A rigorous model close to van Hove's original considerations can be found in Ref. 40. Readers interested in the various physical applications of the van Hove or weak coupling limit are referred to the review articles.[15,16] Starting with (II.12), we rescale the time $\lambda^2 t = \tau$ and go over to the interaction picture

$$\Lambda^0_{-\lambda^{-2}\tau}\Lambda^\lambda_{\lambda^{-2}\tau}\rho$$

$$= \rho + \int_0^\tau d\sigma \left\{ \Lambda^0_{-\lambda^{-2}\sigma} \left[\int_0^{\lambda^{-2}(\tau-\sigma)} dx\, \Lambda^0_{-x} K^\lambda(x) \right] \Lambda^0_{\lambda^{-2}\sigma} \right\} \Lambda^0_{-\lambda^{-2}\sigma}\Lambda^\lambda_{\lambda^{-2}\sigma}\rho$$

$$(\text{III.1})$$

where ρ is the density matrix of the system at time $t = 0$. We recall that Λ^0_t is simply the uncoupled time evolution of the system, which does not depend on λ, $\Lambda^0_t \rho = e^{-iHt}\rho\, e^{iHt}$. Expanding K^λ as in (II.15), we expect that in the limit as $\lambda \to 0$ only the K_0 term will contribute. The integration up to $\lambda^{-2}(\tau-\sigma)$ extends to infinity as $\lambda \to 0$. Therefore, with $\rho(\tau) = \lim_{\lambda\to 0} \Lambda^0_{-\lambda^{-2}\tau}\Lambda^\lambda_{\lambda^{-2}\tau}\rho$,

$$\rho(\tau) = \rho(0) + \lim_{\lambda\to 0} \int_0^\tau d\sigma\, \Lambda^0_{-\lambda^{-2}} K \Lambda^0_{\lambda^{-2}\sigma}\rho(\sigma) \qquad (\text{III.2})$$

where

$$K = \int_0^\infty dx\, K_0(x) \qquad (\text{III.3})$$

The remaining limit is an averaging of K with respect to the free time evolution Λ^0_t. Since the spectrum of H is purely discrete, this time average always exists and is given by

$$L = \lim_{T\to\infty} \frac{1}{2T} \int_{-T}^T dt\, \Lambda^0_{-t} K \Lambda^0_t = \sum_{\omega\in\mathrm{Sp}([H,\,\cdot\,])} Q_\omega K Q_\omega \qquad (\text{III.4})$$

Here Q_ω are the spectral projections of $[H, \cdot]$ considered as an operator on the trace class $T(\mathcal{H})$ corresponding to the eigenvalue ω, i.e.,

$$Q_\omega \rho = \sum_{\varepsilon_m - \varepsilon_n = \omega} P_m \rho P_n$$

where

$$H = \sum_j \varepsilon_j P_j \tag{III.5}$$

is the spectral representation of the Hamiltonian H. Since H^{SR} was assumed to be bounded, K and L are bounded operators on the Banach space $T(\mathcal{H})$. Inserting (III.4) in (III.2), we obtain

$$\rho(\tau) = \rho(0) + \int_0^\tau d\sigma \, L\rho(\sigma) \tag{III.6}$$

or, equivalently,

$$\frac{d}{d\tau}\rho(\tau) = L\rho(\tau), \qquad \rho(\tau) = e^{L\tau}\rho = \Lambda_\tau \rho \tag{III.7}$$

which is the desired Markovian limit of (II.13).

Davies proves that with suitable bounds on the K_n's, our heuristic argumentation is indeed valid.

Theorem 1 (Davies). Let

$$\int_0^\infty dt \, \|K_0(t)\| < \infty \tag{III.8}$$

$$\int_{0 \le t_1 \le \cdots \le t_n \le t} dt_n \ldots dt_1 \|K_n(t, t_n, \ldots, t_1)\| \le a_n(t)$$

and let the $a_n(t)$ be bounded by

$$a_n(t) \le c_n t^{n/2} \tag{III.9}$$

for all $t \ge 0$, where $\sum_{n=1}^\infty c_n z^n$ has an infinite radius of convergence, and by

$$a_n(t) \le d_n t^{n/2 - \varepsilon} \tag{III.10}$$

for some $\varepsilon > 0$, $d_n \ge 0$, and all $t \ge 0$. Then in the norm topology of $T(\mathcal{H})$

$$\lim_{\lambda \to 0} \|\Lambda^0_{-\lambda^{-2}\tau}\Lambda^\lambda_{\lambda^{-2}\tau}\rho - e^{L\tau}\rho\|_1 = 0 \tag{III.11}$$

for all $\rho \in T(\mathcal{H})$ uniformly on every finite interval $[0, \tau_0]$.

If the reservoir is quasifree (e.g., an ideal gas), then all higher-order correlation functions are expressible as sums of two-point correlation functions. As Davies shows, the conditions (III.8) and (III.10) reduce then (at least for a quasifree Fermi system) to a single condition on the two-point function. If H^{SR} is of the form $V \otimes V^R$, then the bound

$$|\mathrm{tr}_R(\omega_R V^R(t) V^R)| \leq a(1+t)^{-(1+\varepsilon)} \tag{III.12}$$

for some $\varepsilon > 0$ and all $t \geq 0$ is sufficient for the theorem to hold. Equation III.12 means that the time-dependent correlation function of V^R evolving under the isolated reservoir time evolution decays faster than t^{-1}. (See Ref. 41 for an extension to reservoirs that are not quasifree.)

For our further investigations we need a more explicit form of the generator L. First, we treat the case of a single reservoir at temperature β^{-1}. We assume that the interaction operator H^{SR} has the form

$$H^{SR} = \sum_\alpha V_\alpha \otimes V_\alpha^R \tag{III.13}$$

where $V_\alpha = V_\alpha \in B(\mathscr{H})$, $V_\alpha^R = V_\alpha^R \in B(\mathscr{H}^R)$, $\sum_\alpha \|V_\alpha\|^2 < \infty$, and, as before, $\mathrm{tr}_R(\omega^R V_\alpha^R) = 0$. $B(\mathscr{H})$ denotes all bounded operators on $\mathscr{H}$. Since K contains the correlation functions $\mathrm{tr}_R(\omega^R V_\gamma^R V_\alpha^R(t))$ and since L is the time average of K, the Fourier transforms

$$h_{\alpha\gamma}(\omega) = \int_{-\infty}^{\infty} dt\, e^{-i\omega t}\, \mathrm{tr}_R(\omega^R V_\gamma^R V_\alpha^R(t)) \tag{III.14}$$

will enter the generator L. We now exploit the fact that ω^R is a thermal equilibrium state to derive an important relation for $h_{\alpha\gamma}$. Thermal equilibrium states of both finite and infinite quantum systems are defined by the Kubo–Martin–Schwinger (KMS) boundary condition.[42,43] We call a state, defined by the density matrix W, a KMS state (thermal equilibrium state) for inverse temperature β if

$$\mathrm{tr}(WAB(t)) = \mathrm{tr}(WB(t-i\beta)A) \tag{III.15}$$

for all observables A, B, and all t. As usual $B(t-i\beta)$ is defined as $\exp(i(t-i\beta)H)B \exp(-i(t-i\beta)H)$, where H is the Hamiltonian of the system under consideration. (We disregard domain questions that can be handled, however.) The reader may check that for a system in a box, (III.15) admits as only solution $e^{-\beta H}/Z(\beta)$. If the system is infinitely extended, there might be (and in some cases definitely is) more than one solution of (III.15) indicating the existence of phase transitions.

Using the KMS boundary condition (III.15) for ω^R and, as a direct consequence, the time invariance of ω^R, we obtain

$$h_{\alpha\gamma}(\omega) = e^{\beta\omega} \int_{-\infty}^{\infty} dt\, e^{-i\omega t}\, \mathrm{tr}_R(\omega^R V_\alpha^R(t) V_\gamma^R)$$

$$= e^{\beta\omega} \int_{-\infty}^{\infty} dt\, e^{i\omega t}\, \mathrm{tr}_R(\omega^R V_\alpha^R V_\gamma^R(t)) = h_{\gamma\alpha}(-\omega)\, e^{\beta\omega} \tag{III.16}$$

that is,

$$h_{\gamma\alpha}(-\omega) = e^{-\beta\omega} h_{\alpha\gamma}(\omega)$$

Since $\{h_{\alpha\gamma}(\omega)\}$ is the Fourier transform of time-dependent correlation functions, it is a positive matrix for all ω. Equation III.16 is essential for deriving the Onsager relations, the principle of minimal entropy production, and other basic postulates of thermodynamics of irreversible processes. To state it in a somewhat oversimplified way: In our model irreversible thermodynamics follows from the fact that the reservoirs are at thermal equilibrium.

We define the Hilbert transform $s_{\alpha\gamma}$ of $h_{\alpha\gamma}$ as

$$s_{\alpha\gamma}(\omega) = i \int_0^\infty dt\, e^{-i\omega t}\, \mathrm{tr}_R\, (\omega^R V_\gamma^R V_\alpha^R(t)) - \tfrac{1}{2} h_{\alpha\gamma}(\omega)$$

$$= \frac{1}{2\pi} P \int_{-\infty}^\infty \frac{h_{\alpha\gamma}(\lambda)}{\lambda - \omega}\, d\lambda \qquad (III.17)$$

where P denotes the principal part, and introduce the operators

$$V_\alpha(\omega) = \sum_{\varepsilon_m - \varepsilon_n = \omega} P_n V_\alpha P_m = V_\alpha(-\omega) \qquad (III.18)$$

If we now perform for the chosen H^{SR} the integration (III.3) and the average (III.4), we obtain for the generator $L : T(\mathcal{H}) \to T(\mathcal{H})$

$$L(\rho) = \sum_{\omega \in \mathrm{Sp}([H,\,\cdot])} \sum_{\alpha\gamma} \{-i s_{\alpha\gamma}(\omega)[V_\gamma(\omega)^* V_\alpha(\omega), \rho]$$

$$+ h_{\alpha\gamma}(\omega)([V_\alpha(\omega)\rho, V_\gamma(\omega)^*] + [V_\alpha(\omega), \rho V_\gamma(\omega)^*])\} \qquad (III.19)$$

$\mathrm{Sp}([H,\,\cdot])$ denotes the spectrum of the Liouville–von Neumann operator $[H,\,\cdot]$. It is the set $\{\varepsilon_m - \varepsilon_n | \varepsilon_m, \varepsilon_n \in \mathrm{Sp}(H)\}$. The generator L is the sum of a Hamiltonian part L_a (the $s_{\alpha\gamma}$-term) and a dissipative part L_s (the $h_{\alpha\gamma}$-term). Here $e^{Lt}\rho$ describes the time evolution of states (Schrödinger picture). As in the quantum mechanics of isolated systems, one is often interested in the time evolution of observables A (Heisenberg picture), which is clearly to be defined by

$$\mathrm{tr}\,(A\, e^{Lt}(\rho)) = \mathrm{tr}\,(e^{L^*t}(A)\rho)$$

$$\mathrm{tr}\,(AL(\rho)) = \mathrm{tr}\,(L^*(A)\rho) \qquad (III.20)$$

The dual L^* of L is defined by (III.20) and acts now on observables. With the help of (III.19) and (III.20) we find

$$L^*(A) = \sum_{\omega \in \mathrm{Sp}([H,\,\cdot])} \sum_{\alpha\gamma} \{i s_{\alpha\gamma}(\omega)[V_\gamma(\omega)^* V_\alpha(\omega), A]$$

$$+ h_{\alpha\beta}(\omega)([V_\gamma^*(\omega), A]V_\alpha(\omega) + V_\gamma^*(\omega)[A, V_\alpha(\omega)])\} \qquad (III.21)$$

Three points should be noted: (*1*) The reservoir temperature enters via $h_{\alpha\gamma}$ through the KMS condition (III.16). (*2*) The "reservoir part" of the interaction H^{SR} is reflected by the numerical coefficients $h_{\alpha\gamma}$ and $s_{\alpha\gamma}$, the "system part" of H^{SR} by the $V_\alpha(\omega)$'s. (*3*) The Hamiltonian H of $\mathscr{S}$ enters indirectly through the $V_\alpha(\omega)$'s by definition (III.18).

We list a few properties of the generator L: .

Property 1. The canonical ensemble $\rho_\beta = Z^{-1} e^{-\beta H}$ is stationary,

$$L(\rho_\beta)=0, \qquad e^{Lt}\rho_\beta = \rho_\beta \tag{III.22}$$

for all $t \geq 0$. To prove (III.22), we insert ρ_β in (III.19) and use (III.16). In general, ρ_β is the only stationary state of $\{\Lambda_t | t \geq 0\}$ and every initial state approaches ρ_β as $t \to \infty$ (see Section VI, Theorem 3).

Property 2. Λ_t commutes with the free time evolution Λ_t^0,

$$e^{Lt}(e^{-iHt}\rho\, e^{iHt}) = e^{-iHt}(e^{Lt}\rho)\, e^{iHt} \tag{III.23}$$

In particular, all functions of the energy in $T(\mathscr{H})$ form an invariant subspace for $\{\Lambda_t | t \geq 0\}$. Equation III.23 follows from the fact that L was defined as the time average over an operator K; see (III.4).

Property 3. Let the spectrum of H be nondegenerate. Then, by property 2, statistical operators diagonal in the energy basis are transformed among themselves. If we denote the diagonal elements of a density matrix in the energy representation by p_j, they satisfy the Pauli master equation

$$\frac{d}{dt}p_j(t) = \sum_k (W_{jk}p_k(t) - W_{kj}p_j(t)) \tag{III.24}$$

$p_j(t) \geq 0$, $\sum_j p_j(t) = 1$. The coefficients W_{jk} satisfy the detailed balance condition

$$W_{jk}\, e^{-\beta\varepsilon_k} = W_{kj}\, e^{-\beta\varepsilon_k} \geq 0 \tag{III.25}$$

Equation III.25 is a consequence of (III.16). Since the spectrum of H is nondegenerate, the $s_{\alpha\gamma}$-term of L vanishes. In the dissipative part L_s of L $2V_\alpha(\omega)\rho V_\gamma(\omega)^*$ gives the first term and $-V_\gamma(\omega)^* V_\alpha(\omega)\rho - \rho V_\gamma(\omega)^* V_\alpha(\omega)$ gives the second term in (III.24).

Property 4. The generator L satisfies the (quantum generalization of the) detailed balance condition with respect to ρ_β

$$L_a(\rho_\beta)=0 \tag{III.26}$$

$$L_s(A\rho_\beta) = L_s^*(A)\rho_\beta \tag{III.27}$$

Equation III.26 is obvious; (III.27) requires some algebraic manipulations, where use is made of (III.16).

Let us define the scalar product $\langle A|B\rangle_\beta = \text{tr}\,[\rho_\beta AB^*]$ on $B(\mathcal{H})$. Then we can rewrite (III.26) and (III.27) as

$$\langle L_a^*(A)|B\rangle_\beta = -\langle A|L_a^*(B)\rangle_\beta$$

$$\langle L_s^*(A)|B\rangle_\beta = \langle A|L_s^*(B)\rangle_\beta \qquad (\text{III.28})$$

Therefore L_a^* is antisymmetric and L_s^* is symmetric with respect to the scalar product $\langle\,\cdot\,|\,\cdot\,\rangle_\beta$. Here (III.28) is the quantum generalization of the detailed balance (III.25) as proposed and studied by Alicki[44] and Kossakowski, Frigerio, Gorini, and Verri.[45] Equation III.28 is characteristic for a system weakly coupled to a single heat reservoir. If $\mathcal{S}$ is coupled in a nontrivial way to two reservoirs at different temperatures, detailed balance is not fulfilled.

Although at first sight the generator L has a rather complicated structure, properties 1 to 3 allow one to visualize the time evolution $\{e^{Lt}|t\geq 0\}$ fairly well. We expand $\rho(t) = e^{Lt}\rho$ in an energy basis. Then, generally, the off-diagonal elements $\langle n|\rho(t)m\rangle$, $n \neq m$, decay exponentially, whereas the diagonal elements $\langle n|\rho(t)n\rangle$, by transitions according to (III.24), finally reach their canonical equilibrium value. Certainly this picture has to be modified in some cases. If the spectrum of H is degenerate, then property 3 is no longer true. One may have then transitions between diagonal and off-diagonal elements. Clearly, also if the coupling is trivial, say $V_\alpha = 1$ for $\alpha = 1$ and $V_\alpha = 0$ otherwise, then $L = 0$ (since the weak coupling limit is obtained in the interaction picture). In Theorem 3 we give a criterion for what is the exception and what is the rule.

Let $\mathcal{S}$ be coupled now to the heat reservoirs $\mathcal{R}^1, \ldots, \mathcal{R}^r$ in equilibrium at inverse temperatures $\beta_1, \ldots, \beta_r$. Since the reservoir parts are independent from each other, the generator L of the weak coupling limit (III.7) is the sum of the generators $L_1, \ldots, L_r$

$$L = L_1 + \cdots + L_r \qquad (\text{III.29})$$

L_k generates the time evolution of $\mathcal{S}$ weakly coupled only to $\mathcal{R}^k$, $k = 1, \ldots, r$. Each L_k has the form given in (III.19) with $h_{\alpha\gamma}^k(\omega)$ satisfying the KMS condition (III.16) with reciprocal temperature β_k.

IV. DETAILED BALANCE AND THE KMS CONDITION

In the preceding section we showed that for a system coupled to a single reservoir at temperature β^{-1}: KMS (of the reservoir)+weak coupling→ detailed balance with respect to the canonical ensemble ρ_β (of the system). The importance of the detailed balance property derives from the fact that, as is seen later, together with the convexity of the entropy production, it implies essentially all postulates of thermodynamics of irreversible processes. In this section we want to argue the converse: detailed balance of the system + weak coupling → KMS (of the reservoir), which may be viewed as a stability property[46,47] that singles out the equilibrium states. This indicates the deep connection between detailed balance (and therefore irreversible thermodynamics) and the KMS boundary condition. The results of this section are due to Kossakowski, Frigerio, Gorini, and Verri,[45] where also the proofs may be found. (The notions used in this section are slightly more abstract than those used otherwise, and it may be omitted without preventing understanding of Sections V to VIII.)

To begin with, we abandon our particular microscopic model, retaining only the assumption that the time evolution of our open quantum-mechanical system $\mathscr{S}$ is Markovian. This leads us to the definition of a dynamical semigroup.[48] A dynamical semigroup is a one-parameter family of linear mappings

$$\Lambda_t : T(\mathscr{H}) \to T(\mathscr{H}), \qquad t \geq 0 \tag{IV.1}$$

which are positive, $\rho \geq 0$ implies $\Lambda_t \rho \geq 0$, are trace-preserving, tr $(\Lambda_t \rho) =$ tr (ρ) form a semigroup

$$\Lambda_{t+s} = \Lambda_t \Lambda_s \tag{IV.2}$$

and are continuous in the trace norm,

$$\lim_{t \to 0_+} \|\Lambda_t \rho - \rho\|_1 = 0 \tag{IV.3}$$

for all $\rho \in T(\mathscr{H})$. One can show that Λ_t is a contraction, $\|\Lambda_t\| = 1$. Positivity and preservation of the trace ensure that Λ_t maps density matrices on density matrices, (IV.2) is the Markovian property, and (IV.3) ensures, by the Hille–Yosida theorem,[49] the existence of a densely defined generator L such that $\Lambda_t = e^{Lt}$. In terms of the Heisenberg picture, defined by tr $(A\Lambda_t(\rho) =$ tr $(\Lambda_t^*(A)\rho)$ for all $\rho \in T(\mathscr{H})$, $A \in B(\mathscr{H})$, we have the following properties:

(i') $\Lambda_t^* B(\mathscr{H})_+ \subset B(\mathscr{H})_+$
(ii) $\Lambda_t^*(1) = 1$
(iii) $\Lambda_{t+s}^* = \Lambda_t^* \Lambda_s^*$
(iv') $\Lambda_t^*(A) \to A$ ultraweakly when $t \to 0_+$
(v) Λ_t^* is ultraweakly continuous

$B(\mathscr{H})_+$ denotes the positive elements in $B(\mathscr{H})$ and a sequence $A_n \in B(\mathscr{H})$ converges ultraweakly to $A \in B(\mathscr{H})$, if tr $(A_n \rho) \to$ tr $(A\rho)$ for all $\rho \in T(\mathscr{H})$.

The reduced dynamics $\Lambda_t^{\lambda^*}$ [cf. (II.5)] has another important property, as first pointed out by Kraus[50] in the context of state changes caused by quantum measurements and subsequently used by several authors.[51,52] To state this property, we recall a definition. Let $\mathcal{A}$ and $\mathcal{B}$ be C^*-algebras, and denote by $M(n)$ (n integer ≥ 1) the algebra of $n \times n$ complex matrices. A linear map $\phi: \mathcal{A} \to \mathcal{B}$ is called n-positive, if the map

$$\phi_n : \mathcal{A} \otimes M(n) \to \mathcal{B} \otimes M(n)$$

$$\phi_n(A \otimes M) = \phi(A) \otimes M, \qquad A \in \mathcal{A}, \quad M \in M(n) \qquad \text{(IV.4)}$$

is positive (i.e., maps a positive operator to a positive operator). ϕ is called completely positive, if it is n-positive for all integers n. The reduced dynamics $\Lambda_t^{\lambda^*}$ is completely positive for all t, since it is the composition of a *-automorphism and a conditional expectation. Therefore it seems physically well motivated to replace the positivity (i') by the stronger property

(i) Λ_t^* is completely positive for all $t \geq 0$

For technical reasons it is convenient to require the continuity property

(iv) $$\lim_{t \to 0_+} \|\Lambda_t^* - 1\| = 0 \qquad \text{(IV.5)}$$

which is stronger than (iv'). The norm continuity of Λ_t^* implies the existence of a bounded, ultraweakly continuous generator $L^*: B(\mathcal{H}) \to B(\mathcal{H})$ of Λ_t^*.

A family of mappings $\Lambda_t^*: B(\mathcal{H}) \to B(\mathcal{H})$, $t \geq 0$, with properties (i) to (v) is called a quantum-dynamical semigroup. (Readers interested in the theory of quantum-dynamical semigroups are referred to the book by Davies[53] and the review article by Gorini et al.[33] The review article by Ingarden and Kossakowski[54] may serve as introduction.) Recently, Lindblad[51] proved that the general form of the generator of a quantum-dynamical semigroup is given by

$$L^*(\mathcal{A}) = i[H, A] + \frac{1}{2} \sum_{j \in I} \{[V_j^*, A]V_j + V_j^*[A, V_j]\} \qquad \text{(IV.6)}$$

where H is a bounded self-adjoint operator, $V_j \in B(\mathcal{H})$ such that $\sum_{j \in I} V_j^* V_j$ converges ultraweakly, and the sum on the right-hand side converges ultraweakly. We note that the generator (III.21) obtained in the weak coupling limit is of the form (IV.6).

In analogy to (III.28) we define detailed balance for a quantum-dynamical semigroup.

Definition 1:

Let $\{\Lambda_t^* | t \geq 0\}$ be a quantum-dynamical semigroup with generator L^*. Let $\rho_0 \in T(\mathcal{H})$ be a density matrix with no zero eigenvalue and invariant under

Λ_t and let a product on $B(\mathcal{H})$ be defined by

$$\langle A|B\rangle_0 = \operatorname{tr}(AB^*\rho_0) \qquad\qquad \text{(IV.7)}$$

Then $\{\Lambda_t^*|t \geq 0\}$ satisfies the detailed balance condition with respect to ρ_0 if L^* can be written as a sum

$$L^* = L_a^* + L_s^* \qquad\qquad \text{(IV.8)}$$

such that

$$L_a^* = i[H, \cdot] \text{ with } H \text{ self-adjoint and } [H, \rho_0] = 0$$

(then L_a^* is antisymmetric $\langle L_a^*(A)|B\rangle_0 = -\langle A|L_a^*(B)\rangle_0$)

$$L_s^* \text{ is symmetric, } \langle L_s^*(A)|B\rangle_0 = \langle A|L_s^*(B)\rangle_0$$

If Λ_t^* satisfies detailed balance with respect to ρ_0 and if the spectrum of ρ_0 is nondegenerate, then Λ_t induces on the states diagonal in the ρ_0-basis a classical Markov process that satisfies the well-known detailed balance of the form (III.25). In this sense Definition 1 generalizes the classical case. The generators of quantum-dynamical semigroups satisfying the detailed balance condition have been completely classified.[44,45]

Let us assume now that the reservoir $\mathcal{R}$ is in an arbitrary state ω^R invariant under the uncoupled time evolution. As long as Davies' conditions (III.8) to (III.10) are satisfied, we can perform the weak coupling limit and obtain again a semigroup with a generator of the form (III.21). However, h_{ij} will not satisfy (III.16), in general. Let us fix for a moment the reservoir part of the interaction $V_1^R = A$, $V_2^R = B$, $V_j^R = 0$, $j \neq 1, 2$, and let us denote h_{12} by h_{AB}. If we assume in addition that the semigroup obtained in the weak coupling limit satisfies the detailed balance condition with respect to some stationary ρ_0, then this imposes certain restrictions on h_{AB} and on h_{BA}. If we assume detailed balance to be satisfied for a whole class of system Hamiltonians (in fact six-level systems suffice), then we obtain a whole class of restrictions on h_{AB} and on h_{BA}. The interesting point is that all these restrictions together admit only the single solution

$$h_{AB}(-\omega) = e^{-\beta\omega}h_{BA}(\omega) \qquad\qquad \text{(IV.9)}$$

for some β independent of A and B. By Fourier transformations we obtain the KMS condition (III.15) for A and B. We repeat the argument for other observables, for which Davies' conditions (Theorem 1) are satisfied. This shows then that ω^R has to be a KMS state on all these observables. Therefore if the state ω^R of the reservoir $\mathcal{R}$ is such that all systems weakly coupled to $\mathcal{R}$ in the state ω^R satisfy the detailed balance condition, then ω^R has to be a thermal equilibrium (KMS) state.

V. ENTROPY PRODUCTION

Entropy production plays a central role in nonequilibrium thermodynamics.[55,56] The entropy production σ is the source term in the (local) balance equation for the entropy density

$$\frac{\partial S}{\partial t} = -\text{div } \mathbf{J}_S + \sigma \qquad (\text{V.1})$$

S being the local entropy density and $\mathbf{J}_S$ being the vector field of entropy flow per unit area per unit time. Here σ is assumed to be always nonnegative, since entropy can only be created, never destroyed; for reversible processes $\sigma = 0$. This is the (local formulation) of the second law of thermodynamics. It turns out that in the phenomenological theory σ can be written as a bilinear form in the fluxes characterizing the irreversible process and in the conjugate thermodynamic forces. If only heat conduction can occur in the system, then the entropy production is given by

$$\sigma = -\frac{1}{T^2}\mathbf{J} \cdot (\text{grad } T) \qquad (\text{V.2})$$

where T is the temperature field and $\mathbf{J}$ is the vector field for the heat flow. (Here $\mathbf{J}$ is the flux and $-1/T^2 \text{ grad } T$ is the conjugate thermodynamic force.)

In our attempt to give a microscopic foundation to the thermodynamics or irreversible processes we should, therefore, in the first place, give a microscopic expression for the entropy production. To begin with, we treat the case of the system $\mathscr{S}$ weakly coupled to a reservoir at a single temperature β^{-1}. The time evolution of $\mathscr{S}$ is given by $\{e^{Lt}|t \geq 0\}$ in the weak coupling approximation (III.7). The entropy S of $\mathscr{S}$ in the state ρ is defined by the well-known expression of von Neumann

$$S(\rho) = -\text{tr } (\rho \log \rho) \qquad (\text{V.3})$$

The entropy flow is $-\beta \, dQ$ (Boltzmann's constant k is set equal to one), where dQ is the heat transfer from the reservoir to the system (i.e., dQ is the change of energy of S that occurs at a reservoir temperature β^{-1} and is reversible as far as the reservoir is concerned).

$$dQ = \text{tr } (\rho(t)L^*(H)) = \text{tr }\left(\rho \frac{d}{dt}H(t) \right)$$
$$\qquad\qquad\qquad\qquad\qquad\qquad\qquad\qquad (\text{V.4})$$
$$\mathbf{J}_S = -\beta \, dQ = \frac{d}{dt} \text{tr } \rho(t) \log \rho_\beta$$

where $\rho_\beta = Z^{-1} e^{-\beta H}$. Therefore, using the balance equation (V.1) in the integrated form, we obtain for the (total) entropy production

$$\sigma(\rho) = \frac{d}{dt}(\operatorname{tr}(\rho(t) \log \rho_\beta) - \operatorname{tr}(\rho(t) \log \rho(t)))|_{t=0} \tag{V.5}$$

$$\sigma(\rho) = -\frac{d}{dt} S(e^{Lt}\rho|\rho_\beta)|_{t=0} \tag{V.6}$$

$S(\rho|\rho_\beta)$ is the relative entropy[57] of ρ with respect to ρ_β. Equation (V.6) is our basic definition.

Equation (V.5) is only a formal expression for the relative entropy, since both summands may be infinite. One way to define the relative entropy properly is

$$S(A|B) = \sum_{i,j} |\langle a_i|b_j\rangle|^2 (a_i \log a_i - a_i \log b_j + b_j - a_i) \tag{V.7}$$

$$(= \operatorname{tr} A \log A - \operatorname{tr} A \log B + \operatorname{tr} B - \operatorname{tr} A)$$

where A, B are statistical operators with eigenvalues $\{a_i\}$, $\{b_j\}$ and eigenvectors $\{|a_i\rangle\}$, $\{|b_j\rangle\}$. Since the second factor is nonnegative, $S(A|B) \geq 0$, possible $S(A|B) = \infty$. We note that $S(A|B) = \infty$, if the range of A is not contained in the range of B. (The range of A is the linear span of all eigenvectors to nonzero eigenvalues of A.)

We collect the essential properties of the entropy production in

Theorem 2. Let dim $\mathcal{H} = N < \infty$ and let L be given by (III.19). Then the entropy production

$$\sigma(\rho) = -\frac{d}{dt} S(\rho|\rho_\beta)|_{t=0}$$

is given by

$$\sigma(\rho) = \operatorname{tr}(L(\rho)(\log \rho_\beta - \log \rho))$$

$$= \sum_{\omega \in \operatorname{Sp}[H, \cdot]} \sum_{j,k=1}^{N} \left\{ \sum_{\alpha,\gamma} h_{\alpha\gamma}(\omega)\langle\Psi_j|V_\alpha(\omega)\Psi_k\rangle\langle\Psi_k|V_\gamma(\omega)^*\Psi_j\rangle \right\}$$

$$\times \{(\rho_k - \rho_j e^{-\beta\omega})(\log \rho_k - \log \rho_j + \beta\omega)\} \tag{V.8}$$

where $\rho = \sum_{j=1}^{N} \rho_j |\Psi_j\rangle\langle\Psi_j|$ is the spectral decomposition of ρ and where $\operatorname{tr}(L(\rho) \log \rho)$ is defined by

$$\operatorname{tr}(L(\rho) \log \rho) = \sum_{j=1}^{N} \langle\Psi_j|L(\rho)\Psi_j\rangle \log \rho_j$$

$$\langle\Psi_j|L(\rho)\Psi_j\rangle \log \rho_j = \begin{cases} -\infty & \text{if } \langle\Psi_j|L(\rho)\Psi_j\rangle \neq 0 \text{ and } \rho_j = 0 \\ 0 & \text{if } \langle\Psi_j|L(\rho)\Psi_j\rangle = 0 \end{cases} \tag{V.9}$$

Here σ has the properties:

(i) $\sigma(\rho) \geq 0$, possibly $\sigma(\rho) = \infty$. If $\{h_{\alpha\gamma}(\omega)\} > 0$ (as a matrix) for all $\omega \in$ Sp$([H, \cdot])$, then $\sigma(\rho) = 0$ if and only if $L(\rho) = 0$.

(ii) σ is a convex function, that is,

$$\sigma(\lambda\rho_1 + (1-\lambda)\rho_2) \leq \lambda\sigma(\rho_1) + (1-\lambda)\sigma(\rho_2) \qquad \text{(V.10)}$$

$0 \leq \lambda \leq 1$, with statistical operators ρ_1, ρ_2.

Proof: Since dim $\mathscr{H} < \infty$ and since $\log \rho_\beta = -\beta H - Z$, we have

$$S(\rho(t)|\rho_\beta) = \text{tr}\,(\rho(t)\log\rho(t)) + \beta\,\text{tr}\,(\rho(t)H) + \beta Z \qquad \text{(V.11)}$$

The derivative of the second term is $\text{tr}\,(L(\rho)\log\rho_\beta)$. Using the spectral decomposition of $\rho(t)$, $\rho(t) = \sum_j \rho_j(t)|\Psi_j(t)\rangle\langle\Psi_j(t)|$, we obtain

$$\frac{d}{dt}\text{tr}\,(\rho(t)\log\rho(t))|_{t=0} = \sum_k \{\rho_k\log\rho_k(\langle\dot\Psi_k(0)|\Psi_k\rangle + \langle\Psi_k|\dot\Psi_k(0)\rangle) + \dot\rho_k(0)\log\rho_k\}$$

$$\text{(V.12)}$$

If $\rho_k = 0$ and $\dot\rho_k(0) > 0$, then $\dot\rho_k(0)\log\rho_k = -\infty$. If $\rho_k = 0$ and $\dot\rho_k(0) = 0$, then using a Taylor expansion of $\rho_k(t)$ ($\rho_k(t)$ is an analytic function) and performing the limit as $t \to 0_+$, one obtains $\dot\rho_k(0)\log\rho_k = 0$. Since

$$\langle\Psi_k|L(\rho)\Psi_k\rangle = \rho_k(\langle\dot\Psi_k(0)|\Psi_k\rangle + \langle\Psi_k|\dot\Psi_k(0)\rangle) + \dot\rho_k(0) \qquad \text{(V.13)}$$

we have

$$\frac{d}{dt}\text{tr}\,(\rho(t)\log\rho(t))|_{t=0} = \text{tr}\,(L(\rho)\log\rho) \qquad \text{(V.14)}$$

with the convention (V.9). We note that for $t > 0$, $\rho_k(t) = 0$ implies $\dot\rho_k(t) = 0$, since $\rho_k(t_-)$, $\rho_k(t_+) \geq 0$.

We compute the entropy production $\sigma(\rho) = -\text{tr}\,(L(\rho)\log\rho) - \text{tr}\,(L(\rho)\beta H)$. We have $\text{tr}\,(L_a(\rho)\log\rho) = 0$, since $[\rho, \log\rho] = 0$, and $\text{tr}\,(L_a(\rho)H) = 0$, since $[H, V_\gamma^*(\omega)V_\alpha(\omega)] = 0$. The contribution of the symmetric part to the entropy production is

$$\sigma(\rho) = 2 \sum_{\omega\in\text{Sp}([H,\,\cdot\,])} \sum_{j,k=1}^{N} \sum_{\alpha,\gamma} h_{\alpha\gamma}(\omega)$$

$$\times \langle\Psi_j|V_\alpha(\omega)\Psi_k\rangle\langle\Psi_k|V_\gamma(\omega)\Psi_j\rangle\rho_k(\log\rho_k - \log\rho_j + \beta\omega) \qquad \text{(V.15)}$$

$$= 2 \sum_{\omega\in\text{Sp}([H,\,\cdot\,])} \sum_{j,k=1}^{N} \sum_{\alpha,\gamma} h_{\alpha\gamma}(\omega)$$

$$\times \langle\Psi_j|V_\alpha(\omega)\Psi_k\rangle\langle\Psi_k|V_\beta(\omega)\Psi_j\rangle e^{-\beta\omega}\rho_j(\log\rho_j - \log\rho_k - \beta\omega)$$

$$\text{(V.16)}$$

where we used (III.16) and (III.18). Adding (V.15) and (V.16) and dividing by 2, one obtains (V.8).

From (V.8) it is obvious that $\sigma(\rho) \geq 0$. If $L(\rho) = 0$, then $\sigma(\rho) = 0$ by (V.8). Let $\sigma(\rho) = 0$. Since $\{h_{\alpha,\gamma}(\omega)\}$ is a positive matrix we can diagonalize it. Let $h_\gamma(\omega) \geq 0$ be its eigenvalues and let $\{\tilde{V}_\gamma(\omega)\}$ be $\{V_\alpha(\omega)\}$ expressed in the eigenbasis of $\{h_{\alpha,\gamma}(\omega)\}$. Then

$$\sigma(\rho) = \sum_{\omega \in \mathrm{Sp}([H,\,\cdot\,])} \sum_{j,k=1}^{N} \sum_\gamma h_\gamma(\omega) |\langle \Psi_j | \tilde{V}_\gamma \Psi_k \rangle|^2 \\ \times (\rho_k - \rho_j)\, e^{-\beta\omega} (\log \rho_k - \log \rho_j + \beta\omega) = 0 \qquad (V.17)$$

implies

$$h_\gamma(\omega) \langle \Psi_j | \tilde{V}_\gamma(\omega) \Psi_k \rangle (\rho_k - \rho_j\, e^{-\beta\omega}) = 0 \qquad (V.18)$$

for all γ, $\omega \in \mathrm{Sp}\,([H, \cdot\,])$ and $j, k = 1, \ldots, N$. Therefore

$$[\tilde{V}_\gamma(\omega), \rho\, e^{\beta H}] = 0 \qquad (V.19)$$

for all γ and $\omega \in \mathrm{Sp}\,([H, \cdot\,])$. Let $\mathcal{N}$ be the set of all operators commuting with all $\tilde{V}_\gamma(\omega)$. Then

$$\rho = A\, e^{-\beta H} \qquad (V.20)$$

with $A \in \mathcal{N}$. We write the generator (III.19) in terms of the $\tilde{V}_\gamma(\omega)$. Then, since $A \in \mathcal{N}$,

$$[\tilde{V}_\gamma(\omega)^* \tilde{V}_\gamma(\omega), A\, e^{-\beta H}] = 0 \qquad (V.21)$$

which implies $L_a(\rho) = 0$. By the detailed balance (III.27)

$$L_s(A\, e^{-\beta H}) = L_s^*(A)\, e^{-\beta H} = 0 \qquad (V.22)$$

since $A \in \mathcal{N}$. Therefore $L(\rho) = 0$.

To prove the convexity of σ we note that $\mathrm{tr}\,(L(\rho) \log \rho_\beta)$ is linear. Therefore we have to show only the convexity of $-\mathrm{tr}\,(L_s(\rho) \log \rho)$. Since $\{h_{\alpha\gamma}(\omega)\}$ is a positive matrix we can diagonalize it for every fixed ω. Then (III.19) takes the form

$$L_s(\rho) = \sum_j [V_j, \rho V_j^*] + [V_j\rho, V_j^*] \qquad (V.23)$$

where the V_j's are certain linear combinations of the $V_\alpha(\omega)$'s. Therefore

$$-\mathrm{tr}\,(L_s(\rho) \log \rho) = 2\sum_j \{\mathrm{tr}\,(V_j^* V_j\rho\, \log \rho) - \mathrm{tr}\,(V_j\rho V_j^*\, \log \rho)\} \qquad (V.24)$$

A theorem due to Lieb[58] asserts that the function

$$\rho \to -\mathrm{tr}\,(\rho^q V_j\rho^{1-q} V_j^*) \qquad (V.25)$$

$0 \leq q \leq 1$, is convex. For $q = 0$ the function is affine. Therefore the derivative at $q = 0$

$$\rho \to \operatorname{tr} (V_j \rho \log \rho V_j^*) - \operatorname{tr} (\log \rho V_j \rho V_j^*) \tag{V.26}$$

is convex. This implies that $-\operatorname{tr} L_s(\rho) \log \rho$ is convex. ∎

COMMENTS

1. The technical restriction dim $\mathscr{H} < \infty$ is annoying and is not overcome easily. Since $S(\rho | \rho_\beta)$ is quite often infinite, it is difficult to prove that $-S(\Lambda_t \rho | \rho_\beta)$ is differentiable and that its derivative is given by (V.8). Of course, we can define the entropy production by (V.8) and then all properties stated are still valid. Our theorem holds also for the classical Markov process with finite state space defined by the Pauli equation (III.24) with detailed balance (III.25). For a general classical Markov process with detailed balance[1,2] one encounters the same difficulties as in the case of an infinite dimensional Hilbert space.

2. Property (ii) is the nonnegativity of the entropy production as required by the phenomenological theory. In general, ρ_β will be the only stationary state. Then $\sigma(\rho) > 0$ for all $\rho \neq \rho_\beta$. This means that $-S(\rho(t) | \rho_\beta)$ is strictly increasing reaching in the limit as $t \to \infty$ the value $-S(\rho_\beta | \rho_\beta) = 0$. At $t = 0$, $-S(\rho(t) | \rho_\beta)$ may have an infinite slope, but at any later time its slope is finite. Typically, this happens if the initial state has, for example, a sharp energy or is otherwise rather pure.

3. Property (iii) is the essential stability property of the entropy production. It guarantees that, as the system is coupled to several reservoirs, there is always a unique state of minimal entropy production. (This statement is too strong. It would be true, if one could prove strict convexity of σ. In general, there is a convex set of states with minimal entropy production.) This situation is analogous to the one in thermodynamics, where the concavity of the entropy ensures stability.

Let us now consider the case of the system $\mathscr{S}$ coupled to reservoirs $\mathscr{R}^1, \ldots, \mathscr{R}^r$ at inverse temperatures $\beta_1, \ldots, \beta_r$. The time evolution of $\mathscr{S}$ is generated by $L = L_1 + \cdots + L_r$. To find a microscopic expression for the entropy production we start again from the balance equation (V.1). The entropy S is still defined by (V.3). The entropy flow is given by

$$J_s = - \sum_{k=1}^{r} \beta_k \, dQ_k \tag{V.27}$$

where dQ_k is the heat transfer from the kth reservoir to the system (i.e., the change of energy of S due to the coupling to the kth reservoir). Therefore

$$dQ_k = \operatorname{tr} (\rho L_k^*(H)) \tag{V.28}$$

and, by (V.14),

$$\sigma(\rho) = \sum_{k=1}^{r} -\beta_k \, \text{tr} \, (\rho L_k^*(H)) - \text{tr} \, (L(\rho) \log \rho)$$

$$= \sum_{k=1}^{r} \{-\beta_k \, \text{tr} \, (L_k(\rho)H) - \text{tr} \, (L_k(\rho) \log \rho)\}$$

$$= \sum_{k=1}^{r} \sigma_k(\rho) \qquad\qquad\qquad (V.29)$$

The entropy production σ (for a specified state of the system) is the sum of the entropy productions σ_k, where σ_k is the entropy production of $\mathscr{S}$ coupled only to the kth reservoir. As a direct consequence of Theorem 2, we note

Corollary 1. Let $\sigma = \sum_{k=1}^{r} \sigma_k$. Then $\sigma \geq 0$. Here $\sigma(\rho) = 0$ if and only if $\sigma_k(\rho) = 0$ for $k = 1, \ldots, r$. If for every L_k the canonical ensemble ρ_β is the only stationary state (in some temperature region), then $\sigma(\rho) = 0$ if and only if $\beta_1 = \beta_2 = \cdots = \beta_r = \beta$ and $\rho = \rho_\beta$.

The entropy production σ is convex.

In the steady state ρ_s, defined by

$$L(\rho_s) = 0, \qquad L = L_1 + \cdots + L_r \qquad\qquad (V.30)$$

the entropy production is a bilinear form $\sigma(\rho_s) = \sum_{k=1}^{r} \beta_k J_k$, where $J_k = \text{tr} \, (\rho_s L_k^*(H))$ is the steady-state heat flow from the kth reservoir to the system. Since ρ_s depends on $\beta_1, \ldots, \beta_r$, J_k does too. We choose some reference temperature β and define the thermodynamic forces by $X_k = -\beta_k + \beta$. Then, using the conservation law $\sum_{k=1}^{r} J_k = \text{tr} \, (L(\rho_s)H) = 0$, σ becomes a bilinear form in the fluxes and the thermodynamic forces

$$\sigma(\rho_s) = \sum_{r=1}^{k} X_k J_k(X_1, \ldots, X_r) \qquad\qquad (V.31)$$

In the *linear* approximation

$$J_k(X_1, \ldots, X_r) = \sum_{j=1}^{r} L_{kj}(\beta) X_j \qquad\qquad (V.32)$$

and the entropy production is a quadratic form

$$\sigma(\rho_s) = \sum_{k,j=1}^{r} L_{kj}(\beta) X_k X_j \qquad\qquad (V.33)$$

The $L_{kj}(\beta)$ are the kinetic coefficients related to the heat conductivities.

From the convexity of σ it follows immediately that

$$\{L_{kj}(\beta)\} \geq 0 \tag{V.34}$$

Since $\{L_{kj}(\beta)\}$ is a matrix on a real vector space, (V.34) does not imply its symmetry $L_{kj} = L_{jk}$ (Onsager relations). As we shall see, however, in Section VII the symmetry essentially follows from the detailed balance property.

VI. APPROACH TO STATIONARITY

On physical grounds one expects that, in general, the reservoir will drive the system to a unique stationary state. Of course, the coupling between the system and the reservoir might be poor (say, $H^{SR} = 0$). Then, since we work in the interaction picture, several stationary states should appear, and, depending on the initial state ρ, $\rho(t)$ should approach one of them as $t \to \infty$.

Definition 2:
A generator L (or the dynamics $\{e^{Lt} | t \geq 0\}$) is called relaxing, if there exists a unique stationary state ρ_s, $L(\rho_s) = 0$, and if

$$\lim_{t \to \infty} e^{Lt}\rho = \rho_s \tag{VI.1}$$

for all initial states ρ.

For a single temperature reservoir, the relative entropy is an efficient tool to study the approach to equilibrium. By Theorem 1 $-d/dt\, S(\rho(t) \| \rho_\beta) > 0$ unless $L(\rho) = 0$. Therefore every initial state ρ tends to a limit as $t \to \infty$,

$$\lim_{t \to \infty} e^{Lt}\rho = A_\rho\, e^{-\beta H} \tag{VI.2}$$

where $A_\rho \in \mathcal{N}$ [cf. (V.19)]. L is relaxing if and only if $\mathcal{N}$ consists only of multiples of the unit operator. In that case the unique stationary state is the canonical ensemble.

One would like to have a condition for relaxation that is more directly related to H and H^{SR}. If we assume that $\{h_{\alpha\gamma}(\omega)\} > 0$ for all $\omega \in \text{Sp}\,([H, \cdot\,])$ and all inverse temperatures β, then $\mathcal{N} = \{V_\alpha(\omega) | \alpha,\ \omega \in \text{Sp}\,([H, \cdot\,])\}'$. $'$ denotes the set of all operators commuting with the ones in $\{\cdot\}$ [i.e., $\mathcal{N}$ is the set of all operators commuting with all $V_\alpha(\omega)$'s]. If, furthermore, $\{H, V_\alpha | \alpha\}' = \{\mathbb{C}1\}$, then, as can be shown,[57] $\mathcal{N} = \{\mathbb{C}1\}$. Hence $\{h_{\alpha\gamma}(\omega)\} > 0$ and $\{H, V_\alpha | \alpha\}' = \{\mathbb{C}1\}$ is a sufficient condition for relaxation. In the case of several reservoirs at different temperatures one can use the fact that, if $L_1, \ldots, L_r$ are relaxing, then also $L = L_1 + \cdots + L_r$ is relaxing (dim $\mathcal{H} < \infty$).

For the case of an infinite dimensional Hilbert space one has to use much more powerful methods than we introduced so far to prove[60] (cf. also Refs. 59, 61–63).

Theorem 3. Let the dynamics of $\mathcal{S}$ coupled to the reservoirs $\mathcal{R}^1, \ldots, \mathcal{R}^r$ be given by the generator $L = L_1 + \cdots + L_r$ as in (III.29). If for given inverse temperatures $\beta_1, \ldots, \beta_r$, $\{h^{(k)}_{\alpha\gamma}(\omega)\} > 0$ for all $\omega \in \mathrm{Sp}\,([H, \cdot])$, $k = 1, \ldots, r$, and if

$$\{H, V^{(k)}_\alpha \,|\, \alpha, k = 1, \ldots, r\}' = \{\mathbb{C}1\} \tag{VI.3}$$

(i.e., only multiples of the unit operator commute with H and the $V^{(k)}_\alpha$'s), then L is relaxing.

Remark: For readers who did not omit Section IV. Theorem 3 is the special case of a very fine theorem.[60] Let $\{\Lambda^*_t \,|\, t \geq 0\}$ be a quantum-dynamical semigroup with generator L^*. Then we define the dissipation function[51] $D: B(\mathcal{H}) \times B(\mathcal{H}) \to B(\mathcal{H})$ by

$$D(A, B) = L^*(A^*B) - L^*(A^*)B - A^*L^*(B) \tag{VI.4}$$

One shows that $D(A, A) \geq 0$. If $D(A, A) = 0$ for all A, then $L^*(A) = i[H, A]$, which means that there is no dissipation at all. Obviously, $D(1, 1) = 0$. The theorem of Frigerio states that if $D(A, A) = 0$ implies $A = c1$, then $\{\Lambda_t \,|\, t \geq 0\}$ is relaxing. In this case we have complete dissipation. Therefore the dissipation function D characterizes the lack of reversibility of L^*. Using the form (IV.6) of the generator L^*, the preceding condition is equivalent to $\{V_j \,|\, j \in I\}' = \{\mathbb{C}1\}$.

Theorem 3 gives us a physically very intuitive criterion for an effective coupling, causing the system to approach a unique steady state. $\{h^{(k)}_{\alpha\gamma}(\omega)\} > 0$ means that the reservoir couples to all relevant frequencies of $\mathcal{S}$ and (VI.3) expresses that, besides the unit operator, there is no other operator that commutes with the Hamiltonian H and the interaction operators $V^{(k)}_\alpha$. To put it somewhat differently: If we form arbitrary linear combinations and products of H and the $V^{(k)}_\alpha$'s (and take weak limits), then we obtain in this way any bounded operator, which means that H and the $V^{(k)}_\alpha$'s together have to be sufficiently incompatible.

VII. ONSAGER RELATIONS

We show the validity of the Onsager relations[64]

$$L_{kj}(\beta) = L_{jk}(\beta) \tag{VII.1}$$

[cf. (V.33)], for the system $\mathscr{S}$ weakly coupled to several reservoirs. In the special case of the Pauli master equation (III.24), Hepp[65,66] proved (VII.1) using the derivation of the Onsager relations for a classical system coupled to stochastic reservoirs as given by Bergmann, Lebowitz et al.[1,67]

Let $\rho_s = \rho_s(\beta_1, \ldots, \beta_r)$ be the unique stationary state of the generator $L = L_{1,\beta_1} + \cdots + L_{r,\beta_r}$, where we indicated explicitly the temperature dependence of the kth generator. Then the kinetic coefficients $L_{kj}(\beta)$ are defined by

$$L_{kj}(\beta) = \frac{\partial}{\partial \beta_i} \operatorname{tr}\left[L_{k,\beta_k}(H)\rho_s(\beta_1, \ldots, \beta_r)\right]\big|_{(\beta_1,\ldots,\beta_r)=(\beta,\ldots,\beta)} \quad \text{(VII.2)}$$

$k \neq j$. By the conservation law $\sum_k \operatorname{tr}(L_k^*(H)\rho_s) = 0$, $L_{jj}(\beta) = -\sum_{k \neq j} L_{kj}(\beta)$. Our proof of (VII.1) consists of two steps: First, we derive a convenient expression for the kinetic coefficients $L_{kj}(\beta)$ and then show their symmetry. The essential ingredient of the proof is the detailed balance property (III.27), which was a direct consequence of the KMS condition for the thermal equilibrium state of the reservoirs.

Lemma 1. Let dim $\mathscr{H} < \infty$. Let the operator for the heat flow from the kth reservoir at inverse temperature β_k to the system be $J_{k,\beta_k} = L_{k,\beta_k}^*(H)$ and let $L_{1,\beta_1} + \cdots + L_{r,\beta_r}$ have a unique stationary state $\rho_s(\beta_1, \ldots, \beta_r)$ in some neighborhood of $(\beta, \ldots, \beta)$. Then the kinetic coefficients $L_{kj}(\beta)$ are given by

$$L_{kj}(\beta) = -\operatorname{tr}\left[\frac{1}{L^*}(J_{k,\beta})J_{j,\beta}\rho_\beta\right] \quad \text{(VII.3)}$$

$$= \int_0^\infty dt \operatorname{tr}[J_{k,\beta}(t)J_{j,\beta}\rho_\beta] \quad \text{(VII.4)}$$

with $J_{k,\beta}(t) = e^{L^*t}J_{k,\beta}$ and $L = L_{1,\beta} + \cdots + L_{r,\beta}$.

Proof: Since the canonical ensemble ρ_β is invariant for $L_{j,\beta}$, we have

$$L_{j,\beta}'(\rho_\beta) = L_{j,\beta}(H\rho_\beta) = J_{j,\beta}\rho_\beta \quad \text{(VII.5)}$$

where $'$ denotes the derivative with respect to β and where we used $L_{a,k}(H) = 0$ and the detailed balance (III.27) for $L_{s,k}$. Since $\rho_s(\beta, \ldots, \beta) = \rho_\beta$

$$(L_{1,\beta} + \cdots + L_{j,\beta+\delta} + \cdots + L_{r,\beta})(\rho_s(\beta, \ldots, \beta+\delta, \ldots, \beta)) - L(\rho_\beta) = 0$$

$$\text{(VII.6)}$$

which implies

$$L_{j,\beta}'(\rho_\beta) = -L(\rho_s'(\beta)) \quad \text{(VII.7)}$$

and, by (VII.5),

$$L(\rho_s'(\beta)) = -J_{j,\beta}\rho_\beta \qquad \text{(VII.8)}$$

Since, by assumption, the kernel of L is one-dimensional (i.e., the eigenvalue zero is nondegenerate), (VII.8) admits the one-parameter family of solutions

$$\rho_s'(\beta) = c\rho_\beta - \frac{1}{L}(J_{j,\beta}\rho_\beta) \qquad \text{(VII.9)}$$

$c \in \mathbb{R}$, which defines $(1/L)(J_{j,\beta}\rho_\beta)$. Since tr $(J_{k,\beta}\rho_\beta) = 0$,

$$L_{kj}(\beta) = -\text{tr}\left[J_{k,\beta} \frac{1}{L}(J_{j,\beta}\rho_\beta) \right] \qquad \text{(VII.10)}$$

ρ_β is the only stationary state of L. By the discussion in Section VI this implies that $\lim_{t\to\infty} e^{Lt}(J_{j,\beta}\rho_\beta) = c\rho_\beta$ and that, except zero, all eigenvalues of L lie in the open left-hand plane. Therefore we can express the inverse in (VII.10) as Laplace transform

$$L_{kj}(\beta) = \int_0^\infty dt \, \text{tr}\,[J_{k,\beta} e^{Lt}(J_{j,\beta}\rho_\beta)] \qquad \text{(VII.11)}$$

where we used again tr $(J_{k,\beta}\rho_\beta) = 0$. The integral (VII.11) is finite. ∎

It is no coincidence that (VII.4) looks like the well-known Kubo formula for transport coefficients. One can derive (VII.4) in the framework of linear response theory adapted to our model.[4,68]

Theorem 4. Let dim $\mathscr{H} < \infty$ and let L_a and L_s be the antisymmetric and the symmetric part of $L = L_{1,\beta_1} + \cdots + L_{r,\beta_r}$. If in some neighborhood of $(\beta, \ldots, \beta)$ the unique stationary state ρ_s of L is also the unique stationary state of L_s, then the Onsager relations (VII.1) are valid.

Proof: Since the stationary state is already defined by $L_s(\rho_s) = 0$, we can repeat the derivation of (VII.4) with L_s instead of L and obtain (VII.4) with L replaced by L_s. By the detailed balance property (III.27)

$$e^{L_s t}(A\rho_\beta) = e^{L_s^* t}(A)\rho_\beta \qquad \text{(VII.12)}$$

Since $[J_{k,\beta}, \rho_\beta] = 0$

$$\text{tr}\,[e^{L_s^* t}(J_{k,\beta})J_{j,\beta}\rho_\beta] = \text{tr}\,[e^{L_s^* t}(J_{j,\beta})J_{k,\beta}\rho_\beta] \qquad \text{(VII.13)}$$

which implies (VII.1). ∎

The conditions of the theorem are certainly met, when $L_a = 0$. They are also fulfilled, whenever the spectrum of H is nondegenerate. Since the free time evolution commutes with e^{Lt} and since ρ_s is unique

$$[H, \rho_s] = 0 \qquad \text{(VII.14)}$$

In general, this does not imply $L_a(\rho_s) = 0$. However, if the spectrum of H is nondegenerate, then ρ_s is a function of H and therefore $L_a(\rho_s) = 0$. In general, one obtains

$$\operatorname{tr}\left[e^{(L_s^* + L_a^*)t}(J_{k,\beta})J_{j,\beta}\rho_\beta\right] = \operatorname{tr}\left[e^{(L_s^* - L_a^*)t}(J_{j,\beta})J_{k,\beta}\rho_\beta\right] \qquad \text{(VII.15)}$$

Without additional assumptions (VII.15) does not imply the Onsager relations (VII.1).

At first sight it is somewhat surprising that the Onsager relations should not be valid in full generality. The reason is to be sought in our construction of the generator L. If we consider L as a function of the Hamiltonian H, then L is discontinuous whenever two eigenvalues meet. (This is so, since the projection P_j on the eigenspace for the energy ε_j changes discontinuously.) There is, however, a unique way to extend L continuously through the points of discontinuity. If we choose this (at finitely many points) modified $\tilde{L}$ as generator of the time evolution, then all our results remain valid and the Onsager relations hold without any restriction.

Remark: The reader will have noticed that we used the assumption $\dim \mathcal{H} < \infty$, which is, in view of physical applications, rather restrictive. The present theorem generalizes to infinite dimensional Hilbert spaces. Since we assumed the partition function to be finite, H is necessarily unbounded. Therefore the heat flow $L^*(H)$ (properly defined) becomes unbounded and one has to be somewhat careful with the integral (VII.4). For clarity and brevity of the argument, we thought it best to avoid such technicalities.

VIII. PRINCIPLE OF MINIMAL ENTROPY PRODUCTION

In the phenomenological theory the steady state may be characterized, under certain conditions, by a minimum of the entropy production compatible with the external constraints imposed on the system.[69-71] This property is valid only if the fluxes are proportional to the thermodynamic forces (i.e., only in the linear regime). As an illustration we consider the case of thermal conduction, where the entropy production as given by irreversible thermodynamics is [cf. (V.2)]

$$\sigma = \int \mathbf{J} \cdot \operatorname{grad} \frac{1}{T} \, dV = \int L_{qq}\left(\operatorname{grad} \frac{1}{T}\right)^2 dV \qquad \text{(VIII.1)}$$

in the linear approximation with $\mathbf{J} = L_{qq} \operatorname{grad} 1/T$. For fixed boundary conditions the minimum of σ is given by

$$\Delta \frac{1}{T} = 0 \qquad \text{(VIII.2)}$$

which implies

$$\frac{\partial}{\partial t} T = 0 \qquad \text{(VIII.3)}$$

by the heat equation $(\partial/\partial t)T = \alpha \, \Delta T$.

In our model the stationary states are defined by $L(\rho) = 0$. They form a convex set (any convex combination of stationary states is stationary). The states of minimal entropy production are defined by $\sigma(\rho)$ minimal. By the convexity of σ they also form a convex set. The external constraints are already specified by the reservoir temperatures appearing in L and σ. In analogy to the phenomenological theory it is tempting to conjecture

$$L(\rho) = 0 \Leftrightarrow \sigma(\rho) = \text{minimal} \qquad \text{(VIII.4)}$$

If all reservoir temperatures are the same, (VIII.4) has been proved in Theorem 2. If the reservoir temperatures are different, one can show by explicit examples (e.g., dim $\mathscr{H} = 2$) that (VIII.4) is not true in general. From our preceding discussion this should come as no surprise, since for any finite temperature difference we are already in the nonlinear regime (the steady-state heat flow is not proportional to the temperature difference, in general). To derive the principle of minimal entropy production we have to proceed in a somewhat different way.

Let $L = L_{1,\beta_1} + \cdots + L_{r,\beta_r}$ have a unique stationary state $\rho_s(\beta_1, \ldots, \beta_r)$ in some neighborhood of $(\beta, \ldots, \beta)$. We then expand

$$\rho_s(\beta_1, \ldots, \beta_r) = \rho_\beta + \sum_{k=1}^{r} \rho_{s,k}(\beta)(\beta_k - \beta) + \cdots \qquad \text{(VIII.5)}$$

Similarly, let $\sigma = \sigma_{1,\beta_1} + \cdots + \sigma_{r,\beta_r}$ have a unique minimum $\rho_m(\beta_1, \ldots, \beta_r)$. By convexity and the smooth dependence of σ_{k,β_k} on β_k, $\rho_m(\beta_1, \ldots, \beta_r)$ depends smoothly on the inverse temperatures. Furthermore, by Theorem 1, $\rho_m(\beta, \ldots, \beta) = \rho_\beta$. We expand

$$\rho_m(\beta_1, \ldots, \beta_r) = \rho_\beta + \sum_{k=1}^{r} \rho_{m,k}(\beta)(\beta_k - \beta) + \cdots \qquad \text{(VIII.6)}$$

Then the principle of minimal entropy production asserts that

$$\rho_{m,k}(\beta) = \rho_{s,k}(\beta) \qquad \text{(VIII.7)}$$

for $k = 1, \ldots, r$ [i.e., (VIII.4) is valid in the linear approximation].

Theorem 5. Let dim $\mathscr{H} < \infty$ and let the spectrum of H be nondegenerate. Let $L_{1,\beta_1} + \cdots + L_{r,\beta_r}$ have a unique stationary state $\rho_s(\beta_1, \ldots, \beta_r)$ and σ_{r,β_r} have a unique minimum $\rho_m(\beta_1, \ldots, \beta_r)$ in some neighborhood of $(\beta_1, \ldots, \beta)$. Then

$$\rho_{m,k}(\beta) = \rho_{s,k}(\beta) \qquad (\text{VIII.7})$$

Proof: In Lemma 1 we showed that $\rho_{s,k}(\beta)$ is defined by

$$L(\rho_{s,k}(\beta)) = -L_{k,\beta}^*(H)\rho_\beta, \qquad (\text{VIII.8})$$

where $L = L_{1,\beta} + \cdots + L_{r,\beta}$ (VII.8).
  Let $U(t) = e^{-iHt}$. Then

$$\sigma_{k,\beta_k}(U(t)\rho U(t)^*) = -\text{tr}\,[U(t)L_{k,\beta_k}(\rho)U(t)^* U(t)(\log \rho + \beta_k H)U(t)^*]$$

$$= \sigma_{k,\beta_k}(\rho) \qquad (\text{VIII.9})$$

Therefore, since the minimal state is unique by assumption,

$$[\rho_m(\beta_1, \ldots, \beta_r), H] = 0 \qquad (\text{VIII.10})$$

Since the spectrum of H is nondegenerate, ρ_m is a function of H. Therefore it suffices to study the minimum of σ amongst the states diagonal in the energy basis $\{\Psi_j | j = 1, \ldots, \dim \mathscr{H}\}$.
  We consider σ as a function of the eigenvalues ρ_j of ρ. Then the minimum of σ is defined by

$$\frac{\partial}{\partial \rho_j}\sigma(\rho) = 0 = -\sum_{l=1}^{r} \left\{ \text{tr}\,(L_{l,\beta_l}(\rho_j)(\log \rho + \beta_l H) + \langle \Psi_j | L_{l,\beta_l}(\rho)\Psi_j \rangle \frac{1}{\rho_j} \right\}$$

$$(\text{VIII.11})$$

where $L_{l,\beta_l}(\rho_j) = L_{l,\beta_l}(\rho_j |\Psi_j\rangle\langle\Psi_j|)$ and $\rho = \rho_m(\beta_1, \ldots, \beta_r)$. By Theorem 1, $\sigma_{k,\beta+\delta}(\rho_{\beta+\delta}) = 0$. Taking the derivative at $\delta = 0$

$$\langle \Psi_j | L_{k,\beta}'(\rho_\beta)\Psi_j \rangle - \langle \Psi_j | L_{k,\beta}'(H\rho_\beta)\Psi_j \rangle = 0 \qquad (\text{VIII.12})$$

for all j, where $'$ denotes the derivative with respect to δ at $\delta = 0$. In (VIII.11) we choose now $(\beta_1, \ldots, \beta_k, \ldots, \beta_r) = (\beta, \ldots, \beta + \delta, \ldots, \beta)$ and we obtain by taking the derivative at $\delta = 0$

$$\sum_{l \neq k} \text{tr}\left[L_{l,\beta}(\rho_{\beta,j})\frac{1}{\rho_\beta}\rho_{m,k}(\beta)\right] + \text{tr}\left[L_{k,\beta}(\rho_{\beta,j})\left(\frac{1}{\rho_\beta}\rho_{m,k}(\beta) + H\right)\right]$$

$$+ \sum_{l=1}^{r} \langle \Psi_j | L_{l,\beta}(\rho_{m,k}(\beta))\Psi_j \rangle \frac{1}{\rho_{\beta,j}} + \langle \Psi_j | L_{k,\beta}'(\rho_\beta)\Psi_j \rangle \frac{1}{\rho_{\beta,j}} = 0 \qquad (\text{VIII.13})$$

for all j, where $\rho_{\beta,j}$ is the jth eigenvalue of ρ_β. Inserting (VIII.12), we obtain

$$\left\langle \Psi_j \middle| \left\{ \sum_{l=1}^r L_{l,\beta}(\rho_{m,k}(\beta)) + L^*_{k,\beta}(H)\rho_\beta \right\} \Psi_j \right\rangle + \rho^2_{\beta,j} \left\langle \Psi_j \middle| \left\{ \sum_{l \neq k} L^*_{l,\beta}\left(\frac{1}{\rho_\beta}\rho_{m,k}(\beta)\right) \right. \right.$$

$$\left. \left. + L^*_{k,\beta}\left(\frac{1}{\rho_\beta}\rho_{m,n}(\beta) + H\right) \right\} \Psi_j \right\rangle = 0 \qquad \text{(VIII.14)}$$

The second term can be rewritten as

$$L^*\left(\frac{1}{\rho_\beta}\rho_{m,k}(\beta)\right) + L^*_{k,\beta}(H) = \frac{1}{\rho_\beta}L(\rho_{m,k}(\beta)) + L^*_{k,\beta}(H) \qquad \text{(VIII.15)}$$

where we used the detailed balance property (III.27) and the fact that $L_a(f(H)) = 0$ for any function f of H. Therefore (VIII.14) vanishes for

$$L(\rho_{m,k}(\beta)) = -L^*_{k,\beta}(H)\rho_\beta \quad \blacksquare \qquad \text{(VIII.16)}$$

Concerning the technical restrictions in Theorem 5, the same remarks as before apply. If we define the entropy production by (V.8), then the theorem is also valid for an infinite dimensional Hilbert space.

References

1. P. G. Bergmann and J. L. Lebowitz, *Phys. Rev.*, **99**, 578 (1955).
2. J. L. Lebowitz, *Ann. Phys.*, **1**, 1 (1956).
3. J. L. Lebowitz, *Phys. Rev.*, **114**, 1192 (1959).
4. J. L. Lebowitz and A. Shimony, *Phys. Rev.*, **128**, 391 (1962).
5. H. Spohn and J. L. Lebowitz, *Commun. Math. Phys.*, **54**, 97 (1977).
6. J. B. Keller, G. C. Papanicolaou, and J. Weilemann, "Heat Conductance in a One-dimensional Random Medium," to appear in *Comm. Pure Appl. Math.*
7. D. Ruelle, *Statistical Mechanics; Rigorous Results*, W. A. Benjamin, New York, 1969.
8. G. G. Emch, *Algebraic Methods in Statistical Mechanics and in Quantum Field Theory*, Wiley, New York, 1972.
9. L. van Hove, *Physica*, **23**, 441 (1957).
10. S. Nakajima, *Prog. Theor. Phys.*, **20**, 948 (1958).
11. I. Prigogine and P. Résibois, *Physica*, **27**, 629 (1961).
12. R. Zwanzig, *J. Chem. Phys.*, **33**, 1338 (1960).
13. R. Zwanzig, *Lect. Theor. Phys. (Boulder)*, **3**, 106 (1960).
14. P. N. Argyres and P. L. Kelley, *Phys. Rev.*, **134**, A98 (1964).
15. F. Haake, *Statistical Treatment of Open Systems by Generalized Master Equations*, Springer Tracts Mod. Phys., **66** (1973).
16. G. S. Agarwal, Progress in Optics, **XI**, 1 (1973).
17. C. Favre and P. Martin, *Helv. Phys. Acta*, **41**, 333 (1968).
18. E. B. Davies, *Commun. Math.* **39**, 91 (1974).
19. H. Narnhofer, *Acta Phys. Austr.*, **36**, 217 (1972).
20. H. Narnhofer, *Acta Phys. Austr.*, Suppl. XI, 527 (1973).
21. D. Robinson and O. E. Lanford III, *Commun. Math. Phys.*, **24**, 193 (1972).
22. D. Robinson, *Commun. Math. Phys.*, **31**, 171 (1973).

23. G. L. Sewell, *Ann. Phys.*, **85**, 336 (1974).
24. C. Radin, *J. Math. Phys.*, **11**, 2945 (1970).
25. G. G. Emch and C. Radin, *J. Math. Phys.*, **12**, 2043 (1971).
26. O. E. Lanford III and J. L. Lebowitz, "Time Evolution and Ergodic Properties of Harmonic Systems," in LNP, Vol. 38, Springer, Berlin, 1975, p. 144.
27. L. Van Hemmen, *Dynamics and Ergodicity of the Infinite Harmonic Crystal*, thesis, University of Croningen, 1976.
28. S. Goldstein, J. L. Lebowitz, and M. Aizenman, "Ergodic Properties of Infinite Systems," in LNP Vol. 38, Springer, Berlin, 1975, p. 112.
29. Z. Rieder, J. L. Lebowitz, and E. Lieb, *J. Math. Phys.*, **8**, 1073 (1967).
30. A. Casher and J. L. Lebowitz, *J. Math. Phys.*, **12**, 1701 (1971).
31. L. van Hove, *Physica*, **21**, 517 (1955).
32. L. van Hove, *Physica*, **23**, 441 (1957).
33. V. Gorini, A. Frigerio, M. Verri, and E. C. G. Sudarshan, *Properties of Quantum Markovian Master Equations*, University of Milan, preprint, 1976.
34. E. B. Davies, *Math. Ann.*, **219**, 147 (1976).
35. E. B. Davies, *Ann. Inst. Poinc.*, **11**, 265 (1975).
36. E. B. Davies, *Commun. Math. Phys.*, **33**, 171 (1973).
37. E. B. Davies, *J. Math. Phys.*, **15**, 2036 (1974).
38. E. B. Davies, *Lett. Math. Phys.*, **1**, 31 (1976).
39. J. V. Pulé, *Commun. Math. Phys.*, **38**, 241 (1974).
40. P. Martin and G. G. Emch, *Helv. Phys. Acta*, **48**, 59 (1975).
41. E. B. Davies and J. P. Eckmann, *Helv. Phys. Acta*, **48**, 731 (1975).
42. R. Haag, N. M. Hugenholtz, and M. Winnink, *Commun. Math. Phys.*, **5**, 215 (1967).
43. N. M. Hugenholtz, in R. F. Streater, Ed., *Mathematics for Contemporary Physics*, Academic, London, 1971.
44. R. Alicki, *Rep. Math. Phys.*, **10**, 249 (1976).
45. A. Kossakowski, A. Frigerio, V. Gorini, and M. Verri, *Commun. Math. Phys.*, **57**, 97 (1977).
46. R. Haag, D. Kastler, and E. Trych-Pohlmeyer, *Commun. Math. Phys.*, **38**, 173 (1974).
47. M. Aizenman, G. Gallavotti, S. Goldstein, and J. L. Lebowitz, *Commun. Math. Phys.*, **48**, 1 (1976).
48. A. Kossakowski, *Rep. Math. Phys.*, **3**, 247 (1972).
49. K. Yosida, *Functional Analysis*, Springer, Berlin, 1972.
50. K. Kraus, *Ann. Phys.*, **64**, 311 (1971).
51. G. Lindblad, *Commun. Math. Phys.*, **48**, 119 (1976).
52. V. Gorini, A. Kossakowski, and E. C. G. Sudarshan, *J. Math. Phys.*, **17**, 821 (1976).
53. E. B. Davies, *Quantum Theory of Open Systems*, Academic, London, 1976.
54. R. S. Ingarden and A. Kossakowski, *Ann. Phys.*, **89**, 451 (1975).
55. S. R. deGroot and P. Mazur, *Non-equilibrium Thermodynamics*, North-Holland, Amsterdam, 1962.
56. I. Prigogine, *Introduction to Thermodynamics of Irreversible Processes*, Wiley, New York, 1967.
57. G. Lindblad, *Commun. Math. Phys.*, **33**, 305 (1973).
58. E. H. Lieb, *Adv. Math.*, **11**, 267 (1973).
59. H. Spohn, *Lett. Math. Phys.*, **2**, 33 (1977).
60. A. Frigerio, *Quantum dynamical semigroups and approach to equilibrium*, University of Milan, preprint.
61. E. B. Davies, *Commun. Math. Phys.*, **19**, 83 (1970).
62. H. Spohn, *Rep. Math. Phys.*, **10**, 139 (1976).

63. D. E. Evans, *Commun. Math. Phys.*, **54,** 293 (1977).
64. L. Onsager, *Phys. Rev.*, **37,** 405; **38,** 2265 (1931).
65. K. Hepp, *Z. Phys.*, **B20,** 53 (1975).
66. K. Hepp, "Results and Problems in Irreversible Statistical Mechanics of Open Systems," in H. Araki, Ed., LNP, Vol. 39, Springer, Berlin, 1975.
67. C. R. Willis and P. G. Bergmann, *Phys. Rev.*, **128,** 391 (1962).
68. H. Spohn, *Linear Response Theory for Weakly Coupled Systems*, in preparation.
69. I. Prigogine, *Etude thermodynamique des phénomènes irreversible*, Desoer, Liege, 1947.
70. P. Mazur, *Bull. Acad. Roy Belg. Cl. Sc.*, **38,** 182 (1952).
71. R. Haase, *Z. Naturforsch.*, **6a,** 522 (1951).

SOLVABLE MODELS FOR UNSTABLE STATES IN QUANTUM PHYSICS

A. P. GRECOS

Faculté des Sciences, Université Libre de Bruxelles, Brussels, Belgium

CONTENTS

I. INTRODUCTION

It is hardly necessary to insist on the importance of the concept of an unstable state in quantum theory. States with a finite lifetime are basic ingredients of most theories in atomic, molecular, and nuclear physics; in elementary particle physics; and in many-body problems. In spectroscopy one must take into account the finite width of spectral lines. Most elementary particles are unstable and, in group-theoretical classification schemes, must be treated on the same footing as stable ones. The notion of a quasiparticle is quite important in the description of interacting many-body systems. These are a few typical cases where decay phenomena, associated with states having a finite lifetime, manifest themselves.

As is well known, the fundamental ideas of the theory of natural line-width for the spectral lines of an atom interacting with the electromagnetic field, go back to the work of Weisskopf and Wigner.[1,2] They studied the exponentially decaying contributions of the corresponding Schrödinger equation and they introduced the approximation of an atom with a finite number of energy levels, which is even today a most useful model when studying decay phenomena.

In the context of a field-theoretical description of unstable particles, Peierls[3] suggested to identify the mass and the (inverse) lifetime of such particles with the real and the imaginary part of complex poles of certain propagators. This implies that, for a Hamiltonian system, unstable states correspond to poles (and more generally to singularities) of the analytic continuation of a class of matrix elements of the resolvent of the Hamiltonian. One should notice here the similarity of this idea with that proposed by Gamow for the description of the α-radiation emitted from radioactive nuclei. In his theory, Gamow[4] introduced (nonnormalizable) "eigenfunctions" that correspond to complex "eigenvalues" of the Hamiltonian of the nucleus. At first sight, it appears that these theories are natural generalizations of the concept of a stable state, defined as an eigenfunction corresponding to a real (point) eigenvalue of the Hamiltonian. However, it is easily realized that, because of the assumed hermiticity of this operator, several difficult physical and mathematical questions arise. In fact, the proper definition of unstable states is still an open problem of theoretical physics and the attempts made for its solution cannot be considered as entirely satisfactory (for a recent review see Fonda[5]).

Although some aspects of the questions involved in the theory of decaying phenomena may be formulated in a general manner, it has been proved very useful to study such phenomena using simple Hamiltonian models that permit explicit calculations. As a matter of fact, the first systematic discussion of the validity of the results of Weisskopf and Wigner was done in a basic paper of Friedrichs,[6] where he analyzed the evolution of a solvable model. He considered a "free" Hamiltonian H_0 having a simple absolutely continuous spectrum and a point eigenvalue embedded in it. The eigenvalue is coupled to the continuum through a bounded interaction λV. Because of the interaction, the eigenvalue disappears and the total Hamiltonian H ($= H_0 + \lambda V$) has no point spectrum (at least for small values of the coupling parameter λ). Friedrichs has shown that the exponentially decaying solution of Weisskopf and Wigner becomes exact in the so-called weak coupling limit, that is, when $|\lambda| \to 0$, $t \to \infty$, while $\lambda^2 t$ is kept finite.

It turns out that several models used in theoretical physics reduce to that considered by Friedrichs, or to some simple variant of it, when one restricts them to some specific subspace. In the next section we present a few examples of such systems. Thus Hamiltonians of the Friedrichs type may be regarded in a certain sense as prototypes for illustrating concepts and testing theories that attempt to deduce, for finite values of λ, phenomenological decay laws from the underlying dynamics.

A dynamical interpretation of dissipative processes, and the decay of an unstable state is certainly such a process, requires a Hamiltonian with a continuous spectrum. Therefore, from a physical point of view, the system must be considered in some asymptotic limit (e.g., infinite volume, thermodynamic limit, etc.). It is only then that qualitatively different phenomena appear. It follows then that, from a mathematical point of view, we need to work in infinite-dimensional (linear) spaces; consequently, the proof of several propositions might be rather difficult. In this respect, solvable models are quite important because we may study their properties using relatively simple mathematical considerations.

Evidently, a solvable model cannot describe properly all aspects of the physical situation. To obtain a mathematically soluble problem we are often forced to introduce approximations that are sometimes physically incorrect. For example, in certain cases we omit in the Hamiltonian the terms that do not conserve the number of particles and therefore the "crossing symmetry" of the system is violated. Nevertheless, the possibility of performing exact calculations justifies the use of such simplified models because we can get a better understanding of the structure of a formalism. Of course, it is necessary to interpret carefully the meaning of the results and, in particular, the generality of the conclusions.

We have already mentioned that the main difficulty in considering exponentially decaying states stems from the fact that the Schrödinger equation cannot admit solutions with such time dependence, given that the Hamiltonian is a bona fide self-adjoint operator. Here we face a problem quite similar to that in nonequilibrium statistical mechanics when we discuss the relation between the irreversible kinetic description and the reversible dynamical equations of motion. During the past few years Prigogine et al. have developed the theory of subdynamics[7] and introduced a certain class of nonunitary transformations[8] that constitute a suitable framework for analyzing this fundamental question of irreversibility (for a recent review see Ref. 9). Models of the Friedrichs type are examples to which the theory of subdynamics may be applied. They may serve to investigate the possibility of using the ideas of nonequilibrium statistical mechanics in order to obtain a consistent description of unstable quantum states.

It should also be noted that some recent work, especially that of Combes,[10] has clarified to a large extent the mathematical problems related to the analytic continuation of the resolvent of an operator. This may lead to a better understanding of the implications of the hypothesis that unstable states are associated to the singularities of propagators.

In this article we review some properties of quantum dynamical systems with a Friedrichs Hamiltonian. We present in the next section a few examples that lead to such models. In the third section we discuss their spectral properties. The structure of exponentially decaying contributions to the evolution of the system is analyzed in the fourth section. Finally, we conclude with some remarks on the generality of these models and we mention some open problems. It should be observed that no attempt has been made to give an exhaustive list of references.

II. SOME SIMPLE QUANTUM SYSTEMS

There are several physical problems leading to a Hamiltonian describing the interaction of a finite subsystem with an infinite reservoir or a field (at zero temperature). In certain cases it is meaningful to restrict the model in a given subspace of the linear space to which belong the states of the system and thus obtain a reduced Hamiltonian with point eigenvalues embedded in a continuous spectrum and coupled to it. In this section we give a few typical examples that fall into this class.

A. The Weisskopf–Wigner Approximation

To study the problem of natural linewidth, we consider a one-electron atom interacting with the radiation field. The Hamiltonian of the system is written as (see, e.g., Heitler[11])

$$H = H_{at} + H_{rad} + H_{int} \tag{II.1}$$

where the different terms are as follows.

The nucleus is assumed to be fixed and therefore the atom is essentially the electron moving in an external (scalar) potential. Its levels ϵ_n and the corresponding eigenstates $\psi_n(\mathbf{x})$ are determined by the Dirac equation. Introducing anticommuting creation and annihilation operators c_n^+ and c_n, the atomic Hamiltonian H_{at} becomes

$$H_{at} = \sum_n \epsilon_n c_n^+ c_n \tag{II.2}$$

Note that the index n denotes a set of quantum numbers characterizing the eigenstate. Here we take into account only discrete levels (with positive energy), although, at least formally, we could include also ionized states (cf. Källen[12]).

For the radiation field, we express the Hamiltonian H_{rad} in terms of bosons

$$H_{rad} = \sum_{\nu=1,2} \int d\mathbf{k}\,\omega(\mathbf{k})\alpha_\nu^+(\mathbf{k})\alpha_\nu(\mathbf{k}) \tag{II.3}$$

where the Bose creation and annihilation operators are determined by the Fourier expansion of the transverse part of the vector potential $\mathbf{A}(\mathbf{x})$

$$\mathbf{A}(\mathbf{x}) = \frac{1}{(2\pi)^{3/2}} \sum_{\nu=1,2} \int d\mathbf{k}\, \frac{\mathbf{e}_\nu(\mathbf{k})}{[2\omega(\mathbf{k})]^{1/2}} [e^{i\mathbf{k}\cdot\mathbf{x}}\alpha_\nu(\mathbf{k}) + e^{-i\mathbf{k}\cdot\mathbf{x}}\alpha_\nu^+(\mathbf{k})] \tag{II.4}$$

We have used the notation $\mathbf{e}_\nu(\mathbf{k})$ for the polarization vectors

$$\mathbf{k}\cdot\mathbf{e}_\nu(\mathbf{k}) = 0 \qquad \mathbf{e}_\nu(\mathbf{k})\cdot\mathbf{e}_\mu(\mathbf{k}) = \delta_{\mu\nu} \qquad (\mu, \nu = 1, 2) \tag{II.5}$$

and $\omega(\mathbf{k})$ is simply $c_0|\mathbf{k}|$, c_0 being the velocity of light.

Finally, the interaction between the electron and the field H_{int} is written as

$$H_{int} = -e \int d\mathbf{x}\, \mathbf{A}(\mathbf{x})\psi^+(\mathbf{x})\boldsymbol{\alpha}\psi(\mathbf{x}) \qquad (\text{II.6})$$

where $\psi(\mathbf{x})$ is the electron field and $\boldsymbol{\alpha}$ are the well-known 4×4 matrices that are expressed in terms of the Pauli spin matrices. As we work in a nonrelativistic approximation, we can take the electron field $\psi(\mathbf{x})$ to be $\sum_n \psi_n(\mathbf{x})c_n$. It is important to notice that H_{int} is a linear combination of terms $\alpha_\nu^+(\mathbf{k})c_n^+ c_m$ and $\alpha_\nu(\mathbf{k})c_n^+ c_m$. Thus, because of the assumed form of $\psi(\mathbf{x})$, the number of electrons $\sum_n c_n^+ c_n$ is conserved and, without loss of generality, it can be taken equal to one.

Suppose now that $\varphi(t)$ is the wave function of the system, which evolves according to the Schrödinger equation

$$i\partial_t\varphi(t) = H\varphi(t) \qquad (\text{II.7})$$

In the Fock representation, $\varphi(t)$ is given by a sequence of functions $\{\varphi_s(k_1, \ldots, k_s; n|t)\}$, where we have denoted by k_m the wave vector $\mathbf{k}_m$ and the polarization ν_m of the mth photon. Then (II.7) becomes a hierarchy of equations, namely,

$$i\partial_t\varphi_s(k_1, \ldots, k_s; n|t)$$

$$= [\omega(\mathbf{k}_1)+\cdots+\omega(\mathbf{k}_s)+\epsilon_n]\varphi_s(k_1, \ldots, k_s; n|t)$$

$$+\sqrt{s+1}\sum_m\sum_\mu \int d\mathbf{k}\, M^+(n, m|k)\varphi_{s+1}(k, k_1, \ldots, k_s; m|t)$$

$$+\frac{1}{\sqrt{s}}\sum_{r=1}^s\sum_m M(m, n|k)\varphi_{s-1}(k_1, \ldots, k_{r-1}, k_{r+1}, \ldots, k_s; m|t), \qquad (\text{II.8})$$

with

$$M^+(m, n|k) = \frac{e\mathbf{e}_\nu(\mathbf{k})}{[2(2\pi)^3\omega(\mathbf{k})]^{1/2}} \int d\mathbf{x}\, e^{i\mathbf{k}\cdot\mathbf{x}}\psi^+(\mathbf{x})\boldsymbol{\alpha}\psi_n(\mathbf{x}) \qquad (\text{II.9})$$

The approximation of Weisskopf and Wigner[1] consists in truncating the preceding hierarchy using the following assumptions: (1) only a finite number of atomic levels are taken into account, and (2) only states with few photons contribute. We will not consider here the domain of the validity of these hypotheses, but we want to point out that Grimm and Ernst[13] have presented a careful and detailed discussion of the possibility of constructing a systematic approximation scheme, based on the preceding ideas, to treat the interaction of a bound electron with the radiation field.

It is evident that the simplest nontrivial model results when considering a two-level atom and, at most, one photon. In this case, (II.8) reduces to the following system:

$$i\partial_t\varphi_0\,(t) = \epsilon_1\varphi_0(t) + \sum_{\nu=1,2}\int d\mathbf{k}\,\bar{M}(\mathbf{k};\nu)\varphi_\nu(\mathbf{k}|t)$$

$$i\partial_t\varphi_\nu(\mathbf{k}|t) = [\omega(\mathbf{k})+\epsilon_0]\varphi_\nu(\mathbf{k}|t) + M(\mathbf{k};\nu)\varphi_0(t) \qquad (II.10)$$

As n takes only two values we have changed slightly the previous notation $[\varphi_0 \equiv \varphi_0(n=2), \varphi_\nu(\mathbf{k}) \equiv \varphi_1(\mathbf{k},\nu;n=1)$, etc.] in order to simplify the resulting expressions. Equation II.10 may be regarded as the Schrödinger equation for a system with a Friedrichs-type Hamiltonian. In fact, it is an easy matter to see that we have one point eigenvalue $\epsilon_1 - \epsilon_0$ coupled, via $M(\mathbf{k},\nu)$ to a (degenerate) continuum represented by the range of the function $[\omega(\mathbf{k})+\epsilon_0]$.

The spectral properties of such a Hamiltonian depend crucially on certain properties of the kernels $M(\mathbf{k},\nu)$. We formulate the necessary requirements in the next chapter. Here we would like to mention that Grimm and Ernst[14] have shown that, for a Dirac hydrogen atom, these kernels are square-integrable and that they satisfy certain analyticity properties that are needed to continue analytically matrix elements of the resolvent of the Hamiltonian. Thus, as might be expected, no divergences occur when the atom has a "finite spatial extension."

To avoid any possible confusion, we should remark that here the approximation of Weisskopf and Wigner is meant to be the set of assumptions that permit to truncate (II.8) and obtain (II.10). However, one often implies something more namely, the approximate exponential solution that (II.10) admits in the weak coupling limit. This aspect is discussed later on.

B. Interaction of an Atom with a One-Dimensional Field

It is often instructive to study the mathematically simpler but physically unrealistic one-dimensional version of a three-dimensional problem. Thus we may consider a quadratic Hamiltonian describing the fictitious case of a two-level atom interacting with a one-dimensional boson field. We write the Hamiltonian of the system as

$$H = \epsilon c^+ c + \int dk\,\omega(k)\alpha^+(k)\alpha(k) + c^+\int dk\,\bar{v}(k)\alpha(k)$$

$$+ c\int dk\,v(k)\alpha^+(k) + c^+\int dk\,\bar{u}(k)\alpha^+(k) + c\int dk\,u(k)\,\alpha(k)$$

$$+ \int\int dk\,dl\,w(k,l)[\alpha^+(k)+\alpha(k)][\alpha^+(l)+\alpha(l)] \qquad (II.11)$$

where c^+ and c are the creation and annihilation operators for the fermion, whereas $\alpha^+(k)$ and $\alpha(k)$ are those for the bosons.

The last three terms in (II.11) do not conserve the number of particles N,

$$N = c^+c + \int dk\, \alpha^+(k)\alpha(k) \qquad (II.12)$$

and are often neglected. For the last term one argues that it is small because it is proportional to the square of the amplitude of the field. Also, the antiresonant terms involving $c\alpha(k)$ and $c^+\alpha^+(k)$ are assumed to be small and, in the so-called rotating wave approximation, are not taken into account. Then the Hamiltonian becomes

$$H = \epsilon c^+c + \int dk\, \omega(k)\alpha^+(k)\alpha(k)$$
$$+ c^+ \int dk\, \bar{v}(k)\alpha(k) + c \int dk\, v(k)\alpha^+(k) \qquad (II.13)$$

Its basic property is that it commutes with N. Consequently, we may restrict H to a subspace, called a sector, where N has some definite value.

In the first nontrivial sector ($N = 1$) either the atom is excited and no photons are present or the atom is in the ground state and one photon is present. Several aspects of this model have been studied in a series of papers by Davidson and Kozak.[15-21] In this sector the wave function of the system can be represented by a column vector $\{\varphi_0(t), \varphi(k|t)\}$. It may be easily seen that the Schrödinger equation takes the form

$$i\partial_t\varphi_0(t) = \epsilon\varphi_0(t) + \int dk\, \bar{v}(k)\varphi(k|t)$$
$$i\partial_t\varphi(k|t) = v(k)\varphi_0(t) + \omega(k)\varphi(k|t) \qquad (II.14)$$

which is again a Friedrichs model. Sometimes one assumes in the definition of the Hamiltonian the so-called dipole approximation (cf. Ref. 15), used frequently in laser theory (see, for example, Ref. 22). Such an approximation leads to an interaction with $|v(k)|^2 \sim |k|$, for large values of $|k|$. A cutoff then is needed in order to avoid divergences.

It should be remarked that the approximations leading from (II.11) to (II.12) might alter qualitatively the decay law. A similar model, due to van Kampen,[23] describing a quantum oscillator interacting with the transverse electromagnetic field has been studied recently by Shirokov[24] and by Rżążewski and Żakowicz.[25] They have concluded that nonexponential terms in the decay of the excited states need not appear when all terms of H are taken into account (for a similar result see Emch and Wolf).[26] Let us finally point out that the Hamiltonian given by (II.13) can be "diagonalized" in all sectors, and this is also possible for the one given by (II.11). In the latter case, certain mathematical questions arise because the transformation is, in general, improper (cf. Ref. 27).

C. The Lee Model

The model introduced by T. D. Lee[28] to illustrate some problems of the renormalization process in quantum field theory may be found in several well-known textbooks (e.g., Refs. 29 to 31). Therefore we limit ourselves here to a very short description of this model.

A spinless nucleon with two internal states, called the N and V particle, interacts with the "θ-particles" of a scalar boson field. The interaction is such that only the process $V \rightleftharpoons N + \theta$ is allowed, whereas the process $N \rightleftharpoons V + \theta$ is forbidden. The nucleon is assumed to be fixed and, with an obvious notation, the Hamiltonian of the system reads

$$H = m_V \psi_V^+ \psi_V + m_N \psi_N^+ \psi_N + \int d\mathbf{k}\, \omega(\mathbf{k}) \alpha^+(\mathbf{k}) \alpha(\mathbf{k})$$

$$+ \frac{g}{(2\pi)^{3/2}} \int d\mathbf{k} \frac{f(k)}{[2\omega(k)]^{1/2}} \{\psi_N^+ \psi_V \alpha^+(\mathbf{k}) + \psi_V^+ \psi_N \alpha(\mathbf{k})\}. \quad (\text{II}.15)$$

It is also assumed that $\omega(k) = \sqrt{k^2 + \mu^2}$ and that the cutoff $f(k)$ tends to zero for large values of $|\mathbf{k}|$.

There are two constants of the motion, namely,

$$B = \psi_V^+ \psi_V + \psi_N^+ \psi_N, \qquad Q = \int d\mathbf{k}\, \alpha^+(\mathbf{k}) \alpha(\mathbf{k}) - \psi_N^+ \psi_N \quad (\text{II}.16)$$

which define a family of sectors for the model. The first nontrivial sector is defined by $B = 1$ and $Q = 0$, that is, it contains the V-particle state and the $N\theta$ scattering states. It may be shown easily that when the Hamiltonian is restricted to this sector, it reduces to a Friedrichs model. The corresponding Schrödinger equation is similar to (II.14) with $\epsilon = m_V - m_N$.

It is not possible to consider here the very interesting questions about ghost states (Källen and Pauli, Ref. 32) that arise when the cutoff is removed, that is, when we take $f(k) = 1$. For the mathematical aspects of this problem we refer to the article of Berezin.[33]

D. Linear Chain of Spins with an Impurity

To study equilibrium and nonequilibrium problems of interacting spin systems, a simplified version of the Heisenberg Hamiltonian, the so-called X-Y model, has been proven very useful.[34,35] A slightly modified model has been considered by Tjon,[36] where at the beginning of an isotropic chain with nearest-neighbor interactions there is an impurity. The Hamiltonian of the model is

$$H = -h_0 s_0^z - h \sum_{j=1}^{N} s_j^z + J \sum_{j=1}^{N} (s_j^X s_j^X + s_j^Y s_j^Y) + g(s_0^X s_1^X + s_0^Y s_1^Y)$$

(II.17)

where s_j^α is the α-component of the spin operator of the jth spin. The coupling parameters J and g, as well as the external magnetic field h, are taken positive.

We define the operators s_j^+ and s_j^- by

$$s_j^X = \tfrac{1}{2}(s_j^+ + s_j^-), \qquad s_j^Y = \tfrac{1}{2}(s_j^+ - s_j^-), \qquad s_j^z = s_j^+ s_j^- - \tfrac{1}{2} \qquad \text{(II.18)}$$

and then the Fermi operators c_j and c_j^+ by the transformation

$$s_j^+ = c_j^+ \exp\left\{ i\pi \sum_{k=0}^{j-1} c_k^+ c_k \right\} \qquad \text{(II.19)}$$

In terms of these fermions the Hamiltonian of the model becomes

$$H = -h_0 c_0^+ c_0 - h \sum_{j=1}^{N} c_j^+ c_j + \tfrac{1}{2} \sum_{j=1}^{N-1} [c_j^+ c_{j+1} + c_j c_{j+1}^+]$$

$$+ \tfrac{1}{2} g[c_0^+ c_1 + c_0 c_1^+] + \text{constant} \qquad \text{(II.20)}$$

The terms in (II.20) that do not involve the impurity can be diagonalized through a unitary transformation. In effect, let η_ν and η_ν^+ be Fermi operators defined by

$$c_j = \left(\frac{2}{N+1} \right)^{1/2} \sum_{\nu=1}^{N} \sin \frac{j\nu\pi}{N+1} \, \eta_\nu \qquad (j = 1, \dots, N) \qquad \text{(II.21)}$$

then, neglecting the constant term, (II.20) becomes

$$H = -h_0 c_0^+ c_0 + \sum_{\nu=1}^{N} \Lambda_\nu \eta_\nu^+ \eta_\nu + \tfrac{1}{2} g \sum_{\nu=1}^{N} \gamma_\nu [c_0^+ \eta_\nu + c_0 \eta_\nu^+], \qquad \text{(II.22)}$$

where $\Lambda_\nu = -h + J \cos(\nu\pi/N+1)$ and $\gamma_\nu = (2/N+1)^{1/2} \sin(\nu\pi/N+1)$. In the limit of an infinite chain we may replace the discrete variable ν by a continuous variable $\theta = (\nu\pi/N+1)$ and the summations by integrations. Thus (II.22) reads

$$H = -h_0 c_0^+ c_0 + \int_{-\pi}^{\pi} d\theta \, (-h + J \cos \theta) \eta^+(\theta) \eta(\theta)$$

$$+ g \int_{-\pi}^{\pi} d\theta \, \sin \theta [c_0^+ \eta(\theta) + c_0 \eta^+(\theta)] \qquad \text{(II.23)}$$

The Hamiltonian that we have obtained is of the same form as that given by (II.12). The only difference is that a fermion field has replaced the boson field. The number operator

$$N = c_0^+ c_0 + \int_{-\pi}^{\pi} d\theta \, \eta^+(\theta)\eta(\theta) \tag{II.24}$$

is a constant of the motion and in the first sector ($N = 1$) the Schrödinger equation is

$$i\partial_t \varphi_0(t) = -h_0 \varphi_0(t) + g \int_{-\pi}^{\pi} d\theta \sin \theta \, \varphi(\theta|t)$$
$$i\partial_t \varphi(\theta|t) = g \sin \theta \, \varphi_0(t) + \cos \theta \, \varphi(\theta|t) \tag{II.25}$$

Thus we have a Friedrichs-type model with a doubly degenerate, bounded, continuous spectrum.

Note that (II.20) may be viewed as the Hamiltonian of a lattice gas of fermions with nearest-neighbor interactions. There is no difficulty in including interactions that involve more lattice sites. This would simply change Λ_ν and γ_ν in (II.22). Moreover, such systems may be considered as a particular case of Fermi lattice systems with bilinear Hamiltonians, studied recently by van Hemmen and Vertogen.[37]

E.   Remarks on Other Models

It is not possible to enumerate in this article all possible physical systems where a Friedrichs Hamiltonian has been used to illustrate some aspects of their behavior. However, we would like to mention briefly two important cases: one from elementary particle physics and the other from physical chemistry.

To describe the decay of the neutral K-meson (or neutral kaon), one assumes that the $K°$ and $\bar{K}°$ states are eigenstates of a Hamiltonian H_0 that includes the strong interactions, whereas an interaction V induces the decay of these states (cf. Lee et al.[38]; for an account of the problem of kaons see Kabir[39] and Bilen'ski[40]). Several questions can be discussed in the framework of an algebraically soluble model of the Friedrichs type, where the kaon is represented by two unstable point eigenvalues of H_0, embedded in its continuous spectrum (Horwitz and Marchand[41,42]).

In order to study radiationless transitions in isolated polyatomic molecules (i.e., electronic relaxation processes that do not involve ionization or breaking of the chemical bonds), Bixon and Jortner[43] have considered a simple model. The mechanism of the transition requires the coupling of an excited electronic state with a "dense" set of vibronic states. In the proposed model one assumes uniformly spaced vibronic states and a constant coupling between the excited state. The background is approximated by a continuum (for a qualitative discussion of the validity of such an approximation see Freed[44]). It is also assumed, for simplicity, that the continuous spectrum extends over the whole real axis. Formally, there is no difficulty in solving the Schrödinger equation of this model (cf. also Ref. 45). Although there is no real physical problem, one should notice that from a mathematical point of view some care is needed because the interaction does not vanish for large values of the energy.

III. SPECTRAL PROPERTIES OF FRIEDRICHS-TYPE MODELS

In this chapter we consider the main properties of Hamiltonians of the Friedrichs type. For most applications the elementary, and to some extent heuristic, considerations that we present here are sufficient. However, one should note that several aspects of such models can be treated in a mathematically general and rigorous manner.

A. Definition of the Hamiltonian

The systems that we have described in the previous chapter lead, when restricted to some sector, to a Hamiltonian of which the general form may be defined as follows: An operator $H = H_0 + \lambda V$ on a (separable) Hilbert space $\mathcal{H}$ is called a Hamiltonian of the Friedrichs type if the following conditions are satisfied:

1. The "unperturbed" Hamiltonian H_0 is a self-adjoint operator and its spectral representation is known.
2. The spectrum of H_0 consists of a continuous part and a point spectrum with a finite number of eigenvalues. We denote by F_c and F_p the projections onto the corresponding invariant subspaces.
3. The "interaction" λV is a bounded self-adjoint operator such that $F_p V F_c \neq 0$, whereas $F_c V F_c = 0$.

With the preceding definition, we assume essentially that H_0 can be represented as a multiplication in some function space. Let us recall here that a functional representation of a linear space is a one-to-one correspondence between the vectors of the space and a set of functions, which preserves the linear structure of the space (for a detailed discussion of this concept we refer to the books of Friedrichs[46,47]). In our case, it suffices to represent the elements of $\mathcal{H}$ by column vectors, which we write as

$$f \leftrightarrow \{f_k; f_\alpha(\omega_\alpha, \eta_\alpha)\}; \qquad k = 1, \ldots, n, \quad \alpha = 1, \ldots, s \quad (f \in \mathcal{H}) \quad \text{(III.1)}$$

Here f_k is a complex number denoting the kth component of an n-dimensional vector and $f_\alpha(\omega_\alpha, \eta_\alpha)$ is a complex-valued function of the real variables ω_α and η_α. We may note that η_α can be a set of variables and that there will be cases where it is not necessary to consider f_α as depending on this accessory variable. Equation (III.1) implies that we have decomposed $\mathcal{H}$ as a (direct) sum of an n-dimensional subspace and s infinite-dimensional ones. The functions are assumed square- integrable and the inner product in $\mathcal{H}$ will be given by the expression

$$\langle f, g \rangle = \sum_{k=1}^{n} \bar{f}_k g_k + \sum_{\alpha=1}^{s} \int_{\mathcal{I}_\alpha} d\omega_\alpha \, d\eta_\alpha \, \bar{f}_\alpha(\omega_\alpha, \eta_\alpha) g_\alpha(\omega_\alpha, \eta_\alpha), \qquad \text{(III.2)}$$

where we have written $\mathcal{I}_\alpha$ for the admitted domain of $(\omega_\alpha, \eta_\alpha)$. Now we assume that this representation of $\mathcal{H}$ is a spectral one for H_0. This means that there exists a set of real numbers $\{\omega_k\}$ $(k = 1, \ldots, n)$ and a set of real-valued functions $\{h_\alpha(\omega_\alpha)\}$ $(\alpha = 1, \ldots, s)$ of the spectral variables $\{\omega_\alpha\}$ such that H_0 may be represented as

$$H_0 f \leftrightarrow \{\omega_k f_k; \, h_\alpha(\omega_\alpha) f_\alpha(\omega_\alpha, \eta_\alpha)\} \qquad \text{(III.3)}$$

The numbers $\{\omega_k\}$ are the point eigenvalues of H_0, whereas the continuous spectrum of this operator is determined by the range of the functions $\{h_\alpha\}$. Clearly, H_0 may be unbounded and then its domain is defined by the requirement that $h_\alpha(\omega_\alpha) f_\alpha(\omega_\alpha, \eta_\alpha)$ must be square integrable.

Because of the hypotheses that we made on the interaction, we can represent V in terms of an $n \times n$ hermitian matrix V_{kl} and a set of functions $\{v_{k|\alpha}(\omega_\alpha, \eta_\alpha)\}$ as follows:

$$Vf \leftrightarrow \left\{ \sum_{l=1}^{n} V_{kl} f_l + \sum_{\beta=1}^{s} \int_{\mathcal{I}_\beta} d\omega'_\beta \, d\eta'_\beta \bar{v}_{k|\beta}(\omega'_\beta, \eta'_\beta) f_\beta(\omega'_\beta, \eta'_\beta); \, \sum_{l=1}^{n} v_{l|\alpha}(\omega_\alpha, \eta_\alpha) f_l \right\}$$

$$\text{(III.4)}$$

Given that V is bounded, the functions $\{v_{k|\alpha}\}$ must be square integrable. Later on we impose certain smoothness conditions on these functions, and when we consider the initial value problem for the Schrödinger (or the von Neumann) equation, some analyticity properties are required. It is convenient to use a matrix representation for H, namely,

$$H \leftrightarrow \begin{bmatrix} \omega_k \delta_{kl} + \lambda \, V_{kl} & \lambda \bar{v}_{k|\beta}(\omega'_\beta, \eta'_\beta) \\ \lambda v_{l|\alpha}(\omega_\alpha, \eta_\alpha) & h_\alpha(\omega_\alpha) \delta(\omega_\alpha - \omega'_\alpha) \delta(\eta_\alpha - \eta'_\alpha) \delta_{\alpha\beta} \end{bmatrix} \qquad \text{(III.5)}$$

In the following we write other operators, in particular the resolvent of H, in the same form.

The finite number of point eigenvalues is a necessary condition for the solvability of the model, because it permits us to reduce the calculation of $[H - z]^{-1}$ to the solution of an algebraic problem. We do not require that all these eigenvalues be embedded in the continuum. We may have bound states as well as resonances, although we will focus our attention on the latter ones. Usually one assumes that the Hamiltonian is bounded from below, that is, that the continuous spectrum is semibounded ($h_\alpha \geq \mu_\alpha$). However, it might be useful to consider also cases with a spectrum extending over the whole real axis or with a bounded continuum.

Without loss of generality, we may suppose that $F_p V F_p = 0$. In fact, we can always transform H by a unitary transformation so that this condition is satisfied; it suffices to solve the eigenvalue problem for $F_p H F_p$. One may modify the Hamiltonian so that a nonvanishing term $F_c V F_c$ is included and thus to introduce scattering of the continuum states. Evidently such a Hamiltonian may be transformed to a Friedrichs type, provided that $F_c H_0 F_c + \lambda F_c V F_c$ can be diagonalized by a unitary transformation. As we have supposed that V is bounded, H is a self-adjoint operator defined on the same domain as H_0. We have already mentioned, in connection with the Lee model, that sometimes unbounded perturbations are also considered. In such a case renormalization techniques might be necessary in order to discuss the properties of the model.

Mathematically rigorous proofs for the general properties of this class of Hamiltonians can be obtained using the results of perturbation theory for linear operators (cf. Kato[48]). In particular, the case of a model with simple continuous spectrum has been studied extensively by Marchand.[49,50]

B. The Resolvent Operator

For a Friedrichs-type Hamiltonian it is an easy matter to compute its resolvent (i.e., to solve the equation)

$$(H - z)f = g \qquad (z \notin \mathrm{sp}\, H) \tag{III.6}$$

More precisely, the solution of (III.6) reduces to an algebraic problem.

In the spectral representation of H_0, (III.6) reads

$$(\omega_k - z)f_k + \lambda \sum_l V_{kl}f_l + \lambda \sum_\beta \int_{\mathscr{I}_\beta} d\omega'_\beta \, d\eta'_\beta \bar{v}_{k|\beta}(\omega'_\beta, \eta'_\beta)f_b(\omega'_\beta, \eta'_\beta) = g_k$$

$$\tag{III.7}$$

$$\lambda \sum_l v_{l|\alpha}(\omega_\alpha, \eta_\alpha)f_l + [h_\alpha(\omega_\alpha) - z]f_\alpha(\omega_\alpha, \eta_\alpha) = g_\alpha(\omega_\alpha, \eta_\alpha) \tag{III.8}$$

From (III.8) we obtain

$$f_\alpha(\omega_\alpha, \eta_\alpha) = -\lambda \frac{1}{h_\alpha(\omega_\alpha) - z} \sum_l v_{l|\alpha}(\omega_\alpha, \eta_\alpha)f_l + \frac{g_\alpha(\omega_\alpha, \eta_\alpha)}{h_\alpha(\omega_\alpha) - z} \tag{III.9}$$

Introducing this expression into (III.7), we have

$$\sum_l M_{kl}(z)f_l = g_k - \lambda \sum_\beta \int_{\mathscr{I}_\beta} d\omega_\beta \, d\eta_\beta \frac{\bar{v}_{k|\beta}(\omega_\beta, \eta_\beta)g_\beta(\omega_\beta, \eta_\beta)}{h_\beta(\omega_\beta) - z} \quad \text{(III.10)}$$

where $M(z)$ is an $n \times n$ matrix with elements

$$M_{kl}(z) = (\omega_k - z)\delta_{kl} + \lambda V_{kl} - \lambda^2 \sum_\alpha \int_{\mathscr{I}_\alpha} d\omega_\alpha \, d\eta_\alpha \frac{\bar{v}_{k|\alpha}(\omega_\alpha, \eta_\alpha)v_{l|\alpha}(\omega_\alpha, \eta_\alpha)}{h_\alpha(\omega_\alpha) - z} \quad \text{(III.11)}$$

Thus by inverting $M(z)$ we obtain the solution of (III.6).

Let $M_{kl}^{-1}(z)$ be the elements of the inverse matrix $M^{-1}(z)$. Then from (III.9) and (III.10) we get for the resolvent of H the representation

$$\frac{1}{H-z} \leftrightarrow \begin{bmatrix} M_{kl}^{-1} & \dfrac{-\lambda \sum_p M_{kp}^{-1} \bar{v}_{p|\beta}(\omega_\beta' \eta_\beta')}{h_\beta(\omega_\beta') - z} \\[2em] \dfrac{-\lambda \sum_p v_{p|\alpha}(\omega_\alpha, \eta_\alpha)M_{pl}^{-1}}{h_\alpha(\omega_\alpha) - z} & X_{\alpha\beta} \end{bmatrix} \quad \text{(III.12)}$$

where

$$X_{\alpha\beta} = \frac{1}{h_\alpha(\omega_\alpha) - z} \delta(\omega_\alpha - \omega_\alpha')\delta(\eta_\alpha - \eta_\alpha')\delta_{\alpha\beta}$$

$$+ \lambda^2 \frac{\sum_{pq} v_{p|\alpha}(\omega_\alpha, \eta_\alpha)M_{pq}^{-1}\bar{v}_{q|\beta}(\omega_\beta', \eta_\beta')}{[h_\alpha(\omega_\alpha) - z][h_b(\omega_\beta') - z]} \quad \text{(III.13)}$$

Here we should point out that $M(z)$ is related to the resolvent of H by

$$M^{-1}(z) = F_p[H - z]^{-1}F_p \quad \text{(III.14)}$$

and that it may be written as

$$M(z) = K(z) - zF_p, \quad K(z) = F_pHF_p - \lambda^2 F_pVF_c \frac{1}{F_cHF_c - z} F_cVF_p \quad \text{(III.15)}$$

The matrix $K(z)$ was introduced by Livsic[51,52] in his investigations of problems in scattering theory. Its properties have been studied recently by Howland.[53]

One may realize that unless we restrict the functions $h_\alpha(\omega_\alpha)$, the discussion of the properties of H becomes rather involved. In most problems of physical interest these functions are monotonic. Then we may change the representation and introduce a new spectral variable $\omega'_\alpha = h_\alpha(\omega_\alpha)$ and replace the interactions by $v'_{k|\alpha}(\omega'_\alpha, \eta_\alpha) = [dh_\alpha^{-1}(\omega'_\alpha)/d\omega'_\alpha]^{1/2}v_{k|\alpha}(h_\alpha^{-1}(\omega'_\alpha), \eta_\alpha)$. The expression for $[H-z]^{-1}$ is similar to that given by (III.12), except that the functions h_α have been replaced by the new spectral variable.

Obviously, the simplest model consists of an unperturbed Hamiltonian H_0 with one point eigenvalue and a simple continuous spectrum. The matrix $M(z)$ is now a complex function $\epsilon(z)$

$$\epsilon(z) = \omega_0 - z - \lambda^2 \int_{\mathscr{I}} d\omega \frac{|v(\omega)|^2}{\omega - z} \tag{III.16}$$

It is an easy matter to show that $\epsilon(z)$ has no complex zeros, in agreement with the fact that H is self-adjoint. In fact, the zeros of $\epsilon(z)$, and more generally the solutions of

$$\sum_l M_{kl}(z)f_l = 0 \tag{III.17}$$

give the point spectrum of H. Of course, (III.17) does not admit solutions for complex values of z. This can be shown using a property of $M(z)$, namely, that it is a dissipative (accretive) operator for Im $z < 0$ (Im $z > 0$).

C.   Eigendistributions in a Special Case

We consider in this section the particular case where H_0 has only one point eigenvalue ω_0 and a continuous spectrum represented as a multiplication by a single spectral variable (i.e., $s = 1$ and $h_1(\omega) = \omega$). This model illustrates most aspects of the problems involved when studying such Hamiltonians.

Here again the matrix $M(z)$ is replaced by a function $\epsilon(z)$ (we use the same notation as previously for a simple continuous spectrum), where

$$\epsilon(z) = \omega_0 - z - \lambda^2 \int_a^b d\omega \frac{1}{\omega - z}|\phi(\omega)|^2; \quad |\phi(\omega)|^2 = \int d\eta |v(\omega, \eta)|^2 \tag{III.18}$$

We have assumed that $a \le \omega \le b$, where one or both endpoints of the interval may be at infinity. The singularities of the resolvent [cf. (III.12)] consist of a cut from a to b and, possibly, poles from the (real) zeros of $\epsilon(z)$.

If ω_0 is a bound state ($\omega_0 \notin [a, b]$), one may prove that $\epsilon(z)$ has always a zero, that is, that H has a point eigenvalue. However, if ω_0 is embedded in the continuum ($\omega_0 \in [a, b]$), then H has no point eigenvalue provided that λ is sufficiently small and

$$\phi(a) = \phi(b) = 0 \tag{III.19}$$

We speak then of an unstable eigenvalue. One may easily check that for the cases that we presented in the second section, (III.19) is satisfied. However when a cutoff is used, as in the paper of Davidson and Kozak,[15] this condition does not hold. Then the Hamiltonian has one or more point eigenvalues (bound states) that depend nonanalytically on the coupling parameter λ, and thus conclusions based on perturbation expansions are, in general, incorrect. From the Plemelj formulas (cf. Ref. 54) we know that, if $\phi(\omega)$ is Hölder continuous, the boundary values of $\epsilon(z)$, as z approaches the real axis, are given by (the integral is a Cauchy principal value)

$$\epsilon_{\pm}(\nu) \equiv \epsilon(\nu \pm i0) = \omega_0 - \nu - \lambda^2 \int_a^b d\omega \frac{|\phi(\omega)|^2}{\omega - \nu} \mp i\pi\lambda^2 |\phi(\nu)|^2$$

(III.20)

Thus H has no point eigenvalue embedded in its continuous spectrum (for any λ) if

$$\phi(\omega) \neq 0, \qquad a < \omega < b$$

(III.21)

Note that the critical value of λ for the nonexistence of an isolated point eigenvalue is

$$\lambda_{cr} = \min\left\{ \frac{(\omega_0 - a)}{\int_a^b d\omega(|\phi(\omega)|^2/\omega - a)}, \frac{(\omega_0 - b)}{\int_a^b d\omega(|\phi(\omega)|^2/\omega - b)} \right\}$$

(III.22)

which may be also considered as a restriction on the position of ω_0 with respect to the end points of the interval (cf. Ref. 14).

The eigenfunction ξ^0 corresponding to a discrete eigenvalue ν_0 of H can be easily calculated. One uses the same steps as for the calculation of the resolvent [it suffices to replace g in the rhs of (III.6) by $\nu_0\xi^0$]. We have

$$\xi^0 = |\epsilon'(\nu_0)|^{-1/2}\left\{ 1, \frac{-\lambda\upsilon(\omega, \eta)}{\omega - \nu_0} \right\}$$

(III.23)

where $\epsilon'(z) = d\epsilon(z)/dz$. The constant $|\epsilon'(\nu_0)|^{-1/2}$ is chosen so that ξ^0 is correctly normalized:

$$\langle \xi^0, \xi^0 \rangle = 1$$

(III.24)

For the continuous spectrum of H we must consider solutions $\xi(\nu)$ of the equation

$$H\xi(\nu) = \nu\xi(\nu), \nu \in [a, b]$$

(III.25)

in the sense of distributions. In the representation that we are working,

(III.24) is written as

$$(\omega_0 - \nu)\xi_{(0)}(\nu) + \lambda \int d\omega' \int d\eta' \, \bar{v}(\omega', \eta')\xi(\omega', \eta'|\nu) = 0 \qquad \text{(III.26)}$$

and

$$\lambda v(\omega, \eta)\xi_{(0)}(\nu) + (\omega - \nu)\xi(\omega, \eta|\nu) = 0 \qquad \text{(III.27)}$$

There are solutions for which $\xi_{(0)}(\nu)$ vanishes and also solutions for which $\xi_{(0)}(\nu) \neq 0$. An easy way to construct them is to consider, for a fixed ω, a set of functions $\{\chi_r(\omega, \eta)\}$ that is orthonormal and complete with respect to η

$$\int d\eta \, \bar{\chi}_r(\omega, \eta)\chi_s(\omega, \eta) = \delta_{r,s}, \qquad \sum_{r=0}^{\infty} \bar{\chi}_r(\omega, \eta)\chi_r(\omega, \eta') = \delta(\eta - \eta')$$

$$\text{(III.28)}$$

and such that

$$\phi(\omega)\chi_0(\omega, \eta) = v(\omega, \eta) \qquad \text{(III.29)}$$

It may be shown, then, that the solutions of (III.26) and (III.27) are given by

$$\xi_0(\nu) = |\epsilon(\nu + i0)|^{-1} \left\{ -\lambda \bar{\phi}(\nu), \chi_0(\omega, \eta) \left[\epsilon_1(\nu) \, \delta(\omega - \nu) + \lambda^2 \frac{\phi(\omega)\bar{\phi}(\nu)}{\omega - \nu} \right] \right\}$$

$$\text{(III.30)}$$

$$\xi_r(\nu) = \{0, \chi_r(\omega, \eta)\delta(\omega - \nu)\} \qquad (r = 1, 2, \ldots) \qquad \text{(III.31)}$$

Similar problems arise in other branches of theoretical physics and, in particular, in linear transport theory.[55] In fact, the preceding construction was used by Felderhof[56] in studying the eigendistributions of the linearized Vlasov operator in plasma physics.

These eigendistributions are orthonormal in the sense that

$$\langle \xi_r(\nu), \xi_s(\nu') \rangle = \delta_{r,s}\delta(\nu - \nu') \qquad \text{and} \qquad \langle \xi^0, \xi_r(\nu) \rangle = 0 \qquad \text{(III.32)}$$

Moreover, together with the eigenfunctions ξ^0 corresponding to the point eigenvalue, if such an eigenvalue exists, they form a complete set, that is,

$$\bar{\xi}^0\xi^0 + \int_a^b d\nu \sum_{r=0}^{\infty} \bar{\xi}_r(\nu)\xi_r(\nu) = \begin{bmatrix} 1 & 0 \\ 0 & \delta(\omega - \omega') \end{bmatrix} \qquad \text{(III.33)}$$

The proof of these relations is based on the solution of a singular integral equation. Other formal proofs can be obtained by relating these eigendistributions to the boundary values of the matrix elements of the resolvent. These properties of the eigendistributions ensure that they can be used to expand vectors of the space with which we are working, provided that the functions in their representation are sufficiently smooth.

It is possible to justify rigorously the use of such eigendistributions by considering the theory of "rigged Hilbert spaces" developed mainly by Gelfand[57] (see also Maurin[58] and Bohm[59]). In this approach one introduces a (topological) space Φ of vectors the representers of which are infinitely differentiable functions and tend to zero as $|\omega| \to \infty$, faster than any power of the variable. We have then the inclusion $\Phi \subset \mathcal{H} \subset \Phi'$, where Φ' is the dual space of Φ (i.e., the space of continuous linear functionals on Φ). The eigendistributions of H belong precisely to Φ', and a consistent theory of expansions of the vectors of Φ can be formulated.

We should remark that the eigenfunction corresponding to the point eigenvalue of H and the eigendistribution corresponding to its continuous spectrum may be viewed as the "matrix elements" of a unitary transformation diagonalizing H. More precisely, they define a transformation U such that U^+HU commutes with H_0. Note that in case it has no discrete spectrum this transformation is a partial isometry, that is, $UU^+ = I$ but $U^+U = F_c \neq I$. Furthermore, one may show (Refs. 47, 49, 50; see also Ref. 60) that up to phase factors the transformation U and its adjoint are the Möller wave operators of scattering theory (see, for example, Ref. 61).

One may generalize the preceding considerations to other cases as well. For example, Marchand[49] has studied the Hamiltonian with a simple continuous spectrum and several embedded eigenvalues. His results can be also extended to the case of a degenerate continuous spectrum. However, the different expressions become rather complicated, and as there is no new basic idea involved, we will not discuss them here.

IV.   TIME-DEPENDENT PROBLEMS

We want now to discuss some aspects of the initial value problem for the Schrödinger and von Neumann equations for systems with Friedrichs-type Hamiltonians. The general solution of the initial value problem can be easily obtained using the eigendistributions or the resolvent of the Hamiltonian. Under certain conditions the solution contains some exponential contributions that are of particular interest. We examine these exponentially decaying terms and discuss briefly certain questions related to their physical interpretation. For simplicity, we limit ourselves to the model with a single point eigenvalue embedded in the continuum.

A.   The Solution of the Initial Value Problem

As we are mainly interested in decay phenomena, we assume that the total Hamiltonian of the system has no point spectrum. To fix the ideas, we generally consider a semibounded continuous spectrum extending over the positive real axis.

In terms of the eigendistributions $\{\xi(\nu)\}$ (see Section III.C), the solution of the Schrödinger equation

$$i\partial_t \varphi(t) = H\varphi(t) \tag{IV.1}$$

is straightforward. It is given by the expression

$$\varphi(t) = e^{-iHt}g = \int_0^\infty d\nu \, e^{-i\nu t}\xi(\nu)g_\nu \tag{IV.2}$$

where $g = \{g_0, g(\omega)\}$ is the initial condition $\varphi(t=0)$ and

$$g_\nu = \langle \xi(\nu), g \rangle = g(\nu) - \lambda \frac{v(\nu)}{\epsilon(\nu+i0)} G(\nu+i0) \tag{IV.3}$$

with

$$G(z) = g_0 - \lambda \int_0^\infty d\omega \frac{\bar{v}(\omega)g(\omega)}{\omega - z} \tag{IV.4}$$

We may calculate now the scalar product of $\varphi(t)$ with a vector f. From (IV.2) we get

$$\langle f, \varphi(t) \rangle - \langle f, e^{-iHt}g \rangle = \int_0^\infty d\nu \, e^{-i\nu t}\tilde{f}_\nu g_\nu \tag{IV.5}$$

where

$$\tilde{f}_\nu = \langle \tilde{f}, \xi(\nu) \rangle = \bar{f}(\nu) - \lambda \frac{\bar{v}(\nu)}{\epsilon(\nu-i0)}\tilde{F}(\nu-i0) \tag{IV.6}$$

and

$$\tilde{F}(z) = \bar{f}_0 - \lambda \int_0^\infty d\omega \frac{v(\omega)\bar{f}(\omega)}{\omega - z} \tag{IV.7}$$

Equation IV.5 can be written as a sum of two terms, namely,

$$\langle f, e^{-iHt}g \rangle = \frac{1}{2\pi i} \int_0^\infty d\nu \, e^{-i\nu t}\left\{ \frac{1}{\epsilon(\nu+i0)}\tilde{F}(\nu+i0)G(\nu+i0) + M(\nu+i0) \right\}$$

$$- \frac{1}{2\pi i} \int_0^\infty d\nu \, e^{-i\nu t}\left\{ \frac{1}{\epsilon(\nu-i0)}\tilde{F}(\nu-i0)G(\nu-i0) + M(\nu-i0) \right\} \tag{IV.8}$$

with

$$M(z) = \int_0^\infty d\omega \frac{\bar{f}(\omega)g(\omega)}{\omega - z} \tag{IV.9}$$

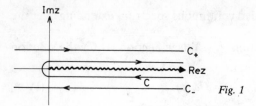

Fig. 1

This formula permits us to make contact with the solution of the initial value problem using the resolvent of H (i.e., a Laplace transformation).

Clearly, the integrands in the rhs of (IV.8) are the boundary values of the analytic function $\epsilon^{-1}(z)\tilde{F}(z)G(z)+M(z)$, as z approaches the cut (Im $z = 0$, Re $z = \nu > 0$) from above or below. Thus we may write

$$\langle f, e^{-iHt}g\rangle = \frac{1}{2\pi i}\int_C dz\, e^{-izt}\left\{\frac{1}{\epsilon(z)}\tilde{F}(z)G(z)+M(z)\right\} \qquad \text{(IV.10)}$$

where C is the contour in Fig. 1. Taking into account the expression for the resolvent of the Hamiltonian [cf. (III.12) and (III.13)], one may see that

$$\frac{1}{\epsilon(z)}\tilde{F}(z)G(z)+M(z)=\left\langle f, \frac{1}{H-z}g\right\rangle \qquad \text{(IV.11)}$$

Moreover, one may note that the contour C in (IV.10) can be replaced by the contours C_+ and C_-, shown in Fig. 1. Then for $t>0$ ($t<0$) the only nonvanishing contribution comes from the integration along C_+ (C_-). In other words, we may extend in the rhs of (IV.8) the integration along the whole real axis and drop the second (first) term for $t>0$ ($t<0$). These considerations show the relation between the solution of the initial value problem for the Schrödinger equation in terms of the eigendistributions of the Hamiltonian and that in terms of its resolvent.

Similar considerations apply to the solution of the von Neumann equation

$$i\partial_t\rho = [H, \rho] = L\rho \qquad \text{(IV.12)}$$

describing the evolution of a (mixed) state, given by a density matrix ρ. In fact, the formal solution of (IV.12) is

$$\rho(t) = \exp\{-iLt\}\rho(0) = \exp\{-iHt\}\rho(0)\exp\{iHt\}$$

$$= \frac{1}{2\pi i}\int_{\tilde{c}} dz\, e^{-izt}\frac{1}{L-z}\rho(0) \qquad \text{(IV.13)}$$

and the resolvent of L is related to that of H by

$$\frac{1}{L-z}\rho = \frac{1}{2\pi i}\int_{\tilde{c}'} dw\, \frac{1}{H-w}\rho\frac{1}{H-w+z} \qquad \text{(IV.14)}$$

For $\text{Im } z > 0$, the contour $\tilde{C}'$ is above the real axis and below z. Using (IV.2), we may write this solution as a (double) Fourier integral, or, because of (IV.14), we may write $\rho(t)$ as the inverse Laplace transform of a function involving the convolution of (matrix elements of) the resolvent of H.[63] We should also recall here that the mean values of observables, which are bounded self-adjoint operators, are given by the trace of their product with the state ρ, which is usually written as

$$\langle A \rangle_\rho = (A, \rho) = \text{tr } A^+ \rho \qquad \text{(IV.15)}$$

In the representation with which we are working, an operator A is given by a matrix

$$A \leftrightarrow \begin{bmatrix} A_{00} & A_{0c}(\omega') \\ A_{c0}(\omega) & A_{cc}(\omega, \omega') \end{bmatrix} \qquad \text{(IV.16)}$$

and (IV.15) becomes

$$(A, \rho) = \bar{A}_{00}\rho_{00} + \int_0^\infty d\omega' \, \bar{A}_{0c}(\omega')\rho_{c0}(\omega') + \int_0^\infty d\omega \, \bar{A}_{c0}(\omega)\rho_{0c}(\omega)$$

$$+ \int_0^\infty d\omega \int_0^\infty d\omega' \, \bar{A}_{cc}(\omega, \omega')\rho_{cc}(\omega', \omega) \qquad \text{(IV.17)}$$

Note, however, that the kernel $A_{cc}(\omega, \omega')$ may contain distributions and, in particular, a Dirac δ-function.

B. Exponentially Decaying Contributions

It has been assumed in this chapter that the eigenvalue ω_0 of H_0 "disappears" because of the interaction and that H has no point eigenvalue. Of course, one expects to find some trace of this unstable eigenvalue. If we suppose that the interaction $v(\omega)$ is analytic in some domain including the positive real axis, then the function $\epsilon(z)$, defined by (III.16), may be continued analytically across the axis. For simplicity, we will suppose that $v(\omega)$ is analytic in some sector $-\theta_0 < \theta < \theta_0$, indicated in Fig. 2. Then $\epsilon_+(z)$

$$\epsilon_+(z) = \omega_0 - z - \lambda^2 \int_{\Gamma_+} d\zeta \frac{\bar{v}(\bar{\zeta})v(\zeta)}{\zeta - z} \qquad \text{(IV.18)}$$

is the analytic continuation of $\epsilon(z)$ below the positive real axis, in the sector determined by some contour Γ_+. In a similar manner we can define $\epsilon_-(z)$ by analytic continuation above the axis in the sector determined by Γ_-.

At least for small, but finite, values of λ, $\epsilon_+(z)$ has a zero z_0,

$$\epsilon_+(z_0) = 0, \quad \text{Im } z_0 < 0, \quad \lim_{\lambda \to 0} z_0(\lambda) = \omega_0 \qquad \text{(IV.19)}$$

Note that this zero does not depend on the precise position of Γ_+, provided

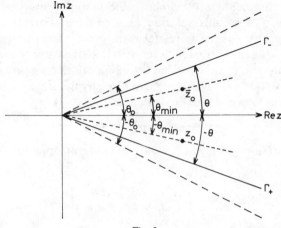

Fig. 2

that $\theta > \theta_{min}$. Also from the symmetry of the problem, it follows that $\epsilon_-(z)$ vanishes at $\bar{z}_0$. In the following we consider the contribution of such a root to the solution of the Schrödinger and von Neumann equations.

Here we should observe that the interaction is represented by a dilation analytic potential. For such cases, a mathematically rigorous theory for resonances in two- and few-body systems has been constructed by Combes[10,64,65] and others,[66,67] and this theory has been applied by Weder[68] to the Lee model in the $N\theta$ sector. Moreover, we should remark that the problem of the perturbation of unstable eigenvalues has been studied extensively in a series of articles by Howland,[69-74] who has also considered the relation of virtual poles of the resolvent of the Hamiltonian to spectral concentration phenomena (see also the work of Baumgärtel and Demuth in Refs. 75 to 80).

If we assume now that the functions $\bar{f}(\omega)$ and $g(\omega)$, which we introduced in the previous section, are analytic in the same sector as the interaction, we can continue analytically the functions $\tilde{F}(z)$, $G(z)$, and $M(z)$ in the second Riemann sheet. Thus (IV.10) may be written as ($t \geq 0$)

$$\langle f, e^{-iHt}g \rangle = -e^{-iz_0t}\frac{1}{\epsilon_+'(z_0)}\tilde{F}_+(z_0)G_+(z_0)$$

$$+\frac{1}{2\pi i}\int_{C'} dz\, e^{-izt}\left\{\frac{1}{\epsilon_+(z)}\tilde{F}_+(z)G_+(z)+M_+(z)\right\} \qquad \text{(IV.20)}$$

where the contour C' is shown in Fig. 3. The second term in the rhs of (IV.20) does not vanish for any choice of f and g, and it decays as an inverse power of t for long times. This nonexponential behavior is a

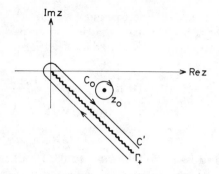

Fig. 3

general feature of Hamiltonians with a bounded or semibounded spectrum. It can be proven either using a theorem of Paley and Wiener[81] on Fourier transforms[82] or a construction due to Sz.-Nagy[83] for the extension of a contraction semigroup to a unitary group in a larger space.[84–87] Note also that if we consider (pure or mixed) states with finite energy, then the decay is necessarily nonexponential for short times, even when the Hamiltonian has a continuous spectrum extending over the whole real axis.[5,88,89]

The situation is quite analogous for the von Neumann equation, as long as we consider observables A for which the kernel $A_{cc}(\omega, \omega')$ [cf. (IV.16)] contains no δ-functions. Corresponding to the first term in the rhs of (IV.20) we have an exponentially decaying term, which we denote by $(A, \rho(t))_0$, and certain other terms that, in general, are not pure exponentials. Let us discuss briefly the exponential contribution that can be written as follows:

$$(A, \rho(t))_0 = \exp\{-i(z_0 - \bar{z}_0)t\} \frac{1}{\epsilon'_+(z_0)\epsilon'_-(\bar{z}_0)} \tilde{\mathscr{A}}(z_0, \bar{z}_0)\mathscr{P}(z_0, \bar{z}_0)$$

(IV.21)

where

$$\mathscr{P}(z_0, \bar{z}_0) = \rho_{00} + \lambda^2 \int_{\Gamma_+} d\zeta \int_{\Gamma_-} d\zeta' \frac{\bar{v}(\bar{\zeta})\rho_{cc}(\zeta, \zeta')v(\zeta')}{(\zeta - z_0)(\zeta' - \bar{z}_0)}$$
$$- \lambda \int_{\Gamma_-} d\zeta' \frac{\rho_{0c}(\zeta')v(\zeta')}{\zeta' - \bar{z}_0} - \lambda \int_{\Gamma_+} d\zeta \frac{\bar{v}(\bar{\zeta})\rho_{c0}(\zeta)}{\zeta - z_0}$$

(IV.22)

and

$$\tilde{\mathscr{A}}(z_0, \bar{z}_0) = \bar{A}_{00} + \lambda^2 \int_{\Gamma_+} d\zeta \int_{\Gamma_-} d\zeta' \frac{v(\zeta)\bar{A}_{cc}(\bar{\zeta}, \bar{\zeta}')\bar{v}(\bar{\zeta}')}{(\zeta - z_0)(\zeta - \bar{z}_0)}$$
$$- \lambda \int_{\Gamma_-} d\zeta' \frac{\bar{A}_{0c}(\bar{\zeta}')\bar{v}(\bar{\zeta}')}{\zeta' - \bar{z}_0} - \lambda \int_{\Gamma_+} d\zeta \frac{v(\zeta)\bar{A}_{c0}(\bar{\zeta})}{\zeta - z_0}$$

(IV.23)

The contours Γ_+ and Γ_- are those in Fig. 2. These formulas, which are somewhat complicated because we work with density matrices, may be cast in a form corresponding to the formalism of subdynamics[7,8] (cf. also Refs. 90 to 92). In fact, the rhs of (IV.21) can be written as

$$(A, \rho(t))_0 = \exp\{-i(z_0 - \bar{z}_0)t\}(A, \Pi\rho)_+ \qquad \text{(IV.24)}$$

where Π has a matrix representation determined easily from (IV.22) and (IV.23). Formally, Π is an idempotent and for $\lambda = 0$ it tends to a projection P, such that $P\rho = \rho_{00}$.

An important point that we want to stress is the following. In the rhs of (IV.24) we have introduced a modified scalar product, as indicated by a $(+)$-subscript. Thus, when we compute $(A, \rho)_+$, we must integrate along Γ_+ and Γ_- instead of the positive real axis as in (IV.17). This means that if we want to isolate the contribution due to a complex pole of the analytic continuation of the resolvent of L, we need to work in a space with a "nonhermitian" metric. We have considered this question using the Friedrichs model as example[9] and we will discuss it in detail in a forthcoming article.[93] We may remark that similar mathematical problems arise with Gamow states in nuclear physics[94-96] and that the idea to associate eigenvectors to complex eigenvalues of the Hamiltonian has been considered also in some recent articles.[97,98] Moreover, we should note that such complex energy eigenvectors may be interpreted as generalized eigenvectors of the Hamiltonian in a suitably chosen rigged Hilbert space.[99,100]

C. Considerations on Nonunitary Transformations

We have indicated in the previous section how, in the framework of the simple Friedrichs model, we can associate an unstable state to an exponentially decaying contribution of the solution of the evolution equation. Similar conclusions can be drawn in more complex models. The whole approach may be viewed as an application of the theory of subdynamics. Of course, the correct interpretation of the formalism requires the use of a nonhermitian scalar product, avoiding thus the objections that can be formulated when working in a normed space.[101]

Closely related to the idea of subdynamics is the theory of nonunitary transformations introduced by Prigogine et al.[8] It predicts that dissipative dynamical systems are characterized by the existence of an invertible transformation Λ, leading to the so-called physical representation, and such that $\Lambda^{-1}L\Lambda$ (L is the Liouville or von Neumann operator) is dissipative, that is,

$$\Phi = \Lambda^{-1}L\Lambda; \qquad \Lambda^{-1} \neq \Lambda^+, \qquad \frac{1}{2i}[\Phi - \Phi^+] \leq 0 \qquad \text{(IV.25)}$$

Furthermore, this transformation has the property

$$\Pi = \Lambda P \Lambda^{-1} \qquad \text{(IV.26)}$$

We do not intend to discuss here this theory, which permits us to look at the problem of irreversibility of dynamical systems from a novel point of view and define a nonequilibrium entropy function (see, for example, Ref. 102). We will make only a few remarks connected with the exponentially decaying contribution given by (IV.24).

As we have already noted, to the pole $z_0 - \bar{z}_0$ we can associate an idempotent Π [cf. (IV.24)], which reduces to a one-dimensional projection P when the coupling parameter goes to zero. These two operators can be related by a transformation Λ as in (IV.26). Obviously, this transformation must be regarded as connecting the space with the hermitian metric, in which the von Neumann equation is defined, to that with the nonhermitian metric, in which the exponentially decaying contribution may be identified.

One cannot define the transformation Λ uniquely from (IV.26). However, this equation determines essentially its component $\chi_0 = P\Lambda P$. Because P is one-dimensional, χ_0 is simply a number, namely,

$$\chi_0 = [\epsilon'_+(z_0)\epsilon'_-(\bar{z}_0)]^{-1/2} \qquad \text{(IV.27)}$$

It may be shown that if $\tilde{\rho}_0 = P\Pi\rho$, then

$$|\chi_0^{-1}\tilde{\rho}_0|^2 = (\Pi\rho, \Pi\rho)_+ \qquad \text{(IV.28)}$$

This relation is quite important for the physical interpretation of the transformation theory. It shows that in order to have a consistent description of the unstable state, we must renormalize the "bare" state by taking into account the correlations, predicted by the theory of subdynamics, between this "bare" state and the field.

V. CONCLUDING REMARKS

We have reviewed here some systems leading to a Friedrichs model, as well as the basic properties of its Hamiltonian. The presence in the initial value of terms decaying slower than the exponential one has caused several discussions as to their meaning[103-105] (cf. also references in Ref. 5). Of course, the exponential decay, and more generally a semigroup property, can be rigorously proven in the limit of weak coupling[6] (see also Refs. 106, 107) or when the system is coupled to a singular reservoir.[108] However, as we have indicated, a correct description of unstable states may be obtained without such unphysical restrictions using methods initially developed for nonequilibrium statistical mechanics. This aspect has been stressed often by I. Prigogine, who has insisted on the necessity of a "thermodynamic" description of elementary particles.[109,110]

One cannot expect that a model of the Friedrichs type will illustrate all interesting problems in decaying phenomena. Perhaps we could get a better insight of such processes by treating the model with the field, represented by the continuum, at finite temperatures.

It should be mentioned that the remarks of the last two sections of the previous chapter apply essentially to observables with discrete spectrum. The extension of these ideas to observables with continuous spectra (e.g., a function of H) is an important open problem. Until now there has not been a satisfactory solution, at least when the continuous spectrum of the Hamiltonian is assumed *ab initio* and not as the result of a limiting process (e.g., the infinite volume limit). We intend to consider this question in a subsequent article.

Acknowledgments

The final version of this paper was written during a visit to the Center for Statistical Mechanics and Thermodynamics of the University of Texas at Austin. The author wishes to thank the members of the Center for their hospitality.

References

1. V. Weisskopf and E. Wigner, Z. Phys. **63,** 54 (1930).
2. V. Weisskopf and E. Wigner, Z. Phys. **65,** 18 (1930).
3. R. Peierls, in E. H. Bellamy and R. G. Moorhouse, Eds., *Proceedings of the 1954 Glasgow Conference on Nuclear and Meson Physics*, Pergamon, London, 1955.
4. G. Gamow, Z. Phys., **51,** 204 (1928).
5. L. Fonda, *Fortschr. d. Physik* **25,** 101 (1977).
6. K. O. Friedrichs, *Comm. Pure. Appl. Math.*, **1,** 361 (1948).
7. I. Prigogine, C. George, and F. Henin, *Physica*, **45,** 418 (1969).
8. I. Prigogine, C. George, F. Henin, and L. Rosenfeld, *Chem. Scripta*, **4,** 5 (1973).
9. A. P. Grecos, in F. C. Auluck, Ed., *Proceedings of the International Conference on Frontiers of Theoretical Physics*, Indian National Science Academy, New Delhi, 1977.
10. J. M. Combes, in the *Proceedings of the International Congress of Mathematicians*, Vancouver, 1974.
11. W. Heitler, *The Quantum Theory of Radiation*, 3rd ed., Clarendon, Oxford, 1954.
12. G. Källen, *Quantum Electrodynamics*, Springer-Verlag, New York, 1972.
13. E. Grimm and V. Ernst, *J. Phys.*, **A7,** 1664 (1974).
14. E. Grimm and V. Ernst, Z. *Phys.*, **A274,** 293 (1975).
15. R. Davidson and J. Kozak, *J. Math. Phys.*, **11,** 189 (1970).
16. R. Davidson and J. Kozak, *J. Math. Phys.*, **11,** 1420 (1970).
17. R. Davidson and J. Kozak, *J. Math. Phys.*, **12,** 903 (1971).
18. R. Davidson and J. Kozak, *J. Math. Phys.*, **14,** 414 (1973).
19. R. Davidson and J. Kozak, *J. Math. Phys.*, **14,** 423 (1973).
20. J. J. Yang, R. Davidson, and J. Kozak, *J. Math. Phys.*, **15,** 491 (1974).
21. R. Davidson and J. Kozak, *J. Math. Phys.*, **16,** 1013 (1975).
22. M. Sargent, M. Scully, and W. Lamb, Jr., *Laser Physics*, Addison-Wesley, Reading, Mass., 1974.

23. N. G. van Kampen, *Kong. Danske Vidensk, Selsk., Mat.-Fys. Meddr.*, **26**(15) (1951).
24. M. Shirokov, *Sov. J. Nucl. Phys.*, **21**, 347 (1975).
25. K. Rżążewski and W. Żakowicz, *J. Phys.*, **A9**, 1159 (1976).
26. G. Emch and J. Wolf, *J. Math. Phys.*, **13**, 1236 (1972).
27. F. A. Berezin, *The Method of Second Quantization*, Academic, New York, 1966.
28. T. D. Lee, *Phys. Rev.*, **95**, 1329 (1954).
29. E. Henley and W. Thirring, *Elementary Quantum Field Theory*, McGraw-Hill, New York, 1962.
30. G. Barton, *Introduction to Advanced Field Theory*, Interscience, New York, 1963.
31. S. Schweber, *An Introduction to Relativistic Quantum Field Theory*, Harper & Row, New York, 1961.
32. G. Källen and W. Pauli, *Kong. Danske Vidensk., Mat. Fys. Meddr.*, **30**(7) (1955).
33. F. Berezin, *Am. Math. Soc. Transl.*, Series 2, **56**, 249 (1966).
34. E. Lieb, T. Schultz, and D. Mattis, *Ann. Phys. (N.Y.)*, **16**, 407 (1961).
35. E. Baruch, in W. E. Brittin, Ed., *Lectures in Theoretical Physics*, Vol. XIV B, Colorado Ass. Univ. Press, Boulder, 1973.
36. J. A. Tjon, *Phys. Rev.*, **B2**, 2411 (1970).
37. J. L. van Hemmen and G. Vertogen, *Physica*, **81A**, 391 (1975).
38. T. D. Lee, R. Oehme, and C. N. Yang, *Phys. Rev.*, **106**, 340 (1957).
39. P. K. Kabir, *The CP Puzzle*, Academic, London, 1968.
40. S. M. Bilen'skii, *Particles and Nuclei*, **1** (Pt. I), 146 (1972).
41. L. P. Horwitz and J. P. Marchand, *Helv. Phys. Acta*, **42**, 1039 (1969).
42. L. P. Horwitz and J. P. Marchand, *Rocky Mt. J. Math.*, **1**, 225 (1971).
43. M. Bixon and J. Jortner, *J. Chem. Phys.*, **48**, 715 (1968).
44. K. F. Freed, *Fortschr. Chem. Forsch.*, **31**, 105 (1972).
45. G. C. Stey and R. W. Gibberd, *Physica*, **60**, 1 (1972).
46. K. O. Friedrichs, *Spectral Theory of Operators in Hilbert Space*, Springer-Verlag, New York, 1973.
47. K. O. Friedrichs, *Perturbation of Spectra in Hilbert Space*, American Mathematical Society, Providence, R.I., 1965.
48. T. Kato, *Perturbation Theory for Linear Operators*, Springer-Verlag, New York, 1966.
49. J. P. Marchand, *Helv. Phys. Acta*, **37**, 475 (1964).
50. J. P. Marchand, in A. O. Barut and W. E. Brittin, Eds., *Lectures in Theoretical Physics*, Vol. XA, Gordon & Breach, New York, 1968.
51. M. S. Livsic, *Sov. Phys. JETP*, **4**, 91 (1957).
52. M. S. Livsic, *Am. Math. Soc. Transl.*, Series 2, **16**, 427 (1960).
53. J. S. Howland, *J. Math. Anal. Appl.*, **50**, 415 (1975).
54. N. I. Muskhelishvili, *Singular Integral Equations*, Noordhoff, Groningen, 1953.
55. K. M. Case and R. F. Zweifel, *Linear Transport Theory*, Addison-Wesley, Reading, Mass., 1967.
56. B. U. Felderhof, *Physica*, **30**, 1171 (1964).
57. I. M. Gel'fand and N. Vilenkin, *Generalized Functions*, Vol. 4, Academic, New York, 1964.
58. K. Maurin, *Generalized Eigenfunction Expansions and Unitary Representations of Topological Groups*, Polish Scientific Publishers, Warsaw, 1968.
59. A. Bohm, in W. Brittin, A. O. Barut, and M. Guenin, Eds., *Lectures in Theoretical Physics*, Vol. IXA, Gordon & Breach, New York, 1967.
60. A. Bohm, *The Rigged Hilbert Space and Quantum Mechanics*, Springer-Verlag, Berlin, 1978.
61. E. C. G. Sudarshan, in *Brandeis Lectures in Theoretical Physics*, Vol. 2., Benjamin, New York, 1962.

62. R. G. Newton, *Scattering Theory of Waves and Particles*, McGraw-Hill, New York, 1966.
63. A. P. Grecos and I. Prigogine, *Physica*, **59**, 77 (1972).
64. J. Aguilar and J. M. Combes, *Commun. Math. Phys.*, **22**, 269 (1971).
65. E. Balslev and J. M. Combes, *Commun. Math. Phys.*, **22**, 280 (1971).
66. B. Simon, *Ann. Math.*, 2nd series, **97**, 247 (1973).
67. Cl. van Winter, *J. Math. Anal. Appl.*, **47**, 633 (1970); **48**, 368 (1974).
68. R. Weder, *J. Math. Phys.*, **15**, 20 (1974).
69. J. Howland, *J. Math. Anal. Appl.*, **23**, 575 (1968).
70. J. Howland, *Pac. J. Math.*, **29**, 565 (1969).
71. J. Howland, *Am. J. Math.*, **91**, 1106 (1969).
72. J. Howland, *Trnas. AMS*, **162**, 141 (1971).
73. J. Howland, *Bull. AMS*, **78**, 280 (1972).
74. J. Howland, *Pac. J. Math.*, **55**, 157 (1974).
75. H. Baumgärtel, *Am. J. Math.*, **95**, 849 (1973).
76. H. Baumgärtel, *Math. Nachr.*, **59**, 265 (1974).
77. H. Baumgärtel, *Math. Nachr.*, **59**, 275 (1974).
78. M. Demuth, *Math. Nachr.*, **64**, 345 (1974).
79. H. Baumgärtel, *Math. Nachr.*, **69**, 107 (1975).
80. H. Baumgärtel and M. Demuth, *J. Funct. Anal.*, **22**, 187 (1976).
81. R. Paley and N. Wiener, *Fourier Transforms in the Complex Domain*, American Mathematical Society, Providence. R. I., 1934.
82. L. A. Khalfin, *Sov. Phys. JETP*, **6**, 1053 (1958).
83. F. Riesz and B. Sz.-Nagy, *Leçons d'Analyse Fonctionnelle*, 4th ed., Gauthier-Villars, Paris, 1965 (appendix).
84. L. P. Horwitz, J. A. La Vita, and J. P. Marchand, *J. Math. Phys.*, **12**, 2537 (1971).
85. D. N. Williams, *Commun. Math. Phys.*, **21**, 314 (1971).
86. K. Sinha, *Helv. Phys. Acta*, **45**, 619 (1972).
87. P. Exner, *Commun. Math. Phys.*, **50**, 1 (1976).
88. M. Havliček and P. Exner, *Czech. J. Phys.*, **B23**, 594 (1973).
89. P. Exner, *Czech. J. Phys.*, B26, 976 (1976).
90. A. P. Grecos and I. Prigogine, in G. Pichon, Ed., *Théories Cinétiques Classiques et Relativistes*, C.N.R.S., Paris, 1975.
91. A. P. Grecos, T. Guo, and W. Guo, *Physica*, **80A**, 421 (1975).
92. A. P. Grecos and M. Theodosopulu, *Acta Phys. Pol.*, **A50**, 749 (1976).
93. A. P. Grecos, to appear in the Proceedings of the VII International Conference on Group Theoretical Methods in Physics, Springer-Verlag.
94. T. Berggren, *Nucl. Phys.*, **A109**, 265 (1968).
95. W. J. Romo, *Nucl. Phys.*, **A116**, 617 (1968).
96. B. Gyarmati and T. Vertse, *Nucl. Phys.*, **A160**, 523 (1971).
97. H. A. Weldon, *Phys. Rev.*, **D14**, 2030 (1976).
98. E. C. G. Sudarshan, C. B. Chiu, and V. Gorini, *Phys. Rev.*, **D**, to appear.
99. H. Baumgärtel, *Math. Nachr.*, **72**, 93 (1976).
100. H. Baumgärtel, *Math. Nachr.*, **75**, 133 (1976).
101. L. Lanz, L. Lugiato, and G. Ramella, *Physica*, **54**, 94 (1971).
102. I. Prigogine and A. P. Grecos, in F. G. Auluck, Ed., *Proceedings of the International Conference on Frontiers of Theoretical Physics*, Indian National Science Academy, New Delhi, 1977.
103. G. Höhler, *Z. Phys.*, **152**, 546 (1958).
104. M. Levy, in E. R. Caianello, Ed., *Lectures on Field Theory and the Many-Body Problem*, Academic, New York, 1961; and references cited there.

105. B. Diu, *Qu'est-ce qu'une Particule Elémentaire?*, Masson, Paris, 1965.
106. E. B. Davies, *J. Math. Phys.*, **15,** 2036 (1974).
107. E. B. Davies, *Quantum Theory of Open Systems*, Academic, London, 1976.
108. V. Gorini, A. Frigerio, M. Verri, A. Kossakowski, and E. C. G. Sudarshan, *Rep. Math. Phys.* to appear.
109. I. Prigogine, in *Fundamental Problems in Elementary Particle Physics*, Proceedings of the Fourteenth Solvay Conference on Physics, Interscience, New York, 1968.
110. I. Prigogine, F. Mayné, C. George, and M. de Haan, *Proc. Nat. Acad. Sci.* (*USA*) **74,** 4152 (1977).

ASPECTS OF KINETIC THEORY

R. BALESCU and P. RÉSIBOIS

Faculté des Sciences, Université Libre de Bruxelles, Brussels, Belgium

CONTENTS

I. INTRODUCTION

More than one century ago Boltzmann introduced his famous equation to describe the approach to equilibrium in dilute gases, which, for spatially homogeneous systems schematically reads:[1,2]

$$\partial_t \varphi_1(\mathbf{v}; t) = n C^B(\varphi_1, \varphi_1) \tag{I.1}$$

Here $\varphi_1(\mathbf{v}; t)$ is the one-particle velocity distribution function (df), n is the particle density, and C^B is the well-known collision operator, whose nonlinear (quadratic) dependence on φ_1 has been explicitly displayed.

This equation has been the keystone to nonequilibrium physics for almost 100 years, because, at least in the simple case of dilute gases, it provides an understanding of the irreversible behavior of many-particle systems (through the celebrated Boltzmann's H-theorem) and, at the same time, it allows us to calculate the macroscopic transport coefficients* in terms of molecular parameters (like the law of interaction between the particles).

Yet, despite its success, this equation was the object of innumerable criticisms.[3,4] These were all rooted in the fact that Boltzmann, being unable to solve the exact dynamics of N particles enclosed in a volume Ω ($N \approx 10^{23}$!):

$$\frac{d\mathbf{r}_a}{dt} = -\{H_N, \mathbf{r}_a\}, \qquad \frac{d\mathbf{p}_a}{dt} = -\{H_N, \mathbf{p}_a\}, \qquad a = 1, \ldots, N \tag{I.2}$$

(H_N is the Hamiltonian of the system and $\{ , \}$ denotes the Poisson

---

* This calculation requires the generalization of (I.1) to spatially inhomogeneous systems.

bracket), faced only the simplest features of this dynamical problem (i.e., the two-body problem) but avoided its many-particle aspects by introducing extramechanical, probabilistic assumptions; these assumptions are made at the level of the distribution functions describing the statistical properties of the gas, not on the individual motion of each molecule. Perhaps the irreversible behavior manifested by (I.1), contrasting with the reversible character of the equations of motion (I.2), was due to this unjustified introduction of probability into the realm of dynamics? Despite the undeniable success of the Boltzmann equation, this dispute was hard to settle, because there was no way to put (I.1) on a sound mathematical basis. Worse than this, besides the successful but extremely empirical proposal by Enskog for dense hard spheres,[5] there was no way to generalize the Boltzmann equation to strongly interacting systems.

A major breakthrough occurred around 1950, through the pioneering work of Bogolubov[6] and, mostly, of Van Hove[7] and Prigogine and co-workers. (The crucial role of the limit $N \to \infty$ was stressed in Ref. 8a for an assembly of oscillators, not displaying a true approach to equilibrium, however. The development of the theory in these early days can be found, for example, in Ref. 8b.) In particular, the latter authors were the first to point out that, *provided that one considers the thermodynamic limit*:

$$N \to \infty, \qquad \Omega \to \infty, \qquad \frac{N}{\Omega} = n : \text{finite} \qquad (I.3)$$

and that suitable assumptions are made on the statistical behavior of the system at the initial time $t = 0$, the observable properties of the system could be described by irreversible equations without any violation of the reversible microscopic equations of motion (I.2).

This idea later culminated in the derivation of the so-called generalized kinetic equation,[9] an exact equation that is schematically written as:*

$$\partial_t \varphi_1(\mathbf{v}; t) = -\int_0^t d\tau G[\mathbf{v}; t | \varphi_1(t - \tau)] + \mathcal{D}(\mathbf{v}; t) \qquad (I.4)$$

Here G is a non-Markovian kernel (nonlinear in $\varphi_1(t - \tau)$) that generalizes the Boltzmann operator nC^B; $\mathcal{D}$ is a term describing the effect of the spatial correlations at the initial time; as such, it is a functional of the *initial* ensemble density $\rho_N(\mathbf{r}_1, \mathbf{v}_1, \ldots, \mathbf{r}_N, \mathbf{v}_N; 0)$ (in the limit (I.3)).

As the derivation sketched in Section II shows, this equation is still rather formal in the sense that the operator G and $\mathcal{D}$ depend in a complicated way on the dynamics of the interacting particles and cannot be

---

* We limit ourselves here to classical systems, though the quantum extension of (I.4) poses no difficulty.

written in explicit form. Nevertheless, it is valid for arbitrary interactions and generalizes thus the Boltzmann equation, to which it reduces in the dilute gas limit. However, it is well known that, with strong interactions, the knowledge of the one-particle df does not suffice to calculate all the observable properties. For example, to calculate the potential energy, one also needs to know the pair correlation function defined by

$$g_2(\mathbf{r}_1, \mathbf{r}_2, \mathbf{v}_1, \mathbf{v}_2; t) = f_2(\mathbf{r}_1, \mathbf{r}_2, \mathbf{v}_1, \mathbf{v}_2; t) - n^2 \varphi_1(\mathbf{v}_1; t)\varphi_1(\mathbf{v}_2; t) \qquad \text{(I.5)}$$

(f_2 is the two-particle df), or its Fourier transform $g_{\mathbf{k},-\mathbf{k}}(\mathbf{v}_1, \mathbf{v}_2; t)$. Yet, in the same way as (I.4) is established, one also gets companion equations for the correlations; for example, one gets (see Section II)

$$g_{\mathbf{k},-\mathbf{k}}(\mathbf{v}_1, \mathbf{v}_2; t) = \int_0^t d\tau\, \mathscr{C}_{\mathbf{k},-\mathbf{k}}(\mathbf{v}_1, \mathbf{v}_2; t|\varphi_1(t-\tau)) + \mathscr{P}_{\mathbf{k},-\mathbf{k}}(\mathbf{v}_1, \mathbf{v}_2; t) \qquad \text{(I.6)}$$

where the non-Markovian operator $\mathscr{C}_{\mathbf{k},-\mathbf{k}}$ expresses part of the correlations as a nonlinear functional of the one-particle df at earlier times, whereas $\mathscr{P}_{\mathbf{k},-\mathbf{k}}$ expresses—in analogy with the function $\mathscr{D}$ in (I.4)—the effect of the initial correlations.

In Section II we present a derivation of equations (I.4, 6), and we briefly indicate how they allow us to study the fundamental questions posed by the irreversibility of macroscopic systems, as opposed to the reversible character of the equations of motion (I.2). However, this is a very difficult problem, because, if one wants to escape the type of criticisms raised earlier against Boltzmann's work, no approximation can be made in the analysis of the (exact) equations (I.4 to I.6). We shall therefore limit ourselves to a few qualitative and rather naive comments, referring the reader to other articles in this book[10] for recent progress made in this field.

Besides these general questions, the generalized kinetic theory—especially when (I.4 to I.6) are supplemented by their quantum extension—allows us to attack an extremely wide variety of specific nonequilibrium problems. To mention just a few, let us cite transport phenomena in dense fluids,[11] the dynamics of critical phenomena,[12] instabilities in plasmas,[13] and even heavy ion collisions in nuclear physics, as was started recently.[14] The physics of these problems, all contained in (I.4 to I.6), turns out to be very rich and quite different from the monotonous, exponential-like relaxation described by the Boltzmann equation in the dilute gas limit. The reason is that in most strongly interacting systems, the naive idea that the collision operator describes processes localized in space and in time (over distances of the order of r_0 and over times of the order $\tau_c = r_0/\langle v \rangle$ where $\langle v \rangle$ is some average velocity) is just wrong: The microscopic processes "excited" in the collision process generally involve long-living excitations that propagate over large distances and long times. These excitations completely alter the simple relaxation picture valid at the Boltzmann level. In particular, much of the recent developments in kinetic theory of neutral fluids is based on the so-called mode-mode coupling theory (see, for examples, the reviews of Ref. 11 and 12) in which the collision is described in terms of hydrodynamic modes, defined at the microscopic level. It is true that many of these mode-mode coupling effects have been originally obtained by more phenomenological approaches than that based on a microscopic kinetic theory. Yet the fact remains that the basic justification for these simpler methods ultimately rests on a detailed analysis of the kinetic equations. In Section III we illustrate this point by discussing two examples that recently received the attention of the authors: First, we indicate how one can analyze the long-time behavior of neutral fluids on the basis of the generalized kinetic equation (I.4); second, we show how a proper description of the turbulent behavior of a plasma comes out of the analysis of (I.6) for the pair correlation function. The detailed mathematics of these two problems is quite difficult and awkward; thus, rather than duplicating recent publications, we prefer here to stress the physics involved and the main theoretical ideas used, skipping the technical details.

These two examples, very different by their field of application as well as by their physical consequences, are quite illustrative of the variety of situations that can be handled by present-day kinetic theory.

II. THE GENERALIZED KINETIC EQUATION

The starting point is the Liouville equation

$$i\,\partial_t \rho_N = i\{H_N, \rho_N\} = L_N \rho_N \tag{II.1}$$

where $\rho_N = \rho_N(r, v; t)$ (we use the abbreviation r, v for the Γ-phase space coordinates $\mathbf{r}_1, \ldots, \mathbf{r}_N, \mathbf{v}_1, \ldots, \mathbf{v}_N$) represents the N-particle distribution function of the system.

The information contained in this equation, which gives us the detailed motion of each particle, is, however, much too rich: the physically relevant questions can all be answered if we know the first few *reduced* df only; for example, limiting ourselves to a spatially homogeneous system, the knowledge of both $\varphi_1(\mathbf{v}_1; t)$ and $g_2(\mathbf{r}_1, \mathbf{r}_2, \mathbf{v}_1, \mathbf{v}_2; t)$ suffices to calculate most of the macroscopic properties of the system, like the pressure tensor, the heat flow, and so on.*

---

* Of course, one can imagine physical properties that depend simultaneously on more than two particles; then we would need also three, four, . . . , l-particle distribution functions. The important point is that, in the thermodynamic limit (I.3), l remains finite.

To illustrate how such a reduced description can be extracted from (II.1), let us consider the case of $\varphi_1(\mathbf{v}_1; t)$. It is convenient to work in two steps:

1. First, derive a *formal master equation* for the N-particle velocity distribution function φ_N:

$$\varphi_N(v; t) = \int_\Omega dr\, \rho_N(r, v; t) \tag{II.2}$$

2. Then integrate φ_N over the $(N-1)$ velocities $\mathbf{v}_2, \ldots, \mathbf{v}_N$ to arrive at the *generalized kinetic equation for $\varphi_1(\mathbf{v}_1; t)$*.

The first step is straightforward: We introduce the projection operators P_N and $Q_N \equiv 1 - P_N$, which are such that

$$P_N g \equiv \Omega^{-N} \int_\Omega dr\, g \tag{II.3}$$

$$P_N^2 = P_N, \qquad Q_N^2 = Q_N, \qquad P_N^+ = P_N, \qquad Q_N^+ = Q_N \tag{II.4}$$

We then multiply the Liouville equation (II.1) by P_N and Q_N, respectively, and we obtain two coupled equations for $P_N \rho_N$ and $Q_N \rho_N$:

$$i\, \partial_t(P_N \rho_N) = P_N L_N(P_N \rho_N) + P_N L_N(Q_N \rho_N) \tag{II.5}$$

$$i\, \partial_t(Q_N \rho_N) = Q_N L_N(P_N \rho_N) + Q_N L_N(Q_N \rho_N) \tag{II.6}$$

Inserting the formal solution of (II.6):

$$Q_N \rho_N(t) = \exp(-iQ_N L_N t)(Q_N \rho_N(0))$$

$$-i \int_0^t d\tau \exp(-iQ_N L_N \tau) Q_N L_N(P_N \rho_N(t-\tau)) \tag{II.7}$$

into (II.5) and using the definitions (II.2 and II.3) together with the readily proved property $P_N L_N P_N = 0$, we arrive at the *formal master equation*:

$$\partial_t \varphi_N(v; t) = - \int_0^t d\tau\, G_N(v; \tau) \varphi_N(v; t-\tau) + \mathscr{D}_N(v; t | Q_N \rho_N(0)) \tag{II.8}$$

Here the non-Markovian operator in velocity-space G_N is defined by

$$G_N(v; \tau) = P_N L_N \exp(-iQ_N L_N Q_N \tau) Q_N L_N \tag{II.9}$$

and $\mathscr{D}_N$ is the following function, depending on the initial value of $Q_N \rho_N(0)$:

$$\mathscr{D}_N(v; t | Q_N \rho_N(0)) = -i\Omega^N P_N L_N \exp(-iQ_N L_N t) Q_N \rho_N(0) \tag{II.10}$$

Despite being an exact identity, valid for any finite N, the formal master equation has very unpleasant features when N becomes large: The objects G_N and D_N have a wild N dependence that make them ill defined in the thermodynamic limit (I.3). This point will not be discussed here in any detail (see, for example, Refs. 3 and 8), but its physical origin is clear: The master equation describes the time evolution of the velocities of the N particles and accounts therefore for the interactions inside groups of particles that are arbitrarily far from one another in space: The number of such processes grows without bound when N and $\Omega = N/n$ increase. Step 2—that is, the reduction from φ_N to $\varphi_1(\mathbf{v}_1; t)$—should eliminate this problem: indeed, the evolution of particle 1 is influenced by its interactions with neighboring molecules only, and the number of these neighbors remains finite even if $N \to \infty$. Accordingly, let us define the functions $\mathbb{B}^N$ and $\mathscr{D}^N$ of the variable $\mathbf{v}_1$:

$$\mathbb{B}^N(\mathbf{v}_1; \tau | \varphi_N(t-\tau)) \equiv \int d\mathbf{v}_2 \ldots d\mathbf{v}_N G_N(v; \tau)\varphi_N(v; t-\tau) \quad \text{(II.11)}$$

$$\mathscr{D}^N(\mathbf{v}_1; t | \rho_N(0)) \equiv \int d\mathbf{v}_2 \ldots d\mathbf{v}_N \mathscr{D}_N(v; t | Q_N \rho_N(0)) \quad \text{(II.12)}$$

which appear after integrating (II.8) over $\mathbf{v}_2, \ldots, \mathbf{v}_N$:

$$\partial_t \varphi_1(\mathbf{v}_1; t) = -\int_0^t dt \mathbb{B}^N(\mathbf{v}_1; \tau | \varphi_N(v; t-\tau)) + \mathscr{D}^N(\mathbf{v}_1; t) \quad \text{(II.13)}$$

Under very general conditions, one can show that $\mathbb{B}^N$ and $\mathscr{D}^N$ remain finite when N becomes large. However, a new difficulty appears with (II.13): The integral defining $\mathbb{B}^N(\mathbf{v}_1; \tau | \varphi_N(t-\tau))$ still involves the complete velocity df for all times and (II.13) is not closed! At this stage the thermodynamic limit (I.3) comes to our help. Suppose indeed, that, in the limit, we asymptotically have*

$$\varphi_N(v; t) \simeq \prod_{i=1}^{N} \varphi_1(\mathbf{v}_i; t) \quad \text{(II.14)}$$

Then we immediately arrive at the generalized kinetic equation (I.4) with:

$$G(\mathbf{v}_1, \tau | \varphi_1(t-\tau)) = \lim_{\infty} \mathbb{B}^N\left(\mathbf{v}_1; \tau \middle| \prod_{i=1}^{N} \varphi_1(\mathbf{v}_i; t-\tau)\right) \quad \text{(II.15)}$$

$$\mathscr{D}(\mathbf{v}_1; t) = \lim_{\infty} \mathscr{D}^N(\mathbf{v}_1; t | \rho_N(0)) \quad \text{(II.16)}$$

($\lim_{\infty}$ denotes the thermodynamic limit).

---

* More rigorously, (II.14) should be replaced by the set of equations

$$\lim_{\infty} \varphi_l(\mathbf{v}_1, \ldots, \mathbf{v}_l; t) = \prod_{i=1}^{l} \lim_{\infty} \varphi_1(\mathbf{v}_i; t), \quad l \text{ finite}$$

It can be shown that these weaker statements are sufficient to justify our later claims.

Of course, the asymptotic validity of the molecular chaos assumption (II.14) is far from obvious and is mathematically very delicate. We shall content ourselves here with a few qualitative remarks:

1. Obviously, a factorization property like (II.14) cannot possibly hold in a finite system; there, velocity correlations—due either to the initial conditions or to the dynamics—can generally not be forgotten.
2. At the initial time, molecular chaos appears as a very reasonable *assumption* in the large system limit. Indeed, φ_N is the integral of ρ_N over the whole volume Ω. If we assume that *the initial correlations between the particles extend over finite distance only*—an assumption to be made again below when studying the function $\mathcal{D}$—the weight of these correlated configurations goes to zero in the limit $\Omega \to \infty$, leading to (II.14) for $t = 0$.
3. The last step toward the justification of (II.14) is to show that this property, if valid at $t = 0$, remains valid at later times. This *persistence of molecular chaos* was rigorously proved by Kac on the basis of a model master equation[15] and, more recently, convincingly demonstrated within the present formalism.[16]

The analysis reported above for φ_1 can be extended without difficulty to establish (I.6) for the pair correlation function. From

$$g_2(\mathbf{r}_1, \mathbf{r}_2, \mathbf{v}_1, \mathbf{v}_2; t) \equiv N(N-1) \int d\mathbf{r}_3 \dots d\mathbf{r}_N Q_N \rho_N(r; t) \qquad (\text{II.17})$$

one readily obtains with the help of (II.3, 7, 14)

$$\mathcal{C}_{\mathbf{k},-\mathbf{k}}(\mathbf{v}_1, \mathbf{v}_2; t | \varphi_1(t-\tau)) = \lim (-i) N(N-1) \Omega^{-N}$$
$$\int dr\, e^{i\mathbf{k}\cdot\mathbf{r}_{12}} \exp(-iQ_N L_N t) Q_N L_N \qquad (\text{II.18})$$

$$\mathcal{P}_{\mathbf{k},-\mathbf{k}}(\mathbf{v}_1, \mathbf{v}_2; t) = \lim_{\infty} N(N-1) \int dr\, e^{i\mathbf{k}\cdot\mathbf{r}_{12}} \exp(-iQ_N L_N t) Q_N \rho_N(0) \qquad (\text{II.19})$$

We have thus far derived the equations governing the time evolution of the one- and two-particle df in a spatially homogeneous system. These equations are exact in the thermodynamic limit, provided that the molecular chaos assumption (II.14) is satisfied at $t = 0$, but they are still quite formal: The whole physics of irreversibility has to be extracted from the complicated objects G, $\mathcal{D}$, $\mathcal{C}$, and $\mathcal{P}$. This is a rather formidable and difficult program and an understanding of these quantities is only slowly growing; we refer the reader to other contributions in this book for some of this progress.[10]

Yet, qualitatively at least, not much mystery is left about the general features of the approach to equilibrium. The following comments illustrate this point of view.

Our remarks are based on a rather straightforwardly demonstrated theorem that:

if

$$\mathscr{D}(\mathbf{v}_1; t) \to 0, \qquad \mathscr{P}_{\mathbf{k},-\mathbf{k}}(\mathbf{v}_1, \mathbf{v}_2; t) \to 0 \quad \text{as } t \to \infty \qquad (\text{II.20})$$

then

$$\lim_{t\to\infty} \varphi_1(\mathbf{v}_1; t) = \varphi_1^{\text{eq}}(\mathbf{v}_1), \lim_{t\to\infty} g_{\mathbf{k},-\mathbf{k}}(\mathbf{v}_1, \mathbf{v}_2; t) = g_{\mathbf{k},-\mathbf{k}}^{\text{eq}}(\mathbf{v}_1, \mathbf{v}_2) \qquad (\text{II.21})$$

are long-term solutions of (I.3,5) (φ_1^{eq} and $g_{\mathbf{k},-\mathbf{k}}^{\text{eq}}$, respectively, denote the Maxwellian distribution and the equilibrium pair correlation). It is tempting to *conjecture* that this theorem is a weak form of the much stronger statement: If and *only if* (II.20) holds, then (II.21) are the *only* long-term solutions of (I.3,5). As the functions $\mathscr{D}$ and $\mathscr{P}_{\mathbf{k},-\mathbf{k}}$ depend on the initial ensemble density $\rho_N(t = 0)$, we conclude that *the approach to equilibrium is an exact and ineluctable consequence of dynamics, for that class of initial conditions satisfying* (II.20).

First of all, let us stress that (II.20) is a condition depending on the spatial correlations at the initial time, expressed in $\rho_N(r, v; 0)$. Indeed both the functions $\mathscr{D}$ and $\mathscr{P}$ depend linearly on $Q_N\rho_N(r, v; 0)$ and, in the absence of spatial correlations, we have identically (see II.2 and II.3):

$$\rho_N(r, v; 0) = \Omega^N \varphi_N(v; 0) \qquad (\text{II.22})$$

or

$$Q_N\rho_N(r, v; 0) = 0 \qquad (\text{II.23})$$

Hence

$$\mathscr{D}(\mathbf{v}_1; t) = 0, \qquad \mathscr{P}_{\mathbf{k},-\mathbf{k}}(\mathbf{v}_1; t) = 0 \qquad (\text{II.24})$$

so that (II.20) is automatically satisfied. Besides such uncorrelated initial conditions (which are very unrealistic for any system involving strong interactions), it seems presently impossible to characterize precisely those initial ensemble densities that satisfy (II.20). Nevertheless, model calculations have thrown some light on this problem (see, for example, Ref. 2); for example, for (II.20) to hold true, one always needs the assumption that, loosely speaking, $\rho_N(r, v; 0)$ describes *smooth correlations, of finite extension in space, between the particles that are going to interact at later times.* For a single system (corresponding to the Dirac-delta initial condition, $\rho_N(r, v; t = 0) = \delta(\mathbf{r}_1 - \mathbf{r}_{10}) \cdots \delta(\mathbf{v}_N - \mathbf{v}_{N0})$ this smoothness condition is obviously not satisfied; therefore *a statistical description appears as a necessary ingredient of the theory*. This, however, does not enter in conflict with the dynamical laws (I.2), because statistical assumptions need to be made at one instant of time only. It also puts little limitation on the physical properties that can be described with the theory, because in many particle systems, the individual motions of given particles in a given experiment generally are of no interest and one looks for a more global information, for which statistical predictions are quite sufficient. Let us also notice that when we speak of correlations of finite extension in space, we mean finite compared to the infinitely large dimension of the system: Here, once more, the thermodynamic limit appears as a necessary ingredient of the theory.

This crude analysis also throws some light on the famous reversibility paradox;[17-19] indeed, let us suppose that we start with an initial condition such that (II.20) is valid. As time goes on, the dynamical evolution of the systems brings in new correlations, which are very sensitive to the detailed motion of each molecule; they extend over larger and larger distances, and involve more and more particles. Suppose now that, after an arbitrary long time t_0 [such that, for example, we already have $\mathscr{D}(\mathbf{v}_1; t | \rho_N(0)) \approx 0$ and a similar property for $\mathscr{P}_{\mathbf{k},-\mathbf{k}}(\mathbf{v}_1, \mathbf{v}_2; t | \rho_N(0))$],* we *exactly* invert the velocity of every particle in the system, preserving these very special correlations. The new "initial" state $\rho_N(r, -v; t_0)$ is a very singular phase-space function, involving long-range and highly singular correlations between the particles that are going to interact in the time interval $(t_0, 2t_0)$. Thus $\mathscr{D}(\mathbf{v}_1; t | \rho_N(r, -v; t_0))$ and $\mathscr{P}_{\mathbf{k},-\mathbf{k}}(\mathbf{v}_1, \mathbf{v}_2; t | \rho_N(r, -v; t_0))$ will not vanish in this interval and no approach to equilibrium will be observed; on the contrary, the system will go back to its initial state at $t = 2t_0$, in agreement with the reversibility of the equations of motion (I.2). (See also Ref. 16b.)

Putting these hand-waving arguments on a sound mathematical basis is an ambitious program of nonequilibrium physics.

III. APPLICATIONS OF THE GENERAL THEORY

A. Long Time Behavior in Neutral Fluids

In the early days of modern kinetic theory it was believed that applications of the general theory would involve computational difficulties only. No surprise was expected with the physics! The reasoning was as follows: Suppose first that the physical initial conditions do indeed satisfy the basic

---

* In this paragraph, we explicitly write the functional dependence of $\mathscr{D}$ and $\mathscr{P}_{\mathbf{k},-\mathbf{k}}$ on the initial ensemble density (see (II.16, II.19)).

requirement (II.20). Then, after some transient time, (I.4) reduces to

$$\partial_t \varphi_1(\mathbf{v}_1; t) = -\int_0^t d\tau\, G[\mathbf{v}_1; \tau | \varphi_1(t-\tau)] \tag{III.1}$$

independent of the initial correlations. Then it was argued that the non-Markovian character of this equation is related to the finite duration of the collision process (of the order $\tau_c = r_0/\langle v \rangle$). Thus in the right-hand side of (III.1), the variable τ only extends over times of the order of τ_c and no great error should be made by assuming $\varphi_1(\mathbf{v}_1; t-\tau) \approx \varphi_1(\mathbf{v}_1; t)$, in which case (III.1) reduces to the Markovian equation

$$\partial_t \varphi_1(\mathbf{v}_1; t) = C[\varphi_1(t)] \tag{III.2}$$

with

$$C[\varphi_1(t)] = \int_0^\infty d\tau\, G[\mathbf{v}_1; \tau | \varphi_1(t)] \tag{III.3}$$

If we tentatively write the following density expansion:

$$C[\varphi_1(t)] = nC^{(2)}[\varphi_1(t)] + n^2 C^{(3)}[\varphi_1(t)] + O(n^3) \tag{III.4}$$

it is readily proved that $C^{(2)}$ is identical to the Boltzmann collision operator C^B. Therefore it was supposed that the generalized kinetic equation was describing essentially the same physics as the Boltzmann equation, though valid for arbitrary densities: a monotonous, exponential-like approach to equilibrium. Of course, the parameters characterizing this relaxation should be hard to calculate, but no difficulty of principle was expected.

However, it soon became clear that this naive approach was wrong and that the physics described by (III.1) is much richer than that of simple relaxation. The first spectacular indication of this was the discovery (see the historical review of Ref. 13)* that the collision operator C—and therefore also the transport coefficients—could not be expanded in power of the density, even for particles interacting through short-range forces. As we shall see soon, this crucial remark led to the idea that, in general, collision processes are *not* localized in space and time: the consequence of this observation is a wealth of interesting results,[11] including the now well-known slow power-law decay of the Green–Kubo integrands.

The nonexistence of the density expansion of the Markovian collision operator, due to processes having a very long duration, can be understood from geometric considerations. The simplest example is the three-body process corresponding to the scattering of a light particle by two fixed

---

* As a matter of fact, this calculation was first done within the framework of Bogolubov's theory, which is essentially equivalent to the Markovian approximation (III.2).

scattering centers (I and II) in two dimensions* (two-dimensional Lorentz gas; see Fig. 1).

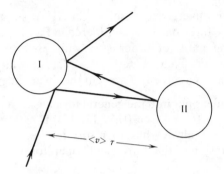

Fig. 1. A diverging three-body process in the two-dimensional Lorentz gas.

Suppose that the collision with I occurs at time 0; the average frequency of this process is $an\langle v\rangle$; in a later time interval $(\tau, \tau + d\tau)$, the probability of a collision with an arbitrary center II is $an\langle v\rangle\, d\tau$; however, if we impose this process to be such that the light particle after its collision with II has to point out into the solid angle under which I is seen from the location of II, this reduces the probability by a factor $a/\langle v\rangle\tau$. Thus the frequency of the three-body process where the intermediate binary collision occurs in the time interval $(\tau, \tau + d\tau)$ is

$$\frac{a^2 n^2 \langle v\rangle\, d\tau}{\tau} \quad (d = 2) \tag{III.5}$$

The total three-body collision frequency—an estimate of $n^2 C^{(3)}$—is obtained by integrating (III.5) over τ (see (III.3)), starting from a lower limit of the order $a/\langle v\rangle$, where our dimensional analysis becomes too rough:

$$n^2 C^{(3)} \approx n^2 a^3 \langle v\rangle \int_{a/\langle v\rangle}^{\infty} d\tau\, \tau^{-1} = \infty \quad (d = 2) \tag{III.6}$$

an expression whose divergence is due to collision processes with an infinitely long duration! We shall not dwell any further here on this divergence of the density expansion of transport coefficients; let us simply make the two following remarks that are helpful to understand how one proceeds in the general case:

1. Because processes of long duration play an important role in the dynamics of dense systems, we are not justified in using the Markovian

---

* The same argument holds in three-dimension for four-particle processes.

approximation (III.3): We have to face the full non-Markovian equation and, in particular, we need to analyze carefully the operator $G[\mathbf{v}_1|\varphi_1(t-\tau)]$.*

2. Of course, the divergence observed in (III.6) should be taken with suspicion: The three particles considered in Fig. 1 are not isolated, and along its long path (of order $\langle v \rangle \tau$ with $\tau \to \infty$), the light particle encounters other fixed centers III, IV,..., which scatter it and presumably act as a screen to prevent the divergence.†

How are these ideas put into the general formalism? The explicit form of the operator $G[\mathbf{v}_1|\varphi_1(t-\tau)]$ (see (II.9, 11, 15)) is extremely complicated, and to study it, one always has to resort to the formal infinite-order perturbation expansion of the "projected" operator of motion:

$$\exp(-iQ_N L_N \tau)$$

$$= \exp(-iQ_N L_N^0 \tau)$$

$$-i \int_0^\tau d\tau' \exp[-iQ_N L_N^0(\tau-\tau')] Q_N \delta L_N \exp(-iQ_N L_N^0 \tau') + \cdots$$

$$= \exp(-iL_N^0 \tau) - i \int_0^\tau d\tau' \exp[-iL_N^0(\tau-\tau')] Q_N \delta L_N \exp(-iL_N^0 \tau') + \cdots$$

$$(\text{III.7})$$

Here we have decomposed the Liouvillian L_N into its kinetic and potential parts, respectively:

$$L_N = L_N^0 + \delta L_N \tag{III.8}$$

$$L_N^0 = -i \sum_{a=1}^N \mathbf{v}_a \cdot \frac{\partial}{\partial \mathbf{r}_a} \tag{III.9}$$

$$\delta L_N = \sum\sum_{a<b} \delta L^{ab} = i \sum\sum_{a<b} \frac{\partial V}{\partial \mathbf{r}_{ab}} \cdot \left(\frac{\partial}{\partial \mathbf{p}_a} - \frac{\partial}{\partial \mathbf{p}_b} \right) \tag{III.10}$$

and we have used the property $Q_N L_N^0 = L_N^0$.

Equation (III.7) involves the free-streaming operator

$$\exp(-iL_N^0 \tau) = \prod_{a=1}^N \exp\left(-\mathbf{v}_a \cdot \frac{\partial}{\partial \mathbf{r}_a} t\right) \tag{III.11}$$

which is related, by a simple Fourier–Laplace transform, to the so-called

---

* One shows that, in general, the correlation term $\mathscr{D}(\mathbf{v}_1; t)$ also decays to zero very slowly with t; it has to be kept in the description if we want quantitative results.

† Yet, except for the Lorentz model discussed earlier, this screening seems insufficient in two-dimensional systems to ensure finite transport coefficients.[11]

one-particle unperturbed propagator:

$$\mathscr{G}_{\mathbf{k}}^{0}(\mathbf{v}_1; z) \equiv \Omega^{-1} \int d\mathbf{r}_1 \int_0^\infty dt\, e^{iz\tau}\, e^{-i\mathbf{k}\cdot\mathbf{r}_1}\, e^{-\mathbf{v}_1\cdot(\partial/\partial\mathbf{r}_1)t}\, e^{i\mathbf{k}\cdot\mathbf{r}_1}$$

$$= \frac{1}{i(\mathbf{k}\cdot\mathbf{v}_1 - z)} \tag{III.12}$$

Combining (II.9) and (III.7, 11, 12), it is easy to understand that the non-Markovian kernel $G(\mathbf{v}_1)$ (we now drop its τ- and φ_1-dependence) can be expressed as a functional of $\mathscr{G}_{\mathbf{k}}^0$ and of δL_N:

$$G(\mathbf{v}_1) = \tilde{G}[\mathbf{v}_1 | \mathscr{G}_{\mathbf{k}}^0, \delta L_N] \tag{III.13}$$

The explicit (perturbative) form of this functional is quite complicated, but we do not need it here; suffice it to say that it represents the dynamics of the collisions in terms analogous to those used in Fig. 1: free motion (straight-line) trajectories interrupted by instantaneous binary interactions.

The preceding argument with the Lorentz gas, however, suggested that physically important collision processes involve the propagation of particles over large distances and long time, or, in Fourier-Laplace language, for small z and $\mathbf{k}$. Clearly, describing the motion of the particles with the help of straight-line trajectories [i.e., with $\mathscr{G}_{\mathbf{k}}^0(\mathbf{v}; z)$] over such scales makes no sense in a dense system and all kind of difficulties [exemplified by the divergence (III.6)] can be expected with such an unrealistic (though formally exact) formalism. What we should rather do is to describe the motion of each particle in terms of a "dressed propagator," which takes into account the presence of the other particles in the fluid. This can be done by a "renormalization procedure," which is schematically depicted in Fig. 2, where free propagation is represented by a thin line and the dots

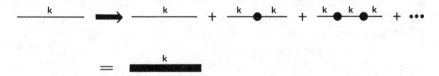

Fig. 2. Schematic representation of the propagator renormalization.

represent collisions with other fluid molecules. As is familiar from field theory (see, for example, Ref. 20), all these contributions can be put together and lead to the dressed propagator $\mathscr{G}_{\mathbf{k}}(\mathbf{v}_1; z)$ represented by the thick line; one finds

$$\mathscr{G}_{\mathbf{k}}(\mathbf{v}_1; z) = \frac{1}{i[\mathbf{k}\cdot\mathbf{v}_1 - z - C_{\mathbf{k}}(\mathbf{v}_1; z)]} \tag{III.14}$$

where $C_\mathbf{k}(\mathbf{v}_1; z)$ generalizes to nonvanishing $\mathbf{k}$ and z the collision-operator (III.3).

It is now possible to rewrite, in an unambiguous way, the non-Markovian kernel $G(\mathbf{v}_1)$ as a new functional of this renormalized propagator

$$G(\mathbf{v}_1) = \overset{z}{\tilde{G}}[\mathbf{v}_1 | \mathcal{G}_\mathbf{k}, \delta L_N] \tag{III.15}$$

To deal with $\overset{z}{\tilde{G}}$ and $C_\mathbf{k}$, we need of course a proper bookkeeping of the infinite number of terms appearing in (III.15) and in the definition of $C_\mathbf{k}$; usually this is done by diagrammatic techniques.

In general, it is almost impossible to extract useful information from (III.15); the main reason is the complication of the exact propagator (III.14). However, if we are confident that low wavenumbers and low frequencies play the major role, the situation drastically simplifies; indeed, one can show (in 3-d at least!)[21] that, for small z and $\mathbf{k}$, $\mathcal{G}_\mathbf{k}(\mathbf{v}_1; z)$ has the following representation:

$$\mathcal{G}_\mathbf{k}(\mathbf{v}_1; z) = \sum_{\substack{\mathbf{k} \to 0 \\ z \to 0}}^{5} \sum_{\alpha=1} |\bar{f}_\mathbf{k}^\alpha\rangle \frac{1}{z + i\Lambda_\alpha^\mathbf{k}} \langle f_\mathbf{k}^\alpha| \tag{III.16}$$

where $\Lambda_\alpha^\mathbf{k}$ represent the *hydrodynamic modes*:

$$\Lambda_{1,2}^\mathbf{k} = \pm ick + \Gamma k^2$$

$$\Lambda_{3,4}^\mathbf{k} = \frac{\eta k^2}{nm} \tag{III.17}$$

$$\Lambda_5^\mathbf{k} = \frac{\kappa k^2}{nmC_p}$$

(c and Γ, respectively, represent the sound velocity and the sound absorption; η is the shear viscosity, κ is the thermal conductivity, and C_p is the specific heat at constant pressure), while the corresponding $|f_\mathbf{k}^\alpha\rangle$, $|\bar{f}_\mathbf{k}^\alpha\rangle$ are simple known functions of the velocity $\mathbf{v}_1$, also depending on the thermodynamic properties of the fluid (like c, C_p, etc., ...). Physically, (III.16) expresses the fact that, in the long term and for long wavelengths, the only surviving excitations in a fluid are the collective hydrodynamic modes, corresponding to the conserved quantities.

Once (III.16) is inserted in (III.15), we get a "mode-mode coupling" description of the collision operator, in which the hydrodynamic part of the propagators $\mathcal{G}_\mathbf{k}$ interact through the δL_N's. This description should validly apply to the long time behavior of the kernel G.

It is out of place to show in detail here how the analysis can be pursued.[11,22,23] Suffice it to say that one can show that, for large times, the dominant contribution to G comes from the contribution schematically depicted in Fig. 3; we first have a "bubble" representing complicated short-range processes, from which two hydrodynamic propagators emerge, with wave numbers $\mathbf{k}$ and $-\mathbf{k}$, and these propagators interact again at time τ. As one can prove that the "bubbles" have a finite limit for $\mathbf{k} \to 0$, it is readily seen from (III.16) that, for $d = 3$:

$$G[\mathbf{v}_1; t|\varphi_1(t)] \sim \sum_\alpha \sum_\beta \int_0^\infty dk\, k^2 \exp\left[-(\Lambda_\alpha^{\mathbf{k}} + \Lambda_\beta^{-\mathbf{k}})t\right] \sim t^{-3/2}, \qquad t \to \infty$$

(III.18)

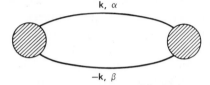

k, α

$-\mathbf{k}$, β

Fig. 3. The dominant mode-mode contribution to G for long times.

The proportionality coefficient in (III.18) cannot be calculated in closed form, but, remarkably enough, one can show that this coefficient is not needed if one wants to compute the behavior of the Green–Kubo integrands from the present kinetic theory formalism.

For example, one finds for the normalized velocity autocorrelation function[23]

$$\Gamma(t) \approx \frac{2}{3n} \frac{1}{[4\pi(D + \eta/nm)t]^{3/2}}, \qquad t \to \infty$$

(III.19)

where D is the self-diffusion coefficient. This result completely agrees with the phenomenological theories[24] and gives them a microscopic foundation.

We shall not dwell any further here on the many applications of the mode-mode coupling theory,[11] which gave rise to a spectacular development in nonequilibrium statistical physics, both from a fundamental and a practical point of view. Let us simply mention here that similar considerations apply to (I.6), describing the pair correlation. To show the richness of the theory, we now discuss, however, this equation for a completely different system, namely, a turbulent plasma.

B. Microstructure of a Turbulent Plasma

One of the most intriguing features of the plasmas of thermonuclear interest is the very weak role played by individual particle collisions. In the relevant conditions of density and temperature, the mean free path of the particles, evaluated on the basis of the appropriate generalization of the Boltzmann equation,[25] may become much longer than the size of the reaction vessel. On the other hand, experiment shows that the motion and evolution of a plasma is governed by laws that are dissipative. One must therefore find an explanation for the dissipation on the basis of mechanisms quite different both from the individual particle collisions (as in Boltzmann's picture) and from the coupling between the hydro-dynamical modes, which, in a neutral fluid, are the only excitations that are only weakly affected by these collisions and therefore long living (as in the mode-mode coupling picture).

Because of the long range of the Coulomb interactions, between charged particles, there exists, however, in a plasma a wide variety of relatively long-living collective excitations, of a quite different type from the hydro-dynamical modes discussed in neutral fluids. The overall dissipation originates now in the interactions between such collective modes, their mutual transformation, their decay, and their growth. A plasma whose evolution is governed by such collective mechanisms is said to be in a *turbulent* state, a rather vague name that is not quite suitable but is of common practice. When viewed from the particle point of view, the existence of collective excitations necessarily implies the existence of large-amplitude and long-range correlations.

A physical effect that has been pretty much discussed in the recent literature is the existence of a spatial microstructure in a turbulent plasma. Numerical simulations[26] very clearly show that such a system has a "granular" appearance. These granules have been called clumps. This concept was introduced simultaneously by Dupree[26] and by Kadomtsev and Pogutse.[27] If such microstructures exist, they would seriously influence the values of the transport coefficients. This can be understood by a very rough argument:[28] If the particles coalesce into macroparticles, the number of such entities in a Debye sphere goes down and therefore the plasma parameter (which is the inverse of that number) goes up: The plasma is more strongly coupled than expected. Although this argument must be considered with care, it is pretty clear that the clumps could provide one mechanism (among many others) by which one would understand the anomalies of the diffusion coefficients, heat conduction, and so on, that one observes both in magnetically confined plasmas (e.g., tokomaks) and in laser-produced plasmas.

A missing point in the initial theoretical treatments of this idea was a clear picture of the mechanism of generation of such correlations. One could not assume that these were simply initial correlations propagated by freely moving particles, because such correlations would die out very quickly, as follows from the discussions of the previous paragraphs. On the other hand, the mechanism of "creation" of correlations from φ_1, appearing in (I.6), is not easy to understand in absence of pure collisions.

Nevertheless, it was shown in a recent work[13] that a careful analysis of the equations of evolution by means of the modern techniques of kinetic theory leads to the conclusion that formula (I.6) is valid even in a "collisionless" plasma. In such a plasma, the particles are indirectly correlated through the long-range interactions with the fluctuating electric field pervading the medium. The "signature" of the turbulence is the fact that all quantities—in particular, the correlations— are functionals of the spectrum of fluctuations of the field, which must be self-consistently related to the particle distribution function by the Poisson equation. Let us further mention that, in agreement with the general discussion of the previous sections, the validity of (I.6) is again subject to an assumption about the initial conditions. These must be of a "self-propagating" type, which means that they are of the type produced by the intrinsic internal interactions within the plasma.

The correlation function (I.6) can be evaluated explicitly in the lowest nontrivial approximation. We cannot explain in detail here the exact nature of this approximation.[13] Let us simply say that the propagators are renormalized [pretty much in the spirit of (III.14)], and moreover the two-body propagators must exhibit (to lowest order) the "trajectory-correlation" property [i.e., they should not be factorizable into a product of two (renormalized) one-body propagators]. In this lowest approximation, the correlation function can be expressed as follows (in the case of a homogeneous and quasistationary turbulence):

$$
g(\mathbf{r}, \mathbf{v}_1, \mathbf{v}_2; t) = \int d\mathbf{k} \int_0^\infty d\tau \exp\left[i\mathbf{k} \cdot \langle \mathbf{r}(t-\tau) \rangle\right]
$$
$$
\times \exp\left[-\tfrac{1}{2}\mathbf{k}\mathbf{k} : \langle \tilde{\mathbf{r}}(t-\tau)\tilde{\mathbf{r}}(t-\tau) \rangle\right] d_{\mathbf{k}}(\mathbf{v}_1, \mathbf{v}_2)
$$
$$
\times \frac{\partial}{\partial \mathbf{v}_1} \cdot \frac{\partial}{\partial \mathbf{v}_2} \varphi_1(\mathbf{v}_1; t)\varphi_1(\mathbf{v}_2; t) \qquad \text{(III.20)}
$$

Here $\varphi_1(\mathbf{v}; t)$ is the one-particle distribution function, which obeys a kinetic equation consistent with the present scheme of approximation: We shall, however, not consider it explicitly here. $d_{\mathbf{k}}(\mathbf{v}_1, \mathbf{v}_2)$ is a diffusion tensor, which is a linear functional of the fluctuation spectrum $S_{\mathbf{k}}$:

$$
S_{\mathbf{k}}(\tau) = \frac{e^2}{8\pi^3 m^2} \int d\mathbf{r}\, e^{-i\mathbf{k}\cdot\mathbf{r}} \langle \mathbf{E}'(\mathbf{x}; t)\mathbf{E}'(\mathbf{x}-\mathbf{r}; t-\tau) \rangle \qquad \text{(III.21)}
$$

where $\mathbf{E}'(\mathbf{x}; t)$ is the fluctuating self-consistent field at point $\mathbf{x}$ and at time t (for a homogeneous quasistationary turbulence, S is independent of $\mathbf{x}$ and t). The interesting feature of (III.20) is the appearance of $\langle \mathbf{r}(t-\tau) \rangle$ and $\langle \tilde{\mathbf{r}}(t-\tau)\tilde{\mathbf{r}}(t-\tau) \rangle$, which are, respectively, the average distance and the mean square distance fluctuation $[\tilde{\mathbf{r}}(t) = \mathbf{r}(t) - \langle \mathbf{r}(t) \rangle]$ at time $t - \tau$, for two particles that are at distance $\mathbf{r}$ and have relative velocity $\mathbf{g} = \mathbf{v}_1 - \mathbf{v}_2$ at time t. The evaluation of the correlation function is thus reduced to a diffusion problem in the turbulent plasma.

If trajectory correlation effects are neglected, one finds for long times τ, in the limit of small initial distance and relative velocity

$$\langle \tilde{r}^2(t-\tau) \rangle^0 = \tfrac{1}{3}\lambda_0^2 \left(\frac{\tau}{\tau_0}\right)^3 \tag{III.22}$$

where λ_0 is a characteristic length of the turbulence and τ_0 the corresponding time. The behavior in τ^3 (which is quite different from the τ behavior in simple diffusion problems) is known as "Dupree damping." If, however, trajectory correlations are included, one finds in the same limit:

$$\langle \tilde{r}^2(t-\tau) \rangle = \tfrac{1}{6}\left(r^2 + \frac{g^2\tau^2}{10}\right)\left(\frac{\tau}{\tau_0}\right)^3 \tag{III.23}$$

Thus the turbulent dispersion tends toward zero for very small initial distances and relative velocities. In this simple fact appears clearly the origin of the clumps. Because of the impeded diffusion of particles that are initially close together in phase space, these particles tend to remain together for a very long time: They form a clump.

In a typical thermonuclear plasma in which λ_0 is of the order of 10^3 Debye lengths, the characteristic dimension of a reasonably long-living clump is of the order of several tens of Debye lengths. For a detailed study of the properties of the clumps we refer the reader to the original literature.[13,26,27]

The examples sketched here were chosen among many possible others in order to illustrate the power and usefulness of the modern methods of kinetic theory in the study of numerous widely different problems.

References

1. L. Boltzmann, *Wien. Ber.*, **66,** 275 (1872); see also L. Boltzmann, *Vorlesungen über Gastheorie*, Leipzig, 1896 (English transl., *Lectures on Gas Theory*, S. Brush, transl., University of California Press, Berkeley, 1964).
2. R. Balescu, *Equilibrium and Nonequilibrium Statistical Mechanics*, Wiley-Interscience, New York, 1975; P. Résibois and M. De Leener, *Classical Kinetic Theory of Fluids*, Wiley-Interscience, New York, 1977.
3. P. and T. Ehrenfest, *Enzyklopaedie math. Wiss.*, Vol. IV, Pt. 32, Leipzig, 1911.
4. S. Brush, *Kinetic Theory*, Pergamon, Oxford, Vol. 1, 1965; Vol. 2, 1966; Vol. 3, 1972 (historical account).
5. D. Enskog, Kungl. Sv. Vetenskapsakad. Handb. 63, n° 4 (1922).
6. N. Bogolubov: *Problemy Dinamicheskoy Teorii v Statisticheskoy Fizike* (Moscow, 1946) (English translation: "*Problems of a Dynamical Theory in Statistical Physics,*" Vol. I (J. de Boer and G. Uhlenbeck, eds.), (North-Holland, Amsterdam 1972)).

7. L. Van Hove, *Physica*, **21**, 517 (1955); **23**, 441 (1957).
8. (a) G. Klein and I. Prigogine, *Physica*, **19**, 1053 (1953); (b) I. Prigogine, *Nonequilibrium Statistical Mechanics*, Wiley-Interscience, New York, 1962, R. Balescu, *Statistical Mechanics of Charged Particles*, Wiley-Interscience, New York, 1963; P. Résibois, in E. Meeron, Ed., *Many Particle Physics*, Gordon & Breach, New York, 1967, and the references quoted there.
9. I. Prigogine and P. Résibois, *Physica*, **27**, 629 (1961); R. Zwanzig, in *Lectures in Theoretical Physics*, Vol. 3 (Summer Inst. Theor. Phys., Univ. Colorado, 1960), Wiley-Interscience, New York 1961.
10. A. P. Grecos, in this volume.
11. Y. Pomeau and P. Résibois, *Phys. Rep.*, **19C**, 64 (1975), and references quoted there.
12. P. Résibois, in L. Garrido, J. Biel, and J. Rae, Eds., *Irreversibility and the Many-Body Problem*, Plenum, New York, 1973, and references quoted there.
13. J. Misguich and R. Balescu, two papers to appear in *Phys. Fluids*; two papers to appear in *Plasma Phys.*
14. W. Nörenberg, *Phys. Lett.*, **52B**, 289 (1974).
15. M. Kac, in *Probability Theory and Related Topics in Physical Science*, Wiley-Interscience, New York 1959.
16. (a) P. Clavin, *C.R. Acad. Sci. Paris*, **274**, 1085 (1972); (b) P. Résibois and M. Maréschal, *Physica* (to appear in 1978).
17. I. Prigogine and P. Résibois, in *Atti del Simposio Lagrangino"*, *Acc. Sc. Torino*, 1964.
18. R. Balescu, *Physica*, **36**, 433 (1967).
19. S. Brush (1972), quoted in Ref. 4.
20. A. Abrikosov, L. Gorkov, and I. Dzyaloshinsky, *Methods of Quantum Field Theory in Statistical Physics*, Prentice-Hall, Englewood Cliffs, N.J., 1963.
21. P. Résibois, *J. Statist. Phys.*, **2**, 21 (1970).
22. P. Résibois, *Physica*, **70**, 413 (1973); P. Résibois and Y. Pomeau, *Physica*, **72**, 493 (1973); M. Theodosopulu and P. Résibois, *Physica*, **82A**, 47 (1976).
23. J. Dorfman and E. Cohen, *Phys. Rev. Lett.*, **25**, 1257 (1970); *Phys. Rev.*, **A6**, 776 (1972); *Phys. Rev.*, **A12**, 292 (1975).
24. B. Alder and T. Wainwright, *Phys. Rev.*, **A1**, 18 (1970); M. Ernst, E. Hauge, and J. Van Leeuwen, *Phys. Rev. Lett.*, **25**, 1254 (1970); *Phys. Rev.*, **A4**, 2055 (1971).
25. R. Balescu (1963), quoted in Ref. 8.
26. T. H. Dupree, *Phys. Fluids*, **15**, 334 (1972); B. H. Hui and T. H. Dupree, *Phys. Fluids*, **18**, 235 (1975); T. H. Dupree, C. E. Wagner, and W. M. Manheimer, *Phys. Fluids*, **18**, 1167 (1975).
27. B. B. Kadomtsev and O. P. Pogutse, *Phys. Rev. Lett.*, **25**, 1155 (1970).
28. F. Engelmann and T. Morrone, *Comments Plasma Phys. Controlled Fusion*, **1**, 75 (1972).

KINETIC THEORY OF PLASMAS

YU. L. KLIMONTOVICH

Moscow State University, Moscow, USSR

CONTENTS

I. KINETIC THEORY OF PLASMAS

The modern kinetic theory of plasmas is based on the well-known work of N. Bogolubov, M. Born, A. Vlasov, H. Green, J. Yvon, J. Kirkwood, L. Landau, and I. Prigogine. In this paper we summarize some new results obtained in the last few years, also based on the classical contributions just mentioned. The first important step was taken by R. Balescu and A. Lenard, who obtained the kinetic equation for a plasma including polarization effects. These papers ended an important stage in the construction of the kinetic equation for a fully ionized plasma. We discuss here only three problems.

1. The kinetic theory of nonideal plasmas.
2. The kinetic theory of fluctuations.
3. The kinetic equation for a partially ionized plasma including the role of inelastic processes.

At present, three kinetic equations are used in the theory of plasmas: the Boltzmann equation, the Landau equation, and the Balescu–Lenard equation. These equations can be written in the following form:

$$\left(\frac{\partial}{\partial t}+\mathbf{v}\,\frac{\partial}{\partial \mathbf{z}}+\mathbf{F}_a\,\frac{\partial}{\partial \mathbf{p}}\right)f_a(\mathbf{z},\mathbf{p},t)=I_a(\mathbf{z},\mathbf{p},t) \tag{I.1}$$

where f_a is the one-particle distribution function, a is the component

193

index, I_a is the collision integral, and F_a is the average force. We write the Boltzmann integral in the form proposed by Bogolubov, namely,

$$I_a(x, t) = \sum_b n_b \int \frac{\partial \phi_{ab}}{\partial \mathbf{z}} \frac{\partial}{\partial \mathbf{p}} f_a(\mathbf{z}, \mathbf{P}(-\infty), t) f_b(\mathbf{z}, \mathbf{P}'(-\infty), t) \, dx',$$

$$x = (\mathbf{z}, \mathbf{p}) \tag{I.2}$$

where $\mathbf{P}(-\infty)$, $\mathbf{P}'(-\infty)$ are the initial momenta of the two particles that collide at time t, and ϕ_{ab} is the interaction energy of the particles. The Landau collision integral is

$$I_a(x, t) = \sum_b 2e_a^2 e_b^2 n_b \frac{\partial}{\partial p_i} \int \frac{k_i k_j}{k^4} \delta(\mathbf{kv} - \mathbf{kv'})$$

$$\times \left(\frac{\partial}{\partial p_j} - \frac{\partial}{\partial p_j'} \right) f_a(\mathbf{z}, \mathbf{p}, t) f_b(\mathbf{z}, \mathbf{p'}, t) \, d\mathbf{k} \, d\mathbf{p'} \tag{I.3}$$

where the integration over the vector $\mathbf{K}$ is performed using

$$\frac{1}{z_{\min}} \equiv k_{\max} > k > k_{\min} \equiv \frac{1}{Z_D} \tag{I.4}$$

and Z_D is the Debye radius. The Balescu–Lenard collision integral is

$$I_a(x, t) = \sum_b 2e_a^2 e_b^2 n_b \frac{\partial}{\partial p_i} \int \frac{k_i k_j}{k^4} \frac{\delta(\mathbf{kv} - \mathbf{kv'})}{|\mathscr{E}(\omega, \mathbf{k})|^2}$$

$$\times \left(\frac{\partial}{\partial p_j} - \frac{\partial}{\partial p_j'} \right) f_a(\mathbf{z}, \mathbf{p}, t) f_b(\mathbf{z}, \mathbf{p'}, t) \, d\mathbf{k} \, d\mathbf{p'} \tag{I.5}$$

This expression differs from the Landau integral as it accounts for the polarization of the plasma. The dielectric constant is determined by the expression

$$\mathscr{E}(\omega, \mathbf{k}) = 1 + \sum_a \frac{4\pi e_a^2 n_a}{k^2} \int \frac{\mathbf{k}(\partial f_a / \partial \mathbf{p})}{\omega - \mathbf{kv} + i\Delta} \, d\mathbf{p}, \qquad \Delta \to 0 \tag{I.6}$$

The three collision integrals displayed have the following common properties.

1. The collision integrals vanish when the Maxwell distribution

$$f_a(\mathbf{p}) = \frac{1}{(2\pi m_a k_B T)^{3/2}} \exp \left[-\frac{\mathbf{p}^2}{2 m_a k_B T} \right] \tag{I.7}$$

is inserted into them.

2. The integral

$$I(\mathbf{z}, t) = \sum_a n_a \int \varphi_a(\mathbf{p}) I_a(x, t) \, d\mathbf{p} = 0$$

when

$$\varphi_a = 1, \mathbf{p}, \frac{\mathbf{p}^2}{2m_a} \qquad (I.8)$$

3. The gas or plasma entropy given by

$$S(\mathbf{z}, t) = -k_B \sum_a n_a \int \ln f_a \cdot f_a \, d\mathbf{p} \qquad (I.9)$$

does not decrease with time. That is

$$\frac{dS(t)}{dt} \geq 0, \qquad S(t) = \int S(\mathbf{z}, t) \, d\mathbf{z} \qquad (I.10)$$

In the Boltzmann, Landau, and Balescu–Lenard equations the correlations between particles are not completely taken into account. These correlations are found to influence the dissipative processes, but they make no contribution to the thermodynamic functions of the plasma (Chaps. 2, 9, 10 in Ref. 11). This is seen, in particular, from (I.9) for the entropy, which is valid only for the ideal gas and ideal plasma (Section 14 in Ref. 11.)

Let us examine the assumptions under which it is possible to obtain the kinetic equation (I.1) with collision integrals (I.2), (I.3), (I.5). We begin by discussing these assumptions for an ordinary gas.

1. The Boltzmann equation for a monoatomic gas corresponds to the first term in the density expansion, that is, to the binary collision approximation. In this approximation the first two equations of the BBGKY hierarchy become closed equations for the distribution functions f_a and f_{ab}.

2. The first additional assumption used in deriving the Boltzmann equation from the equations for the functions f_a and f_{ab} is the loss of initial correlations, suggested by Bogolubov. However, in the derivation of the kinetic equation one can neglect only those correlations for which the correlation times τ_{COR} are less than τ_{COL}. The assumed loss of the initial correlations corresponds to saying that the long-lived correlations (with $\tau_{COR} > \tau_{COL}$) play no essential role in the kinetic theory.

3. The second assumption used is the complete neglect of temporal delay, that is, $f_a(t-\tau) \to f_a(t)$. In the same approximation we can neglect the spatial variation of the functions f_a over distances of the order of z_0.

4. Usually (but indirectly) it is assumed that the collision dynamics that determines the collision integral is continuous. This implies neglect of random interactions between particles in the infinitely small volume l_{ph}^3. The volume l_{ph}^3 is defined in the following manner (Sec. 16 in Ref. 11):

$$nl_{ph}^3 \sim \frac{1}{\sqrt{\mathscr{E}}}, \qquad \mathscr{E} = nz_0^3 \ll 1,$$

where $\mathscr{E}$ is the reduced density, and z_0 is the interaction radius. From this definition it follows that $l_{ph} \sim z_{av}/\sqrt{\mathscr{E}} \gg z_{av}$. But $l_{ph} \sim \sqrt{\mathscr{E}} l_{path} \ll l_{path} \equiv l_{COL}$ (the mean free path), which implies that the number of particles in the physically infinitesimal volume l_{ph}^3, namely, $n l_{ph}^3$, is so big that it is possible to neglect fluctuations. In this limit the distribution function in the kinetic equation is not a random function.

Using for the distribution function f_{ab} an expression that takes into account terms linear in the reduced density, we can obtain a more accurate expression for the collision integral. This was done by Choh and Uhlenbeck.[4,11] It is possible to construct a kinetic equation for a nonideal gas in the triple-collision approximation (Chap. 4 in Ref. 11).

It appears, superficially, that there exists a method of constructing a kinetic equation for a gas in any approximation (i.e., to any power in the density). In each level of approximation only one assumption is made, namely, loss of the initial correlations on a time scale much shorter than τ_{COL}. However, fundamental difficulties are encountered when it comes to realization of this program. The investigations carried out by Weinstock, Goldman, and Frieman and by Dorfman and Cohen have shown that the contributions made to the collision integral by the second and higher terms in the density expansion contain divergent integrals. This shows that the construction of the kinetic equation for a dense gas by direct utilization of the method of successive approximation in powers of the density is impossible. Note, however, that the divergences can be eliminated by taking into account the contribution of the most divergent diagrams that are produced when expansion in terms of the density is used (Kawasaki, Oppenheim).

The problem of constructing a kinetic equation for a dense gas can be solved in another manner, without summing divergent diagrams in a density expansion. To do so, however, it is necessary to dispense with the condition that the initial correlations vanish for $t > \tau_{COL}$ (Chap. 3 in Ref. 11).

In the kinetic theory of plasmas we make use of the polarization approximation (the first-order term in the plasma parameter)

$$g_{ab} \ll f_a f_b, \qquad g_{abc} = 0$$

where now

$$l_{ph} \sim z_D, \qquad \tau_{ph} \sim \frac{1}{\omega_L}, \qquad n l_{ph}^3 \sim \frac{1}{\mu} \gg 1$$

and μ is the plasma parameter. We define as small scale those fluctuations for which

$$z_{COR} < l_{ph} \ll l_{COL}, \qquad \tau_{COR} < \tau_{ph} \ll \tau_{COL}$$

and shall define as large scale those for which

$$z_{COR} > l_{ph} \sim z_D, \qquad \tau_{COR} > \tau_{ph} \sim \frac{1}{\omega_L}$$

The condition that the initial correlations vanish corresponds to the assumption that the large-scale correlations (with $z_{COR} > z_D$) play no essential role in the kinetic theory of plasma.

In the general case there is only partial loss of correlation in the plasma. The large-scale fluctuations do not attenuate rapidly and should, consequently, be taken into account when a kinetic theory is constructed.

This article is not intended to give a detailed discussion of the whole range of plasma kinetic problems. Our main purpose is to display some of the resources of the modern kinetic theory and to demonstrate how all results can be derived from the microscopic plasma equations.

There is a brief list of references at the end of the article. We have included, for the most part, books dealing with the problems in question. In these books one can find a more complete discussion of the kinetic theory of plasmas.

II. KINETIC EQUATIONS FOR IDEAL AND NONIDEAL PLASMAS

To start, we use the closed system of equations[10]

$$\frac{\partial N_a}{\partial t} + \mathbf{V} \frac{\partial N_a}{\partial \mathbf{z}} + e_a \left(\mathbf{E}^M(\mathbf{z}, t) + \frac{1}{c} [\mathbf{v} \mathbf{B}^M(\mathbf{z}, t)] \right) \frac{\partial N_a}{\partial \mathbf{p}} = 0$$

$$\text{rot } \mathbf{B}^M = \frac{1}{c} \frac{\partial \mathbf{E}^M}{\partial t} + \frac{4\pi}{c} \sum_a e_a \int \mathbf{v} N_a(x, t) \, d\mathbf{p}$$

$$\text{rot } \mathbf{E}^M = -\frac{1}{c} \frac{\partial \mathbf{B}^M}{\partial t} \tag{II.1}$$

$$\text{div } \mathbf{B}^M = 0$$

$$\text{div } \mathbf{E}^M = 4\pi \sum_a e_a \int N_a(x, t) \, d\mathbf{p}$$

for the microscopic phase densities

$$N_a(x, t) = \sum_{1 \le i \le N} \delta(x - x_{ia}(t)), \qquad x = (\mathbf{z}, \mathbf{p})$$

of each component of the plasma and the microscopic electric and magnetic field strengths $\mathbf{E}^M$, $\mathbf{B}^M$.

For given experimental conditions these functions are random functions.

For the Coulomb plasma the system of equations for the random functions N_a, $\mathbf{E}^M$ is simpler:

$$\frac{\partial N_a}{\partial t} + \mathbf{v}\frac{\partial N_a}{\partial \mathbf{z}} + e_a\mathbf{E}^M\frac{\partial N_a}{\partial \mathbf{p}} = 0$$

$$\text{rot } \mathbf{E}^M = 0, \qquad \text{div } \mathbf{E}^M = 4\pi \sum_a e_a \int N_a(x, t)\, d\mathbf{p} \tag{II.2}$$

We shall now study the equations for the Coulomb plasma.

Direct averaging of (II.2) gives for the first moments

$$\bar{N}_a = n_a f_a, \qquad \mathbf{E} = \overline{\mathbf{E}^M} \tag{II.3}$$

where f_a is the one-particle distribution function. Using the definition

$$\overline{\mathbf{E}^M N_a} = \mathbf{E} n_a f_a + \overline{\delta\mathbf{E}\,\delta N_a} \tag{II.4}$$

we can write the following equations for f_a and E:

$$\left(\frac{\partial}{\partial t} + \mathbf{V}\frac{\partial}{\partial \mathbf{z}} + e_0\mathbf{E}\frac{\partial}{\partial \mathbf{p}}\right)f_a = -\frac{e_a}{n_a}\frac{\partial}{\partial p}\overline{\delta N_a\,\delta\mathbf{E}} \equiv I_a(x, t) \tag{II.5}$$

$$\text{rot } \mathbf{E} = 0, \qquad \text{div } \mathbf{E} = 4\pi \sum_a e_a n_a \int f_a\, d\mathbf{p} \tag{II.6}$$

The collision integral I_a is representable in the form

$$I_a = -\frac{e_a}{n_a}\frac{1}{(2\pi)^4}\frac{\partial}{\partial \mathbf{p}}\int Re\,(\delta N_a\,\delta\mathbf{E})_{\omega,\mathbf{k},\mathbf{z},\mathbf{p},t}\, d\omega\, d\mathbf{k} \tag{II.7}$$

from which we see that the collision integral is expressed in terms of the correlation of the fluctuations of the phase density and the electric field.

Consider now the kinetic equations for ideal and nonideal plasmas in the polarization approximation.

For the ideal plasma it is possible to use assumptions 2 to 4. In this case the collision integral is the Balescu–Lenard integral (I.5). We shall use another form of this collision integral, namely,

$$I_a(\mathbf{z}, t) = \frac{e_a^2}{16\pi^3}\frac{\partial}{\partial p_i}\int\frac{k_i}{k^2}\delta(\omega - \mathbf{kv})\left[(\delta\mathbf{E}\,\delta\mathbf{E})_{\omega,\mathbf{k}}\frac{\partial f_a}{\partial \mathbf{p}} + \frac{8\pi Im\mathscr{E}(\omega, \mathbf{v})}{|\mathscr{E}(\omega, \mathbf{k})|^2}f_a\right]d\omega\, d\mathbf{k} \tag{II.8}$$

where

$$(\delta\mathbf{E}\,\delta\mathbf{E})_{\omega,\mathbf{k}} = \sum_b\frac{(4\pi)^2 e_b n_b}{k^2}\int\frac{2\pi\delta(\omega - \mathbf{kv})f_b\, d\mathbf{p}}{|\mathscr{E}(\omega, \mathbf{k})|^2}$$

$$(\delta\mathbf{E}\,\delta\mathbf{E})_k = \frac{1}{2\pi}\int(\delta\mathbf{E}\,\delta\mathbf{E})_{\omega,\mathbf{k}}\, d\omega \tag{II.9}$$

are the spectral densities of the fluctuations of the electric field.

The dielectric constant is determined by (I.6). Equation (II.8) assumes the following form in the equilibrium state:

$$(\delta \mathbf{E}\, \delta \mathbf{E})_{\omega,\mathbf{k}} = \frac{8\pi}{\omega} \frac{Im\, \mathscr{E}(\omega, \mathbf{k})}{|\mathscr{E}(\omega, \mathbf{k})|^2} k_B T \qquad (II.10)$$

The kinetic equation for a plasma with the collision integral (I.5) or (II.8) does not fully account for the interactions between particles. The interactions determine the dissipative processes, for example, the process of establishing equilibrium. But from (I.8) it follows, in particular, that the total energy is not conserved in the process of establishing the equilibrium state; only the kinetic energy of the particles is conserved. The equation of state is that of the ideal plasma.

If we take into account retardation effects and the nonuniformity in the collision integral, it is possible to obtain a kinetic equation for the nonideal plasma (Chaps. 9 and 10 in Ref. 11). In the kinetic equation for a nonideal plasma the properties (I.8) hold only when $\varphi_a = 1$. When $\varphi_a = p^2/2m_a$, the case of a spatially uniform plasma, we have

$$\sum_a n_a \int \frac{\mathbf{p}^2}{2m_a} I_a(\mathbf{z}, \mathbf{p}, t)\, d\mathbf{p} = -\frac{\partial}{\partial t} \int \frac{(\delta \mathbf{E}\, \delta \mathbf{E})_{\omega,\mathbf{k}}}{8\pi} \frac{d\omega\, d\mathbf{k}}{(2\pi)^4} \qquad (II.11)$$

which properly accounts for conservation of total energy:

$$\frac{\partial}{\partial t}\left[\sum_a n_a \int \frac{\mathbf{p}^2}{2m_a} f_a\, d\mathbf{p} + \int \frac{(\delta \mathbf{E}\, \delta \mathbf{E})_{\omega,\mathbf{k}}}{8\pi} \frac{d\omega\, d\mathbf{k}}{(2\pi)^4}\right] = 0 \qquad (II.12)$$

The spectral density is determined by (II.9).

III. INCLUSION OF THE AVERAGED DYNAMICAL POLARIZATION IN THE BOLTZMANN KINETIC EQUATION FOR A NONIDEAL PLASMA

The Boltzmann collision integral for a Coulomb plasma contains a divergence at large distances, and the Balescu–Lenard integral diverges at small distances. In many papers (see Section 56 in Ref. 11) different forms of the collision integral simultaneously taking into account binary collision processes and polarization processes have been proposed.

The simplest form proposed is a combination of three integrals: the Boltzmann I_a^B, Landau I_a^L, and Balescu–Lenard I_a^{BL} integrals:

$$I_a = I_a^B - I_a^L + I_a^{BL} \qquad (III.1)$$

In this expression the integral I_a^L compensates for the divergence of the Boltzmann integral at large distances and the divergence of the integral I_a^{BL} at small distances. Such a generalization, although attractive because of its relative simplicity, is not completely satisfactory, since the thermodynamic functions of the nonideal plasma (Sec. 56 in Ref. 11).

We shall consider another model, in which the dynamical character of the plasma polarization is taken into account approximately.

The collision integral (II.7) can be written in the form

$$I_a = \sum_b n_b \frac{\partial}{\partial p_i} \int k_i \, \Phi_{ab}(\mathbf{k}) \, Im \, g_{ab}(\mathbf{k}, \mathbf{p}, \mathbf{p}', t) \frac{d\mathbf{k}}{(2\pi)^3} \, d\mathbf{p}'$$

$$\Phi_{ab}(\mathbf{k}) = \frac{4\pi e_a e_b}{k^2}$$

(III.2)

If we substitute into this the solution of the equation for the correlation function $g_{ab}(\mathbf{k}, \mathbf{p}, \mathbf{p}', t)$ in the polarization approximation, we obtain the expression (I.5). The expression in the integrand in (I.5) is proportional to the square of $\Phi_{ab}^2(\mathbf{k})$. One of the factors in (III.2) remains unchanged; the second changes when the polarization is taken into account:

$$\Phi_{ab}(\mathbf{k}) \rightarrow \frac{\Phi_{ab}(\mathbf{k})}{|\mathscr{E}(\omega, \mathbf{k})|^2}$$

(III.3)

We shall take the averaged effect of the dynamical polarization into account in the following way. In place of (III.3) we shall use the following effective potential:

$$\tilde{\Phi}_{ab}(\mathbf{k}) = \frac{\Phi_{ab}(\mathbf{k})}{\sum_d e_d^2 n_d} \sum_c e_c^2 n_c \int \frac{f_c(\mathbf{p})}{|\mathscr{E}(\mathbf{k}\mathbf{v}, \mathbf{k})|^2} \, d\mathbf{p} = \frac{e_a e_b}{\sum_c e_c^2 n_c} \frac{(\delta \mathbf{E} \, \delta \mathbf{E})_\mathbf{k}}{4\pi}$$

(III.4)

We use here the expression (II.9) for the spectral density of the field fluctuations.

In the local equilibrium approximation

$$\tilde{\Phi}_{ab}(\mathbf{k}) = \frac{\Phi_{ab}(\mathbf{k})}{1 + z_D^2 k^2} \qquad \text{and} \qquad \tilde{\Phi}_{ab}(\mathbf{z}) = \frac{e_a e_b}{z} e^{-z/z_D}$$

(III.5)

We can use now the following system of equations for the distribution functions:

$$\left(\frac{\partial}{\partial t} + \mathbf{v} \frac{\partial}{\partial \mathbf{z}} + e_a \mathbf{E} \frac{\partial}{\partial \mathbf{p}}\right) f_a = \sum_b n_b \int \frac{\partial \Phi_{ab}}{\partial \mathbf{z}} \frac{\partial g_{ab}(x, x', t)}{\partial \mathbf{p}} \, dx' \equiv I_a(x, t)$$

$$\left(\frac{\partial}{\partial t} + \mathbf{v} \frac{\partial}{\partial \mathbf{z}} + \mathbf{v}' \frac{\partial}{\partial \mathbf{z}'} - \frac{\partial \tilde{\Phi}_{ab}}{\partial \mathbf{z}} \frac{\partial}{\partial \mathbf{p}} - \frac{\partial \tilde{\Phi}_{ab}}{\partial \mathbf{z}'} \frac{\partial}{\partial \mathbf{p}'}\right) f_{ab}(x, x', t)$$

$$= \left(\frac{\partial}{\partial t} + \mathbf{v} \frac{\partial}{\partial \mathbf{z}} + \mathbf{v}' \frac{\partial}{\partial \mathbf{z}'}\right) f_a(x, t) f_b(x', t)$$

where

$$f_{ab} = f_a f_b + g_{ab} \qquad \text{(III.6)}$$

The collision integral can be displayed as the sum of three terms:

$$I_a = I_{a(1)} + I_{a(2)} + I_{a(3)} \qquad \text{(III.7)}$$

The expression

$$I_{a(1)} = \sum_b n_b \int \frac{\partial \phi_{ab}}{\partial \mathbf{z}} \frac{\partial}{\partial \mathbf{p}} f_a(\mathbf{z}, \tilde{\mathbf{P}}(-\infty), t) f_b(\mathbf{z}, \tilde{\mathbf{P}}'(-\infty), t) \, d\mathbf{z}' \, d\mathbf{p}' \qquad \text{(III.8)}$$

corresponds to (I.2), but here $\tilde{P}(-\infty)$, $\tilde{P}'(-\infty)$ are the initial momenta of the particles interacting under $\tilde{\Phi}_{ab}$.

$$I_{a(2)} = -\sum_b n_b \int_0^\infty \int \frac{\partial \Phi_{ab}}{\partial \mathbf{z}} \frac{\partial}{\partial p} \left(\frac{\partial}{\partial t} + \frac{(\mathbf{v} + \mathbf{v}')}{2} \frac{\partial}{\partial \mathbf{z}} \right) \tau \frac{\partial}{\partial \tau} f_a(\mathbf{z}, \tilde{\mathbf{P}}(-\tau), t)$$
$$\times f_b(\mathbf{z}, \tilde{\mathbf{P}}'(-\tau), t) \, d\tau \, d\mathbf{z}' \, dp' \qquad \text{(III.9)}$$

$$I_{a(3)} = -\sum_b \frac{n_b}{2} \int \frac{\partial \phi_{ab}}{\partial \mathbf{z}} \frac{\partial}{\partial \mathbf{p}} \left(z_{ab} \frac{\partial}{\partial \mathbf{z}} \right) [f_a(\mathbf{z}, \tilde{\mathbf{P}}(-\infty), t) f_b(\mathbf{z}, \tilde{\mathbf{P}}'(-\infty), t)$$
$$- f_a(\mathbf{z}, \mathbf{p}, t) f_b(\mathbf{z}, \mathbf{p}', t)] \, d\mathbf{z}' \, d\mathbf{p}' \qquad \text{(III.10)}$$

The complete collision integral (III.7) has the properties (I.8) only for $\varphi_a = 1$. For $\varphi_a = p_i$ we have

$$\sum_b n_a \int p_i I_a \, d\mathbf{p} = -\sum_a \frac{\partial}{\partial z_j} \Delta P_{ij}^a \qquad \text{(III.11)}$$

where ΔP_{ij}^a is the contribution, because of the interaction between particles to the stress tensor P_{ij}^a.

In the state of local equilibrium we have

$$\Delta P_{ij}^a = \delta_{ij} \, \Delta p^a$$
$$\Delta p_a = -\sum_a \frac{2\pi n_a n_b}{3} \int_0^\infty z \Phi_{ab}' \, e^{-\Phi_{ab}/k_B T} z^2 \, dz \qquad \text{(III.12)}$$

For $\varphi_a = p^2/2m_a$ in the local equilibrium approximation

$$\sum_a n_a \int \frac{p^2}{2m_a} I_a \, dp = -\sum_a \left(\frac{\partial \Delta U_a}{\partial t} + \frac{\partial}{\partial \mathbf{z}} [\mathcal{U}(\Delta U_a + \Delta p_a)] \right) \qquad \text{(III.13)}$$

where Δp_a is defined by (III.12) and

$$\Delta U_a = \sum_a 2\pi e_a^2 n_a \int_0^\infty \Phi_{ab} \, e^{-\Phi_{ab}/k_B T} z^2 \, dz \qquad \text{(III.14)}$$

is the contribution, resulting from the interactions between particles to the internal energy density. The full energy density is

$$U = \sum_a n_a \tfrac{3}{2} k_B T + \tfrac{1}{2} \sum_{a,b} n_a n_b \int \Phi_{ab}(z) f_{ab}(z)\, dz \qquad (\text{III.15})$$

where

$$f_{ab}(z) = \exp\left[-\frac{e_a e_b}{2 k_B T} \exp\left(-\frac{z}{z_0} \right) \right] \qquad (\text{III.16})$$

At short distances this function goes over into the distribution function of a Boltzmann gas and at large distances it coincides with the Debye distribution function, corresponding to the first approximation in the plasma parameter μ. The corresponding limiting properties also hold for the thermodynamic functions.

Allowance for the average dynamic polarization and for the nonideality effects leads to a change in the kinetic coefficients. It can be shown[11] that the average dynamic polarization leads to a change in the form of the Coulomb logarithm. The effective electric field E_{eff} has been calculated (see Ref. 11). Both effects decrease the value of the electrical conductivity σ. It is possible that this explains why the experimental conductivities are lower than those calculated by Spitzer's theory.

IV. KINETIC THEORY OF FLUCTUATIONS IN A GAS AND A PLASMA

We recall that in the derivation of the kinetic equation the condition of loss of initial correlations is imposed. Actually, as already noted, only a partial decay of small-scale correlations occurs. Thus in the derivation of the kinetic equation it is actually assumed that large-scale fluctuations do not play any role in the kinetic theory.

The large-scale fluctuations, in the general case, do not have time to attenuate within the relaxation time of the distribution functions. As a result, the BBGKY hierarchy does not lead to the kinetic equation when account is taken of the large-scale fluctuations. In this case we obtain only a system of equations for the functions f_a and the correlation functions of the large-scale fluctuations $\tilde{g}_{ab}$, $\tilde{g}_{abc}$. This system of equations, in the simplest case, can be replaced approximately by the Langevin equations for random distribution functions with random sources $y_a(x, t)$. Such a replacement naturally raises the problem of determining the statistical characteristics of the random source. For a gas at equilibrium this problem was first solved by Kadomtsev (see Ref. 11).

For the development of the kinetic theory of fluctuations we can start with the system of equations for the large-scale random phase densities $\tilde{N}_a$ and large-scale random functions $\tilde{\mathbf{E}}$, $\tilde{\mathbf{B}}$. For an ideal plasma this system is (Section 61 in Ref. 11)

$$\left(\frac{\partial}{\partial t}+\mathbf{v}\frac{\partial}{\partial \mathbf{z}}+e_a\left(\tilde{\mathbf{E}}+\frac{1}{c}[\mathbf{v}\tilde{\mathbf{B}}]\right)\frac{\partial}{\partial \mathbf{p}}\right)\tilde{N}_a(x,t)$$

$$=\sum_b\int\frac{\partial\Phi_{ab}}{\partial\mathbf{z}}\frac{\partial}{\partial\mathbf{p}}\tilde{N}_a(\mathbf{z},\tilde{\mathbf{P}}(-\infty),t)\tilde{N}_b(\mathbf{z},\tilde{\mathbf{P}}'(-\infty),t)\,dx' \qquad \text{(IV.1)}$$

$$\operatorname{rot}\tilde{\mathbf{B}}=\frac{1}{c}\frac{\partial\tilde{\mathbf{E}}}{\partial t}+\frac{4\pi}{c}\sum_a e_a\int\mathbf{v}\tilde{N}_a\,d\mathbf{p}$$

$$\operatorname{rot}\tilde{\mathbf{E}}=-\frac{1}{c}\frac{\partial\tilde{\mathbf{B}}}{\partial t}$$

$$\operatorname{div}\tilde{\mathbf{B}}=0 \qquad \text{(IV.2)}$$

$$\operatorname{div}\tilde{\mathbf{E}}=4\pi\sum_a e_a\int\tilde{N}_a\,d\mathbf{p}$$

The right-hand side of (IV.1) describes the dissipative processes that are defined by the small-scale fluctuations.

The averaging of (IV.1) and (IV.2) gives equations for the first moments:

$$\bar{\tilde{N}}=n_a f_a, \qquad \bar{\tilde{\mathbf{E}}}=\mathbf{E} \qquad \text{(IV.3)}$$

Using the definition

$$\langle\tilde{N}_a\tilde{\mathbf{E}}\rangle=n_a f_a\mathbf{E}+\langle\delta\tilde{N}_a\,\delta\mathbf{E}\rangle$$

we can write for the Coulomb plasma the following equations:

$$\left(\frac{\partial}{\partial t}+\mathbf{v}\frac{\partial}{\partial \mathbf{z}}+e_a\mathbf{E}\frac{\partial}{\partial \mathbf{p}}\right)f_a=I_a+\tilde{I}_a$$

$$\operatorname{rot}\mathbf{E}=0 \qquad \text{(IV.4)}$$

$$\operatorname{div}\mathbf{E}=\sum_a 4\pi e_a n_a\int f_a\,d\mathbf{p}$$

The collision integral I_a for the ideal plasma is defined by (III.8) or by (I.3). In (IV.4)

$$\tilde{I}_a=-\frac{e_a}{n_a}\frac{\partial}{\partial \mathbf{p}}\overline{\delta\tilde{N}_a\,\delta\tilde{\mathbf{E}}} \qquad \text{(IV.5)}$$

is the part of the collision integral that is determined by large-scale fluctuations.

In the polarization approximation for large-scale fluctuations,

$$\tilde{g}_{ab} \ll f_a f_b, \qquad \tilde{g}_{abc} = 0$$

and the equations for the random functions $\delta \tilde{N}_a$, $\delta \tilde{E}$ become

$$\left(\frac{\partial}{\partial t} + \mathbf{v}\frac{\partial}{\partial \mathbf{z}} + e_a \mathbf{E}\frac{\partial}{\partial \mathbf{p}} + \delta \hat{I}_a\right)\delta \tilde{N}_a + e_a\, \delta \tilde{E}\frac{\partial n_a f_a}{\partial \mathbf{p}} = y_a(x, t)$$

(IV.6)

$$\operatorname{rot} \delta \mathbf{E} = 0, \qquad \operatorname{div} \delta \mathbf{E} = 4\pi \sum_a e_a \int \delta \tilde{N}_a\, d\mathbf{p}$$

The operator $\delta \hat{I}_a$ is defined by the expression

$$\delta \hat{I}_a F_a(\mathbf{p}) = -\sum_b n_b \int \frac{\partial \Phi_{ab}}{\partial \mathbf{z}}\frac{\partial}{\partial \mathbf{p}}\,[F_a(\check{\mathbf{P}}(-\infty))f_b(z, \check{\mathbf{P}}'(-\infty), t)$$

$$+ f_a(\mathbf{z}, \check{\mathbf{P}}(-\infty), t)F_b(\check{\mathbf{P}}'(-\infty))]\, dz'\, dp'$$

(IV.7)

The moments of the random source are

$$\langle y_a \rangle = 0, \qquad \langle y_a(x, t)y_b(x', t)\rangle = A_{ab}(x, x', t)\,\delta(t - t')$$

(IV.8)

The intensity of the random source can be displayed in the form of a sum of two parts:

$$A_{ab} = A_{ab}^B + \tilde{A}_{ab}$$

(IV.9)

The first part of (IV.9) is defined by small-scale fluctuations (by the collision integral I_a^B)

$$A_{ab}^B(x, x', t) \equiv A_{ab}(\mathbf{z}, t, \mathbf{p}, p')\,\delta(\mathbf{z} - \mathbf{z}')$$

$$= [(\delta \hat{I}_a + \delta \hat{I}_b) - (\delta \hat{I}_a + \delta \hat{I}_b)_0]n_a\,\delta_{ab}\,\delta(\mathbf{z} - \mathbf{z}')\,\delta(\mathbf{p} - \mathbf{p}')f_a(\mathbf{z}, \mathbf{p}, t)$$

(IV.10)

The subscript 0 in the second term means that operators act only on the functions f_a, but not on the δ functions. The result (IV.10) is for the case that the number of particles in the volume $l_{ph}^3 \sim z_D^3$ is not infinite.

In the equilibrium state Kadotzev's result follows (IV.10):

$$A_{ab}(x, x') = (\delta \hat{I}_a + \delta \hat{I}_b)n_a\,\delta_{ab}\,\delta(x - x')f_a(\mathbf{p})$$

(IV.11)

The second term in (IV.9) is determined by the collision integral $\tilde{I}_a$

$$\tilde{A}_{ab}(x, x', t) = n_a\,\delta_{ab}[\delta(x - x')\tilde{I}_a(x, t) - \frac{1}{V}(\tilde{I}_a(x, t)f_b(x', t) + \tilde{I}_b(x', t)f_a(x, t))]$$

(IV.12)

We see that in the general case the intensity of the Langevin source is not determined by the one-particle distribution function f_a. Indeed, in the general case characteristic relaxation times for distribution functions and fluctuations are of the same order. Therefore without simplifications it is impossible to express the collision integral $\tilde{I}_a$ in terms of distribution functions f_a. This is possible, however, if the kinetic stage of the relaxation process is finished and the system is near the local equilibrium state. In this case it is possible to obtain the following expression for the collision integral $\tilde{I}_a$ (Sec. 64 in Ref. 11)

$$\tilde{I}_a(x, t) = \frac{e_a^2}{(2\pi)^4} \frac{\partial}{\partial p_i} \int \frac{K_i}{K^2} \operatorname{Re} i \hat{L}_{a,\omega,\mathbf{K},\mathbf{p}}^{-1} \otimes \left[(\delta \mathbf{E} \, \delta \mathbf{E})_{\omega,\mathbf{K}} \mathbf{K} \frac{\partial f_a}{\partial \mathbf{p}} \right.$$
$$\left. + \frac{8\pi \operatorname{Im} \tilde{\mathscr{E}}(\omega, \mathbf{K})}{|\mathscr{E}(\omega, \mathbf{K})|^2} f_a(x, t) \right] d\omega \, d\mathbf{K} \tag{IV.13}$$

The operator $\hat{L}^{-1}$ is defined by the expression

$$\hat{L}_{a,\omega,\mathbf{K},\mathbf{p}}^{-1} = \frac{1}{(\omega - \mathbf{K}\mathbf{V} + i \, \delta \hat{I}_a)} \tag{IV.14}$$

$\tilde{\mathscr{E}}(\omega, \mathbf{K})$ is the dielectric constant for the large-scale fluctuations

$$\tilde{\mathscr{E}}(\omega, \mathbf{K}) = 1 + \sum_a \frac{4\pi e_a^2 n_a}{K^2} \int \hat{L}_{a,\omega,\mathbf{K},\mathbf{p}}^{-1} \mathbf{K} \frac{\partial f_a}{\partial \mathbf{p}} \, d\mathbf{p} \tag{IV.15}$$

The spectral density of the large-scale fluctuations of the electric field is defined by the expression

$$(\delta \tilde{\mathbf{E}} \, \delta \tilde{\mathbf{E}})_{\omega,\mathbf{K}} = \frac{\sum_a ((4\pi)^2 e_a^2 n_a / K^2) 2 \operatorname{Re} \int i \hat{L}_{a,\omega,\mathbf{K},\mathbf{p}}^{-1} f_a \, d\mathbf{p}}{|\tilde{\mathscr{E}}(\omega, \mathbf{K})|^2} \tag{IV.16}$$

The integration over the values of the vector $\mathbf{K}$ in (IV.13) is performed using

$$K < \frac{1}{z_D} \tag{IV.17}$$

If

$$\frac{1}{(\omega - \mathbf{K}\mathbf{V} + i \, \delta \hat{I}_a)} \to \frac{1}{(\omega - \mathbf{K}\mathbf{V} + i \, \Delta)}, \qquad \Delta \to 0 \tag{IV.18}$$

the expression (IV.13) coincides (for $K < 1/z_D$) with the Balescu–Lenard collision integral in the form (II.8).

Thus the collision integral for the ideal plasma is the sum of two terms:

$$I_a = I_a^B + \tilde{I}_a$$

The first term is defined by the small-scale fluctuations with $z_{COR} < z_D$. It can be presented in the form of a Boltzmann collision integral taking into account the polarization effects, or in the Balescu–Lenard form. The second term is defined by the large-scale fluctuations.

V. THE KINETIC EQUATION FOR A PARTLY IONIZED PLASMA WITH INELASTIC PROCESSES TAKEN INTO ACCOUNT

We have considered the kinetic equations for a gas and a fully ionized plasma. These are the two limiting cases of the more general partly ionized plasma. The kinetic theory of a partly ionized plasma is much more complicated than what we have considered so far. This is due to the need to take into account intramolecular motions, and ionization and recombination processes. At the present time only the kinetic theory of an ideal partly ionized plasma has been developed.

We have described elsewhere[15] (see Chap. 14 in Ref. 11) a method for describing nonequilibrium processes in a partly ionized plasma. The starting equations introduce four operator density matrices that describe the states of free and bound charged particles as well as the transition from bound states to free states and vice versa. These equations are considered simultaneously with those for the microscopic electromagnetic field strengths, and one obtains the kinetic equations for the electron, ion, and atom distribution functions in which inelastic processes are taken into account.

The collision integrals I_a for the electrons and ions $(a = e, i)$, and for the atoms, can be presented as the sum of two terms:

$$I_a(\mathbf{p}_a, t) = [I_a(\mathbf{p}_a, t)]_1 + [I_a(\mathbf{p}_a, t)]_2 \qquad (V.1)$$

The first term is defined by the expression

$$[I_a(\mathbf{p}'_a, t)]_1 = \frac{e_a^2}{(2\pi)^3 \hbar} \int d\mathbf{p}''_a \, d\omega \, d\mathbf{K}$$

$$\times \frac{1}{K^2} \delta(\hbar \mathbf{K} - [\mathbf{p}'_a - \mathbf{p}''_a]) \delta\left(\hbar\omega - \left[\frac{\mathbf{p}'^2}{2m_a} - \frac{\mathbf{p}''^2}{2m_a}\right]\right)$$

$$\times \left\{ (\delta\mathbf{E}\,\delta\mathbf{E})_{\omega,\mathbf{K}} [f_a(\mathbf{p}''_a, t) - f_a(\mathbf{p}'_a, t)] \right.$$

$$\left. - \frac{4\pi\hbar \,\mathrm{Im}\,\mathscr{E}(\omega, \mathbf{K})}{|\mathscr{E}(\omega, \mathbf{K})|^2} [f_a(\mathbf{p}''_a, t) + f_a(\mathbf{p}'_a, t)] \right\} \qquad (V.2)$$

The spectral density of the field fluctuations and the dielectric constant are the sums of four terms:

$$(\delta\mathbf{E}\,\delta\mathbf{E})_{\omega,\mathbf{K}} = (\quad)^{ff} + (\quad)^{fb} + (\quad)^{bf} + (\quad)^{bb} \qquad (V.3)$$

$$\mathscr{E}(\omega, \mathbf{K}) = 1 + 4\pi(\alpha_{ff} + \alpha_{fb} + \alpha_{bf} + \alpha_{bb}) \qquad (V.4)$$

The indices f and b denote the free and bound states. Equation V.2 also can be presented as the sum of four terms that describe the same four processes.

The collision integral (V.2) has the property

$$\frac{V}{(2\pi\hbar)^3} \int [I_a(p_a, t)]_1 \, dp_a = 0 \qquad (V.5)$$

so that the concentrations c_a are unchanged by the relaxation processes.

The second term in (V.1) is defined by the expression

$$[I_a(\mathbf{p}_a, t)]_2 = \frac{1}{(2\pi)^3} \frac{V}{\hbar(2\pi\hbar)^3} \sum_m \int d\mathbf{p}_b' \, dP'' \, d\omega \, d\mathbf{K} \frac{1}{K^2} \left| \frac{P m_b \mathbf{p}_a' - m_a \mathbf{p}_b'}{m_a + m_b}, m \right|^2$$

$$\otimes \, \delta(\hbar\mathbf{K} - [\mathbf{p}_a' + \mathbf{p}_b' - \mathbf{P}'])\delta\left(\hbar\omega - \left[\frac{\mathbf{p}_a'^2}{2m_a} + \frac{\mathbf{p}_b'^2}{2m_b} - E_m - E\mathbf{p}'' \right] \right)$$

$$\otimes \left[(\delta\mathbf{E} \, \delta\mathbf{E})_{\omega, \mathbf{K}}(f_m(\mathbf{P}'', t) - N f_a(\mathbf{p}_a', t) f_b(\mathbf{p}_b', t)) \right.$$

$$\left. - \frac{4\pi\hbar \operatorname{Im} \mathscr{E}(\omega, \mathbf{K})}{|\mathscr{E}(\omega, \mathbf{K})|^2}(f_m(\mathbf{P}'', t) + N f_a(\mathbf{p}_a, t) f_b(\mathbf{p}_b', t)) \right] \qquad (V.6)$$

Here the $P_{\alpha\beta}(K)$ are matrix elements

$$P_{\alpha\beta}(K) = \int \left[e_e \exp\left(i \frac{m_i}{m_e + m_i} \mathbf{K}\mathbf{z} \right) \right.$$

$$\left. + e_i \exp\left(-i \frac{m_e}{m_e + m_i} \mathbf{K}\mathbf{z} \right) \right] \Psi_\alpha^*(\mathbf{z}) \Psi_\beta(\mathbf{z}) \, d\mathbf{z}$$

and the Ψ_α are the eigenfunctions of the isolated atom; $\alpha = n$ for the discrete spectrum and $\alpha = \mathbf{p}$ for the continuous spectrum.

The integral (V.6) also is describable in terms of four processes, but now

$$\frac{V}{(2\pi\hbar)^3} \int [I_a(\mathbf{p}_a, t)]_2 \, d\mathbf{p}_2 \neq 0 \qquad (V.7)$$

so that these processes do change the concentrations. For the fully ionized plasma the second term of the collision integral $[I_a]_2 = 0$ and the first term $[I_a]_1$ is the same as the Balescu–Lenard quantum collision integral.

The integral I_n in the kinetic equation for atoms also can be presented in the form of a sum of two terms:

$$I_n(\mathbf{P}', t) = [I_n(\mathbf{p}', t)]_1 + [I_n(\mathbf{p}', t)]_2 \qquad (V.8)$$

Each term describes four processes. For the first term

$$\sum_n \int [I_n(\mathbf{P}', t)]_1 \frac{V}{(2\pi\hbar)^3} \, d\mathbf{P}' = 0 \qquad (V.9)$$

but for the second term

$$\sum_n \int [I_n(\boldsymbol{P}', t)]_2 \frac{V}{(2\pi\hbar)^3} d\boldsymbol{P}' \neq 0 \qquad (V.10)$$

The properties of the collision integrals are such that they ensure conservation of the total number of charged particles, total momentum, and energy. The Maxwell–Boltzmann distribution is the equilibrium solution of the kinetic equation under condition of chemical (ionization) equilibrium.

The approximation in which the kinetic equation has been derived corresponds to the Born approximation for elastic and inelastic processes, providing that the atomic motions and the polarization of the plasma have been taken into account.

The polarization of the plasma arises from four processes: the motion of free charged particles, variation of the internal states of the atoms, transitions of charged particles from the bound states to free ones, and transitions of charged particles from the free states to bound ones.

In Ref. 15 (see also Section 84 in Ref. 11) we obtained the kinetic equations for the electron, ion, and atom distribution functions of a partly ionized plasma, in which all processes due to interaction of particles with a transverse electromagnetic field are taken into account. The kinetic equations take into account all the usual processes of photoionization, photorecombination, emission, absorption, and so on, and also take into account all the so-called anomalous effects. These include, besides the Cerenkov effect, the anomalous Doppler effect, anomalous *bremsstrahlung*, effects of spontaneous and stimulated ionization, recombination, and so on. The pressing problem that remains is to develop the kinetic theory of a nonideal partly ionized plasma.

References

1. N. N. Bogolubov, *Problems of a Dynamical Theory in Statistical Mechanics*, Moscow, 1946.
2. A. A. Vlasov, *Zh. Eksp. Teor. fiz.*, **8**, 291 (1938); *Theory of Many Particles*, Gordon & Breach, New York, 1950.
3. M. Born and H. Green, *Proc. Roy. Soc.*, **A188**, 10 (1946).
4. G. E. Uhlenbeck and G. W. Ford, *Lectures in Statistical Mechanics*, American Mathematical Society, Providence, R.I.
5. I. R. Prigogine, *Nonequilibrium Statistical Mechanics*, Interscience, New York, 1963.
6. L. D. Landau, *Zh. Eksp. Teor. Fiz.*, **7**, 203, 1937.
7. L. Kadanoff and G. Baym, *Quantum Statistical Mechanics*, Benjamin, New York, 1962.
8. R. Balescu, *Phys. Fluids*, **4**, 85 (1960).
9. A. Lenard, *Ann. Phys.*, **10**, 390, 1960.
10. Yu. L. Klimontovich, *Statistical Theory of Nonequilibrium Processes in Plasma*, Moscow, 1964; Pergamon, New York, 1966.

11. Yu. L. Klimontovich, *Kinetic Theory of the Nonideal Gas and of Nonideal Plasma*, Moscow, 1975, Pergamon, 1978.
12. V. P. Silin, *Introduction to the Kinetic Theory of Gases*, Moscow, 1971.
13. J. H. Fertiger and H. G. Kaper, *Mathematical Theory of Transport Processes in Gases*, Amsterdam–London, 1972.
14. G. Ecker, *Theory of Fully Ionized Plasmas*, New York, 1972.
15. Yu. L. Klimontovich, *Zh. Eksp. Teor. Fiz.*, **52,** 1233 (1967) (*Sov. Phys. JETP*, **25,** 820 (1967)); *Zh. Eksp. Teor. Fiz.*, **54,** 136 (1967) (*Sov. Phys. JETP*, **27,** 75 (1968)).

HOW DOES INFORMATION ORIGINATE? PRINCIPLES OF BIOLOGICAL SELF-ORGANIZATION*

MANFRED EIGEN

CONTENTS

Abstract

Information as a structural correlate of function originates in a dynamical process of self-evaluation. The prerequisites for such self-organization can be derived from suitable game models. The degree to which these prerequisites for self-organization can be fulfilled by defined material systems is investigated, and their adequacy as explanatory models is discussed. Two essential principles of organization are derived.

1. The principle of natural selection, as formulated by Darwin, is established as an "extremal principle," valid for material systems showing self-, complementary, or cyclic reproduction under suitable boundary conditions. The amount of information that can be collected in such a system is limited by the quality of information transfer, determined by the physical parameters of the system.
2. The hypercyclic integration of replicative systems is demonstrated. This yields new mechanisms for the expansion of information capacity.

The importance of both these principles for evolution from molecular to cellular systems is examined using realistic examples. Conclusions are drawn whose relevance should extend beyond the confines of molecular biology.

I. INTRODUCTION

In our everyday speech, the concept *information* is associated with the content and meaning of an item of news. To inform somebody means to

* Translation by Christopher J. Holloway, based on a paper in *Berichte der Bunsen-Gesellschaft* **80**(11), 1059–1081 (1976).

supply him with knowledge. For the person informed, the news takes "gestalt," in the psychological sense.[1] That is, in fact, the linguistic root of the word *information*.

The exact sciences have adopted the concept of information and have assigned to it an absolute meaning, free of subjective values.[2,3] Here information is a uniquely defined measure of probability, which specifies how many yes-no decisions one needs on average in order to be able to achieve a definite result within a given number of alternatives. Starting from this number of alternatives, one could designate information as a measure of complexity. If one thinks, on the other hand, of the choice available, the complementary notation, measure of scarcity, or surprise would be apt.

What actually separates this absolute concept of information from its semantic pendant? In principle, can the meaning of the content of an item of news not be made objective? The question of such an objective evaluation of content and meaning of news is posed, when we wish to understand *how* information—such as the blueprint of a living entity or speech-transfixed thought—can originate in a "natural" way.

Let us for a moment consider information that "originates" through the deciphering of a piece of news. The news specialist starts decoding from the properties of the language; that is, from the frequency of use of certain linguistic symbols, from the probability of their sequence, from statistics of word length, the rules of syntax, and other objective assumptions. Actually, there is no sharp transition from objective to subjective. Thus it is possible to establish objectively some predictions of the sense of the news—at least, that which presents a single meaning. Finally, what happens exclusively in the brain of the news receiver is subjective, dependent on his previous history, from which, however, his completed experience and the possibilities of evaluation of his organ of thought are "objectively" derivable, and which (at least in principle) can be represented by some type of probability distribution. Of course, we must view here the system, news *and* receiver, as a nonseparable unit. The origin of information is an environment-drawn process, dependent on certain starting and boundary conditions; in fact, a self-organizational process, which occurs in the brain of a person, namely, the receiver of the news.

Is the self-organizational generation of information an accidental, historically unique coincidence of events, or is it a "regularity among events"?[4] Only then would we be justified in making it the object of physical consideration. The underlying "physical" problem associated with this question is, above all, complexity. It is essential to determine whether a mechanism could legitimately bring about a defined choice, in spite of the fact that the feasible states of matter constitute such a vast number that the expected value for the emergence of a *particular* state within the spatial and temporal limits of our real world is a priori practically zero.

Let us now examine some examples of complexity. A cubic centimeter of a rarefied gas contains about 10^{18} molecules at room temperature and 1 to 30 atmospheres pressure. Ludwig Boltzmann[5] has estimated the time required for the reproduction of a particular microstate. He defined here the microstate by the stipulation of a spatial position for each individual molecule to within 10 Å, and its speed within 1 to 3% of each momentary value. Boltzmann reckoned this (Poincaréic) recurrence time to be $10^{10^{19}}$ years. This figure is of a dimension beyond the bounds of our imagination and lies outside any physical materialization. A detailed knowledge of the microstructure of the gas is required for a macroscopic state only insofar as one can derive predictions from it, by an assessment of statistical weights with respect to the average state and mean deviations. Quantum mechanics, which makes no distinction between particles of the same type in the same quantum state, has altered the assessment of the microstate quantitatively. Inasmuch as it concerns a measure of complexity, and thereby the effective uniqueness of each microstate within our spatial and temporal borders, Boltzmann's considerations are nevertheless definitely relevant.

A further example is represented by the snowflakes depicted in Fig. 1. The effect of the diverse structural details attracted the attention of Thomas Mann.[6] In *The Magic Mountain* he writes, "Gems, diadems, diamond brooches—the most skilled jeweller could have created no richer and more delicate work . . . , and in all the myriads of enchanting stars with their secret miniature splendour, too small for the naked eye of man to see, was not one like unto another." That is in fact correctly observed. If one wanted to know precisely how many detailed structures there really are, one would have to determine how many of the approximately 10^{18} H_2O molecules in a snowflake are incorrectly oriented and in how many ways these misorders can be distributed within the lattice. The result is a number that absolutely supports Thomas Mann's statement. Nevertheless, the microscopic detail is, to a certain extent, portrayed in the macroscopic. That we can observe this multitude of forms lies in the fact that during crystal growth, a definite "historical" misordered proportion of the lattice was "frozen," thereby determining the macroscopic picture.

Such a "historically" stipulated detail of the microscopic confronts us even more obviously in the macrostructure of a living entity. The genetic pattern of a colibacterium, transmitted from generation to generation in the form of a single giant DNA molecule, consists of about 4 million ordered symbols of a molecular four-letter alphabet in a linear chain. Transferred to the letter symbols of our language, such a sentence would

Fig. 1. Snow crystals (photographed by W. A. Bentley and W. J. Humphreys, Ref. 7.)

have the scale of a book, about 1000 printed pages long. The microscopic detail of the molecular symbol sequence determines uniquely the macroscopic characteristics of a colibacterium, its ability to metabolize certain substances; that is, to be able to exploit their (chemically bound) free energy for self-maintenance and for reproduction. Alterations of this molecular detail would have far-reaching macroscopic consequences (e.g., death and decay of the individual). The symbol order is only one of over $10^{2,000,000}$ alternative sequences. Again this complexity is so great that we would have no chance of achieving, by coincidence, the correct sequence synthetically. In the special succession of the symbols is reflected the detail of a historic evolutionary process. Among all the possible sequences of equal length, a considerable number (which, however, still represents a diminishingly small fraction) would certainly possess the capability to code for a living system. Nevertheless, here we have the crucial macroscopic property—namely, "to live"—decided by the microscopic detail. Because of this property, "information" came into being in the course of evolution and is represented by the arrangements of molecular building blocks.

II. PRINCIPLES OF MATERIAL SELF-ORGANIZATION

A. Information Originates Through "Natural" Selection

The idea proposed in the title is actually not new. We can attribute this statement to Charles Darwin, although he formulated it somewhat differently[9]: "Just what is 'natural' selection?" Here it is the word *natural* that is important. It immediately precludes the question, "*Who* undertakes the selection?" It signifies that the process of choice is effected by definite criteria of evaluation, inherent to the system. Straightaway one is forced to pose another question: "Do the criteria of evaluation (if they exist) provide for a deterministic result for the selection process, and is the result already determined by the initial conditions?" If this were so, one could actually only speak of a "revealing" of the information. Equilibration within a closed system belongs in this category. It is characterized by an extremal principle, which for given initial distributions, and under fixed boundary conditions, results in a definite prediction for the terminal distribution. The final state is thus independent of the way in which it is reached.

The work of the great schools of population genetics, Ronald A. Fisher,[10] John B. S. Haldane,[11] and Sewell Wright[12] have, during the first half of this century, produced work that first led to the concept of selection through the properties of the living entity, as, for example, its self-reproduction, its finite lifetime, and through stipulation of certain boundary conditions (selectional pressure). One has also been able to substantiate this concept in mathematical form. The consequence of this was that obvious properties of living entities proved to be "sufficient" to explain their selective and evolutive behavior.

If we want to employ this concept later in a general way regarding the behavior of material systems, then we must ask beyond this about the *necessary* preconditions. Which properties have to accompany a material system, so that it organizes independently by natural selection?

B. "Glass Bead Games"

The principle of self-organization of a system should next be investigated on the basis of a series of models. These are so formulated that they illustrate which properties of matter are especially concerned with self-organization through natural selection. In this case we are concerned with "glass bead games," which have been developed in their present form by Ruthild Winkler.[13]

We start with a board of $8 \times 8 = 64$ squares, which are clearly defined by a coordinate plan (Fig. 2). In these squares we place glass beads of various colors. For simplicity, we commence with just two species (e.g., blue and yellow beads). To ascertain the fate of individual beads, a pair of octa-

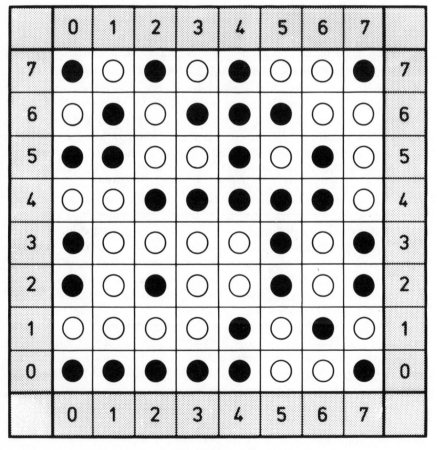

Fig. 2. Game board (Ref. 13) with an arbitrary distribution of two sorts of beads.

hedral dice, whose sides are marked according to the coordinate plan, is thrown. What actually happens then to the individual beads is determined by the different rules of play.

1. Ehrenfest's Urn Game. Let us commence with a game, which had already been described at the turn of the century by Paul and Tatjana Ehrenfest. It is well known to physicists under the name *Ehrenfest's urn model*.[14] At the start, the board is completely occupied with beads, whereby the proportion and distribution of the two bead types plays no role. Now we define the following rules of play: Every bead, whose coordinates turn up by throwing the pair of dice, must be removed from that square and replaced by a bead of the other color. The result of this game is totally independent of the initial distribution. It always results in equal proportions of both bead colors. It also happens very quickly, in fact almost within one generation, which is represented by 64 throws. (At 64 throws each bead has on average the chance to be exchanged once.) We have drawn the course of such a game by means of a computer. It is depicted in Fig. 3. The ordinate is labeled according to the occupation number of the beads, and the abscissa according to the number of throws. The subdivisions on the abscissa represent the number of generations.

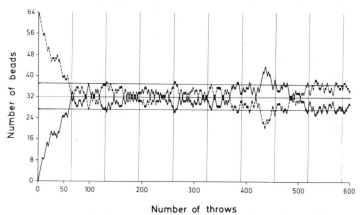

Number of throws

Fig. 3. Computer simulation of Ehrenfest's urn game. (Two sorts of beads are involved, the population numbers of which are represented by the solid and dashed curves respectively.)

The Ehrenfest game simulates generally the establishment of an equilibrium. In the present case, it is characterized by a genuine equipartition of both bead colors. Naturally, one can formulate the rules such that any desired equilibrium proportion is reached, just as one can employ as many bead sorts as desired. The *mean value* of the occupation number achieved simulates quite realistically the equilibrium position expected for an equivalent reaction system, indeed actually as one would calculate by the law of mass action. It is a characteristic of this equilibrium proportion that the deviations remain on average within certain limits, which are determined by the square root of the particle numbers. The probability distribution takes the form of a Gaussian bell curve with a maximum at the equilibrium proportion (here $N/2$), and a half-width $\sim\sqrt{N}$.

The limitation of fluctuations is particularly clearly shown in the game. Whenever a variation away from equal distribution occurs, the probability for a reduction in that deviation increases—in fact, proportionally to the extent of this fluctuation. The bead type present in excess is more likely to come up as the result of a throw and is therefore removed more quickly to the advantage of the underrepresented type. The system thus regulates itself continuously to its stable distribution value. This behavior is valid for all systems in equilibrium, independent of how complicated the reaction mechanisms are in detail. Initially, it seems amazing that the simple rules of the Ehrenfest game apply also to complicated reaction systems. However, the reason is that, near equilibrium, every step can be linearized with respect to small variations. The fading of the deviation is then always linearly proportional to the magnitude of the fluctuation. Because of the relationships of microscopic reversibility, valid near to equilibrium, it cannot come to true oscillations (i.e., to strict periodic variations of the bead distribution). This statement is based generally on Onsager's relationships.[15]

All behavior simulated by a game is, of course, a consequence of the applied rules of play. Now we need to modify these rules systematically. What happens, for example, when we simply reverse the rule of the Ehrenfest game and define: Each bead resulting from a throw is not exchanged by one of the other color, but is doubled at the expense of the other bead color. In detail it appears thus: If we throw, for example, a blue bead, we remove any one of the yellow beads and place in the free square a further blue bead.

2. The Catastrophe Game. In Fig. 4 we see again a typical run of the game simulated by the computer. Here it is useful to start with an equal distribution. The game shows that the equipartition is in no way stable under these conditions. First of all, there are fluctuations, as in the Ehrenfest game, around the equal distribution value. However, they demonstrate no tendency to self-regulation. Rather, there is (as a consequence of the rules applied) very soon an amplification of the deviations. In spite of equal chances for both bead types, the equal distribution is unstable. A large variation from the equal distribution value occurring accidentally quickly decides the outcome of the game. The deviation amplifies itself; in fact, the

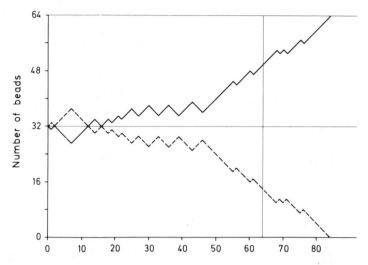

Fig. 4. Computer simulation of the Catastrophe game (two sorts of beads as in Fig. 3).

greater the deviation from the equal distribution, the greater the amplification. As fast as the equal distribution value is established in the Ehrenfest game—namely, after about one generation—so quickly is the catastrophe completed here for one of the bead types. For that reason we have christened this game the Catastrophe game. Although it results in selection, since selective advantage is solely based on the excess of population numbers, it is of little application to an evolutionary process. The species established initially is no longer displaceable by newly emerging mutants (small fluctuations).

3. "Random Walk." We have thus become acquainted with two macroscopic basic behavior patterns of a statistical distribution: stability and instability. Of course, there must also be a third fundamental type of deviation response, which occupies an intermediary position between the two extremes.

In both the preceding games, the reactional behavior of the systems was dependent on the respective quantities of occupation. In the first case, this led to an independent reduction of the variation; in the second case, to its amplification. The dependence of the reactional behavior on the occupation distribution was guaranteed by the throwing of game coordinates. Now, in the third game, the behavior of the beads should be independent of their respective amounts of occupation. Thus as a "dice" we simply use a coin, and define: For "heads" one of the blue beads is removed and replaced by a yellow one. For "tails" it is exactly the opposite, yellow being replaced by blue. The uncertainty of the elementary event can, at this stage, directly affect the macroscopic distribution. There is in no way "mass action," either in the sense of a reduction of the deviation or in the sense of an amplification. The chance, even after many "heads" throws, to throw "tails" remains at 50%.

Figure 5, again a computer simulation, shows the "random walk" type of drift in its macroscopic behavior. No proportion of occupation is preferred to any other. The probability is constant for all values of occupation. To differentiate this from the Gaussian bell curve of the Ehrenfest game or from the singularity of the Catastrophe game, we obtain here a right-angle distribution curve.

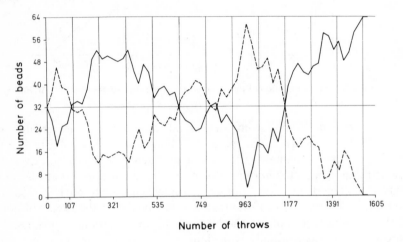

Fig. 5. Computer simulation of the Random Walk game. (Two sorts of beads as in Fig. 3.)

The difference with respect to the first two games is demonstrated especially drastically in the probability value for the extreme occupation conditions—complete covering of the field of play by one bead color or another. In the "random walk" game, it requires on average (in agreement with the known Einsteinian quadratic displacement relationship for diffusion) very nearly thirty-two generations (32^2, or about 1000 throws) until one reaches an extreme condition starting from an equal distribution. In the Catastrophe game this situation establishes itself effectively after only one generation (64 throws). In the Ehrenfest game, on the other hand, one would have to throw on average about 10^{19} times, since each of the two extreme conditions is only one of 2^{64} ($\sim 10^{19}$) possible states. The equal distribution, though, which in the Ehrenfest game is practically achieved from the extreme state within one generation, is about 10^{18} times more likely than any of the extreme situations. It constitutes around one-tenth of all microstates.

The games described here simulate in reality natural-law behavior of matter. The proportions are chosen so that deviations are noticeable. The chemist, who does not play with glass beads but with molecules, has to deal with quantities of totally different orders of magnitude (e.g., that of Avogadro's number, 10^{24}). Deviations here lie generally below the limits of perception. Thus the chemist can, without hesitation, work with the laws of thermodynamics, as if these were deterministic statements. Here one should note that our limited board is also suitable for the representation of complexity. The number of possible microstates is so great that with certainty no game would ever reproduce itself exactly. The story of the chess board with the grains of rice is known well enough. Similarly, in our most rudimentary games, every square has two possibilities for occupation. This yields, for 64 squares (as already mentioned), 10^{19} possible microstates. Even if one made a move every second, the age of the Earth would not suffice to produce a particular microstate with certainty.

4. Rational Explanation of the Games. Naturally, we can consider combining the various basic types of game with one another. However, we must take care not to sever the link with reality. The exchange of one bead with another actually represents two different processes, which in general are independent of each other. These are a formation reaction (birth) and a decomposition reaction (death). In the preceding games the retroactiveness of the formation and decomposition reactions on the variations of bead distribution is analogous: It is stabilizing, amplifying, or indifferent. But we can also imagine the situation where the formation and decomposition processes are not coupled with one another, and act against the effect of the distribution. In this way they could compensate their effects, or create a dependence of the quantitative correlation on the occupation proportions. Further, if we state that the rate changes for decomposition and formation of the fundamental reaction types are independent of one another, we obtain nine possible combinations. They are presented in Table I in the form of a matrix, as known from John von Neumann's games theory.[16] Here we call the type of deviation response a strategy. If the response to the fluctuation is an increase in rate in the same direction as the fluctuation, then we employ the symbol S_+. If the rate change has opposite sign to the deviation, we call the strategy S_-. Finally, indifferent (i.e., deviation independent) relationships we label with S_0. Care has to be taken that S_+ and S_- apply as strategies for the formation and decomposition processes opposing each other with respect to the deviations. S_+ always means an increase with increasing number of particles, and a decrease of rate with decreasing number of

TABLE I

Payoff Matrix for the Statistical Glass Bead Games

Formation / Decomposition	S_+	S_o	S_-
S_+	+ + variable	+ o stable	+ − stable
S_o	o + unstable	o o indifferent	o − stable
S_-	− + unstable	− o unstable	− − variable

particles. As far as formation is concerned, S_+ *also* gives a further increase in the number of particles for an increase in rate. An increase in decomposition rate (S_+), however, has the opposite effect; that is, it is then not deviation amplifying, but reducing. A similar logic applies to S_-, but here the relationships are inverted.

Let us now examine the matrix more closely. Seven of the system relationships are identical with the fundamental types discussed previously—stability, instability, and indifference. In only two cases does one obtain a variable relationship. These typical "results of the game" that occupy corner parts of the matrix on the main diagonal command our further attention.

Naturally, one can encompass simple combinations in such a matrix. In nature, transitions between the various strategies are actually possible for more complicated reaction mechanisms. Actually, as it turns out, we require only simple fundamental mechanisms for self-organizational behavior of matter. This must be realistic for the whole class of substance in wide concentration ranges, in fact commencing with a single particle per total volume right up to concentrations of macroscopically comprehensible orders of magnitude.

Now what does a certain strategy really mean in the language of the reaction kineticist?

S_+ is to be considered as a universal strategy for a chemical decomposition. In the simplest case, S_+ represents a reaction of first order, in which the individual particles is characterized by a certain lifetime. Already the use of the word *lifetime* approaches the idea that such an application is in general to be considered for a living entity as well.

For a decomposition reaction, how can one apply the strategy S_0 or even S_- to reality? The biochemist will, above all, have a ready answer. One decomposes the substance in question with the aid of an enzyme. With sufficient excess of substrate, the rate of decomposition is independent of the substrate concentration. Here all the enzyme molecules are saturated by substrate, and the maximal rate is achieved, which is dependent on the amount of enzyme alone, not on the amount of substrate (Michaelis–Menten mechanism).[17] In exactly the same way, the strategy S_- can be applied to reality. In this case, however, one requires two binding sites for the substrate. One is the catalytic site and accounts for substrate decomposition. The other is simply a regulating site. The binding of a substrate molecule at this site effects a conformational change of the enzyme, which can thereby regulate (e.g., totally switch off) the catalytic function.[18] Such "allosteric enzymes" are well known to the biochemist. We see immediately that both the previously mentioned strategies cannot claim a universal validity. They will only be applicable to specialized structures, and then only in certain concentration ranges, that is, at relatively high concentrations.

On the other hand, different conditions apply to the formation processes. In this case, S_0 is the most frequently applicable strategy, since in an open system the formation rate of a certain product will in general depend on the concentrations of certain reactants, but not on the concentration of the reaction product. This is, of course, not true in a closed system. Here an equilibrium is very quickly established. Then each variation of the product concentration produces an opposing effect on the reactant concentration. We have already become acquainted with such behavior in the Ehrenfest game. The strategy S_- finally is applicable to the formation reaction, whichever mechanism is in action in detail. In the open system (and only in this) the strategy S_+ may also be universally applicable to reality for the formation reaction. S_+ then represents an inherent type of autocatalysis. There are whole classes of substance (e.g., the nucleic acids) that generally distinguish themselves by such behavior in a suitable environment. However, this can only be effected in the sense of S_+ in an open system with a continuous influx of energy. At equilibrium, the effect would be completely compensated by a respective term in the rate of decomposition, owing to the microscopic reversibility. For the formation rate, then, only the preceding term (S_-) is left.

5. Selection. We now examine somewhat more closely the two situations labeled "variable" in the games matrix, particularly with reference to the previously stated question concerning the mechanism of selection.

A process of natural selection should take as its basis an inherent mechanism of evaluation. The qualifications for this are

1. To stabilize certain (i.e., "adapted") states of matter.
2. To destabilize other states (namely, those not or badly adapted).
3. To permit, with indifference, the coexistence of equivalent variants as the starting point for the development of new mutants.

It is important that this inherent capability for evaluation becomes effective with the emergence of a single individual particle or species. Whenever a mutant with suitable properties makes its appearance (and then, because of the probability relationships, it can only concern a single copy), the evaluation mechanism must become effective and, if need be, permit an amplification to a macroscopic population. Of the two situations labeled "variable," only one conforms in this way, since only the combination of strategies S_+/S_+ fulfills the stated preconditions. It alone makes use of a universal decomposition strategy and requires for the decomposition an equally universal complying strategy, namely, inherent self-reproduction. From the point of view of the origin of information, the latter prerequisite is particularly important. When the defined state of matter always represents a definite informational content, within the finite lifetime of this state, then the information can be preserved or evolve further only if the respective state reproduces itself continuously.

We call a material system whose basic reaction mechanism is characterized by this combination of strategies a Darwinian system. It must fulfill the following conditions:

1. The system must possess a metabolism. This means that the system must build each individual species from energy-rich matter and decompose into energy-deficient products. Metabolism is thus a conversion of free energy, a continuous compensation for the entropy generated by the completed reaction process. Only in this way can the system be prevented from going over to an equilibrium state with its unsuitable combination of strategies. This essential precondition for every animate system was clearly recognized more than 30 years ago by Erwin Schrödinger.[19]

2. The system must be self reproductive. On the one hand, the necessary competitive behavior for selection can only generally be effected in this way. Also, within the finite lifetime of individuals, the evolutively acquired informational content can only be preserved by continual reproduction. The self-reproductive formation must at least counterbalance the decomposition of the system.

3. Self-reproduction does not have to be absolutely exact. Owing to miscopying, mutations must appear continuously. They are the only source of new information. The system learns from its mistakes. The major cause of mutations is disturbances arising from thermic movements, which are superimposed on the physical transfer process and cause this process to be "blurred."

Of these three conditions, the first and second are determined simply by the combination of strategies. The second condition is in addition partly derived, and the third condition entirely derived, from the support of evolutive behavior. The justification for this is provided in the fourth section. First of all, we want to study the consequences of these three conditions, again based on a glass bead game.

In this game we employ four bead types (e.g., blue, yellow, red, and green). Formation and decomposition here are two processes, independent of one another. We have to throw the dice separately for them, and we shall do this strictly alternately. Thus the total number of beads on the board will always remain constant. The decomposition rate obeys the Ehrenfest rules (i.e., every bead whose coordinates are thrown must be removed from the field of play). For simplicity we set the probability of decomposition for each individual bead as equal. For each type there is then a gross decomposition rate, which is proportional to the respective representation on the field of play. The rules for the formation rate we shall take from the Catastrophe game. Every bead thrown in this phase is doubled (i.e., a bead of the same color is placed on the square that has become free as a result of the previous throw). However, we apply here different rates of formation for the various bead types. Every time a blue bead is involved, a further throw is made with a cube (normal dice), and the process of doubling may only be carried out if the result was in the range of 1 to 5. The yellow bead type is only doubled for 1 to 4, and finally the green bead only for 1 to 3. The formation rate for the latter is thus on average half as great as that of the blue bead type. It is important that in the formational phase the dice are thrown until a duplication can be carried out. This represents a regulation, which maintains a constant total population.

Figure 6 shows once again the course of a typical game. Here the green beads are actually the first to die out, followed by the yellow bead type. The red and blue bead types compete now for a while, but finally the higher selective value of the blue type asserts itself. It remains as the only surviving type from the selective competition.

This is selection in the Darwinian sense. What is missing here is purely the finite quota of errors, and thereby mutation. We would like to examine their effects in a further experiment. Here we revert to the use of two bead

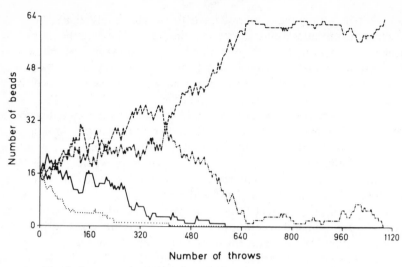

Fig. 6. Computer simulation of a Selection game (four sorts of beads).

types: blue and yellow. We provide the blue beads with a selective advantage: The blue bead type may duplicate itself at a rate four times that of the yellow type. In a real selectional experiment, a superior species competes with a multitude of less suitable species. All errors, which appear in the superior and stably reproducing form, represent for these a loss of copies. All error copies, on the other hand, which appear in the reproduction of interior competitors, do not really represent a new sort, since almost all these error copies still belong to the pool of the inferior competitors. Our game simulates this case, namely, the competition between a superior species (blue) and many less suitable mutants (grouped together as "yellow"). In order to simulate realistically the effect of the error rate on the selection, we give only the superior, the blue bead type, a finite error rate. The yellow bead type is reproduced exactly, which simply means that (nearly) every mistake made there does not lead to a loss from the pool of the competitors. The probability that a blue bead comes into being by misreproduction of a yellow bead is, with sufficient complexity of the species represented, negligibly small.

The courses of four games are depicted in Fig. 7. The precisions of reproduction of the blue type in these four experiments amount to 100%, 75%, 50%, and 25%. We see that the yellow bead type gains the upper hand with an error rate of only 50%. Since the advantage of the blue bead types constitutes a factor of 4, one might wonder at first that this should result in a reproductive rate twice as high for the yellow type in spite of the 50% error rate of the blue type. This, though, is not correct, since all mistakes of the blue type are to the benefit of the pool represented by the yellow color. For an error quota of 75%, the blue bead type dies out finally, although it possesses an advantage with respect to each individual bead type in the yellow pool. This game demonstrates very clearly that for a given selective advantage, a definite threshold exists for the error rate.

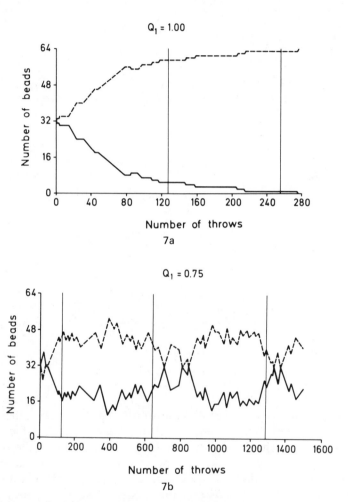

Fig. 7. Computer simulation of a Selection game involving two sorts of beads. The "blue" beads—represented by the dashed curve—reproduce four times as rapidly as the "yellow" ones, which are represented by the solid line. "Blue" furthermore has an error rate (i.e., they produce "yellow" instead of themselves) of (a) 0%, (b) 25%, (c) (*next page*) 50%, and (d) 75%, while "yellow" [representing all inferior (nonblue) species] has a mutation rate of zero (i.e., they nearly all remain nonblue when they mutate).

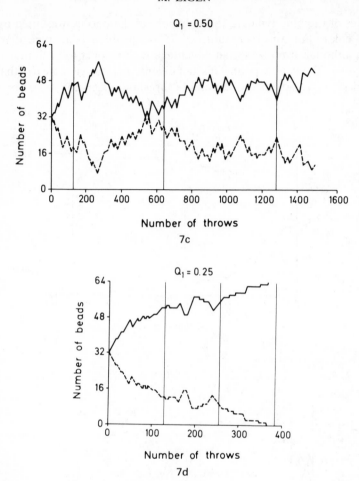

Fig. 7. *Continued.*

Let us summarize once again the lessons to be drawn from the various glass bead games: Darwinian selectional behavior is related by the fulfillment of certain preconditions. These become unalterable once the complexity of the system has become so great that all the alternatives cannot be represented at the same time. As long as we start with simple preconditions to be fulfilled for large classes of substance in unlimited concentration ranges, only one definite combination of strategies comes into question for selectional values in the sense of Darwin, which has metabolism, inherent self-reproductivity, and mutageneity as pre-requisites. By compliance with these *necessary* preconditions, "natural" selection results as a quality, stipulated by the properties of matter. It is already founded on the mechanism, since, even for selectively equivalent species, the result of competition is always a unique selectional decision. For this, though, the adherence to certain constraints is prerequisite. If the species are not equivalent, a preference is shown for the better-suited component and the process of selection becomes deterministically predictable. Of course, these predictions are precisely valid only in the limiting case of sufficiently great numbers. For finite occupation, on the other hand, one can only make probability predictions. Finally, mutation is, in detail, a completely undetermined stochastic process. Since selective amplification of individual mutants can appear as a macroscopic phenomenon, only the direction of the evolutionary processes and not their individual routes are determined.

III. QUANTITATIVE TREATMENT OF THE DARWINIAN SYSTEM

A. Formalization of the Problem

Now we want to examine selectional behavior in more detail from a quantitative viewpoint.

The mathematical treatment of the Darwinian system is presented in a condensed form in Tables II to IV.

The Ansatz takes account of the three *necessary* preconditions named earlier: metabolism, self-reproductivity, and mutageneity. The first condition requires separate terms for the formation and decomposition rates of each species taking part in the competition (index i). Formation and decomposition rates are *not* linked to one another by the condition of microscopic reversibility. For the maintenance of metabolism, a continuous influx of energy-rich matter (or of energy in a form capable of executing work) is necessary. The state labeled by the respective index i is intermediary. It forms spontaneously (i.e., with positive affinity) from the energy-rich building blocks, and is equally spontaneously decomposed into an energy-deficient end state. The accompanying rates are designed by

$$A_i Q_i x_i \qquad \text{formation}$$

and

$$-D_i x_i \qquad \text{decomposition}$$

In the case of the nucleic acids, the only class of substance known to us that fulfills these prerequisites, the energy-rich building blocks are the nucleoside triphosphates and the products of decomposition are the respective monophosphates. They result from hydrolysis or enzymatic cleavage of the nucleic acid strands.

TABLE II
Mathematical Representation of the Darwinian System[a]

Given: n distinguishable "species" i with their population numbers (or concentrations) x_i

Prerequisite: Each "species" i has a metabolism, is self-reproductive and mutagenic

Rate equation (explanation in the text, $\dot{x} \equiv dx/dt$):

$$\dot{x}_i = (A_i Q_i - D_i)x_i + \sum_{k=i} \varphi_{ik} x_k + \frac{\Phi_0 x_i}{\sum_{k=1}^{n} x_k} \tag{1}$$

Constraints (constant overall organization):

$$A_i Q_i = \text{const}; \quad \sum_{k=1}^{n} x_k \equiv c_n = \text{const} \tag{2}$$

Conservation of error copies:

$$\sum_{l=1}^{n} A_l(1 - Q_l)x_l = \sum_{l=1}^{n} \sum_{k=l} \varphi_{lk} x_k \tag{3}$$

Consequently,

$$\sum_{k=1}^{n} (A_k - D_k)x_k \equiv \sum_{k=1}^{n} E_k x_k = -\Phi_0 \tag{4}$$

Hence the rate equation reduces to:

$$\dot{x}_i = (W_i^0 - \bar{E}(t))x_i + \sum_{k=i} \varphi_{ik} x_k \tag{5}$$

with

$$W_i^0 = A_i Q_i - D_i \tag{6}$$

$$\bar{E}(t) = \frac{\sum_{k=1}^{n} E_k x_k}{\sum_{k=1}^{n} x_k} \tag{7}$$

[a] Cf. Refs. 20 and 21.

The terms A_i, Q_i, and D_i by no means need to be independent of x_i or the other x_k, although this is usually the case. Above all, A_i contains a functional term, which is dependent on the concentration of the energy-rich building blocks. If one buffers the building blocks to constant values of concentration, then one can incorporate this function into the rate constant for the formation reaction; otherwise it has to be taken into account explicitly.

The decomposition rate is, as usual, representative of the application of the strategy S_+. It assigns to each individual state i a finite lifetime $1/D_i$, which on its part can depend on the concentration of the other decomposition-catalyzing or state-stabilizing reaction partners.

The respective expression of a proportionality to x_i for the formational term takes account of the second prerequisite stated earlier: the inherent *self-reproductivity*. Each matrix effect expresses itself in such a proportionality to x_i, which, however, in no way excludes additional concentration dependences of A_i.

TABLE III
Solutions of Rate Equations for Constant Selective Values W_i^0 (at Constant Overall Organization)

Equation 5 from Table II can be transformed to (cf. Ref. 21):

$$\dot{\xi}_l(t) = (\lambda_l - \bar{E}(t))\xi_l(t) \tag{1}$$

where $\bar{E}(t)$, as defined by (4) in Table II, and c_0 are conserved and can be written as:

$$\bar{E}(t) = \sum_{l=1}^{n} \frac{\lambda_l \xi_l(t)}{c_n} \tag{2}$$

$$c_n = \sum_{k=1}^{n} x_k = \sum_{l=1}^{n} \xi_l = \text{const} \tag{3}$$

The "normal modes" ξ_i can be represented as linear combinations of the x_k. The coefficients follow from the components of the eigenvectors and their respective eigenvalues λ_k. All species i contribute to a given quasispecies (e.g. the selected one), reflecting their "kinship" relations.

The solutions of the system (1) show the following behavior for $t \to \infty$. All ξ_i for which $\lambda_i < \bar{E}(t)$ approach zero, thereby shifting the threshold for selection $\bar{E}(t)$ steadily [according to (2)], until finally only one quasispecies (usually the "wild type" plus a mutant distribution) survives. It is characterized by the maximum eigenvalue λ_{max}.

Hence "selection" can be characterized by an extremal principle:

$$\bar{E}(t) \to \lambda_{max} \tag{4}$$

The average excess production of the system tends toward the maximum eigenvalue.

Finally, the third precondition is expressed in the explicit consideration of the term Q_i, also in the error flux rate $\varphi_{ik}x_k$. Q_i is a quality factor, yielding which fraction of the preceding copies from the replication regulated by the template is absolutely free from error. $Q_i = 1$ means completely precise (i.e., error-free) reproduction. The quantity $(1 - Q_i)$ then naturally encompasses a multitude of different miscopies, which can be assigned into classes (with, respectively, one, two, three, etc., errors) by partition functions. Everything that is lost through a misreproduction in the class i appears again (with complete account being taken of all classes) in some other class k (usually closely related to i). In the same way, the class i continuously receives additional copies through misreproduction of its related classes k. This is accounted for by the term $\sum_{k \neq i}^{n} \varphi_{ik}x_k$. Here the law of conservation applies, which results from the summation of all classes of species:

$$\sum_{l=1}^{n} A_l(1 - Q_l)x_l = \sum_{l=1}^{n} \sum_{k \neq l} \varphi_{lk}x_k$$

Finally, the expression also contains a flux term Φ_i. For this the assumption is made that on the basis of a general flux Φ_0, individuals i are proportionately (i.e., $\sim x_i/c_n$) lost from, or else delivered to, the system. Further, a dilution by influx of solvent can occur.

TABLE IV
Error Threshold and Maximal Reproducible Information

Perturbation theory, applied to the system treated in Tables II and III, yields for the dominant species "m" (the "master copy") the condition:

$$W_m > \bar{E}_{k \neq m} \tag{1}$$

where

$$W_m = W_m^0 + \sum_{k \neq m} \frac{\varphi_{mk} \varphi_{km}}{W_m - W_k} \tag{2}$$

$$W_m^0 = A_m Q_m - D_m \tag{3}$$

and

$$\bar{E}_{k \neq m} = \sum_{k \neq m} E_k x_k \Big/ \sum_{k \neq m} x_k$$

If the sum term of W_m is negligible with respect to W_m^0 (as usually is the case), it follows that

$$Q_m > \sigma_m^{-1} \tag{4}$$

where

$$\sigma_m = \frac{A_m}{\bar{E}_{k \neq m} + D_m} \tag{5}$$

Then we have

$$\frac{x_m}{\sum_{k=1}^n x_k} \approx \frac{Q_m - \sigma_m^{-1}}{1 - \sigma_m^{-1}} \tag{6}$$

and

$$\frac{x_k}{x_m} \approx \frac{\varphi_{mk}}{W_m - W_k} \tag{7}$$

The relations (4) to (7) can be generalized for higher approximations of W_m.

For any set of ν_m symbols, of which each can be reproduced with an accuracy q_{mk} we have

$$Q_m = \prod_{k=1}^{\nu_m} q_{mk} \equiv \langle q_m \rangle^{\nu_m} \tag{8}$$

The geometric mean $\langle q_m \rangle$ can be replaced by the arithmetic mean, if the q-values are close to one (as usually is the case).

Substitution of (8) into (4) yields the important threshold relation:

$$\nu_{max} < \frac{\ln \sigma_m}{1 - \langle q_m \rangle} \tag{9}$$

Since for any self-reproductive system a threshold relation of the form of (4) exists, the form of (9) is of general validity. This relation states that there is an absolute upper limit for the information content of any stable self-producing system.

In nature there are *certain* flux and boundary conditions, which (provided they are known) can be taken into consideration for an integration of the equations. In experiments, one will attend to the compliance with definite and expediently chosen conditions. This is just the same in thermodynamics. Natural processes, such as the origin of the weather, hardly adhere to simple boundary conditions (e.g., constant temperature or constant entropy), nor does a hurricane endeavor to maintain a controlled condition of pressure or flux.

The introduction of the condition "constant organization" has proved to be particularly appropriate for the execution of experiments. Generally it means that we hold the proportion of the entirety of nonorganized symbols (e.g., mononucleotides) constant in relation to organized symbols (polynucleotides). In the special case of an evolution experiment it represents a buffering of the energy-rich building blocks, whose concentrations can then be included as constants in the A-terms. Furthermore, the sum of the concentrations x_i can also be held constant, so that the additional condition $\sum_{k=1}^{n} x_k = c_n = $ constant is valid. In this case the flux terms are variable, since the constancy of the total concentrations can only be regulated by adjustment of the flux. The flux term Φ_0 in particular must be chosen, such that the total excess production $\sum_{k=1}^{n} (A_k - D_k)x_k$ is compensated by removal. Each individual species contributes to this flux a fraction proportional to its presence x_i/c_n. Thus one may substitute $\Phi_0/n = \bar{E}(t)$, if $\bar{E}(t)$ denotes the mean excess production: $\sum_{k=1}^{n} (A_k - D_k)x_k/\sum_{k=1}^{n} x_k$. As long as the system is not in the stationary state, $\bar{E}(t)$ appears as a function of time.

When one employs this additional condition, one achieves a dramatic simplification of the expression. Here it is noted that the chosen additional condition "constant organization" (as was shown elsewhere[20]) leads to totally similar results, as for the alternative boundary condition of constant fluxes. For the natural prerequisites, constant (average) fluxes are perhaps more typical primarily of the later stages of evolution (after which the initially concentrated energy-rich starting materials had been exhausted). Both are limiting cases of "simple" boundary conditions, analogous to the constancy of intensive or extensive variables in equilibrium thermodynamics, or the constancy of generalized forces or fluxes in the thermodynamics of irreversible processes.

The form of the reaction equations, simplified on the basis of boundary conditions, can generally be solved for the case of constant W_i^0 and φ_{ik}.[21,22] The explicit expressions resulting from the general solution by application of perturbation theory are identical with previously reported approximate solutions.[20]

The physical interpretation of the results can be summarized as follows: Selection in a Darwinian system, characterized by the preconditions stated, can (in formal analogy to equilibrium in thermodynamics) be characterized by an extremal principle. The mean (excess) productivity $\bar{E}(t)$ tends for $t \to \infty$ to the maximal eigenvalue λ_{max} of the entire system. The "quasi-species" denoted by this eigenvalue is selected—all the remaining "quasispecies" (i.e., those denoted by all other eigenvalues) die out. The selection of the dominant "quasispecies" succeeds at the total expense of the others.

The appearance of a mutant with a selective advantage at constant flux values is associated with a fluctuation term in the entropy production, which Ilya Prigogine and Paul Glansdorff[24] might have called a "dangerous" contribution. The substitution of a less advantageous by a more advantageous mutant results in rate and affinity changes of opposite signs—similar to those calculated for a simple autocatalytic process far from equilibrium (cf. Ref. 24). The instability associated with the selection process and resulting in the breakdown of the formerly (meta)-stable quasispecies and its substitution by a new one is hence in accordance with expectations drawn from the Prigogine-Glansdorff principle.

B. What is a Quasispecies?

For the case where, within the population, one species is distinguished above all others by a notably higher selective value $W^0 > W^0_{k \neq m}$, and further, where all individual error production rates φ_{mk} and φ_{km} are small compared with W^0_m, the species m determines the nature of the quasispecies. The quasispecies, however, includes also all the mutants of this dominant form, corresponding to their rates of generation. The normal mode ξ_m assigned to the eigenvalue λ_{max} is a linear combination of the concentrations of the dominant species and also its mutants. In this linear combination, the x_m term dominates. The condition $\varphi_{km} \ll W^0_m$ can always then be taken as fulfilled, if m characterizes the dominant species, and the informational content of this is so large that many different mutants, each having one single error, exist. For any mutant l, the term $\varphi_{lm} x_m$ particularly is *not* generally small compared with $W_l x_l$. These owe their very existence above all to the constant production of errors from the dominant species.

If the precondition of the unique dominance of *one* species is not fulfilled (i.e., some $W_k \approx W_m$), then perturbation theory as applied earlier falls down. The selection with respect to mutants is far from sharp, whenever $(W^0_m - W^0_k) \to 0$, however only for finite φ_{mk} and φ_{km}-terms. Indeed, in the limiting case, one can combine the species degenerated ($W_m = W_k$) in their selectional values into an indistinguishable class in the quasispecies.

Admittedly, all members must be able to diverge by mutation ($\varphi_{mk}, \varphi_{km}$ finite); that is, they must be "related" to one another. The selection between nonrelated species ($\varphi_{ik}, \varphi_{ki} \approx 0$) remains sharp; these are characterized by different eigenvalues. The structure of the equations $\dot{\xi}_i = (\lambda_i - \bar{E}(t))\xi_i$ shows that only the quasispecies with the greatest eigenvalue can survive under selectional pressure, which may by all means be temporally variable (e.g., periodic).

C. The Principle of Selection

A system characterized by the extremum principle given in Table IV we designate as a Darwinian system. It should not thereby be claimed that Darwin himself had limited his principle of natural selection in this way. Nonetheless, from him comes the characterization of "survival of the fittest" (which in fact is a term he adopted in later editions of his work from Herbert Spencer). *Fittest* is the consequence of the effect of the extremum principle. As distinguished from the extremum principle, which characterizes equilibrium behavior in thermodynamics, it is strictly valid only for linear rate expressions (constant W_k^0). Nevertheless, it can be demonstrated that for a series of (biologically relevant) nonlinear rate expressions (with terms of higher order in x_i or x_k), a similar optimum criterion remains valid, if certain boundary conditions are fulfilled. In nonlinear systems, however, a coexistence of difference species, and particularly the growth of new mutants, is no longer possible, since the selective value at this stage becomes a function of the concentration of the species concerned (or of others). Just that, which is what Darwin wanted to explain—namely, the diversity of the species and their descendence—would no longer be explicable for such a system.

Even if additional terms of a nonlinear type cannot be totally eliminated for the development of the species from single-celled entities, selection in the sense of Darwin, as well as the evolution of the species based on it, is still best described by the system dealt with in Tables II to IV. It is due to this that we call it Darwinian. In addition, not only do the directly self-reproductive species belong in this class of systems, but also every species having linearly complementary or cyclically reproductive systems (as it is, for example, represented by the nucleic acids), or the so-called Mendelian population systems based on sexual recombination (although only for sufficiently large population numbers). The theory sketched in Table II has been extended for such cases. An analogous solution scheme can also be found for these systems through suitable transformations.

Actually, the expressions in population genetics were purely derived from linear autocatalytic propagational terms. One is therefore justified in asking, "What has actually come out of these studies which is new, as compared to the statements of population genetics?" The answer is, "Above all, it is the displacement of emphasis from 'sufficient' to 'necessary' prerequisites, which will yet stand out as significant for the molecular field." In population genetics, one simply started from the properties of living entities, as, for example, from their reproduction, their finite lifetimes, from the existence of a certain rate of mutation, and so on. One has attempted to grasp the dynamic behavior arising therefrom by a system of differential equations. From this resulted the fact that, given certain constraints, the properties named are "sufficient" to account for the phenomenon of natural selection. Here the nature of the Darwinian principle has already emerged. Under certain preconditions it is a valid, deducible principle, and therefore not an axiom of life, which could lay claim to general validity, independent of these preconditions. It seemed, then, to be something specifically linked to the properties of the living being. Hence the conjecture that it is applicable as a characteristic for this and *only* for this was introduced.

In our study, on the other hand, the *necessary* prerequisites stand in the foreground. We could not classify matter according to whether it "lives" or came out of the chemist's retorts. It is exclusively decisive which preconditions are needed so that a system can evolve by natural selection. Under suitable conditions even lasermodes demonstrate the phenomenon of natural selection—wholly in the Darwinian sense. Yet one cannot form a living entity from it (in the generally accepted sense).

The significant difference with respect to the statements of population genetics, however, is first and foremost expressed in the consideration of the informational content of the species.

IV. THE INFORMATIONAL ASPECT

A. Maximal Information Content of the "Species"

In the games described previously a species i was characterized simply by a glass bead of a particular color. We could then assign to it a population variable x_i and describe its behavior by means of a deterministic sequence or a stochastic statement for the turnover probability. However, a living organism is evidently characterized only *very* inadequately by an index i. At least, we have to ask, "How much information is contained in each reproductive component unit?" or, "How much information is necessary for the characterization of all the integrated functions of the unit?"

Modern molecular biology has already provided substantial answers to this question. Our question is really, "How does this information come into being? If genetic information arises selectively by self-organization of a material system, what is the correlation between the important physical parameters for the selection, and the resultant quantity of information from the selection?"

An answer to this question is compiled in Table IV. The decisive quantity is the accuracy, the quality of the reproductive informational transcription Q_i. It is dependent on the accuracy of transfer of information for each component symbol in the genome of the organism.

The quality of transfer of the component symbol q_{ik} is determined by molecular properties, that is, by the essential physical (or chemical) interactions of the copying process (e.g., the base pairing interactions of nucleotides).

The relationship derived in Table V,

$$\nu_{max} = \frac{\ln \sigma_m}{1 - \langle q_m \rangle}$$

is of fundamental significance for the question of the origin of information. The mean value $\langle q_m \rangle$ of the accuracy of transfer of the component symbol is defined initially as a geometric mean. However, if the q_{mk} values do not differ greatly from unity, the geometric and arithmetic means become identical. It is interesting that the kinetics parameter appears only in the form of the quantity $\ln \sigma_m$, that is, as a logarithmic term. The specific form of σ_m—in which the reproduction and decay parameters of the selected species, relative to the corresponding mean values for the distribution of competitors ($\bar{E}_{k \neq m}$), enters—is defined unequivocally by the mechanism dealt with in Table II. Indeed, the relationship is formally valid for *any* other mechanism of informational transfer. If the information is not to "melt away," *at least* one, but preferably a much larger number of *completely* correct copies, must be transmitted from generation to generation, and these must be able to compete favorably with their errors copies. Selective advantages, with regard to an expansion of the informational content, can only to a certain extent be drawn as a consequence of this logarithmic dependence. Even systems that have not yet achieved the stationary state are governed by this criterion. The requirements of symbol accuracy could be, in fact, even more restricting.

The mean value $\langle q_m \rangle$ is decisive for the amount of symbols that can be transmitted from generation to generation. Let us consider some examples.

For an enzyme-free reproduction of nucleic acids, with $\langle q_m \rangle = 0.99$, just as much information can be accumulated as is contained in transfer nucleic acids (composed of about 80 nucleotides). This is not even enough for the coding of a single protein molecule, if one adheres to the yardsticks valid in nature. A polypeptide of 30 amino acid residues, however, may well have been adequate as an initial functional carrier (catalyst). But: *One* protein does not yet constitute a translational apparatus, and without this the proteins could not have evolved. The interactions between nucleotides, based on the hydrogen bonding of complementary base pairs ($A=U$ or $G\equiv C$), definitively suffice for a value of the order 0.90–0.99, but not more.

If a (highly developed) replicational enzyme were once available, somewhat as it is coded in a bacteriophage, then units of about 100 to 10,000 symbols could be transferred reproducibly without more ado.

Q_β-phage, for instance, contains 4500 nucleotides, and its average single-digit quality factor $\langle q_m \rangle$ has been determined to 3×10^{-4} (with a σ_m of about 4).[26,41] The coli bacterium, with its DNA polymerases, including a "proofreading" exonuclease function, possesses a system with whose help the informational content can be enlarged to several millions of symbols. Finally, in the human genome, several thousand millions of nucleotides are combined, of which only a fraction contain the information for the functional properties, which, however, is transmitted in toto from generation to generation. Nonetheless, the enzymes of a coli bacterium are just as efficient as those of the human organism. The difference between the two systems lies in the amount of symbols and in the functions represented by this information.

B. Information Games

We would like to examine the relationship between quantity of information and quality of transfer by means of a concrete example. For this we lay down a target sentence, which represents a particular information content (see Fig. 8). This target sentence should result from an initial sentence, which is not related in its meaning, via an evolutionary mechanism. In our example the initial sentence (which is not especially philosophically profound) reads, "Make adventure of my cake." There is only one phenotypical level at which the sentence can ever be tested. At this level the target sentence "Take advantage of mistake" (German original: *Lern aus den*

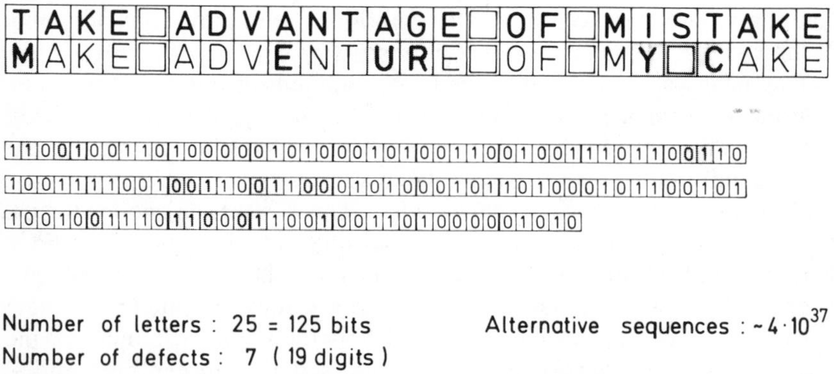

T	A	K	E	□	A	D	V	A	N	T	A	G	E	□	O	F	□	M	I	S	T	A	K	E
M	A	K	E	□	A	D	V	E	N	T	U	R	E	□	O	F	□	M	Y	□	C	A	K	E

1 1 0 0 1 0 0 1 1 0 1 0 0 0 0 0 1 0 1 0 1 0 0 1 0 1 0 0 1 1 0 0 1 0 0 1 1 1 1 0 1 1 0 0 1 1 0

1 0 0 1 1 1 1 1 0 0 1 0 0 1 1 1 0 0 1 1 0 0 0 1 0 1 0 0 0 1 1 0 1 1 1 0 1 0 0 0 1 0 1 1 0 0 1 0 1

1 0 0 1 0 0 1 1 1 0 1 1 0 0 0 1 1 0 0 1 0 0 1 1 1 0 1 0 0 0 0 0 0 1 0 1 0

Number of letters : 25 = 125 bits   Alternative sequences : ~$4 \cdot 10^{37}$
Number of defects : 7 (19 digits)

Fig. 8. Word game simulating the self-reproductive transfer of information. The first sentence has the target structure (usually unknown in advance). It is *given* to the computer only to determine the selection value ("meaning") of each individual sequence (which in nature is established by the dynamic selection parameters of each mutant). For the evolution process the computer uses only the second sentence, or its mutant offsprings. Each individual sentence has a finite lifetime and a finite error rate of reproduction.

Fehlern) must be known. Also in the living cell, the genotypically laid down information is mainly checked for functional efficiency at some phenotypical level. Advantage there appears to be expressed by the properties of the entities. Thus, without violating the object of the presentation, we can say, "Every sentence that is in better agreement with the target sentence by one bit of the encoded binary sequence reproduces itself by a factor $e \approx 2.7$ faster than the original copy." Progress depends then solely on the differential advantage within each distribution, and for this it is unimportant whether or not the final sentence is decided from the start. In nature the "arm" of evolution only becomes apparent through the evolutionary route. We define further a finite lifetime for each sentence and a mean precision of reproduction $\langle q_i \rangle$ per binary symbol and run the experiment on a computer. The sentences propagate themselves according to a reproduction rate typical for the number of errors, whereby we alternately allow growth to 100 copies and then a reduction to 10 copies. Note that the only way of conserving and evolving information is by reproduction of otherwise unsuitable sentences. The target sentence, stored in the computer, is only used for evaluation, not for conserving the information.

Figure 9 shows this sentence in three different phases—namely, in the first, tenth, and thirtieth generation of reproduction. (In the computer this evolutionary process is completed within seconds.) In Fig. 10 one sees the result for various error quotas. The experiment supports the validity of the relationship for the threshold of error introduced in Table V. As in a dice game, where the sequence of certain throws is subject to considerable fluctuations, so here the number of generations in which the evolutionary target is achieved varies. However, the frequency distribution of errors is precisely fixed. For $1 - \langle q_i \rangle = 1/\nu_i$ (where ν_i represents the number of symbols in the sentence—in the present case, 125 binary units), Q_i should become equal to e^{-1}; that is, on average 30% of the sentences should be reproduced correctly. This theoretical prediction is largely fulfilled, as one can deduce from the third column of Fig. 10. The selective advantage per corrected bit amounts to 2.7. This means that the more correct copies reproduce 2.7 times faster than those that are in error by one more bit. Within one generation, therefore, only such copies that on average differ from one another by about 1 or 2 errors are likely to coexist. The value of $\ln \sigma_i$ for this lies close to 2.

For our sentence this means that as soon as the error rate exceeds 1.6%, the information must "melt away." A higher σ_i has little effect on this. Figure 11 demonstrates that (with respect to the proportionality of $\ln \sigma_i$) a tenfold increase of the selective advantage only shifts the error threshold by a factor $\leqslant 2$.

Initial :　　　　MAKE ADVENTURE OF MY CAKE

Goal :　　　　TAKE ADVANTAGE OF MISTAKE

v = 125

mutation rate (1 - q) : 0.8 %

selective advantage : 2.7

sentence = replicative unit

```
1 )   YAKE  ADVENTERE  OF  MYPSAKE  /   UAOE  ADZENTERE  OF  MYPSAKE  /
      MA:E  ADVENTCRE? OF  MY TAKE   /   MAKE  ADVENTERF  OE  MY SAKE   /
      UAKE  ADVENTERE  OF  MYPSAKE  /   MAKE  ADVFNTCRF  OF  MY TMKE   /
      MAKE  ADFENIERE  OF  MYPSAKE  /   MAKE  ADVENDERE  OF  MYPSAKE  /
      MAKE  ADFENTERE  OF  MYPKAKF  /   MAKC  ADVENTURE  OF  MYPSAKF  /
```

```
10 )  XAOE  BDVINTAIE · OF·MY:TAKE  /   TAPE  BDVINTAIE   OF·UYKTAKE  /
      TAOE  BDVINTAIE   OF·MYKTAKE  /   TAOE  BDVINTAIE   OFAMYKTAOE  /
      TAOE  BDVINTAIE   OF·MYKTAKE  /   TAOE  BDVINTAIE   OF·KYKXAKE  /
      TAOE  BDVGNTAIE   OF·MYKTAKE  /   TAOE  BDVINTAIE   OF·MYKTAKE  /
      TAOE  BDTINTAIE   OF·MYKTAKF  /   TAOE  BDVINTAIE   OF·MYKTASE  /
```

```
44 )  TAKE  ADVANTACE  OD  MISTAKE  /   TAKE  ADVANTAGE  OF  MISTAKE  /
      TAKE  ADVANTAGE  OF  MISTAKE  /   TAKE  ADVANTAGE  OF  MISTAKE  /
      TAKE  ADVANTIGI  OF  MISTAKE  /   LAKE  ADVALTAGE  OD  MISTAKE  /
      TAKE  ADVANTAGE  OF  UISTAKE  /   TAKE  ADVANLAGE  OF  MISTAKE  /
      TAKE  ADVBNTAGE  OF  MISIAKE  /   TAKE  ADVALTAGE  KF  MISTAKE  /
```

Fig. 9. Examples of sentences selected in the information transfer game.

Selective Advantage :　　2.7

Error Rate	Final Sentence	Representation	Generation
0.1 %	SAKE ADV!NTMOE OF MYCSAKE	9 : 1	> 55
0.3 %	TAKE ADVANTAGE OF MISTAKE	7 : 3	~ 50
0.5 %	TAKE ADVANTAGE OF MISTAKE	5 : 5	~ 40
0.8 %	TAKE ADVANTAGE OF MISTAKE	3 : 7	~ 40
1.2 %	TAKE ADVANTAGE OF MISTAKE	1 : 9	~ 65
1.5 %	TAKE ADVANTAGEBQF NISTAKE	0	65
2.0 %	SQQE AFV.NTA.E OF !ISTAKE	0	65
5.0 %	K . : ! AA,EARVAEFHODPUWWTA:C	0	10

Fig. 10. Evolution of sentences at different error rates. The first number in the column "representation" refers to the number of error-free copies (out of a total of 10). Selective advantage: 2.7.

This computer experiment makes the following quite clear:

1. Error rates that are too low make the evolutionary process protracted; the rate of progress stays small.
2. Error rates that are too high lead to dispersion of the information. This happens as soon as a certain threshold of error, determined by the quantity of symbols, is exceeded.
3. The ideal evolutionary condition is situated just below this threshold.

Selective Advantage : 27

Error Rate	Final Sentence	Representation	Generation
0.8 %	TAKE ADVANTAGE OF MISTAKE	4 : 6	stable
1.0 %	TAKE ADVANTAGE OF MISTAKE	3 : 7	stable
2.0 %	TAKE ADVANTIGE OF MISTAKE	1 : 9	13
5.0 %	? OE ADU!NLASE O , NJSDBUE WJAM? ?CNHMLEUF? K. ASG! .TV.		9

Fig. 11. Evolution of sentences of different error rates. Selective advantage: 27.

What possibilities of extended self-organization has a system, whose amount of information is already accommodated to the quality of transfer of symbols? What is meant here is a system that optimally exploits existing mechanisms of transfer by available physical interactions. Apparently, the system would have to develop a new improved reproductional content. However, since the one determines the other, we must enquire after a new principle of organization. The answer to such a question is of decisive importance for an appreciation of evolution, for the formation of molecular replicational and translational apparatus of the cell, and thereby for the "origin of life."

The problem can be clarified once again by the word game discussed earlier. Let us assume that the accuracy of transfer of the reproductional mechanism given by the computer program is just sufficient to ensure conservation of the informational content of single words. Further progress is then attainable solely by the compilation of words into a meaningful sentence. For predetermined error rates, how can such an expansion of the informational content be achieved? Here we cannot simply increase the informational content, so that the whole sentence is reproduced as a unit, since we already know that the informational content would then "melt away."

With regard to this we wish to try out four possibilities:

1. We assume that all words of the sentence can come into being simultaneously, so that they possess an accuracy of transfer adapted to their symbol content and identical selective values. The individual words then represent self-contained replicative units. We hope that in sufficiently large quantities (e.g., a total of 100 copies), all words needed for a meaningful sentence are concurrently and reproducibly conserved, and that by their presence (in the meaning uniquely manifesting itself on the phenotypical level), the entire information is stably maintained. Figure 12 shows the closing stages of such an experiment. For identical selective values of all words (degeneracy of selectivities) only one of the four words will survive. With sufficiently frequent repetition of the experiment, each word has around 25% probability of being the survivor. The fact is that, owing to the reproductional mechanism, the words must hold their own selectively against miscopies and are forced to compete also with one another. That is also true if one assumes selective degeneracies (an assumption being extremely unrealistic for natural processes). As it emerged from the mechanistic analysis, selection is primarily a consequence of the inherent mechanism. The selection parameter, characterized by the word *fittest*, only determines the preferred direction.

2. We enclose the words, which again represent self-replicative units, into compartments and ensure that at sufficiently high concentration, a proliferation into new compartments occurs. At once, one is faced by a difficulty, which can only be solved with the aid of a complicated mechanism (which would hardly be available in nature). This is because, under conditions ideal for a competition of the limited space, a strong competition between the different words would immediately arise (even more so than in case 1). Thus there must be a mechanism that takes care that proliferation is set up for all four words if ever one copy (or a few copies) exists. Apart from the difficulties associated with realization, the same problem is encountered in the limitation of information, preventing the unification of the words into a sentence. The compartment could then effectively be viewed as a replicative unit, and for this, a significantly greater accuracy of transfer would be required, as it is delivered from the individual words. The information would be lost quite rapidly.

3. By means of a phenotypical level (e.g., meaning), we introduce couplings between the words. These ensure that, for example, the reproduction of a word is encouraged by the presence of the preceding word in the sentence (i.e., proportional to the number of copies of this word). The selective value is thus in a certain way dependent on the concentration of the respective preceding word. [We employ an expression in the form

Word Competition

	Mutation-Rate
TAKE	3.1 %
ADVANTAGE	1.4 %
OF	6.3 %
MISTAKE	1.8 %

4 Typical results : (in 10 experiments 2 : 4 : 3 : 1)

Generation 10 :   TAK!, TAKE, TAK!, TAKE

T.KE, LAKE, TAK!, TAKE

TALU

Generation 12 :   .DVANTAGE, ADVANTAGE, ADVANTAGE,

ADVANTAGE, ADVANTAGE, ADVANTAGE,

ADVANTA!E, A?VALTAGE, ADVAMTAGE,

IFVANTAGE.

Generation 22 :   PF, OF, KF, OV, OF, OD

OF, OF, OF, OU

Generation 21 :   UISTAKE / MISTAKE / MISTAKE

MESTAME / MISTAKE / MISTAKE

MISTAKE / MGST.KE / MGKTAME

MISTAKE

Fig. 12. Typical results of word competitions with equal selection values of the different words.

$\dot{x}_i = (A_i + B_i x_{i-1}) x_i$, where x_i is the concentration of the word i, x_{i-1} is that of the preceding word, and A_i and B_i are reproduction parameters to be adjusted.] In nature such an interaction can very easily be realized on the basis of physical or chemical interactions—for example, through complex formation (stabilization) or by a certain regulating or catalytic action. The surprising result of this experiment is that this and all similar types of simple "coupling," as well as branch-confined types of couplings, do *not* lead to stabilization of the item of news. This would only be expected in the first instance if the rate parameter of the coupled propagation reaction (*A*) were greatest for the first word. Otherwise, during the initial competition, in which the couplings play a significant role, the first word dies out and no further stabilization can occur. In nature the adherence to such an order, in which a correlation between the magnitude of the rate parameter and the logical sequence defined by the coupling exists, would in any case present problems. However, it is apparent that even with compliance to this order in the presence of strong coupling, the entire advantage takes effect solely on the last word of the chain. (This can be substantiated mathematically.) The result of a large series of experiments was then similarly always the exclusive selection either of the word *take* or (in the overwhelming number) of the word *mistake*. This latter type of forwarding the advantage suggests another sort of coupling.

4. Actually, this is the only remaining possibility that could achieve the target. We shall try to stabilize the whole sentence by a cyclic coupling of the self-reproductive word units (of which "complex formation" between all four words is an "extreme" case). Symbolically, it is obvious in the

Hypercycle

Selective advantage per error : 2.7

	Mutation - Rate	A	B
TAKE	3.1 %	1.0	4.0
ADVANTAGE	1.4 %	1.0	3.8
OF	6.3 %	1.0	4.3
MISTAKE	1.8 %	1.0	4.2

20 th generation :

```
TAME,  TAKE,  TAKE,  TAKE,  TAKE,  TAKE,
TAKE,  TAKC,  TAKE,  T KE,  TQKI,  TASE,
ADVANTAGE,  ADVAMTACE,  ADVANTAGE,  ADVANTAGI,
ADVANTAGE,  ADVANTAGE,  ADVBNTAGE,  ADVANTAGE
ADVANTAGE,  IDVANTAGE,  ADAANTQGE
PF,  OF,  OF,  :F,
MISTAKE,  MISTAKE,  UISTAKE,  MESTAKE,  MISTAQE
```

Fig. 13. Hypercyclic coupling among self-replicative words. The result is a stabilization of the information content of the whole sentence despite its high average error rate (which would be above threshold if the total sentence were the replicative unit).

sentence at hand that the four terminal letters of the last word are identical to the initial word. If one programs the computer to take account of the consideration of coupling only in the presence of this combination of letters, then for this sentence, cyclic coupling automatically results. The typical product of such an experiment is reproduced in Fig. 13. Every expression (even after many generations) preserves the total information (i.e., all four words), whose sequence results also from the temporal sequence of the oscillation pattern. In agreement with the analytical solution of the differential equation system belonging to this, an oscillation of the presence of words ensues (see Fig. 14), which clearly expresses the cyclic nature of the coupling. In the deterministic case, for small amplitudes, it is a genuine harmonic oscillator. The fluctuations in the game have their origins in the stochastic nature of the process.

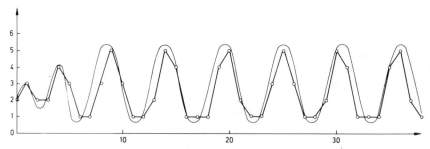

Fig. 14. Oscillation of the abundance of a single word in the hypercyclically coupled sentence, which on average is nearly harmonic (thin line). The abscissa denotes the number of generations.

V. THE HYPERCYCLE AS AN ORDERING PRINCIPLE

We have recognized metabolism, self-reproductivity, and mutageneity as "necessary" prerequisites of material self-organization. These properties guarantee living entities the stable conservation of their genetic information. This places certain demands on the cell's machinery for the processing of information. In population genetics one can start from the empirically assured fact that the necessary machinery for the preservation of information is an integral and inherent constituent of every cell. This assumption is not, however, necessarily valid when dealing with the *origin* of the first cell as the smallest self-productive unit. Also, one cannot proceed on the assumption that Darwinian behavior (in the stricter sense) is "sufficient" for the origin of life. On the contrary, the examples described in the preceding section demonstrate that additional principles for the origin and integration of a sufficiently extensive capacity or information are of significance.

A. Growth and Limitations in the Hypercyclic Systems

Self-reproductive behavior expresses itself, in the simplest cases, in the form of exponential growth curves (provided that the "birth rate" exceeds the "death rate") (see Fig. 15). In the limited system (e.g., the total number of the competing species is held constant), this means selection in the Darwinian sense (see Fig. 16). The inferior type dies out, the superior can preserve itself. Important for this kind of selection is the fact that each mutant can grow at any time, starting from a single copy, provided it excels only through a sufficiently selective advantage. Here the growth parameter is a constant (independent of the population number) that is based on the physical properties of the respective mutant. This is characteristic for Darwinian behavior, namely, that advantageous mutants can emerge at

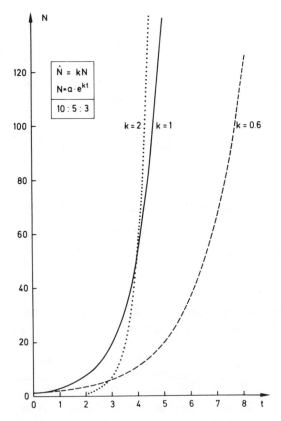

Fig. 15. Exponential growth. The k values are in the ratio $10:5:3$. The species with highest growth rate ($k = 2$) does not originate until at $t = 2$.

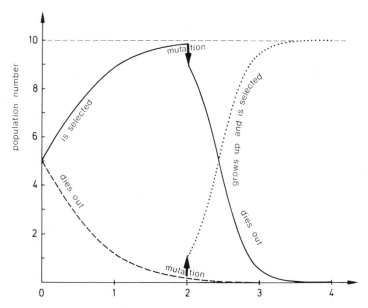

Fig. 16. Effect of limitation on exponential growth (as represented in Fig. 15). Result: selection with reinforcement of advantageous mutants.

any time. Only thus can the profusion of the divergent species be understood, which represented for Darwin the inspiration for thoughts, which connected the principle of descendence with the principles of evolution and selection.

Coexistence of competitors requires some sort of stabilization, which confines the exponential growth law, or the formation of niches that uncouple the competition. Let us imagine, for example, that two species draw their nourishment from mutually independent sources and adjust themselves to a constant rate of production of foodstuffs in their growth behavior. Their population number would then increase only linearly with time instead of exponentially (Fig. 17). A limitation in the total number (e.g., through a death rate dependent on the population number) would then result in coexistence and only manifest itself, for the selective advantage, in a certain increase in the population density (see Fig. 18). Such behavior is certainly to be deduced from the prerequisites of the Darwinian system. It only requires a suitable adjustment of the boundary conditions.

We examine now the rather different case of a "hypercycle," somewhat as we did in the cyclically coupled unit of production of the previously mentioned example. Its growth curve would take the form of a hyperbola (Fig. 19), owing to the greater than linear dependence of the rate on the

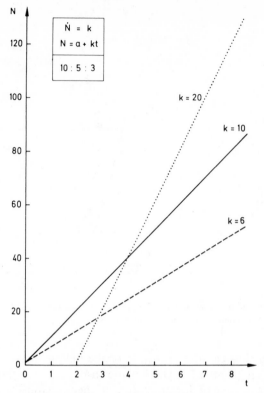

Fig. 17. Linear growth. The ratios of the constants are the same as in Fig. 15.

population count. The significant difference from the exponential function lies in a singularity (i.e., the curve disappears to infinity within a finite time). Such a case can naturally never occur, since limitations would come into play, which result from the limitations of the source of the building material, braking the growth. Next, one could predict that there is actually no great difference between the exponential and hyperbolic growth laws, particularly since one can always approximate a hyperbolic curve in a (finitely) limited region by an exponential function. Figure 20 presents, nonetheless, a significant difference. With limitation, one obtains once again selection, indeed sharper and more discriminating than in the case of pure exponential growth: In the case of exponential growth, mutants, newly emerged and expelled by a selective advantage, could increase. In the case of hyperbolic growth this is not usually possible. Let us imagine approximating, in the critical region of the limitation, the hyperbola by an exponential function. It would be $\exp(k_1' N t)$. An advantage $k_2' > k_1'$ would

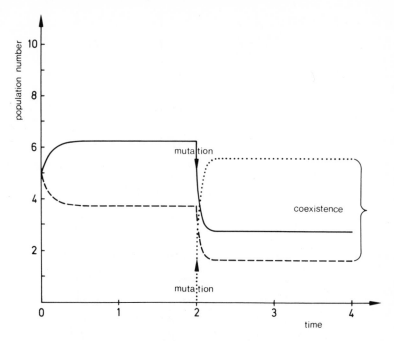

Fig. 18. Effect of limitation on linear growth (as shown in Fig. 17). Result: coexistence in the ratios of growth rates.

be practically of no more use to the mutant,[2] as soon as the system 1 has once established itself and grown to macroscopic dimensions. An order of magnitude $N_1 \approx 10^{10}$ particles/cm^3 would, in the case of molecules, still represent a very dilute solution, namely, about $10^{-11} M$. No selective advantage in the physical parameters that determine k'_2 would be thinkable that could make up for these 10 orders of magnitude for the advantage.

A hyperbolic system thus exhibits a behavior that differs significantly from the Darwinian system. No multiplicity of species would be possible. It would give a universal once = for = ever decision, and only this could optimize itself by yet further integration of information. Darwin had started directly from the multiplicity of species. For that reason, one should not call the once = for = ever selection behavior encountered in hyperbolic systems Darwinian.

The hypercycle is self-reproductive in two senses. It is made up of self- or cyclic-reproductive units, which themselves are connected by a cyclic coupling into a "hyperreproductive" unit. Naturally, there are manifold possibilities for the connection, so that we have to distinguish between the various classes of hypercycles, characterized by the reaction order of the

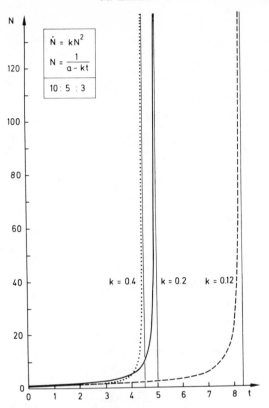

Fig. 19. Hyperbolic growth. The ratios of k are again the same as in Figs. 15 and 17.

coupling terms. The extreme of coupling would be complex formation among all partners (as mentioned earlier) being the prerequisite of any function.

B. Stability of the Hypercycles

Mathematically, one cannot represent the hypercycles in a similarly closed form as the Darwinian system. In general, the solutions of the differential equation system cannot be solved explicitly. Nevertheless, many of the questions of direct interest to us can be answered in a general manner. It is interesting, for instance, whether definite solutions *exist*, above all whether the system is stable, or whether the significant constituents for the existence and function of the hypercycle are coexistent. Of course, they must not fall victim to the once = for = ever selection. Many general investigations of related problems have been undertaken by René

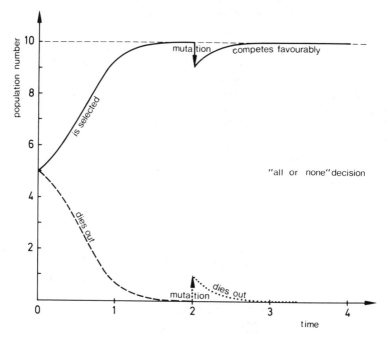

Fig. 20. Effect of limitation on hyperbolic growth (as shown in Fig. 19). Result: "all-or-none" selection. No amplification of later mutants is seen even though these may be distinguished by larger growth parameters k.

Thom.[23] Ilya Prigogine[24] and his school have been foremost in considering the stability viewpoints of chemical reaction systems. We (including Peter Schuster and his group in Vienna) have primarily employed fixed-point as well as more general trajectorial analysis. Especially the behavior in the environment of the stationary solutions in the multidimensional coordinate system x_1 to x_n is examined. The fixed points can be classified by "sinks," "sources," "limit cycles," and so on, or by their pendants in the multidimensional space. The decisive question for the evolutionary ability of an information-integrating system is that of the stability of the respective coupled reaction system, whose components must be coexistent, at the same time behaving in toto selectively with respect to other competitors. For the mathematical analysis this means looking for a stable fixed point, a limit cycle, or an analogous space curve whose coordinates for the population count of all components of the coupled system to be selected are different from zero. At this stage we should not go into more detail of the analysis. It can be found in an exhaustive paper published together with Peter Schuster.[26]

The result of this analysis is that conclusion drawn from the special example in the preceding chapter: The self-replicative components significant for the integration of information reproduce themselves only in a coexistent form when they are connected to one another through cyclic coupling. The mutual stabilization of the components of hypercycles succeeds for more than four partners in the form of nonlinear oscillations, as depicted in the example in Fig. 21 (cf. the oscillation of words in the example of the self-reproductive sentence, Fig. 14). The competitive behavior of the whole hypercycle with respect to all other systems is characterized by once = for = ever selection.

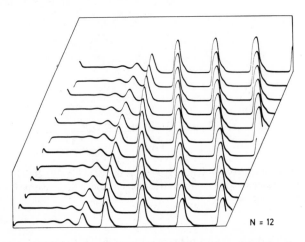

Fig. 21. Oscillation in a hypercyclic system consisting of six self-replicating RNA-cycles (complementary plus and minus strands) and six proteins. The latter are the translation products of the six RNA-plus strands.

VI. MODELS

For an appreciation of biological self-organization, we have to start from concrete models. A general theory can (if it is correct) only lay down minimum conditions. Concrete models, on the other hand, can go beyond this but can also (even if they are consistent and correct within themselves) be irrelevant. Only observations or experiments can decide their relevance.

A. The First Replicative Units

A realistic model for a prebiotic hypercycle must be constituted such that it permits expansion of the limited informational capacity of the nucleic acids, as determined by the physicochemical properties of this class of substance, by integration of a second class of substances with greater functional capacity. On the basis of experimental investigations that have been carried out primarily by Dietmar Pörschke,[27] data have been provided that permit an elucidation of the parameters for the transfer of information and for the significant interactional parameters (formation energies, cooperativity parameters, rate constants, and so on). With these data one can estimate or give limits for the mean single-digit quality parameter $\langle q_i \rangle$, which is responsible for the reproduction of single nucleic acid molecules. In ideal cases one arrives at values that allow a reproductive conservation of information for molecules consisting of up to 100 nucleotide residues. Underlying this estimation is the following mechanism of transfer: The complementary building block (A=U or G≡C) for a segment (A, U, G, or C) of the template strand is stored up as a nucleoside triphosphate (ATP, UTP, GTP, CTP). The differentiation between "correct" and "false" letters follows on the basis of the difference of the energies of interaction. These are sufficiently large only when there is "cooperative" pair formation (e.g., by the storing up of energy-rich building blocks onto a fragment, already completed and attached to the template strand). With cleavage of the pyrophosphate, the nucleoside monophosphate now becomes covalently bonded to the pattern fragment. This terminal is, however, particularly susceptible to hydrolysis. A noncomplementary terminal segment is cleaved more easily again than a complementary one. This effect can, in special cases, be increased by fitting into certain forcing structures. The process of unfolding of secondary or tertiary structures can be assisted by the surfaces of basic proteins (not necessarily of reproducibly fixed structure). However such a reaction may have proceeded (model systems are available in sufficient abundance), the quality of transfer of the single symbol is limited to a great extent by the relatively insignificant difference in interaction energies of complementary and noncomplementary base pairs, which can be enhanced only to a limited degree through the formation of superstructures. In any case, those structures that could significantly protect themselves against hydrolytic decomposition, employing spatial folding, would be advantageous. It is probable that the first structures possessing this reproducibility were precursors of the transfer nucleic acids known today. Many properties, which have come to light in the meantime, possessed by this (probably) oldest stable reproductive class of substance render this assumption justified.

B. Hypercyclic Organization of a Translational Apparatus

An advancement beyond this stage could only be achieved by the incorporation of *new* interactions, and these new interactions had to be just as reproducible as the replicative *t*-RNA-type nucleic acid units. Such a requirement is most suitably accomplished by a translation of the information stored in the replicative units, and this in turn requires a new kind of alphabet distinguished by a greater functional capacity than the original. On the basis of the detailed knowledge available today, which has been accumulated in about 30 years of molecular-biological research, we can take it that the second class of substance is comprised of the proteins, which (historically) were able to develop even before the nucleic acids. Protein molecules or protenoids were probably existent in great numbers as environment factors, but they were not *inherently* reproducible, which would be the prerequisite for any further optimization of any coincidental correlations. Of the approximately 20 *natural* building blocks of the proteins known today (the natural amino acids), in all certainty relatively few were originally around in great quantity, so that they could offer themselves for the development of an original alphabet. For the development of a class of substance having large functional efficiency, units of polar and nonpolar, acidic and basic natures were necessary. The minimal requirements of a prebiotic hypercycle as described earlier had to start from nucleic acid and protein adaptors, having corresponding coupling properties, and for further improvement of the quality of reproduction from a polymerizing factor (i.e., a replicase). The hypercycle depicted in Fig. 22 satisfies in principle these minimum requirements. We have contrasted such a hypercycle with other alternatives, which exhibit the same functional character (translational adaptors, replicase, etc.). The result of a thorough analysis (see the work with Peter Schuster already cited[26]) confirms that the hypercyclic connection alone is capable of integrating the current system of replicative subunits and of stabilization as a functional unit (in competition with units that do not contribute to the translational or replicational function). The simple connection via functions like repli-

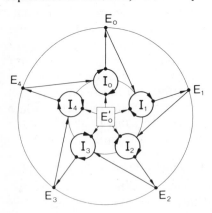

Fig. 22. Scheme of a hypercycle of five self-replicating RNA-cycles (I_0 to I_4) and their translation products (E_0 to E_4), which mediate the couplings. The proteins E_0 to E_4 at the same time provide for the replicase and synthetase functions.

cation, translation, and so on, would not be sufficient for cooperative stabilization of the components.

It is to be noted here that the equation system that represents the hypercycle shown in Fig. 22 contains terms of higher than second order. Thus for the replication of nucleic acids, for example, not only the particular templates (plus and minus strands), but also a general replicase or specific ones (acting as coupling factors) are essential. For the translation of gene information one employs, apart from the respective nucleic acid templates, the complete translational machinery. The fact that hypercycles of different order can be distinguished has already been mentioned. A hypercycle of nth order would require the presence of all n factors for every reaction in the system. Here one thinks immediately of cooperative reaction steps, which demand the presence of a complex made up of all components. The difficulty with such n-fold complexes, though, arises from the fact that their nucleation occurs only at high concentrations. That would mean: *All* factors would have to be present in rather high concentration, before they could distinguish and amplify themselves with respect to other structures. The lower the order, the easier it is to effect stepwise formation.

We notice that the hypercyclic order is a theoretically justifiable, *essential* requirement for the integration of subsystems capable of replication into a unit of greater informational content. *How* this connection appeared historically in detail, however, cannot be predicted by theory. Models can only be checked by their experimental verification. Any optimization of a system requires trying out alternatives; it cannot be achieved purely by chance in a single step. The genetic code appears to be a product of optimization. The information for the whole translational machinery must therefore have originated in a stepwise manner, and by the testing of various combinations. If this information had ever become united into a single replicative nucleic acid unit, it would have to have resulted, as in the evolution of the species, in a multitude of fine-structured codes. The effective process of selection in a hypercyclic system, on the other hand, permits no branching, but optimizes the system emerging first to the exclusion of all other alternatives. The result of this process is in every case a universal code that actually owes its origin to a chance event but whose final outcome (namely, as a product of an optimization process) can no longer be thought of as the mere result of chance.

C. From Hypercycle to Cell

Why do we find today only living entities that propagate as replicative units? The genomal content of a cell is integrated into a unit, which as such is reproduced only when the cell divides.

Even if the hypercycle turns out to be an essential organizational form for the transition from replicative molecules to reproductive, multi-molecular machinery, this does not mean the hypercyclic organizational form represents the optimal *final* form of organization.[28] Already at the first mention of the hypercycles[20] the following was clearly expressed:

From a multitude of replicative units, the hypercycle can discover the one appropriate to one another and, if the combination offers some advantage, amplify them selectively. In this manner, a totally novel function (e.g., translation) can come into being. Furthermore, the hypercycle can optimize all its phenotypical characteristics. What it is not capable of doing, however, is selectively evaluating purely genotypical changes. What is meant here is the selective amplification of a mutation, which demonstrates its advantages only after translation into the phenotype. This advantage takes effect on the successive reactions of the cycle but is not able to couple back selectively to the mutated genotype and to improve this with respect to its forerunners. The genotypical mutant does propagate itself *sympathetically*, but not *selectively*.

A proper evolution of interplay between genotypical and phenotypical functions can finally develop only in a compartment. To differentiate from homogeneous distribution, a compartmentalized hypercycle would be able to exploit selectively *genotypical* advantages in competition with other compartments, owing to preferential neighborhood effects. A favorably mutated compartment would develop more strongly, and thus finally evolve more quickly. Compartmentalization is, however, only of advantage *after* completed hypercyclic organization; otherwise it simply amplifies the competition between the genotypes that are phenotypically independent of one another. The compartmentalization is accomplished by the individualization of all replicative units represented by the hypercycle. This event can be viewed as the birth of the *cell*, the smallest living entity known to us today. Individualization could, however, only occur *after* the proteins had optimized themselves enough, so that, with their aid, the total quantity of symbols of the genome was reproduced sufficiently free of error. Probably this conclusive step of individualization was initiated by an enzyme that connected the open nucleic acid chains to one another and that we today call a ligase. Through the action of such ligase, the whole genome in all probability developed a cyclic structure, just as we encounter it today in the microbes.

VII. CONCLUSIONS

Our reflections on the source of information in "living" systems, or rather, on the self-organization of a material system, which we define as the origin of life, are summarized schematically in Fig. 23.

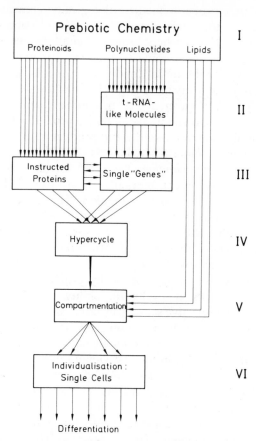

Fig. 23. Scheme of prebiotic evolution.

It begins with a prebiotic phase in which the chemical prerequisites for the origin of life must first be satisfied. Here in a reducing atmosphere, under the action of light, heat, energy-rich radiation, and electrical discharges, the *chemical* "building blocks" were formed. Where they appeared in an energy-rich form, they could also combine to give higher molecular structures. No limitation, other than the law of mass action, was operative. Where reactants with a positive affinity appeared, their reaction could always occur. This phase, which set in after the earth's crust had cooled sufficiently, more than 4000 million years ago, must have lasted for a relatively long time. The starting materials for the synthesis of macromolecules had to accumulate sufficiently, so that the complicated reactions of higher order, which are characteristic of the self-organizational phase, could occur at an appreciable rate.

The proteins were probably the first biologically relevant self-organizational molecules to come out of the macromolecular mass. The experiments of Sidney Fox[29] and others (compiled citations, see Ref. 30) reveal the wide-ranging functional capacity that the spontaneously emerging structures possess. For the initial conditions, this is of great significance; the environment made available a large choice of catalytic functions right from the start. These, however, could not optimize themselves systematically. The route from the spontaneous initial product, determined in all cases by structural advantages, to the end product, selected only from the functional viewpoint, is fantastically long. It consists of innumerable steps. This was where *information originated*, through a selective process in which the majority of noncompetent alternatives were excluded from the start because of a steady progressive evaluation. This process of evolution, toward optimal function of an inherently unstable structure, requires an intrinsic self-reproductivity of the total evolving molecular class. With certain exceptions limited to special structures, this property was not satisfied by the proteins, owing to the principles of their formation. On the other hand, the nucleic acids, historically probably subordinate to the proteins, had the capacity to develop this ability.

The first phase of a selective self-organization was completed at the level of the nucleic acids, in which the presence of proteins may have assisted. The evolution of the first replicative molecular units could be achieved by the laws of natural selection in the Darwinian sense; that is, selective and specified coexisting alternatives permitted. From this phase resulted "quasispecies" consisting of a large number of sufficiently stable and reproducible (t-RNA-type) structures, each of which possessed only the most limited informational content. These individually were much too small to allow the development of a new function. The Darwinian phase of the "creation of molecular carriers of information" therefore must have been superseded during an integrational phase, in which many individual carriers of information were united into a new functional superstructure, which again had to be able to compete, and thereby be selectively optimizable. In this phase the liaison of proteins and nucleic acids was effected.

Proteins and proteinoids were present in sufficient quantity that couplings could develop, which led to translation, and whose functions were amplified through the products formed (particular models are described in Ref. 36). It is important that only the functions of the released structures, and not the structures themselves, had to be amplified by back-coupling. When a definite structure was formed through translation, and also distinguished itself by the function applicable, it could be favored selectively through the self-propagation cycle.

The function had become reproducible, and thereby optimizable. Theory demonstrates that a minimum requirement for the integrating form must have been a hypercyclic organization, although how the detailed structure of the historical hypercycle looked remains open. The consequence of hypercyclic organization is once = for = ever selection. The translational system that first comes into existence will itself constantly improve but will tolerate no new translational system with an alternative configuration. Thus a single code, universal in its fine structure, and universal machinery with a chirality uniform for the whole molecular class could evolve. This was not, however, because everything was unique (in which case it could not have been optimized), but because a selectional mechanism was at work that differed fundamentally from the simple Darwinian formalism. This integrational phase led inevitably to compartmentalization. Because of proximity effects, occurring in the organized systems, this made possible the selective utilization of genotypical changes. It accelerated the optimization and finally permitted the individualization of the compartments into a new superordered replicative unit, the primitive cell. Moreover, this process was necessary, as soon as the prerequisites for a sufficiently precise symbol transfer were fulfilled. It abolishes the strict all-or-none limitation of the hypercyclic phase and allows evolution of a great variety of differential alternatives. The evolution of the species in the sense of Darwin could then begin (even if further discoveries, such as "genetic recombination" were required for its full development).

If we now ask how foolproof such a system is, we have to differentiate rigorously between a theoretically based necessity from assumptions and mere model ideas. The sequence proposed for the events is deducible logically; it cannot be reversed at any stage. Both phases of evolution of individual carriers of information, both on the molecular and on the cellular level, require a selective behavior in the Darwinian sense, with its specifiable "essential" preconditions. Both phases, however, can only become united with one another via a hypercyclic integrating intermediary phase. Here self-reproductive molecules become self-reproductive "cell machineries." Clearly, the difference between integrating and differentiating phases in the picture of the evolutionary tree (Fig. 24) leads to the expression "The roots converge to the stem, which diverges into the branches." The universal code is not the primary reason for the assumption of this convergent phase; rather, it appears now as a consequence of logical reasoning.

The logical necessity for the named steps simply prescribes their "existence" as a minimal requirement. The number of additional steps required for the model actually to be realizable remains open.

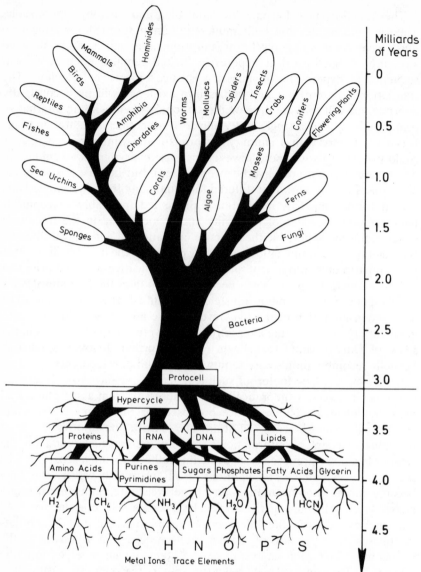

Fig. 24. The "Tree of Evolution."

Has the question posed in the title of this paper—"How does information originate?"—been answered sufficiently? The discussion has shown that several very *general* definable preconditions need to be satisfied, so that with certainty our ideas are not limited to the area of genetics. On the other hand, it has also been shown that in the early phases of evolution, totally variant types of mechanisms appeared that were specially adapted to the particular problem. Investigations in connection with the origin of information in the immune system,[35-38] about the determination of cells in the differentiation of organs,[39] and ideas on self-organization in the central nervous system[40,41] show that, despite all the differences in the problems, structures, and boundary conditions, the thoughts expressed in this article should be of significance beyond the realm of molecular genetics.

References

1. W. Köhler, *Die Aufgabe der Gestaltspsychologie*, Walter de Gruyter, New York, 1971.
2. C. E. Shannon and W. Weaver, *The Mathematical Theory of Communication*, University of Illinois Press, Urbana, 1971.
3. L. Brillouin, *Science and Information Theory*, Academic, New York, 1962.
4. E. Wigner, Proc. of the Robert A. Welch Found.
5. L. Boltzmann, *Ann. Phys.*, **57**, 773 (1896); **60**, 392 (1897).
6. Th. Mann, *Der Zauberberg*, Fischer-Verlag, Berlin, 1924.
7. W. A. Bentley and W. J. Humphreys, *Snow Crystals*, Dover, New York, 1931.
8. J. D. Watson, *The Molecular Biology of the Gene*, Benjamin, New York, 1970.
9. C. Darwin, The Origin of Species, London, 1859, 6. Aufl. 1872.
10. R. A. Fisher, *Proc. Roy. Soc.* **B141**, 510 (1953).
11. J. B. S. Haldane, *Proc. Cam. Phil. Soc.*, **23**, 838 (1927).
12. S. Wright, *Genetics*, **16**, 97 (1931).
13. (a) M. Eigen and R. Winkler, *Das Spiel*, Piper-Verlag, Munich, 1975; (b) M. Eigen and R. Winkler, in *Mannheimer Forum 73/74*, Studienreihe Boehringer Mannheim, 1974.
14. P. a. T. Ehrenfest, *Phys. Z.*, **8**, 311 (1907).
15. L. Onsager, *Phys. Rev.*, **37**, 405 (1931); **38**, 2265 (1931).
16. J. von Neumann and O. Morgenstern, *Theory of Games and Economic Behavior*, Princeton University Press, Princeton, N.J., 1953.
17. L. Michaelis and M. L. Menten, *Biochem. Z.*, **49**, 333 (1913).
18. M. Eigen, *Nobel Symp. No. 5*, S. 334, Wiley-Interscience, New York, 1966.
19. E. Schrödinger, *What Is Life?*, Cambridge University Press, New York, 1944.
20. M. Eigen, *Naturwissenschaften*, **58**, 465 (1971).
21. B. L. Jones, R. H. Enns, and S. S. Rangnekar, *Bull. Math. Biol.*, **38**, 15 (1976).
22. C. J. Thompson and J. L. McBride, *Math. Biosci.*, **21**, 127 (1974).
23. R. Thom, *Stabilité Structurelle et Morphogénèse*, Benjamin, Reading, Mass., 1972.
24. P. Glansdorff and I. Prigogine, *Thermodynamic Theory of Structure, Stability, and Fluctuations*, Wiley-Interscience, New York, 1971.
25. M. W. Hirsch and S. Smale, *Differential Equations, Dynamical Systems, and Linear Algebra*, Academic, New York, 1974.
26. M. Eigen and P. Schuster, *Naturwissenschaften*, **64**, 541 (1977), **65**, 7 (1978) and **65** (1978) in press.
27. (a) D. Pörschke, *Biopolymers*, **10**, 1989 (1971); (b) M. Eigen and D. Pörschke, *J. Mol. Biol.*, **53**, 123 (1970); (c) D. Pörschke, O. C. Uhlenbeck, and F. H. Martin, *Biopolymers*, **12**, 1313 (1973).
28. E. U. von Weizsäcker, in *Offene Systeme*, Klett-Verlag, Stuttgart, 1974.
29. S. W. Fox, in *The Origin of Prebiological Systems and of Their Molecular Matrices*, Academic, New York, 1965.

30. K. Dose and H. Rauchfuss, *Chemische Evolution und der Ursprung lebender Systeme*, Wiss. Verlagsges, Stuttgart, 1975.
31. S. Spiegelman, *Q. Rev. Biophys.*, **4**, 213 (1971).
32. M. Sumper and R. Luce, *Proc. Natl. Acad. Sci. (USA)*, **72**, 162 (1975).
33. B. Küppers and M. Sumper, *Proc. Natl. Acad. Sci. (USA)*, **72**, 2640 (1975).
34. C. K. Biebricher and L. E. Orgel, *Proc. Natl. Acad. Sci. (USA)*, **70**, 934 (1973).
35. N. K. Jerne, *Ann. Immunol. (Inst. Pasteur)*, **125C,** 373 (1974).
36. P. H. Richter, in *Theoretical Immunology*, Dekker, New York, 1976.
37. G. W. Hoffmann, *Eur. J. Immunol.*, **5,** 638 (1975).
38. G. Adam and E. Weiler, in *The Generation of Antibody Diversity*, Academic, London, 1976.
39. H. R. Wilson and J. D. Cowan, *Kybernetik*, **13,** 55 (1973).
40. Ch. v. d. Malsburg, *Kybernetik*, **14,** 85 (1973).
41. E. Domingo, R. A. Flavell, and C. Weissmann, *Gene* **1,** 3 (1976).

PATTERN FORMATION IN REACTING AND DIFFUSING SYSTEMS

G. NICOLIS, T. ERNEUX,
and
M. HERSCHKOWITZ-KAUFMAN

*Faculté des Sciences de l'Université Libre de Bruxelles,
Brussels, Belgium*

CONTENTS

One of the most striking and intriguing aspects of natural phenomena is that complex systems, involving a large number of strongly interacting elements, can form and maintain *patterns of order* extending over a macroscopic space and time scale. From the most "elementary" level of temporal organization of simple chemical networks to the most "macroscopic" level of development and functioning of multicellular organisms or even of societal systems, concepts such as regulation, information, and communication play a prominent role. They are associated with a state of high *coherence*, which is perhaps the most striking feature of complex systems.

What are the driving forces that induce such a coherent behavior and how common can this phenomenon be? Imagine we were able to reconstitute a complex biochemical pathway and put all reactants in a vessel that is closed to mass and energy transfer from the outside world. Then, after a sufficient lapse of time, the system would settle to a final state characterized by the complete absence of dynamical phenomena at a macroscopic scale. The reason is that each "forward" process tending to produce a certain chemical substance in excess at a given point or at a given instant would be canceled by an equally probable "backward" process. This statement, known as *detailed balance*, characterizes the states near thermodynamic equilibrium.

Conversely, a system may be constrained to remain away from thermodynamic equilibrium and detailed balance if it is subject, permanently or transiently, to a flux of matter and/or energy. In this case a permanent, or transient, difference between local and background concentrations can be achieved and give rise to different kinds of dynamical behavior, including the emergence of spatial patterns or of rhythmic phenomena.

A most striking result of thermodynamic theory of irreversible processes pioneered by Ilya Prigogine[1,2] was to prove that the transition to ordered behavior requires a critical distance from equilibrium, as well as nonlinear kinetics. Under these conditions the branch of steady states corresponding to the extrapolation of equilibriumlike behavior as the distance from equilibrium increases may become unstable. The system then evolves to a new regime that may correspond to a spatially or temporally organized state. These regimes have been called by Prigogine *dissipative structures*. It has been shown that the transition to a dissipative structure is governed by a thermodynamic stability criterion involving excess entropy production (i.e., the amount of dissipation introduced by the disturbance causing the instability). In this way, pattern formation is connected with thermodynamic functions of direct experimental interest.

It is hardly necessary to convince the reader that systems undergoing nonlinear kinetics and subject to a flux of matter and/or energy are extremely common: Nonlinearity seems to be a general rule in such diverse fields as biochemistry (e.g., through regulatory steps involving allosteric enzymes) or fluid dynamics (e.g., through Reynolds stresses). Plant life or large-scale atmospheric phenomena are directly influenced by the flux of electromagnetic energy from the sun; cell membranes are subject to quite appreciable composition and electrical potential differences; and so forth. It is therefore legitimate to expect that rhythmic phenomena and pattern formation should be a *generic property* of a large class of physicochemical systems.

The purpose of the present paper is to analyze some general mechanisms and principles underlying pattern formation. Part I deals with problems amenable to a deterministic description. We first discuss briefly the status of reaction-diffusion equations. In Section I.B the type of information obtained from linear stability analysis is summarized. Section I.C deals with bifurcation theory applied to the first instability leading to pattern formation. Secondary bifurcations are analyzed in Sections I.D to I.F, with particular emphasis on time-periodic solutions. Section I.G is devoted to general comments. Section II deals with the role of fluctuations in the onset of self-organization phenomena. After a brief survey of recent results we discuss, in Sections II.B and II.C, some analogies with equilibrium phase transitions. The role of external noise in bifurcation phenomena is outlined in Section II.D, whereas Section II.E, the final section, summarizes the principal viewpoint adopted throughout the paper.

We do not discuss experimental examples or biological problems, as these are covered elsewhere in this volume.

I. DETERMINISTIC ANALYSIS

A. Reaction-Diffusion Equations: Search for Archetypes

We consider a reacting mixture in a volume V. The mixture comprises a number of "major" chemicals $\{A_i\}$, $\{E_i\}$, whose concentrations (generally in excess) are taken to be controlled externally. In addition, the reactions give rise to N variable intermediates $X_1, \ldots, X_N$.

The system is taken to be isothermal, without convection, and subject to time-independent boundary conditions. Moreover, diffusion within V is approximated by Fick's law and the diffusion coefficient matrix is taken to be diagonal as well as space and time independent. Let X_i be the composition variables. Their time evolution is given by:[3]

$$\frac{\partial}{\partial t} X_i = f_i(\{X_j\}, \lambda) + D_i \nabla^2 X_i \quad \text{(reaction-diffusion equations)} \quad \text{(I.1)}$$

The source term $f_i(\{X_j\}, \lambda)$ describes the effect of chemical reactions. According to well-known laws of chemical kinetics, it is a *nonlinear* (usually polynomial) function of X_j values. λ is a set of characteristic parameters like rate constants or major reactant concentrations. Note that, with a suitable change of vocabulary, (I.1) become identical to the evolution equations encountered in such diverse situations as laser physics, superradiance, population dynamics, or even sociology.

We want to analyze the type of behavior predicted by (I.1). Whatever the answer to this question might be, we require that it satisfy *structural stability*. In other words: a slight change of the right-hand sides of (I.1)— introduced, for instance, by varying some of the parameters λ by $\delta\lambda$— should lead to solutions remaining in a neighborhood of $0(\delta\lambda)$ of the solutions of the original system. The breakdown of this requirement will signal the transition to a qualitatively new behavior and can be used to provide the boundaries of sets of parameter values that, once crossed, lead to pattern formation.

This program has been applied successfully to the elucidation of the generic behavior of dynamical systems described by certain classes of ordinary differential equations.[4] Unfortunately, the complexity of the nonlinear partial differential system (I.1) has so far compromised the extension of these methods to reaction-diffusion systems. Hence we adopt a different procedure, which may be summarized as follows.

1. Suppose for a moment that the diffusion terms are absent in (I.1). Since $f_i(\{X_j\}, \lambda)$ does not depend explicitly on space and time, the relations

$$f_i(\{X_j^0\}, \lambda) = 0 \qquad (I.2a)$$

define a set of numbers $\{X_j^0\}$ that are space and time independent. If, moreover, X_j^0 are physically acceptable:

$$X_j^0 \geq 0 \qquad (I.2b)$$

they will represent uniform and steady-state solutions of the (truncated) differential system (I.1). If now diffusion is switched on, these numbers may no longer be solutions of the enlarged system, as they will not satisfy in general the boundary conditions. To avoid such spurious phenomena that have nothing to do with self-organization and pattern formation, we limit ourselves to systems subject to either of the following types of boundary conditions:

$$\{X_j^\Sigma\} = \{X_j^0\} \qquad \text{(Dirichlet condition)} \qquad (I.3a)$$

or

$$\{\mathbf{n} \cdot \nabla X_j^\Sigma\} = \{0\} \qquad \text{(Neumann or zero-flux conditions)} \qquad (I.3b)$$

where $\mathbf{n}$ is the normal to the surface Σ (see Fig. 1). We are thus certain that under (I.3a) or (I.3b) the system is allowed to be in state $\{X_j^0\}$, which is obviously the most symmetrical state allowed by the outside world.* As such, it is the prototype of disorder. In this respect it is reminiscent of equilibriumlike behavior and will be referred to hereafter as *thermodynamic branch*.

2. Bearing these points in mind, we now view pattern formation as a *transition* from this disordered configuration on the thermodynamic branch to a new type of solution of the rate equations (I.1). On physical grounds, we require the latter to be stable and to be generated spontaneously by small macroscopic fluctuations. The simplest situation compatible with

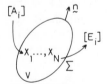

Fig. 1. An open system subject to the flow of initial and final chemicals $\{A_i\}$ and $\{E_i\}$, respectively, and containing n intermediates $X_1, \ldots, X_n$ within the reaction volume V. Here $\mathbf{n}$ is the outward directed normal to the surrounding surface Σ.

* The reader has realized by now that the boundary conditions express the constraints acting on the system from the external world, as discussed in the introduction.

these requirements is one where the disordered branch of states loses its stability. This problem can be tackled by the powerful tools of *stability theory*, which are summarized in Section I.B.

3. If such an analysis confirms that the most symmetrical state is unstable, and thus untenable, in what direction will the system evolve? If the equations admit new branches of stable solutions, from where do these branches emanate? Again the simplest situation is one where various types of solution coalesce when one of them, in our case the disordered branch, loses its stability. This is the phenomenon of branching, or *bifurcation*, and is depicted in Fig. 2. Note that in both stability and bifurcation one essentially deals with the parametric dependence of certain types of solution.

4. Suppose now that bifurcation can indeed take place. The "critical" values λ_c of the parameters defining both the loss of stability of the "reference" solution and the emergence of new solutions provide, then, the conditions for pattern formation to occur. Note the similarity with the viewpoint pioneered by Thom[4] in the context described earlier in this section. In both cases, pattern formation is associated with a singular behavior of the solutions. This is obvious from Fig. 2, where branches (a)–(d) and (b)–(c) cannot be joined smoothly. As we see in subsequent sections, this is reflected by a generally nonanalytic dependence of bifurcating branches on the parameter λ.

Having reduced pattern formation to a stability and bifurcation problem, we can dispose of the powerful tools of bifurcation theory[5] to tackle the question of the behavior of the systems past the instability of the disordered branch of solution. This is the object of Sections I.C. to I.E. A final remark is, however, in order.

In the vicinity of a bifurcation point a system is necessarily structurally unstable. It is conceivable that small perturbations (resulting, for instance, from imperfections or impurities) play an important role and can even compromise the existence of bifurcation.[6] In the latter case the disordered branch and the pattern will not be joined in the vicinity of λ_c but will be separated by a finite distance. How typical are then the results obtained from bifurcation theoretical analysis?

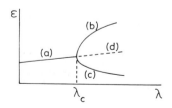

Fig. 2. A bifurcation diagram illustrating the branching of solutions as a function of a bifurcation parameter λ. Here ε is a characteristic function of the amplitude of the available solutions. Full and broken lines denote, respectively, stable and unstable solutions. (*a*), (*d*), thermodynamic branch; (*b*), (*c*), stable dissipative structures.

We do not attempt to answer this difficult question. Rather, we propose to regard the mechanism of pattern formation via instability and bifurcation as an *archetype* of the behavior of complex systems. We intend to focus on some qualitative aspects of the solutions and to show that they bear strong similarities to the patterns observed in actual physical situations where chemical reactions and diffusion are present. The question of quantitative *modeling* of concrete physicochemical problems by reaction-diffusion equations leading to instability and bifurcation is not raised in our paper. The reader is referred to a recent monograph by Nicolis and Prigogine[3] and to the paper by Hess, Goldbeter, and Lefever in this volume for a detailed presentation of this aspect of self-organization phenomena.

B.   Linear Stability Analysis

The "principle" of linearized stability[5] permits us to obtain a great deal of information about the solutions of the nonlinear system (I.1) by studying the linearized form of these equations around some reference state. In this section we consider the behavior around the thermodynamic branch $\{X_i^0\}$, postponing until Section I.D the analysis of the stability properties of dissipative structures.

Let ϕ_m, $-k_m^2$ be the eigenfunctions and eigenvalues of the Laplace operator within the volume V and under the boundary conditions (I.3):

$$\nabla^2 \phi_m = -k_m^2 \phi_m \tag{I.4}$$

Setting

$$X_i = X_i^0 + x_i \tag{I.5}$$

and linearizing (I.1) with respect to x_i we obtain

$$\frac{\partial}{\partial t} x_i = \sum_j \left(\frac{\partial f_i}{\partial X_j}\right)_0 x_j + D_i \nabla^2 x_i \tag{I.6}$$

As the coefficients $(\partial f_i / \partial X_j)_0$ are space and time independent, (I.6) admits solutions of the form:

$$x_i(r, t) = a_i e^{\omega_m t} \phi_m(r) \tag{I.7}$$

Inserting into (I.6), we find that ω_m must satisfy the *characteristic equation*[3]

$$\det \left| \left(\frac{\partial f_i}{\partial X_j}\right)_0 - (\omega_m + D_i k_m^2)\delta_{ij}^{kr} \right| = 0 \tag{I.8}$$

This relation implies, in particular, that ω_m is an eigenvalue of the differential-matrix operator:

$$L \equiv \{L_{ij}\}$$

$$L_{ij} = \left(\frac{\partial f_i}{\partial X_j}\right)_0 + D_i \nabla^2 \delta_{ij}^{kr}$$

(I.9)

Solving (I.8), we obtain the values of ω_m in terms of k_m and the system's parameters λ, D_i:

$$\omega_m = \omega_m(k_m^2, \lambda, \{D_i\})$$

(I.10)

From this relation one can obtain the following sort of information.

1. One may first compute how the real part Re ω varies in terms of λ for a given k_m. If for a certain λ_c, Re ω switches from negative to positive values, the reference state will switch from asymptotic stability to instability [cf. (I.7)]. The situation is depicted in Fig. 3.

We expect that the evolution past the instability will lead to different kinds of solutions according to whether the imaginary part Im ω is nonzero or vanishes. Actually, a clearcut answer to this question is available when ω_m is a simple eigenvalue of L. One then shows[5] that for Im $\omega = 0$ one has bifurcation of steady-state solutions, whereas for Im $\omega \neq 0$ one has bifurcation of time periodic solutions.

2. For a given λ above λ_c one may also compute the rate of amplification of various modes, as shown in Fig. 4.

As the spectrum of k_m is discrete for a finite system, there will be a finite number of unstable modes. Moreover, one expects that the fastest-growing modes will determine the properties of the solution the system will evolve to beyond the instability. Computer simulations confirm this point in the case of steady-state solutions.[7]

3. At the onset of the instability, Re $\omega_m = 0$, (I.10) provides a relation between the bifurcation parameter λ and the eigenvalue k_m which in fact determines the wavenumber of the spatial mode:

$$\lambda = \lambda(k_m^2)$$

(I.11)

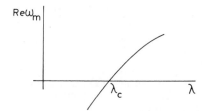

Fig. 3. Behavior of the real part of the (simple) eigenvalue of the operator L in (I.9), in the vicinity of a bifurcation point λ_c.

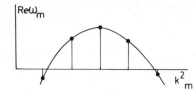

Fig. 4. Real part of the eigenvalues ω_m corresponding to the unstable modes emerging from the thermodynamic branch.

This is referrred to as *marginal stability* relation. A few typical situations are depicted in Fig. 5. Let us focus on curve *b*. As the value of λ is increased starting from the stable region, the first instability will take place at a set of values (λ_c, k^2_{mc}) that is both compatible with the boundary conditions and close to the minimum of the curve. In simple situations the critical wavenumber k^2_{mc} will be a well-isolated eigenvalue of the Laplacian. In other cases, notably in symmetric spatial domains, the spacing between eigenvalues will be small or even zero (multiple eigenvalues). Some interesting phenomena arising when this happens are discussed in Sections I.D and I.E.

4. In general, k^2_m will depend on the dimensions of the domain. For instance, for a one-dimensional system of length l (I.4) implies

$$k^2_m = \frac{m^2 \pi^2}{l^2}, \qquad m = 0, 1, \ldots \qquad (I.12)$$

The stability diagrams (Fig. 5) can therefore also be expressed in terms of the length l. An important conclusion that comes from this is that instability cannot take place unless the length l exceeds a critical value l^*. The biological implications of this result are numerous and are discussed in detail elsewhere.[3]

In addition to the type of information discussed so far in this section, linear stability analysis gives also some indications about the behavior that can be expected from systems involving one, two, or several variable intermediates.

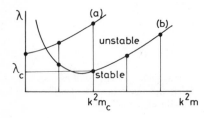

Fig. 5. Linear stability diagram associated with the bifurcation of time-periodic solutions (curve a) and steady-state solutions (curve b).

For one intermediate, the characteristic equation (I.8) becomes

$$\omega_m = \left(\frac{\partial f}{\partial X}\right)_0 - Dk_m^2 \qquad (I.13)$$

This relation implies, of course, that time-periodic solutions are not possible, since ω_m is necessarily real. Moreover, as diffusion plays always a stabilizing role, the *first* instability from the thermodynamic branch for boundary conditions of the type (I.3) will necessarily lead to space-independent solutions.

For two intermediates the characteristic equation is of second degree and is, therefore, compatible with an oscillatory instability and the bifurcation of time-periodic solutions. There are, however, some restrictions that have to be fulfilled by the kinetics: the reaction sequence must comprise a trimolecular or a higher-order step.[8,9] As regards space dependencies, one can show that diffusion may play a destabilizing role and therefore be responsible for the emergence of new branches of space-dependent solutions at the first bifurcation point.

The situation is qualitatively similar for systems involving three or more intermediates, except that a trimolecular step is no longer necessary. For this reason we shall illustrate in the sequel pattern formation on a simple trimolecular model involving two variables:

$$A \rightarrow X$$

$$B + X \rightarrow Y + D$$

$$2X + Y \rightarrow 3X \qquad (I.14)$$

$$X \rightarrow E$$

A, B, D, E are the major reactants and X, Y stand for the variable intermediates. In suitably rescaled variables[10] the rate equations take the form

$$\frac{\partial X}{\partial t} = A - (B + 1)X + X^2 Y + D_1 \nabla^2 X$$

$$\frac{\partial Y}{\partial t} = BX - X^2 Y + D_2 \nabla^2 Y \qquad (I.15)$$

where D_1, D_2 are, respectively, the diffusion coefficients of X and Y.

This model has been extensively studied in the literature (see Ref. 3 for a survey). Section I.C gives a short account of the most important results pertaining to the first bifurcation from the thermodynamic branch.

C. Symmetry-Breaking: The First Bifurcation from
the Thermodynamic Branch

In this and the subsequent two sections we deal with the analytic study of bifurcating solutions of the trimolecular model (I.14) and (I.15). We focus on the asymptotic regimes, both stable and unstable, that may arise in *bounded media*. The knowledge of such regimes is of primary importance because of their influence on the general evolution as well as on the possible transient regimes that may arise from various initial conditions.

We begin with a one-dimensional system of length l, subject to Dirichlet or to zero-flux boundary conditions. The uniform steady-state solution on the thermodynamic branch is

$$X_{st} = A, \qquad Y_{st} = \frac{B}{A} \tag{I.16}$$

The linearized operator L [see (I.9)] is also easily found to be

$$L \equiv \begin{pmatrix} B - 1 + D_1 \nabla^2 & A^2 \\ -B & -A^2 + D_2 \nabla^2 \end{pmatrix} \tag{I.17}$$

Its eigenvectors $\begin{pmatrix} u_m \\ v_m \end{pmatrix}$ are of the form

$$\begin{pmatrix} u_m \\ v_m \end{pmatrix} = \begin{pmatrix} 1 \\ C_m \end{pmatrix} \sin \frac{m\pi r}{l}, \qquad m = 1, 2, \ldots \tag{I.18a}$$

for the boundary conditions (I.3a), and

$$\begin{pmatrix} u_m \\ v_m \end{pmatrix} = \begin{pmatrix} 1 \\ C_m \end{pmatrix} \cdot \cos \frac{m\pi r}{l} \tag{I.18b}$$

for the boundary conditions (I.3b). The eigenvalue ω_m can be computed straightforwardly and the explicit form of the marginal stability condition (I.11) can be derived. Using B as bifurcation parameter, the results are as follows.[3]

1. A complex eigenvalue has a positive real part if

$$B > \tilde{B}_m = 1 + A^2 + (D_1 + D_2) \frac{m^2 \pi^2}{l^2} \tag{I.19}$$

The (purely imaginary) eigenvalues are simple along the marginal stability curve $B = \tilde{B}_m$ [Fig. 5, curve (a)].

2. One (real) eigenvalue becomes positive if

$$B > B_m = 1 + \frac{D_1 A^2}{D_2} + \frac{A^2 l^2}{D_2 m^2 \pi^2} + D_1 \frac{m^2 \pi^2}{l^2} \tag{I.20}$$

The shape of the marginal stability curve $B = B_m$ is as in Fig. 5, curve b. Here B_c is the value of B corresponding to the value of k_m^2 closest to the minimum of this curve.

3. As B increases, the first instability will take place when B crosses the value

$$B = \min (B_c, \tilde{B}_0) \qquad (\text{I.21a})$$

for zero-flux conditions, and

$$B = \min (B_c, \tilde{B}_1) \qquad (\text{I.21b})$$

for Dirichlet conditions.

From these relations one can derive conditions on A, D_1, D_2 ensuring that the first instability will involve real or complex conjugate eigenvalues ω_m. Let us first consider the case of real ω_m values and assume that there is no degeneracy.

We insert the decomposition (I.5) into the full nonlinear rate equations (I.15). Moreover, we decompose the operator L (I.17) as follows:

$$L \equiv L_B = L_c + (L - L_c) = L_c + \begin{pmatrix} (B - B_c) & 0 \\ -(B - B_c) & 0 \end{pmatrix} \qquad (\text{I.22})$$

where L_c is the operator L evaluated at B_c. As we deal with simple eigenvalues, we know from the preceding section that B_c is necessarily a bifurcation point of steady-state solutions.

We obtain in this way the differential system (setting also $\partial x/\partial t = \partial y/\partial t = 0$, since we are interested in the steady-state solutions):

$$L_c \begin{pmatrix} x \\ y \end{pmatrix} = \begin{pmatrix} -1 \\ 1 \end{pmatrix} h(x, y) \qquad (\text{I.23})$$

where

$$h(x, y) = 2Axy + \frac{B}{A} x^2 + x^2 y + (B - B_c)x \qquad (\text{I.24})$$

Close to the bifurcation point the corrections to the thermodynamic branch $(A, B/A)$ should be small. We express this by expanding both $\begin{pmatrix} x \\ y \end{pmatrix}$ and $\gamma = B - B_c$ in terms of a small parameter ϵ:

$$\begin{pmatrix} x \\ y \end{pmatrix} = \epsilon \begin{pmatrix} x_0 \\ y_0 \end{pmatrix} + \epsilon^2 \begin{pmatrix} x_1 \\ y_1 \end{pmatrix} + \cdots$$

$$\gamma = B - B_c = \epsilon \gamma_1 + \epsilon^2 \gamma_2 \cdots \qquad (\text{I.25})$$

The second relation in particular enables one to determine the leading singularity reflecting the $B - B_c$ dependence of the solution from the differential system itself, rather than impose it *a priori*.

Introducing the expansion into (I.23) and identifying equal powers of ϵ, we obtain a set of relations of the form

$$L_c\begin{pmatrix} x_k \\ y_k \end{pmatrix} = \begin{pmatrix} -1 \\ 1 \end{pmatrix} a_k, \qquad k = 0, 1, \ldots \tag{I.26}$$

together with the boundary conditions

$$x_k(0) = x_k(l) = y_k(0) = y_k(l) = 0$$

or

$$\frac{dx_k(0)}{dr} = \frac{dx_k(l)}{dr} = \frac{dy_k(0)}{dr} = \frac{dy_k(l)}{dr} = 0 \tag{I.27}$$

The first few coefficients a_k are

$$a_0 = 0$$

$$a_1 = \gamma_1 x_0 + \frac{B_c}{A} x_0^2 + 2A x_0 y_0 \tag{I.28}$$

$$a_2 = \gamma_2 x_0 + \left(\gamma_1 + 2 \frac{B_c}{A} x_0 + 2A y_0 \right) x_1 + 2A x_0 y_1 + \gamma_1 \frac{x_0^2}{A} + x_0^2 y_0$$

The first relation implies that

$$L_c\begin{pmatrix} x_0 \\ y_0 \end{pmatrix} = 0 \tag{I.29}$$

Recalling (I.17) and (I.18) as well as the results of Section I.B, we conclude that $\begin{pmatrix} x_0 \\ y_0 \end{pmatrix}$ is proportional to the zero eigenvector $\begin{pmatrix} u_{m_c} \\ v_{m_c} \end{pmatrix}$ of L_c. Thus L_c is not invertible. To solve the *inhomogeneous* equations (I.26) for $k > 1$ we need therefore a *solvability condition*. The latter, known as the Fredholm alternative, is given by the following theorem:[5]

Theorem. The vector $\begin{pmatrix} x_k \\ y_k \end{pmatrix}$ is a solution of (I.26) provided the right-hand side $\begin{pmatrix} -a_k \\ a_k \end{pmatrix}$ is orthogonal to the null eigenvector of the adjoint operator L_c^*.

Here *orthogonality* is defined by means of a scalar product that is a combination of the scalar product familiar from vector analysis and the scalar product defined in function spaces such as Hilbert spaces. This, together with the observation that the space dependence of the eigen-vectors of the adjoint operator L_c^* is the same as that of $\begin{pmatrix} u_{m_c} \\ v_{m_c} \end{pmatrix}$ leads to the following quantitative expression of the theorem (depending on the choice of boundary condition):

$$\int_0^l dr \begin{vmatrix} \sin \dfrac{m_c \pi r}{l} \\ \cos \dfrac{m_c \pi r}{l} \end{vmatrix} a_k(\{x_{k-m}(r), y_{k-m}(r)\}) = 0, \quad 0 < m \le k, \ k = 1, \ldots \quad (I.30)$$

These relations, together with (I.28), determine entirely the coefficients γ_i. Next, from the second relation (I.25), one determines ϵ as a function of $(B - B_c)$. Inserting this calculated ϵ into the first relation (I.25) and solving the inhomogeneous equations (I.26), one has an explicit expression for the solution $\begin{pmatrix} x \\ y \end{pmatrix}$.

As an example, we give the first two terms of the perturbation series of $\begin{pmatrix} x(r) \\ y(r) \end{pmatrix}$ for zero-flux conditions[7]

$$\begin{pmatrix} x(r) \\ y(r) \end{pmatrix} = \pm \left(\frac{B - B_c}{\gamma_2} \right)^{1/2} \begin{pmatrix} 1 \\ C_{m_c} \end{pmatrix} \cos \frac{m_c \pi r}{l}$$

$$+ \left(\frac{B - B_c}{\gamma_2} \right) \begin{vmatrix} p_0 + p_2 \cos \dfrac{2 m_c \pi r}{l} \\ q_0 + q_2 \cos \dfrac{2 m_c \pi r}{l} \end{vmatrix} + \cdots \quad (I.31a)$$

with

$$\gamma_2 = -2 A q_0 - A q_2 - A C_{m_c} p_2 - \frac{B_c}{A} p_2 - \frac{3}{4} C_{m_c}$$

$$C_{m_c} = \frac{\left(\dfrac{D_1 m_c^2 \pi^2}{l^2} + 1 - B_c \right)}{A^2}$$

$$p_0 = 0, \qquad q_0 = -\frac{1}{A^2} \left(A C_{m_c} + \frac{B_c}{2A} \right) \quad (I.31b)$$

$$\begin{pmatrix} p_2 \\ q_2 \end{pmatrix} = \frac{(A C_{m_c} + B_c/2A)}{3\left(A^2 - 4 D_1 D_2 \dfrac{m_c^4 \pi^4}{l^4} \right)} \begin{vmatrix} -4 D_2 \dfrac{m_c^2 \pi^2}{l^2} \\ 1 + 4 D_1 \dfrac{m_c^2 \pi^2}{l^2} \end{vmatrix}$$

The $\pm$ in front of the dominant term reflects the fact that $\gamma_1 = 0$ in this problem. It implies a *twofold multiplicity* of the dissipative structures arising at the first bifurcation point. Stated differently, beyond the transition the system has equal *a priori* probability to evolve to two different solutions, depending on the initial conditions. The coefficient γ_2 depends on A, D_1, D_2, l and may be positive or negative. In the first case the bifurcating branches exist in the supercritical region $B > B_c$. A theorem of birfucation theory[5] guarantees then their stability in the vicinity of B_c.

The most important property of the dissipative structures described by (I.31) is, of course, their *symmetry-breaking* character. When the critical value B_c is crossed, the most symmetrical solution $(A, B/A)$ of the rate equations ceases to be stable and the system evolves to a regime with a lesser spatial symmetry. In the case described by (I.31) this symmetry-breaking is accompanied by a double degeneracy translated by the "critical exponent" $\frac{1}{2}$ describing the way the "order parameter"—the amplitude of the solution $\begin{pmatrix} x(r) \\ y(r) \end{pmatrix}$—varies in terms of $B - B_c$.

The bifurcation of time-periodic solutions follows the same lines. One has, of course, to keep the time derivatives in the rate equations. Introducing the scaled variable

$$\tau = \Omega t \qquad (I.32)$$

where Ω is the (unknown) frequency of the solution, we write these equations in the form

$$\mu_m \frac{\partial}{\partial \tau} \begin{pmatrix} x_n \\ y_n \end{pmatrix} - L_{\tilde{B}_m} \begin{pmatrix} x_n \\ y_n \end{pmatrix} = \sum_{k:1}^{n} \tilde{w}_k \frac{\partial}{\partial \tau} \begin{pmatrix} x_{n-k} \\ y_{n-k} \end{pmatrix} + \begin{pmatrix} a_n \\ -a_n \end{pmatrix} \qquad (I.33)$$

where we have set

$$\begin{pmatrix} x \\ y \end{pmatrix} = \epsilon \begin{pmatrix} x_0 \\ y_0 \end{pmatrix} + \epsilon^2 \begin{pmatrix} x_1 \\ y_1 \end{pmatrix} + \cdots$$

$$B = \tilde{B}_m + \epsilon \tilde{\gamma}_1 + \epsilon^2 \tilde{\gamma}_2 + \cdots \qquad (I.34)$$

$$\Omega = \mu_m + \epsilon \tilde{\omega}_1 + \epsilon^2 \tilde{\omega}_2 + \cdots$$

$$\mu_m = Im\omega_m$$

The terms $\tilde{B}_m$ and ω_m are given by (I.19) and by (I.10) of the linear stability analysis, and $\begin{pmatrix} a_n \\ -a_n \end{pmatrix}$ has the same structure as earlier in this section.

To solve the inhomogeneous equation (I.33) we require an extended solvability condition, whereby an integral over the period is taken in the definition of the scalar product. For zero-flux boundary conditions we finally obtain for the first bifurcation[11]

$$\binom{x(t)}{y(t)} = \left(\frac{B-1-A^2}{\tilde{\gamma}_2}\right)^{1/2} \binom{\cos \Omega t}{\tilde{B}_0 \rho \cos (\Omega t + \lambda)}$$

$$+ \left(\frac{B-1-A^2}{\tilde{\gamma}_2}\right) \left[\begin{matrix} 2a_2 \cos (2\Omega t + \psi) \\ b_0 + 2b_2 \cos (2\Omega t + \varphi) \end{matrix}\right] + \cdots \qquad (\text{I.35a})$$

$$\Omega(B) \simeq A + \left(\frac{B-1-A^2}{\tilde{\gamma}_2}\right) \tilde{\omega}_2$$

The coefficients ρ, λ, ψ, φ, $\tilde{\gamma}_2$, a_2, b_2, b_0, $\tilde{\omega}_2$ are explicit functions of the parameter A defined by the following relations:

$$\tilde{B}_0 \rho \, e^{i\lambda} = \frac{1}{A}(-A+i); \quad b_0 = \frac{1}{2A^3}(A^2-1)$$

$$b_2 \, e^{i\varphi} = \frac{1}{12A^3} - \frac{5}{12A} + i\left(\frac{1}{3A^2} - \frac{1}{6}\right); \quad a_2 \, e^{i\psi} = \frac{1}{3A} + \frac{i}{6A^2}(A^2-1) \quad (\text{I.35b})$$

$$\tilde{\gamma}_2 = \frac{1}{2A^2}\left(1 + \frac{A^2}{2}\right); \quad \tilde{\omega}_2 = -\frac{A}{6} - \frac{1}{6A^3} + \frac{7}{24A}$$

It turns out that $\tilde{\omega}_2$ is always negative. The remarkable point is that in the model considered in the present section—and indeed in all models involving only two variable intermediates—the first bifurcation of time-periodic solutions leads necessarily to *homogeneous* oscillations. Such solutions are known as limit cycles and play an important role in biology as discussed in the paper by Hess, Goldbeter, and Lefever in the present volume.

Computer simulations confirm the existence of bifurcations from the thermodynamic branch. Moreover, they suggest a surprisingly large multiplicity of patterns as one moves away from the first bifurcation point. To interpret this important feature we need to explore further the bifurcation diagram of our model reaction. This is the object of Sections I.D to I.F.

D. Successive Primary Bifurcations

In the supercritical region, beyond the first bifurcation, the system exhibits a multitude of qualitatively different patterns. The appearance of these new solutions can be understood on the basis of the marginal stability relations (I.19) and (I.20). Each time B is crossing a point $B = B_m$ (or $B = \tilde{B}_m$) corresponding to one of the marginal stability curves a further branching from the thermodynamic branch takes place. Depending on the parameter values, the successive bifurcating solutions may be all steady-state space-dependent solutions (case of real eigenvalues), either all time-periodic solutions (case of complex eigenvalues) or alternatively steady state or time periodic. In addition to these *primary* bifurcations emerging from the "trivial" reference state one can also get *secondary* bifurcations emerging from a previously established pattern, and arising from the interactions between the various possible solutions.

Let us briefly indicate the analytical expressions describing the first few primary branches in each of the cases mentioned earlier.

1. Succession of Steady-State Patterns.   The null eigenfunctions of the linear operator L are all time independent and the new solutions bifurcating at $B = B_m$ will be similar to those bifurcating at $B = B_c$, which have been described in Section I.C. However, instead of having a dominant spatial dependence given by mode $\cos m_c \pi r / l$ (or $\sin m_c \pi r / l$, depending on the boundary conditions), they will now contain the fundamental mode $\cos m \pi r / l$ (or $\sin m \pi r / l$). In the case of zero-flux boundary conditions described by (I.31) the new branches will again have the property of double degeneracy characterizing the first bifurcating branch.

2. Succession of Time-Periodic Solutions.   When the system is subject to zero-flux (or periodic) boundary conditions, the first bifurcating solution, starting at $\tilde{B}_0$, is always a uniform limit cycle given by relation (I.35). In contrast, the new branches appearing at $\tilde{B}_m$, $m = 1, 2, \ldots$, are characterized by a nonzero wavenumber. For zero-flux boundary conditions the analytic calculations for the second branch, following the method outlined in (I.C), lead to the following expressions in the neighborhood of the bifurcation point $\tilde{B}_1$:

$$\begin{pmatrix} x \\ y \end{pmatrix} = \begin{pmatrix} X_1 - A \\ Y_1 - B/A \end{pmatrix} = \epsilon \begin{pmatrix} \cos \Omega t \\ \tilde{B}_1 \rho' \cos (\Omega t + \lambda') \end{pmatrix} \cos \frac{\pi r}{l}$$

$$+ \epsilon^2 \sum_{k : 0, 2} \begin{pmatrix} a_{k0} + 2a_{k2} \cos (2\Omega t + \psi_k) \\ b_{k0} + 2b_{k2} \cos (2\Omega t + \varphi_k) \end{pmatrix} \cos \frac{k \pi r}{l} \ldots \quad \text{(I.36)}$$

where

$$\epsilon = \left(\frac{(B - \tilde{B}_1)}{\tilde{\gamma}_2'} \right)^{1/2} \quad \text{and} \quad \Omega = \mu_1 + \epsilon^2 \tilde{\omega}_2' + 0(\epsilon^4)$$

The coefficients appearing in (I.36) are explicit functions of A, D_1, and D_2, and are, in the limit of small D_1 and D_2, conveniently related to the coefficients (I.35b) by the following relations:

$$\rho' \to \rho \qquad \lambda' \to \lambda$$

$$a_{00} \to 0 \qquad a_{20} \to 0 \qquad a_{02} \to \frac{a_2}{2} \qquad a_{22} \to \frac{a_2}{2}$$

$$b_{00} \to \frac{b_0}{2} \qquad b_{20} \to \frac{b_0}{2} \qquad b_{02} \to \frac{b_2}{2} \qquad b_{22} \to \frac{b_2}{2}$$

$$\psi_0 = \psi_2 = \psi \qquad\qquad \varphi_0 = \varphi_2 = \varphi$$

$$\tilde{\gamma}_2' \to \frac{3}{4}\tilde{\gamma}_2 \qquad \mu_1 \to A \qquad\qquad \tilde{\omega}_2' \to \frac{3}{4}\tilde{\omega}_2$$

(I.37)

An interesting situation appears when the second bifurcation point is one of multiple branching arising from multiple eigenvalues. This is the case for periodic boundary conditions. Consider a closed ring of length l:

$$X(0, t) = X(l, t)$$
$$Y(0, t) = Y(l, t)$$

(I.38a)

At $\tilde{B}_0 = 1 + A^2$, the uniform steady state (I.16) becomes unstable and a stable limit cycle solution appears. The next bifurcation, at $\tilde{B}_1' = 1 + A^2 + (D_1 + D_2)4\pi^2/l^2$ is characterized by a fourfold multiplicity. Indeed, according to the linear stability analysis, the purely imaginary eigenvalue of the operator $L(\tilde{B}_1')$ is now degenerate and there are four corresponding critical eigenvectors:

$$\phi_1 = \begin{pmatrix} 1 \\ \alpha \end{pmatrix} e^{i\tau} \cos\frac{2\pi r}{l}, \qquad \phi_2 = \begin{pmatrix} 1 \\ \bar{\alpha} \end{pmatrix} e^{-i\tau} \cos\frac{2\pi r}{l}$$

$$\phi_3 = \begin{pmatrix} 1 \\ \alpha \end{pmatrix} e^{i\tau} \sin\frac{2\pi r}{l}, \qquad \phi_4 = \begin{pmatrix} 1 \\ \bar{\alpha} \end{pmatrix} e^{-i\tau} \sin\frac{2\pi r}{l}$$

(I.38b)

where

$$\alpha \simeq \tilde{B}_0 \rho\, e^{i\lambda} = \frac{i - A}{A} \quad \text{for } D_1, D_2 \to 0$$

This means that more than one time-periodic solution may be expected to emerge from $\tilde{B}_1'$ as pictured in Fig. 6. In the same way as for zero fluxes at the boundaries, the asymptotic branches can be evaluated around $\tilde{B}_1'$ by

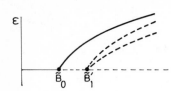

Fig. 6. Bifurcation diagram displaying the successive primary bifurcations that may arise from the thermodynamic branch. Here ϵ is a characteristic function of the amplitude of the solutions available ($\epsilon = 0$ corresponds to the uniform reference state (I.16)). Thin and thick lines correspond to steady-state and time-periodic solutions respectively, whereas full and broken lines denote, respectively, stable and unstable solutions.

expanding x, y and Ω in power series of ϵ:

$$\begin{pmatrix} x \\ y \end{pmatrix} = \begin{vmatrix} X - A \\ Y - \dfrac{B}{A} \end{vmatrix} = \epsilon \begin{pmatrix} x_0 \\ y_0 \end{pmatrix} + \epsilon^2 \begin{pmatrix} x_1 \\ y_1 \end{pmatrix} + \cdots$$

$$\Omega = \mu'_1 + \epsilon \tilde{\omega}'_1 + \epsilon^2 \tilde{\omega}'_2 + \cdots$$

with

$$T = \frac{2\pi}{\Omega} \quad \text{and} \quad B - \tilde{B}'_1 = \epsilon \tilde{\gamma}'_1 + \epsilon^2 \tilde{\gamma}'_2 + \cdots$$

The first approximation $\begin{pmatrix} x_0 \\ y_0 \end{pmatrix}$ to the solution vector has now to be constructed out of the four eigenvectors given in (I.38b). This yields

$$\begin{pmatrix} x_0 \\ y_0 \end{pmatrix} = \beta_1 \phi_1 + \bar{\beta}_1 \phi_2 + \beta_2 \phi_3 + \bar{\beta}_2 \phi_4$$

The complex coefficients β_1, β_2 are determined in the perturbation procedure in the same way as $\{\tilde{\gamma}'_n\}$ and $\{\tilde{\omega}'_n\}$ by means of two orthogonality conditions and given initial conditions. The detailed calculations show that two distinct spatiotemporal regimes, differing in their qualitative properties, bifurcate from $\tilde{B}'_1$. One of these solutions is completely analogous to the solution given by (I.36) for impermeable boundaries and has the general form:

$$\begin{pmatrix} x \\ y \end{pmatrix} \simeq \epsilon \begin{pmatrix} \cos \Omega t \\ \tilde{B}_0 \rho \cos (\Omega t + \lambda) \end{pmatrix} \left(c \cos \frac{2\pi r}{l} + c' \sin \frac{2\pi r}{l} \right) + 0(\epsilon^2)$$

with

$$\Omega \simeq A + 0(\epsilon^2); \qquad \epsilon = \left(\frac{(B - \tilde{B}'_1)}{\tilde{\gamma}'_2} \right)^{1/2}$$

and

$$\tilde{\gamma}'_2 \simeq \tfrac{3}{4}(c^2 + c'^2)\tilde{\gamma}_2 \tag{I.39}$$

c, c' are two constants fixed by the initial conditions. The second solution is given in first approximation by

$$\begin{pmatrix} x \\ y \end{pmatrix} \simeq \epsilon \begin{pmatrix} \cos\left(\Omega t + \dfrac{2\pi r}{l}\right) \\ \tilde{B}_0 \rho \cos\left(\Omega t + \lambda + \dfrac{2\pi r}{l}\right) \end{pmatrix} + 0(\epsilon^2) \qquad (\text{I.40})$$

where again

$$\Omega \simeq A + 0(\epsilon^2); \qquad \epsilon = \left(\frac{(B - \tilde{B}'_1)}{\tilde{\gamma}'_2}\right)^{1/2}$$

but

$$\tilde{\gamma}'_2 \simeq \tilde{\gamma}_2$$

It corresponds to a traveling wave solution with a characteristic propagation velocity (see also Ref. 11). These solutions are further discussed in the paper by Hanusse, Ross, and Ortoleva in this volume.

3. Succession of Steady-State or Time-Periodic Solutions. When the bifurcation parameter crosses a value $B = B_m$ (or $\tilde{B}_m$), the corresponding eigenfunctions of the differential-matrix operator L are either time independent or time periodic. An interesting situation, which we consider later, corresponds to the alternative branching of a steady, space-dependent solution given by (I.31) and a space-independent limit cycle oscillation like that in (I.35).

The stability of these successive primary bifurcating branches is not guaranteed *a priori*, as in the case of the first supercritical branch, in the vicinity of its bifurcation point. Indeed, the stability properties of the new solutions are related to the eigenvalues of operator L, which for $B > B_m(\tilde{B}_m)$, $B_m \neq B_c(\tilde{B}_m \neq \tilde{B}_0)$ has already at least one eigenvalue with positive real part. To know in what region, if any, these new branches can be stabilized, one has thus to perform explicit stability calculations of the primary branches. In this respect, secondary bifurcations have a particular significance in that they can accompany an exchange of stability between two distinct nontrivial solutions. This situation is of special interest in a system with only two variables when periodic or zero-flux boundary conditions apply. Since the first time-periodic solution is always uniform in space, it is only through the stabilization of the subsequent time-periodic branches by a secondary bifurcation that spatiotemporal solutions can appear when situation (2) applies. In the case (3) of mixed steady-state or time-periodic branching, stable, spatiotemporal regimes may also result from a secondary bifurcation appearing on the first primary solution. This is further discussed in Section I.E.

E.   Stability of the Successive Primary Solutions and Secondary Bifurcations

The problem of stability of the successive primary branches can be formulated in a general form as follows:

Let (X_m, Y_m) be a primary solution—stationary or time periodic—bifurcating from the homogeneous state $(A, B/A)$ at $B = B_m(\tilde{B}_m)$; the (small) perturbation $\begin{pmatrix} u \\ v \end{pmatrix} = \begin{pmatrix} X \\ Y \end{pmatrix} - \begin{pmatrix} X_m \\ Y_m \end{pmatrix}$ is solution of the linearized set of equations around (X_m, Y_m):

$$\frac{\partial}{\partial t}\begin{pmatrix} u \\ v \end{pmatrix} = \begin{pmatrix} (2X_m Y_m - B - 1)u + X_m^2 v + D_1 \dfrac{\partial^2}{\partial r^2}u \\[2ex] (B - 2X_m Y_m)u - X_m^2 v + D_2 \dfrac{\partial^2}{\partial r^2}v \end{pmatrix} \tag{I.41}$$

with

$$0 \le r \le l = 1$$

and

$$\left[\frac{\partial}{\partial r}\begin{pmatrix} u \\ v \end{pmatrix}\right]_{r=0} = \left[\frac{\partial}{\partial r}\begin{pmatrix} u \\ v \end{pmatrix}\right]_{r=1} = 0$$

for zero flux boundary conditions. We want to know whether the perturbations are bounded for all times, or whether they grow in time.

According to the general stability theory,[5] the solutions of (I.41) are linear combinations of solutions of the form

$$\begin{pmatrix} u \\ v \end{pmatrix} = e^{\nu t}\Gamma(r)$$

if (X_m, Y_m) is stationary, or:

$$\begin{pmatrix} u \\ v \end{pmatrix} = e^{\nu t}\Gamma(r, \tau)(\tau = \Omega t)$$

where $\Gamma(r, \tau)$ is 2π-periodic in τ, if (X_m, Y_m) is 2π-periodic in τ. This gives us the following eigenvalue problems associated with the differential equations (I.41): When (X_m, Y_m) is stationary:

$$\nu\Gamma = \begin{pmatrix} 2X_m Y_m - B - 1 + D_1 \dfrac{\partial^2}{\partial r^2} & X_m^2 \\[2ex] B - 2X_m Y_m & -X_m^2 + D_2 \dfrac{\partial^2}{\partial r^2} \end{pmatrix}\Gamma$$

with

$$0 \le r \le 1, \qquad \left(\frac{\partial}{\partial r}\Gamma\right)_{r=0} = \left(\frac{\partial}{\partial r}\Gamma\right)_{r=1} = 0 \tag{1.42a}$$

When (X_m, Y_m) is 2π-periodic in τ:

$$\nu\Gamma + \frac{\partial}{\partial t}\Gamma = \begin{vmatrix} 2X_mY_m - B - 1 + D_1\dfrac{\partial^2}{\partial r^2} & X_m^2 \\ B - 2X_mY_m & -X_m^2 + D_2\dfrac{\partial^2}{\partial r^2} \end{vmatrix} \Gamma$$

$$\tag{I.42b}$$

$$0 \le r \le 1; \qquad \left(\frac{\partial}{\partial r}\Gamma\right)_{r=0} = \left(\frac{\partial}{\partial r}\Gamma\right)_{r=1} = 0$$

We know from Section I.C that (X_m, Y_m) converges to an analytical series in a small parameter ϵ when B approaches B_m, its bifurcation point. Under these conditions ν and Γ may also be expanded in convergent power series of the same small parameter

$$\nu = \nu^0 + \epsilon\nu^1 + \epsilon^2\nu^2 + \cdots \tag{I.43}$$

$$\Gamma = \Gamma_0 + \epsilon\Gamma_1 + \epsilon^2\Gamma_2 + \cdots \tag{I.44}$$

Introducing (I.43) and (I.44) into the equations (I.42), we obtain at each order in ϵ a set of linear equations with constant coefficients of the form:

$$M\Gamma_0 = 0 \tag{I.45}$$

$$M\Gamma_1 = \rho_1(\Gamma_0, \nu^0) \tag{I.46}$$

$$M\Gamma_2 = \rho_2(\Gamma_0, \Gamma_1, \nu^0, \nu^1) \tag{I.47}$$

and in general:

$$M\Gamma_n = \rho_n(\Gamma_0, \Gamma_1, \ldots, \Gamma_{n-1}, \nu^0, \nu^1, \ldots, \nu^{n-1}) \tag{I.48}$$

The operator M is defined by

$$M \equiv \begin{vmatrix} -B_m + 1 - D_1\dfrac{\partial^2}{\partial r^2} + \nu^0 & -A^2 \\ B_m & A^2 - D_2\dfrac{\partial^2}{\partial r^2} + \nu^0 \end{vmatrix}$$

for stationary (X_m, Y_m), or

$$M \equiv \begin{vmatrix} -\tilde{B}_m + 1 - D_1\dfrac{\partial^2}{\partial r^2} + \nu^0 + \dfrac{\partial}{\partial t} & -A^2 \\ \tilde{B}_m & A^2 - D_2\dfrac{\partial^2}{\partial r^2} + \nu^0 + \dfrac{\partial}{\partial t} \end{vmatrix}$$

for time-periodic (X_m, Y_m).

The systems of inhomogeneous systems equations (I.46) to (I.48) will admit a solution if, and only if, each ρ_n, $n = 1, 2, 3, \ldots$ satisfies some solvability conditions;[5] in general, this is only possible for particular values of the unknown parameters such as $\nu^1, \nu^2, \ldots$. The homogeneous system (I.45), on the other hand, can present two types of solutions:

a. $\text{Re}\,(\nu^0) = 0$ (with $\text{Im}\,(\nu^0)$ different or equal to zero)

b. $\text{Re}\,(\nu^0) \neq 0$ (with $\text{Im}\,(\nu^0)$ different or equal to zero)

In the first case, calculations at the next order give in first approximation

$$\text{Re}\,(\nu) = -a(B - B_m) + 0((B - B_m)^2)$$

where a is a positive real number only depending on the parameters describing the reference solution (X_m, Y_m). $\text{Re}\,(\nu)$ is negative if (X_m, Y_m) is defined supercritically $(B > B_m)$, which will be assumed to be the case in the sequel. In the second case, $\text{Re}\,(\nu^0)$ can be either positive or negative. All $\text{Re}\,(\nu^0)$ are negative when (X_m, Y_m) is the first bifurcating solution; one $\text{Re}\,(\nu^0)$ is positive when (X_m, Y_m) is the second primary branch, two $\text{Re}\,(\nu^0)$ are positive when (X_m, Y_m) bifurcates in the third place, and so forth.

To get more information on the stability properties of the different primary solutions one has now to consider specific situations of successive primary branching. For simplicity we limit ourselves to values of the bifurcation parameter B only involving the first two primary branches and follow the behavior of the eigenvalue with smallest (but nonzero) real part at $\epsilon = 0$ when ϵ is different from zero but still small. We suppose furthermore that in the small vicinity of the two first bifurcation points, the other eigenvalues remain negative.

1. Succession of Two Stationary Solutions.
From Section I.C we have for $i = m$ or m'

$$\begin{pmatrix} X \\ Y \end{pmatrix} = \begin{pmatrix} A \\ \dfrac{B}{A} \end{pmatrix} + \epsilon \begin{pmatrix} 1 \\ C_i \end{pmatrix} \cos i\pi r + 0(\epsilon^2)$$

$$\epsilon = \left(\frac{B - B_i}{\gamma_2} \right)^{1/2}$$

(I.49)

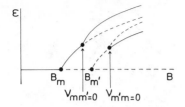

Fig. 7. Bifurcation diagram displaying the successive primary and (conjectured) secondary bifurcations.

where the coefficients are defined in (I.31b). By solving (I.45) to (I.48) we obtain:

$$\nu_{ii'} = \nu_{ii'}^0 + \epsilon^2 F_1(A, D_1, D_2, m) \tag{I.50}$$

where $F_1(A, D_1, D_2, m)$ is a complicated function of the parameters which may be either positive or negative depending on the parameter values; $i = m$ (or m') denotes that the reference solution of which the stability is considered bifurcates at $B = B_i$; $i' = m'$ (or m) stands for the fact that the corresponding eigenvector has a basic wavenumber i':

$$\mathbf{\Gamma}_{ii'} = \begin{pmatrix} 1 \\ C'_{i'} \end{pmatrix} \cos i'\pi r + 0(\epsilon)$$

$$C'_{i'} = (D_1 i'^2 \pi^2 - B_i + 1 + \nu_{ii'}^0)/A^2 \tag{I.51}$$

$\nu_{ii'}^0$ is the smallest solution of the quadratic equation

$$(\nu_{ii'}^0)^2 - \nu_{ii'}^0[B_i - 1 - A^2 - (D_1 + D_2)i'^2 \pi^2] + D_2 i'^2 \pi^2 (B_{i'} - B_i) = 0 \tag{I.52}$$

In the limit of $\delta = (B_{i'} - B_i) \to 0$ it is given by

$$\nu_{ii'}^0 = \frac{-D_2 i'^2 \pi^2 \delta}{|B_i - 1 - A^2 - (D_1 + D_2)i'^2 \pi^2|} \tag{I.53}$$

so that $\nu_{mm'}^0$ is negative and $\nu_{m'm}^0$ is positive.

When the sign of the coefficient of ϵ^2 in (I.50) is opposite to that of $\nu_{ii'}^0$, $\nu_{ii'}$ tends to zero when ϵ grows, thus indicating a possible stability change of the primary solution (X_i, Y_i). This situation can be realized for different sets of values of the parameters. The problem of the validity of the expansions (I.43) to (I.44) when $\nu_{ii'} = 0$ will be considered in more detail in subsection 4.

2. Succession of a Steady-State and a Time-Periodic Regime (the Homogeneous Limit-Cycle Solution).

Two cases have to be considered: The structured steady-state and uniform time-periodic solutions are given, respectively, by (I.31) and (I.35). By solving (I.45) to (I.48) we now obtain

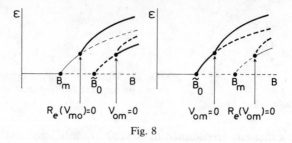

Fig. 8

for the stationary solution, bifurcating at B_m:

$$\text{Re}\,(\nu_{m0}) = \text{Re}\,(\nu_{m0}^0) + \epsilon^2 F_2(A, D_1, D_2, m) \qquad (I.54)$$

$F_2(A, D_1, D_2, m)$ is here again a complicated function of the parameters. $\text{Re}\,(\nu_{m0}^0)$ can be rewritten in terms of $\delta = B_m - \tilde{B}_0$:

$$\text{Re}\,(\nu_{m0}^0) = \frac{\delta}{2}$$

The corresponding eigenvector is

$$\Gamma_{m0} = \begin{pmatrix} 1 \\ C_0' \end{pmatrix} + 0(\epsilon) \qquad (I.55)$$

For the time-periodic solution, bifurcating at $\tilde{B}_0$, one has a similar expression for ν_{0m}:

$$\nu_{0m} = \nu_{0m}^0 + \epsilon^2 F_3(A, D_1, D_2, m) \qquad (I.56)$$

ν_{0m}^0 is the smallest solution of

$$(\nu_{0m}^0)^2 + \nu_{0m}^0(D_1 + D_2)m^2\pi^2 + D_2 m^2\pi^2(B_m - \tilde{B}_0) = 0 \qquad (I.57)$$

and the related eigenvector is given by

$$\Gamma_{0m} = \begin{pmatrix} 1 \\ C_m'' \end{pmatrix} \cos m\pi r + 0(\epsilon) \qquad (I.58)$$

$$C_m'' = -\frac{(B_0 - 1 - D_1 m^2 \pi^2 - \nu_{0m}^0)}{A^2} \qquad (I.59)$$

If $\delta = (B_m - \tilde{B}_0) \to 0$ we obtain for ν_{0m}^0:

$$\nu_{0m}^0 = \frac{-D_2\delta}{(D_1 + D_2)} + 0(\delta^2) \qquad (I.60)$$

Here again, if the parameters are such that the sign of the coefficient of ϵ^2 in the expressions (I.54) or (I.56) is opposite to that, respectively, of $\text{Re}\,(\nu_{m0}^0)$ or ν_{m0}^0, a stability change of the limit cycle solution ($\text{Re}\,(\nu_{m0}) = 0$) or of the space-dependent structure ($\nu_{m0} = 0$) is conceivable.

3. Succession of Two Time-Periodic Solutions. We now consider the two time-periodic solutions bifurcating respectively, at $\tilde{B}_0$ and $\tilde{B}_1$. The first solution is the homogeneous limit cycle solution calculated in (I.3), and the following branch, which exhibits a basic wavenumber of one, is given by (I.36).

By solving again (I.45) to (I.48) we obtain for the homogeneous time-periodic solution:

$$\text{Re}\,(\nu_{01}) = -\frac{1}{2}(D_1 + D_2)\pi^2 - \epsilon^2\frac{\tilde{\gamma}_2}{2} + 0(\epsilon^4)$$

$$\Gamma_{01} = \begin{pmatrix} 1 \\ C \end{pmatrix} e^{i\tau}\cos\pi r + 0(\epsilon)$$

(I.61)

with

$$C(D_1 = 0, D_2 = 0) = \frac{1}{A}(-A + i)$$

(I.62)

and the following for the inhomogeneous time-periodic solution:

$$\text{Re}\,(\nu_{10}) \simeq \frac{1}{2}(D_1 + D_2)\pi^2 - \frac{\epsilon^2}{8}\tilde{\gamma}_2 + 0(\epsilon^4)$$

$$\Gamma_{10} = \begin{pmatrix} 1 \\ C \end{pmatrix} e^{i\tau} + 0(\epsilon)$$

(I.63)

For $D_1, D_2 \to 0$ and C given by (I.62).

One sees immediately from the expressions (I.61) and (I.63) that the first tendency for the homogeneous solution is to remain stable and for the spatiotemporal regime to undergo a stabilization.

It is tempting in the second case to use expression (I.63) to determine for $D_1, D_2 \to 0$ the first approximation to the critical value $\epsilon_c^2 = (\tilde{B}_{1c} - \tilde{B}_1)/\tilde{\gamma}_2'$ where Re (ν_{10}) becomes zero. The validity of this conjecture is investigated below.

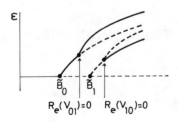

Fig. 9

4. Validity Range of the Perturbation Expansions (I.43) and (I.44) at the Point of Secondary Bifurcation. The problem of stability changes accompanying secondary bifurcation has been recently investigated in the situation described in Section I.E.1 by Mahar and Matkowsky[12] and in the situations of Sections I.E.1 and 2 by Keener.[13] Both analyses are carried out for fixed boundary conditions and in the limit of ϵ_c and $\delta = B_m - B_{m'}$ small. In a forthcoming paper Erneux and Herschkowitz-Kaufman[14] analyze the existence of secondary critical points for a system subject to zero-flux boundary conditions, whatever the values of ϵ_c and δ.

The analysis shows that, provided $\delta \to 0$, ϵ_c and Γ_c are in first approximation equal to the values obtained when $\nu_{ii'} = 0$ [or Re $(\nu_{ii'}) = 0$] by the perturbation expansions (I.43) and (I.44), which are valid for $\epsilon \to 0$. Thus in this limit, the behavior of the critical eigenvalues $\nu_{ii'}$ is well described by (I.50) for the situation of Section I.E.1 and (I.54 to I.56) for that of Section I.E.2, when ϵ goes from zero to ϵ_c.

The problem of two interacting time-periodic solutions, described in Section I.E.3, is more complex. Erneux and Herschkowitz-Kaufman[14] have developed a convergent perturbation procedure for ϵ_c and Γ_c when $D_1, D_2 \to 0$:

$$\epsilon_c = D^{1/2}b_0 + Db_1 + \cdots$$

$$\Gamma_c = \Gamma_{0c} + D^{1/2}\Gamma_{1c} + \cdots$$

with

$$D = D_1 \quad \text{and} \quad \theta = \frac{D_2}{D_1} = 0(1)$$

This analysis confirms the persistence of stability for the homogeneous limit-cycle solution and the stabilization of the spatiotemporal regime. However, it also shows the inadequacy of the expansion (I.43) and (I.44) to produce the correct critical value ϵ_c [corresponding to Re $(\nu_{ii'}) = 0$] even in the limit $D_1, D_2 \to 0$. Therefore, contrary to the cases of Sections I.E.1 and 2, expression (I.55) for Re (ν_{10}) remains good in the small vicinity of $\epsilon = 0$ only.

The eigenvalue problem (I.41) has also been approached numerically by a Galerkin-type procedure. We set

$$\Gamma = \sum_{j:1}^{N} \binom{p_j}{q_j} \phi_j$$

using for the N functions ϕ_j, the functions appearing in the perturbation procedure up to the order ϵ^3:

$$\{\phi_j\} = \{1; e^{ij\tau}, e^{-ij\tau}, j = 1, 2, 3\}$$

$\left\{\binom{p_j}{q_j}\right\}$ are the unknown coefficients to determine.

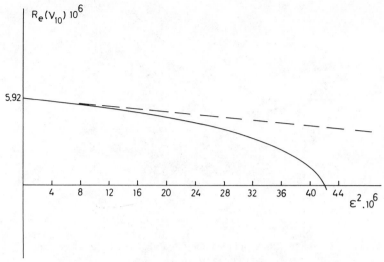

Fig. 10. A Galerkin-type of procedure is used with 11 eigenfunctions. Here Re (ν_{10}) is plotted as function of $\epsilon^2 = (B - \tilde{B}_1)/\tilde{\gamma}_2'$. It diverges rapidly from the calculated value given by (I.63), valid for B sufficiently near $\tilde{B}_1$. $A = 2$, $D_1 = 8 \times 10^{-7}$, $D_2 = 4 \times 10^{-7}$.

Figure 10 illustrates for very small values of D_1 and D_2 how the behavior of Re (ν_{01}) deviates from that of expression (I.63) as ϵ grows.

On the other hand, the stability of the uniform limit cycle in the neighborhood of the first two bifurcation points $\tilde{B}_0$ and $\tilde{B}_1$ can also be confirmed numerically using the property[15]

$$2 \operatorname{Re} (\nu_{01}) = -(D_1 + D_2)\pi^2 + \theta \qquad (I.64)$$

where

$$\theta = \frac{1}{T} \int_0^T (2X_0 Y_0 - B - 1 - X_0^2)\, dt \qquad (I.65)$$

T is the period of (X_0, Y_0). Figure 11 gives the behavior of θ for $A = 2$ and values of B between 5.0 and 5.4. One observes that θ remains negative for these values, thus establishing the stability of the periodic solution (I.35) in this range.

F. Numerical Results

The time evolution of X and Y was also analyzed numerically on a CDC 6500 computer for parameter values corresponding to different situations described in Sections I.C to I.E. The simulations provide a striking confirmation of the *multiplicity of simultaneously stable dissipative structures* in the supercritical region $B > B_c$ (or $\tilde{B}_c$). Some examples follow.

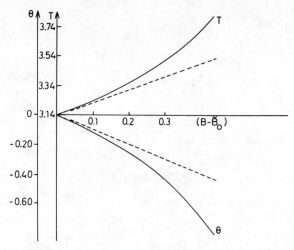

Fig. 11. Here θ, given by (I.65), and the period T of the limit-cycle solution are plotted as function of $(B - \bar{B}_0)$ for $A = 2$.

We used for the integration of (I.15) a Runge-Kutta method with variable stepsize combined with a finite difference approximation for the Laplacian operator.

1. Bifurcation of Steady-State Patterns. The numerical values chosen correspond to the conditions prescribed by (I.20). Figure 12 illustrates

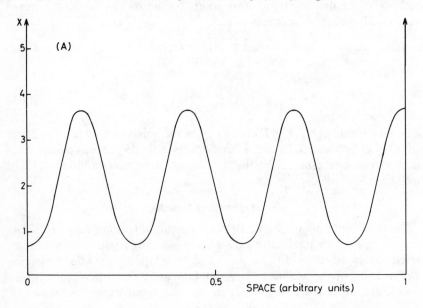

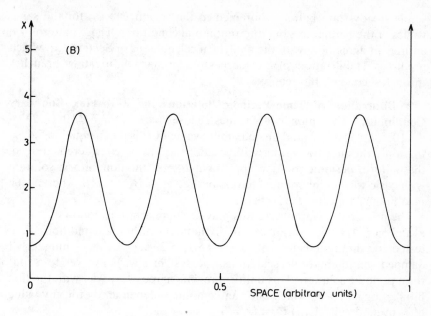

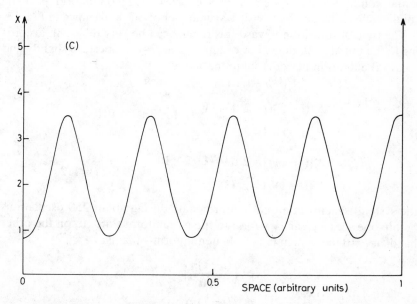

Fig. 12. Steady-state profiles obtained for the same parameter values but different initial conditions: $A = 2$, $D_1 = 1.6 \times 10^{-3}$, $D_2 = 8 \times 10^{-3}$, $B = 4.6$. Zero fluxes of X and Y are maintained at the boundaries.

stable steady-state patterns obtained in the region $B \gg B_c$, for the same values of the parameters but different initial conditions. These patterns are entirely in agreement with the analytical calculations of Section I.E, which predict a stabilization of the successive primary bifurcating branches through secondary bifurcations.

2. Bifurcation of Time-Periodic Solutions for Zero-Flux Boundary Conditions. The numerical values chosen are $D_1 = 8 \times 10^{-3}$, $D_2 = 4 \times 10^{-3}$, $l = 1$, $A = 2$, and the maximum value of B is taken equal to 5.4, so that only the first two primary bifurcations are to be considered. Indeed, the uniform periodic solution appears at $\tilde{B}_0 = 5$, the nonuniform solution with basic wavenumber equal to one bifurcates at $\tilde{B}_1 = 5.118$, whereas the next branch appears at $\tilde{B}_2 = 5.48$.

The principal asymptotic results may be summarized as follows:

a. For $5 < B < 5.4$ various, spatially nonuniform initial conditions were tested for different values of B ($B = 5.1$, 5.2, 5.3). After a number of damped spatiotemporal oscillations, the system always evolves to a uniform periodic solution. This uniform regime agrees very well with the first bifurcating solution, which one can compute independently for any value of B by setting in (I.15) $D_1 = D_2 = 0$.

b. As B is increased to the value $B = 5.4$ and with spatially nonuniform initial conditions the system always evolves to a nonuniform periodic solution characterized by a basic wavenumber of one and shown in Fig. 13.

The time of transition toward this regime can be very different, according to the initial conditions. For example, if at $t = 0$ we perturb slightly and locally the uniform periodic solution, that is:

$$h = 0.01$$

$$X[(i-1)h] = 2.03471 \qquad \text{for } 3 \le i \le 101$$
$$Y[(i-1)h] = 2.33479$$

and

$$X[(i-1)h] = 2.03471 + 2 \times 10^{-5} \qquad \text{for } i = 1, 2$$
$$Y[(i-1)h] = 2.33479$$

the system remains in a quasiuniform state during about 250 oscillations before one sees a clear evolution to the nonuniform state. If, on the other hand, at $t = 0$ the system is sharply nonuniform—that is:

$$X = 2.5 + 2 \cos\left(\frac{(i-1)\pi}{100}\right)$$
$$Y = 3 - 1.5 \cos\left(\frac{(i-1)\pi}{100}\right) \qquad 1 \le i \le 101,$$

the nonuniform periodic regime is rapidly reached after a few oscillations.

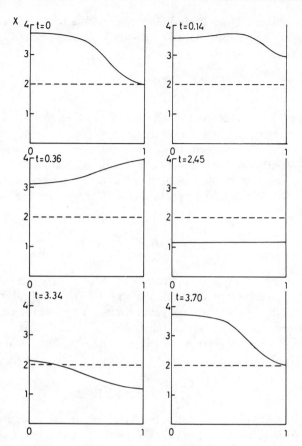

Fig. 13. Spatiotemporal regime for a one-dimensional system subject to zero flux boundary conditions. The characteristic stages of evolution of the spatial distribution of X during one period are shown. One observes a short progressive stage from the left to the right. Time $t = 0$ is taken after the system has presented a regular periodicity ($A = 2$; $D_1 = 8 \times 10^{-3}$, $D_2 = 4 \times 10^{-3}$; $B = 5.4$; and $l = 1$).

These results suggest two remarks: There are many arguments in favor of correlating the nonuniform solution obtained numerically to the second primary bifurcating solution (periodicity, basic wavenumber of one). For this latter an analytical expression has been constructed for B in the neighborhood of $\tilde{B}_1$ [see, for example, (I.36)]. On the other hand, we know that the uniform periodic solution remains linearly stable when B increases from 5 to 5.4. The numerical results show that at $B = 5.4$ it has small chances to be maintained when slightly perturbed. In this situation it becomes important to analyze the global stability of the uniform periodic regime.

3. Bifurcation of Time-Periodic Solutions on a Ring. The numerical simulations have been performed by dividing the ring into M equal intervals. Conditions (I.38a) become

$$X_{M+1} = X_1, \qquad Y_{M+1} = Y_1$$

$$X_M = X_0, \qquad Y_M = Y_0$$

The values of the parameters are the same as in subsection I.F.1 except for the length of the ring, which was chosen to be $l = 2$.

In addition to the uniform limit cycle solution, for values of $B > \tilde{B}_1'$ $= 1 + A^2 + (2\pi/l)^2(D_1 + D_2)$, the two types of spatiotemporal asymptotic regimes expected from the theoretical predictions were also obtained, starting from different initial conditions. At $B = 5.4$ with the initial conditions:

$$\binom{X}{Y} - \begin{pmatrix} A \\ \dfrac{B}{A} \end{pmatrix} \sim \cos \frac{2\pi r}{l}$$

we observe asymptotically a space-dependent, time-periodic regime similar to the first type of solution given by (I.39). It is of interest to note that this periodic solution has the same characteristics as the one shown in Fig. 13 for zero-flux boundary conditions.

The second type of asymptotic solution, the traveling wave, exists at $B = 5.8$.* The initial condition used to obtain the wave shown in Fig. 14 is

$$\binom{X}{Y} = \Psi\left(\frac{2\pi r}{l}, B\right)$$

where $\Psi(2\pi r/l, B)$ corresponds to the limit cycle solution $\Psi(\tau, B)$ computed from (I.15) by setting $D_1 = D_2 = 0$ and unfolded along the ring.

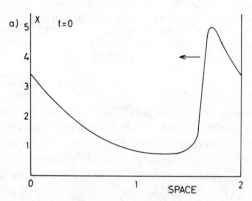

* At $B = 5.4$, with the same initial conditions, the system evolves to the uniform limit cycle solution.

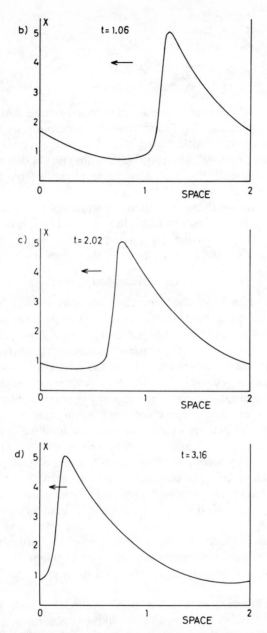

Fig. 14. Progressive spatiotemporal solution for a one-dimensional system subject to periodic boundary conditions ($X_{M+1} = X_1$, $Y_{M+1} = Y_1$). Time $t = 0$ is chosen after the system has settled down to a periodic regime ($B = 5.8$, $l = 2$, and other parameter values are as in Fig. 13).

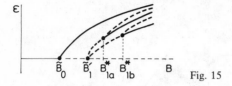

Fig. 15

The behavior of this asymptotic solution was followed during a substantial number of periods. We found that the maximum amplitude of the "wavefront" is constant within ±0.5%, and so is its propagation velocity, given by $v = \Delta r/\Delta t \simeq 0.48$. The propagation direction is determined by the initial condition and the time taken by the propagating front to travel once around the ring is $T = 4.28$.

The appearance of these two stable spatiotemporal regimes with the characteristics given in Section I.D.2 indicates a possible stabilization of the second primary branches (see, for example, Fig. 7), as is already substantiated analytically for zero fluxes at the boundaries.

G. Discussion

The analysis outlined in the preceding sections shows that reaction–diffusion systems can display a surprising variety of patterns, which in some cases can be stable simultaneously. Of special interest is the possibility of increasing the complexity of the pattern via successive bifurcations, as the bifurcation parameter moves away from the first point where the thermodynamic branch becomes unstable. This property of reaction-diffusion systems makes them suitable for the modeling of complex phenomena such as chemical evolution in the prebiotic era, embryonic development and morphogenesis, the activity of the nervous and immune systems, or the dynamics of interacting populations. Some examples are outlined in the recent monograph by Nicolis and Prigogine.[3] See also Ref. 16. It is also to be noted that reaction-diffusion equations are being actively studied by many authors from different standpoints. Detailed references may be found in a review paper by Fife[17] and in a proceedings volume of a recent meeting on the subject.[17]

Despite these advances, the full exploration of the bifurcation diagram of a reaction-diffusion system remains an open problem. It is our opinion that new qualitative insight can be obtained from two types of investigation.,

1. The development of exactly solvable models. Such a model has recently been analyzed by Lefever, Herschkowitz-Kaufman, and Turner.[18] It is a simplified version of the trimolecular model—(I.14) and (I.15)—whereby the entry term $A \rightarrow$ and the decay term $X \rightarrow$ are deleted. To allow for an open system, diffusion is taken into consideration in such a way that

the fluxes of X and Y remain nonvanishing on the boundaries. The inhomogeneous steady states of this model can be computed exactly in terms of elliptic functions. Secondary bifurcations are also possible under certain conditions.

2. A second fascinating possibility would be to extend the analytical work described in Sections I.C to I.E to tertiary and higher bifurcations. The situation is particularly interesting in the case of time-periodic solutions. Indeed, the appearance of a third period (in case of a tertiary bifurcation) not commensurable with those of the primary and the secondary branches would suggest transition to chaotic behavior.[19] Such a behavior is believed to take place in fluid dynamics[20] and corresponds to the emergence of "strange attractors" in the sense of Ruelle and Takens.[21] It is also shown to occur in finite-difference equations describing interacting populations.[22] For reaction-diffusion systems a similar behavior can be expected,[23] although the mechanisms responsible for it have not yet been elucidated. In this respect, therefore the investigation of tertiary bifurcations appears to be a promising line of approach.

Finally, the existence of cascading bifurcations illustrates the important role of initial conditions in the evolution. To a large extent these conditions have a stochastic character, as they are partly determined by (random) external disturbances as well as by the internal fluctuations generated by the system itself. These remarks suggest that in the vicinity of bifurcation points it is necessary to incorporate fluctuations into the description. This analysis is carried out in Section II of this paper.

II. STOCHASTIC ASPECTS OF SELF-ORGANIZATION

A. General Survey

To assess the influence of fluctuations in the behavior analyzed in Section I, we assume that the dynamical processes considered define a Markov process in some appropriate space of state variables. The reasonableness of this assumption depends on the choice of these variables. For instance, lumping all space and momentum degrees of freedom and dealing only with such variables as total numbers of particles might well compromise the Markovian character of the process. On the other hand, by incorporating explicitly all phase space variables, one expects the Markovian assumption to be secured, but at the same time one is faced with the complex task of solving the kinetic equations of nonequilibrium statistical mechanics.

A reasonable "intermediate" choice is possible in reaction-diffusion systems that are subject to constraints varying on a macroscopic scale, larger than the scales characterizing molecular motion (mean free path, relaxation time, etc.). Let such a system be divided into spatial cells, whose dimensions are of the order of a few mean free paths. Being relatively small, such cells do not feel directly the influence of boundary conditions. Yet they contain enough particles for the notion of statistical equilibrium to make sense. Because of the scale separation assumed, this equilibrium will be established to a good approximation within the individual cells, although the characteristic equilibrium parameters—numbers of particles, temperature, convection velocity—will vary from cell to cell.

If this picture is adopted, the central quantity to evaluate becomes the probability $P(X_{\alpha i}, t)$ to have given numbers of particles of type α in cell i ($i = 1, \ldots, n$) at time t. The time evolution of P is determined by two types of process:

1. The chemical reactions going on in each cell. Their effect is to create and destroy particles of different species within the cell.* Hence they will be assimilated to a *birth and death* process.[24]
2. The exchange of particles between adjacent cells through diffusion. This does not affect the total number of particles of a particular species, but tends to redistribute these particles among the n cells. Hence it can again be assimilated to a birth and death process. The transition probabilities for both (1) and (2) can be constructed unambiguously.[3]

One may now write a multivariate *master equation* describing the combined effect of (1) and (2). It will be convenient to work in the generating function representation:

$$F(\{s_{\alpha i}\}, t) = \sum_{X_{\alpha i}=0}^{\infty} \prod_{i,\alpha} s_{\alpha i}^{X_{\alpha i}} P(\{X_{\alpha i}\}, t) \tag{II.1}$$

where α runs for the chemical species and i for the spatial cells. One obtains in this way:

$$\frac{\partial F}{\partial t} = \sum_i \left(\frac{\partial F}{\partial t}\right)_{i, \text{chem}}$$

$$+ \sum_{\alpha i} d_\alpha \ (s_{\alpha, i+1} + s_{\alpha, i-1} - 2s_{\alpha i}) \frac{\partial F}{\partial s_{\alpha i}} \tag{II.2}$$

The term $(\partial F/\partial t)_{i,\text{chem}}$ describes the effect of chemical kinetics within cell i and is system dependent. For instance, in the trimolecular model (I.14), it

---

* We assume that we deal with a dilute mixture, such that the explicit effect of inter-molecular forces in the reaction kinetics can be neglected.

takes the following form:

$$\left(\frac{\partial F}{\partial t}\right)_{i,\text{chem}} = A(s_{X_i} - 1)F + B(s_{Y_i} - s_{X_i})\frac{\partial F}{\partial s_{X_i}}$$

$$+ (s_{X_i}^3 - s_{X_i}^2 s_{X_i})\frac{\partial^3 F}{\partial s_{X_i}^2 \partial s_{Y_i}}$$

$$+ (1 - s_{X_i})\frac{\partial F}{\partial s_{X_i}} \tag{II.3}$$

In contrast, the diffusion term has a "universal" form that is affected only by the boundary conditions. If the cell size is of the order of the mean free path R, the diffusion rates d_α will be related to the Fickian diffusion coefficients D_α through

$$d_\alpha = \frac{D_\alpha}{R^2} \tag{II.4}$$

Having an explicit equation for the probability function, one may envisage computing such quantities as local variances $\langle \delta X_{\alpha i}^2 \rangle$ or correlation functions between cells $\langle \delta X_{\alpha i} \delta X_{\beta j} \rangle$. This analysis has been carried out for several models and the principal results are as follows (see Ref. 3 for a more complete survey).

The first point to be realized is that the multivariate master equation (II.1) leads to an *infinite hierarchy* of moment equations. To obtain information on the first few moments, one usually truncates this hierarchy. Although one may develop several plausibility arguments in favor of such a truncation, it is fair to say that this procedure introduces approximations that are difficult to control, owing to the absence of a systematic perturbation parameter.

Let the hierarchy be truncated at the level of second-order moments. The latter, together with the averages, then give rise to a closed set of equations from which correlation functions can be computed explicitly. As an example consider the trimolecular model below the instability point. Setting

$$\langle \delta X_i \, \delta X_j \rangle = \langle X \rangle \, \delta_{ij}^{kr} + \frac{1}{l} G_{ij}^{XX} \tag{II.5}$$

where $\langle X \rangle$ refers to the thermodynamic branch, l is the size of the system, one has the following expression for the doublet spatial correlation function

G_{ij} in a one-dimensional system of length l:

$$G_{ij} = \sum_{k=1}^{n} \left[\cos (i+j) \frac{k\pi}{n+1} - \cos (i-j) \frac{k\pi}{n+1} \right]$$

$$\times \left[\left(\Gamma - \lambda_k \frac{n^2}{l^2} d \right) \otimes I + I \otimes \left(\Gamma - \lambda_k \frac{n^2}{l^2} d \right) \right]^{-1} \epsilon \qquad (II.6)$$

Here Γ and ϵ contain the influence of chemical kinetics, I is the unit matrix, $\otimes$ denotes tensor product, and λ_k is the eigenvalue of the linearized stability problem.

Expression (II.6) can be analyzed as the bifurcation point B_c, for the onset of a spatial dissipative structure is approached from below. For B well below B_c one finds that G_{ij} is practically zero beyond a small number of neighbors $|i-j|$. Correspondingly, the variances of fluctuations within each cell remain of the order of magnitude of the average values $\langle X \rangle, \langle Y \rangle$. This result, which is reminiscent of the behavior of fluctuations in equilibrium systems away from points of phase transition, implies that there is a clearcut distinction between macroscopic behavior and fluctuations.

The situation changes when $B \leq B_c$, as depicted in Fig. 16. The system develops *long-range correlations* that, moreover, display a spatial pattern identical to the macroscopic pattern that will emerge beyond the bifurcation point. Note that this long-range order is a consequence of nonlinearity and nonequilibrium, since the elementary interactions within the system are all short range.

In analogy with equilibrium phase transitions, one can identify a *correlation* length ξ, which diverges as $B \to B_c$ as follows:

$$\xi \sim |B - B_c|^{-1/2} \qquad (II.7)$$

Moreover, the long-range character of the correlation functions induces a divergence of the variances $\langle \delta X_i^2 \rangle$ within each cell, according to the power law

$$\langle \delta X_i^2 \rangle \sim |B - B_c|^{-1} \qquad (II.8)$$

The divergence exponents in both (II.7) and (II.8) turn out to be those characterizing a "mean field" or Landau-type theory of critical phenomena.[25] Such *classical exponents* could therefore be attributed to the truncation of moment equations that was carried out in order to obtain the explicit expression for the correlation function (II.6).

The purpose of the subsequent two sections is to discuss the generality of the classical exponent behavior. In Section II.B we consider a model that can be solved exactly in the absence of diffusion. We show that classical behavior is recovered in the thermodynamic limit. Section II.C is devoted

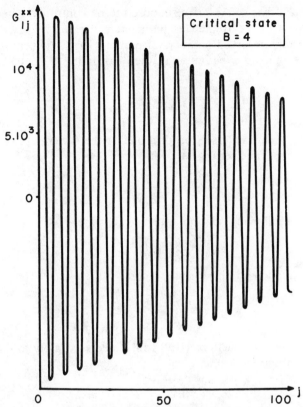

Fig. 16. Generation of long-range spatial correlations in the trimolecular model for $A = 2$, $B = 4$, $D_1 = 1$, $D_2 = 4$. The correlation function features both an overall damping as well as regular spatial oscillations.

to the role of diffusion. As the analytic work pertaining to this aspect is rather preliminary, we place the main emphasis of our arguments on the results of computer simulations.

B. An Exactly Solvable Model

1. Macroscopic Behavior. In order to simplify the analysis as much as possible we consider a reacting system involving a single-variable intermediate X:[26]

$$A + 2X \underset{k_2}{\overset{k_1}{\rightleftharpoons}} 3X$$

$$X \underset{k_4}{\overset{k_3}{\rightleftharpoons}} B \tag{II.9}$$

The concentrations of A, B are supposed to be controlled externally and will from now on be treated as parameters. Let n_X, A, B denote the macroscopically observed numbers of particles of X, A, B in the system. The macroscopic rate equation is then (assuming a spatially uniform system):

$$\frac{dn_X}{dt} = -k_2 n_X^3 + k_1 A n_X^2 - k_3 n_X + k_4 B \tag{II.10}$$

Note that k_1, k_2 are inversely proportional to the (size of system)^2. Dividing through by k_2 and setting

$$a = \frac{k_1 A}{k_2}, \qquad b = \frac{k_4 B}{k_2}, \qquad k = \frac{k_3}{k_2} \tag{II.11}$$

we obtain the following relation at the steady state:

$$n_X^3 - a n_X^2 + k n_X - b = 0 \tag{II.12}$$

This cubic equation admits a triple root at

$$a = (3k)^{1/2}, \qquad b = \frac{a^3}{27} = \frac{ak}{9} \tag{II.13}$$

Note that if the two reactions (II.9) conformed separately to the detailed balance condition one would have

$$a_{eq} n_X^2 = n_X^3, \qquad b_{eq} = k_{eq} n_X$$

or, eliminating n_X:

$$b_{eq} = a_{eq} k_{eq}$$

This is incompatible with the second condition (II.13). In other words, the phenomena discussed in the sequel are due to the fact that the system operates away from the state of chemical equilibrium.

As is well known, the behavior in the vicinity of the triple root (II.13) is related to the so-called *cusp catastrophe*.[4] To study this in detail, we introduce two parameters δ, δ' through

$$k_2 = \frac{1}{A^2}, \qquad k_1 = \frac{3}{A^2}, \qquad k_4 = 1$$

$$k_3 = 3 + \delta, \qquad B = A(1 + \delta') \tag{II.14}$$

Equation (II.12) becomes

$$n_X^3 - 3A n_X^2 + (3 + \delta)A^2 n_X + (1 + \delta')A^3 = 0 \tag{II.15}$$

For $\delta = \delta' = 0$ this equation admits the triple root

$$n_X = A \tag{II.16}$$

The behavior in the vicinity of this root is found by setting

$$n_X = A(1 + x) \tag{II.17}$$

We find

$$\delta' = x^3 + \delta(x + 1) \tag{II.18}$$

This relation is of the same form as the Van der Waals equation near the critical point,[27] provided δ' is interpreted as the "pressure" and δ as "the temperature" difference from the critical point values. If the instability point is approached along the path

$$\delta = \delta' \tag{II.19}$$

then, for $\delta > 0$, $x = 0$ is the only real solution of (II.18). If, on the other hand, $\delta < 0$ (i.e.; if we are above the instability point), then in addition to $x = 0$ we have two real roots:

$$x_\pm = \pm\sqrt{-\delta} \tag{II.20}$$

Obviously, $|x_\pm|$ can be interpreted as the "order parameter" of the problem.[27] We see that for $\delta < 0$ the system displays multiple steady states, as illustrated in Fig. 17.

Note the resemblance with the bifurcation diagrams of Section I. As a matter of fact, $\delta = 0$ is a bifurcation point of the new branches (b), (c), whereas branch (a) is continued across the bifurcation point. A linearized stability analysis of the time-dependent analogue of (II.18) shows that branch (a') is unstable, whereas both (b) and (c) are asymptotically stable.

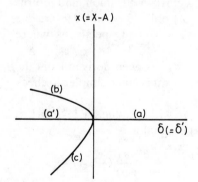

Fig. 17. Bifurcation diagram associated with multiple steady-state transitions in model (II.9).

2. Fluctuation Behavior. We now consider the effect of fluctuations. We adopt the assumption that the diffusion of species A, B, X is very efficient and spreads instantaneously throughout the system any localized fluctuation that may arise randomly. This restricts us to fluctuations whose range is *hydrodynamic* (i.e., comparable to the size of the system itself). The master equation in the generating function representation is easily found to be

$$s^2\left(\frac{d^3F}{ds^3} - 3A\frac{d^2F}{ds^2}\right) + (3+\delta)A^2\frac{dF}{ds} - (1+\delta')A^3F = 0 \qquad \text{(II.21)}$$

where from now on we limit ourselves to the steady-state behavior. We recall that once F is calculated, the moments of the probability distribution $P(X)$ can be computed in terms of derivatives of F at $s = 1$.[24]

We are interested in the properties of the steady-state solution of (II.21) as the size parameter $A \to \infty$ (the thermodynamic limit) and the bifurcation parameter $\delta = \delta'$ goes to zero. Now the generating function $F(s)$ obeys [see (II.21)] a third-order differential equation with coefficients of a second degree or less. A Laplace transform on the variable s produces, therefore, a second-order equation with coefficients of a third degree or less. Differentiating with respect to the new variable leads to a homogeneous third-order equation of the type satisfied by hypergeometric functions of the kind $_3F_2$. On obtaining the solution, which is regular at $s = 1$, taking the inverse Laplace transform and inserting a Barnes type of integral representation for the integrand,[28] one finally shows that[29]

$$F(s) = P(0)\frac{\Gamma(2-\beta)\Gamma(2-\gamma)}{\Gamma(2-\theta)\Gamma(2-\epsilon)}$$

$$\times \frac{1}{2\pi i}\int_c d\omega\,\Gamma(\omega)\frac{\Gamma(2-\epsilon+\omega)\Gamma(2-\theta+\omega)}{\Gamma(2-\beta+\omega)\Gamma(2-\gamma+\omega)}(-3As)^\omega \qquad \text{(II.22)}$$

where C starts at $+\infty$, circles the origin in a clockwise fashion, and returns to $+\infty$, enclosing the singularities of $\Gamma(-\omega)$ only; β, γ and θ, ϵ are the roots of $\Omega^2 - 5\Omega + \lambda + 6 = 0$ and $\Omega^2 - 5\Omega + \mu + 6 = 0$, respectively, where $\lambda = A^2(3+\delta)$ and $\mu = (A^2/3)(1+\delta')$.

The way the integral representation of $F(s)$ has been derived from the differential equation has only been sketched, for it is much simpler to check the result by inspection.

As $\Gamma(-\omega)$ has poles at $w = 0, 1, 2, \ldots, n$ with residue $(-1)^{n+1}/n!$, it follows that the coefficient C_p of s^p in $F(s)$ is equal to

$$\frac{\Gamma(2-\beta)\Gamma(2-\gamma)}{\Gamma(2-\theta)\Gamma(2-\epsilon)}\frac{(-1)^p}{p!}\frac{\Gamma(2-\epsilon+p)\Gamma(2-\theta+p)}{\Gamma(2-\beta+p)\Gamma(2-\gamma+p)}(-3A)^p$$

and the ratio C_{p+1}/C_p is given therefore by

$$\frac{3A}{p+1}\frac{(2-\epsilon+p)(2-\theta+p)}{(2-\beta+p)(2-\gamma+p)}=\frac{p^2-p+\mu}{p^2-p+\lambda}\frac{3A}{p+1}$$

$$=\frac{A}{p+1}\frac{3p(p-1)+A^2(1+\delta')}{p(p-1)+A^2(3+\delta)}$$

Now the master equation for the probability distribution is easily seen to have as stationary solution:[30]

$$P(X)=P(0)\frac{1}{X!}\prod_{J=1}^{X}\frac{k_1(J-1)(J-2)+k_4}{k_2(J-1)(J-2)+k_3} \qquad (\text{II.23})$$

hence

$$\frac{P(X+1)}{P(X)}=\frac{1}{X+1}\frac{(3/A)X(X-1)+A(1+\delta')}{(1/A^2)X(X-1)+3+\delta}$$

As $C_0=P(0)$ the proof is complete.

As the singularities of $\Gamma(2-\epsilon+\omega)$ and $\Gamma(2-\theta+\omega)$ have an imaginary part of order A, the contour can be expanded and a new variable $t=\omega/A$ introduced. The integral representation for $F(s)$ can now be put in the form

$$F(s)=C\int_c dt\,\varphi(t)\,e^{Af(t,s)}$$

where

$$f(t)=t-\ln t+t\ln 3+(t-ik)\ln(t-ik)$$
$$+(t+ik)\ln(t+ik)-(t-i\lambda)\ln(t-i\lambda)$$
$$-(t+i\lambda)\ln(t+i\lambda)+t\ln s$$

$$k^2=\frac{1+\delta'}{3}-\frac{1}{4A^2} \qquad (\text{II.24})$$

$$\lambda^2=3+\delta-\frac{1}{4A^2}$$

$$\varphi(t)=t^{-1/2}\frac{k^2}{t^2+k^2}\frac{t^2+\lambda^2}{\lambda^2}$$

+ (terms that will prove negligible for our purpose when $A\to\infty$), and C is a constant.

This form is suitable for asymptotic evaluation when A is very large. It turns out that as we shall be primarily interested in the asymptotic behavior of the variance $\langle \delta X^2 \rangle$, only weak asymptotic results will be required. Indeed, let

$$G(s, \gamma) = C \int_c dt\, \varphi(t)\, e^{A[f(t)+vt]}$$

then clearly

$$\frac{\langle \delta X^2 \rangle}{A} = \frac{1}{A} \frac{\partial^2 \ln G(s, v)}{\partial v^2}\bigg|_{\substack{s=1 \\ v=0}}$$

When $\delta > 0$, $f(t)$ has a single maximum and by standard steepest descent methods[31] one finds

$$\lim_{A \to \infty} \frac{\langle \delta X^2 \rangle}{A} = \frac{4}{\delta} + 1, \qquad \delta > 0 \tag{II.25}$$

However, when $\delta < 0$, $f(t)$ has two maxima. The dominant contribution leads to

$$\lim_{A \to \infty} \frac{\langle \delta X^2 \rangle}{A} = -\frac{2}{\delta} - \frac{3}{\sqrt{-\delta}} + 1, \qquad -1 \le \delta < 0 \tag{II.26}$$

This behavior is associated with the macroscopic branch $x_- = -\sqrt{-\delta}$ [see (II.20)], for which the probability distribution has an absolute maximum. Stochastically, therefore, the other branch $x_+ = \sqrt{-\delta}$ does not exist in the thermodynamic limit $A \to \infty$, even though the macroscopic analysis predicts asymptotic stability for this branch!

At the bifurcation point, the results of the analysis outlined above diverge as $\langle \delta X^2 \rangle$ is not of order A. This stems from the fact that at its maximum, $f(t)$ has vanishing second and third derivatives as well. In this case the variance has to be calculated directly as $\langle X^2 \rangle - \langle X \rangle^2$. By a suitable modification of the method of steepest descent, one can show that[29]

$$\lim_{A \to \infty} \frac{\langle \delta X^2 \rangle}{A^{3/2}} = 4 \frac{\Gamma(\tfrac{3}{4})}{\Gamma(\tfrac{1}{4})}, \qquad \delta = 0 \tag{II.27}$$

In summary, we find that as the system approaches the bifurcation point on either side, the variance diverges according to a *classical exponent*:

$$\langle \delta X^2 \rangle \sim |\delta|^{-1}$$

This is an *exact result*, no matter how small $|\delta|$ is. Although not obvious *a priori*, this result can nevertheless be understood intuitively: By imposing spatial fluctuations in the hydrodynamic range one effectively realizes the conditions of validity of a "mean-field" type of theory. It is therefore not overly surprising that one recovers the classical result familiar from equilibrium phase transitions.

3. Singular Perturbation Analysis. The master equation (II.21) can also be analyzed from an alternative viewpoint, which is particularly well suited for developing systematic approximation schemes of solutions. Such schemes are necessary in more complex systems, where an exact solution of the master equation is not available.

We define the cumulant generating function $\phi(s)$ by

$$F(s) = e^{A\phi(s)} \tag{II.28}$$

Substituting into (II.21), we find the following nonlinear equation for $\phi(s)$:

$$\epsilon^2 u'' + 3\epsilon u'(u-1) + u^3 - 3u^2 + \frac{3+\delta}{s^2} u - \frac{1+\delta}{s^2} = 0 \tag{II.29}$$

We have set

$$u = \frac{d\phi}{ds} \equiv \phi'$$

$$\delta = \delta' \tag{II.30}$$

$$\epsilon = \frac{1}{A}$$

In a macroscopic system $(A \to \infty)$ is a small quantity. Thus (II.29) defines a *singular perturbation* problem.[32] Setting $\epsilon = 0$, one can find the first approximation of a solution in power series in ϵ, the so-called outer solution. One can show[33] that this leads to the same results as the exact solution, and in particular to a classical exponent divergence of the variance, as long as δ is not strictly zero.

At the bifurcation point, $\delta = 0$, the outer solution breaks down completely. This fact was recently emphasized in the context of bifurcation theory,[6] where it has been shown that the least "imperfection" may lead to qualitatively new behavior. In the present context, the higher derivative terms in (II.29) constitute such an "imperfection." One usually deals with this difficulty by working out an *inner expansion* valid in the immediate vicinity of ($\epsilon = 0$, $\delta = 0$). The idea is to stretch the vicinity of the point $s = 1$, which is related to the macroscopic behavior, by introducing a new variable η through

$$s = 1 + \epsilon^a \eta, \qquad a > 0 \tag{II.31}$$

Similarly, one expands the solution $\omega(\eta) \equiv u(s)$ in (generally fractional) powers of ϵ as follows:

$$\omega = 1 + \epsilon^c \omega_1 + \epsilon^{2c} \omega_2 + \cdots, \qquad c > 0 \tag{II.32}$$

The master equation (II.29) becomes for $\delta = 0$:[29]

$$\epsilon^{2(1-a)} \omega'' + 3\epsilon^{1-a}(\omega - 1)\omega' + (\omega - 1)^3$$
$$+ 3(-2\epsilon^a \eta + 3\epsilon^{2a}\eta^2 + \cdots)\omega + (2\epsilon^a \eta - 3\epsilon^{2a}\eta^2 + \cdots) = 0 \tag{II.33}$$

with

$$\omega' \equiv \frac{d\omega}{d\eta}$$

In order to satisfy this equation order by order it is necessary to have

$$a \geq \min\{2(1-a)+c, 1-a+2c, 3c\} \tag{II.34}$$

Moreover, to avoid singular behavior in ϵ we require

$$a \leq 1 \tag{II.35}$$

These *exponent inequalities* still leave us with considerable freedom in the choice of a and c. However, one can see that any choice not corresponding to an equality in (II.34) would result in *homogeneous* equations for $\omega_1, \omega_2, \ldots$, and hence a large number of undetermined constants. To avoid this we require all dominant contributions of the various terms in (II.33) to be of the same order of magnitude:

$$a = 2(1-a)+c = 1-a+2c = 3c \tag{II.36}$$

This implies

$$a = \frac{3}{4}, \qquad c = \frac{1}{4} \tag{II.37}$$

We thus obtain, to order $\epsilon^{3/4}$:

$$\omega_1'' + 3\omega_1\omega_1' + \omega_1^3 - 4\eta = 0 \tag{II.38}$$

Or, in terms of a function $\bar{\phi}$ such that

$$\bar{\phi}' = \omega \tag{II.39a}$$

$$\bar{\phi}''' + 3\bar{\phi}'\bar{\phi}'' + \bar{\phi}'^3 - 4\eta = 0 \tag{II.39b}$$

Setting

$$\bar{F} = e^{\bar{\phi}} \tag{II.40a}$$

we can further transform (II.39b) to

$$\bar{F}''' - 4\eta\bar{F} = 0 \tag{II.40b}$$

It can be verified (see also Ref. 28) that a positive solution of this equation that is nonvanishing in the limit $\eta \to \infty$ is*

$$\bar{F}(\eta) = \mathcal{N} \int_{-\infty}^{\infty} dk\, e^{k\eta}\, e^{-k^4/16} \tag{II.41}$$

where $\mathcal{N}$ is the normalization constant.

Returning to the relation between ω_1 and $\bar{F}$, (II.39) to (II.40), we obtain

$$\omega_1 = \frac{F'}{F} = \frac{\displaystyle\int_{-\infty}^{\infty} dk\, k e^{k\eta}\, e^{-k^4/16}}{\displaystyle\int_{-\infty}^{\infty} dk\, e^{k\eta}\, e^{-k^4/16}} \tag{II.42}$$

In the vicinity of $\eta = 0$ this function behaves as

$$\omega_1 \approx \eta \frac{\displaystyle\int_{-\infty}^{\infty} dk\, k^2\, e^{-k^4/16}}{\displaystyle\int_{-\infty}^{\infty} dk\, e^{-k^4/16}} = \eta^4 \frac{\Gamma(\tfrac{3}{4})}{\Gamma(\tfrac{1}{4})} \tag{II.43}$$

Utilizing (II.31), we conclude that, to order $\epsilon^{1/4}$:

$$\omega \simeq 1 + \epsilon^{1/4} \frac{s-1}{\epsilon^{3/4}} \frac{4\Gamma(\tfrac{3}{4})}{\Gamma(\tfrac{1}{4})} \tag{II.44}$$

From this relation we can compute the variance by differentiating once. We find

$$\frac{\langle \delta X^2 \rangle - \langle X \rangle}{A} = \left(\frac{d\omega}{ds}\right)_{s=1}$$

$$= \epsilon^{-1/2} 4 \frac{\Gamma(\tfrac{3}{4})}{\Gamma(\tfrac{1}{4})} = A^{1/2} 4 \frac{\Gamma(\tfrac{3}{4})}{\Gamma(\tfrac{1}{4})} \tag{II.45}$$

In other words, at the instability point the (reduced) variance diverges as the square root of the size parameter A or, alternatively, $\langle \delta X^2 \rangle$ diverges as $A^{3/2}$. This result is in agreement with that of subsection 2, (II.27). The

---

* A generating function cannot go to zero for $\eta \to \infty$, as one can see already from the Poisson distribution, for which

$$F \propto \exp(s-1) = \exp(\epsilon^a \eta).$$

method can be extended straightforwardly to $\delta \neq 0$. One recovers again[29] a divergence of the variance in agreement with both the outer solution and the exact result, (II.25) and (II.26).

C.   Effect of Diffusion

When diffusion is taken into account, a new time and length scale is introduced in the system beyond those associated with the pure chemical kinetics. As a result, the range of fluctuations need not only be determined by extrinsic parameters (of the order of the size of the system), as in the preceding section, but could also depend on intrinsic parameters of molecular origin. On purely dimensional grounds we would expect the latter to be suitable combinations of the diffusion coefficients and of the rate constants. This might well modify the behavior of long-range correlations in the vicinity of the instability with respect to the behavior predicted by "classical" theories of the Landau type.

A striking confirmation of the role of diffusion in the onset of instabilities is reported by Hanusse.[34] This author carries out a Monte Carlo simulation of a variety of chemical models. Consider first a system presenting multiple steady-state transitions of the type discussed in Section II.B. Let the initial value of the probability distribution be peaked around one of the (macroscopically) stable branches. When the values of the diffusion coefficients are small relative to the chemical rates, one observes a rapid transition to a single branch. In other words, multiple steady-state and hysteresis effects are destroyed by the fluctuations. When, on the other hand, the values of the diffusion coefficients are large, two stationary distributions may appear, one for each macroscopic state. The transition between the two states involves a *nucleation process*. The latter is initiated preferentially near the boundaries, where the probability of appearance of a critical nucleus seems to be larger than that in the bulk.

A second striking group of results refers to systems undergoing bifurcation of time-periodic or space-dependent solutions around a single uniform steady-state solution (as, for instance, the trimolecular model). Suppose a limit-cycle oscillation appears under macroscopically homogeneous conditions. One may then simulate a diffusion coupling of variable range by coupling a given cell with its n-first neighbors. This amounts to realizing on the computer the "mean field" picture that has been developed recently by Nicolis, Malek-Mansour, Kitahara, and Van Nypelseer.[35] The simulations reveal that for given values of reaction parameters and of diffusion coefficients, there exists a critical number of neighbors above which the instability of the steady state does not show up. This is a direct evidence of the modification of macroscopic behavior by local fluctuations.

We turn now to the analytical work on this topic. The mathematical complexity of the problem is considerable, and for this reason little is known beyond the results based on the truncation procedures outlined in Section II.A. Nevertheless, it is very tempting to argue as follows. If correlations become indeed long-ranged, then the initially chosen size of the cells in the multivariate formalism [see (II.2)] becomes immaterial. Similar results would thus be obtained by considering a partition of space in cells of larger size, provided the latter remains smaller than the correlation length. This idea, which is reminiscent of the *scaling hypothesis* familiar from equilibrium phase transitions, suggests that it would be of interest to analyze the transformation properties of the stochastic equations under a change in length scale.

A preliminary analysis of this aspect has been reported by Mou, Nicolis, and Mazo.[33] It follows closely the ideas of renormalization group formulation of equilibrium critical phenomena (see Ma, Ref. 25, for a recent survey). Consider the case of a single chemical variable X, and let i label a particular spatial cell. If λ is the bifurcation parameter, d the diffusion coefficient of X, and $\bar{u}$ a coupling constant representing the effect of nonlinearities, one can write the steady-state equation for the average of X_i as follows (diffusion is taken to operate along a single space dimension):

$$uN(\langle X_i^2 \rangle, \langle X_i^3 \rangle, \dots) + (2-r)\langle X_i \rangle = \langle X_{i+1} \rangle + \langle X_{i-1} \rangle \qquad (II.46)$$

where

$$u = \frac{\bar{u}}{d}, \qquad r = \frac{\lambda - \lambda_c}{d} \qquad (II.47)$$

The function N expresses the contributions to be added to the linearized part of the equation. Naturally, (II.46) is coupled with those for $\langle X_i^2 \rangle$, $\langle X_i^3 \rangle$, and so on.

The next step is to vary (for instance, double) the cell size. Having X_i' stand for the variables within the enlarged cells, we require the equation for $\langle X_i' \rangle$ to have the same structure as (II.46):

$$u'N(\langle X_i'^2 \rangle, \langle X_i'^3 \rangle, \dots) + (2-r')\langle X_i' \rangle = \langle X_{i+1}' \rangle + \langle X_{i-1}' \rangle \qquad (II.48)$$

with

$$u' = f(u, r)$$
$$r' = g(u, r) \qquad (II.49)$$

The passage from (II.46) to (II.48) has been carried out explicitly, but a number of assumptions had to be adopted. First, the "initial" coupling constant u was taken small. Second, interactions other than nearest-neighbor ones, which appear through the equations for the higher moments, have been neglected. And third, the evolution of higher moments was taken to depend principally on diffusion. Under these conditions, (II.49) can be written down explicitly. They express the transformation of coupling and bifurcation parameters under a scale change. The fixed points of this transformation and their stability give information about the location of the instability point, the law of divergence of the correlation length, and so forth. A very interesting step in a similar direction was undertaken recently by Dewel, Borckmans and Walgraef[39] using the ideas underlying the renormalization group approach to critical phenomena. Further work using this technique as well as the singular perturbation analysis outlined in the preceding section is necessary in order to elucidate the onset of an instability through fluctuations in the presence of diffusion.

D. Influence of External Noise

So far we have been concerned with the dynamics of the *internal fluctuations*, which are generated spontaneously by the system itself. In most of the unstable transitions analyzed in the present paper, the bifurcation parameter is an externally controlled parameter and is subject, in general, to fluctuations. In the vicinity of the bifurcation point this *external noise*, even if it has a small variance, can influence deeply the macroscopic behavior of the system. In this section we outline an approach to this problem. The interesting question of coupling between internal fluctuations and external noise will not be addressed. Rather than remain general, let us illustrate the main ideas on a simple example.[36] We consider a modified Schlögl model [see (II.9)]:

$$A + 2X \rightleftarrows 3X$$

$$B + 2X \rightarrow C \qquad\qquad (II.50)$$

$$X \rightarrow D$$

We set all rate constants equal to one and

$$r = A - 2B \qquad\qquad (II.51)$$

The phenomenological rate equation reads as follows:

$$\frac{dn_X}{dt} = -n_X^3 + rn_X^2 - n_X \qquad\qquad (II.52)$$

It is straightforward to see that at $r = 2$ there is one stable and one unstable steady-state solution emerging, as shown in Fig. 18.

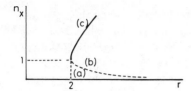

Fig. 18. Steady-state values versus the parameter
r in model (II.50).

In addition, $n_x = 0$ [branch (a) in Fig. II.3] is always a solution that remains infinitesimally stable. Thus, $r = 2$ is not a bifurcation point, although Fig. 18 bears some strong similarities with the diagram of Fig. 17.

We now consider r to be a random variable, and the simplest assumption is that it corresponds to a Gaussian white noise with

$$\langle r \rangle = p$$
$$\langle r^2 \rangle = \sigma^2 \tag{II.53}$$

Instead of (II.52) we write a *stochastic differential equation*[37] of the form

$$dX_t = [(\sigma^2 - 1)X_t^3 + rX_t^2 - X_t]\,dt + \sigma X_t^2\,dW_t \tag{II.54}$$

where W_t is a Wiener process. Its derivative—in the sense of generalized stochastic processes—is a Gaussian white noise with zero mean and a delta function quadratic correlation.

With (II.54) one can associate a Fokker–Planck equation for the probability density $P(X, t)$:

$$\frac{\partial P}{\partial t} = -\frac{\partial}{\partial X}[(\sigma^2 - 1)X^3 + rX^2 - X]P$$

$$+ \tfrac{1}{2}\sigma^2 \frac{\partial^2}{\partial X^2} X^4 P \tag{II.55}$$

It is straightforward to see that $X = 0$ remains a stationary solution of (II.54) and that it is always stable. In contrast, the Fokker–Planck equation (II.55) does not admit any stationary probability distribution. In other words, in the stochastic model *the transition point $r = 2$ of the deterministic case disappears*. A close inspection of the reaction sequence shows that this phenomenon is caused by the step $X \to D$, which gives rise to a "drift" toward zero. After a certain time, which might tend to infinity, the process certainly reaches the boundary zero and subsequently remains there, since zero is a stationary point.

Further examples indicating the important role of external noise in nonequilibrium phenomena have been analyzed recently. In particular Arnold, Horsthemke, and Lefever[40] consider the effect of white and "colored" external noise in certain types of nonlinear systems and show the possibility of new transitions not predicted by the phenomenological equations when the variance of the noise exceeds some critical value. Moreover Horsthemke and Stucki[38] show that the behavior of the Lotka–Volterra oscillator under external noise may be twofold. A white noise tends to disrupt the macroscopically predicted periodicity and spread the system's trajectory throughout the phase space. In contrast, a "real noise" displaying a finite correlation time may give rise to *preferred trajectories* which are maintained during the simulation time. This is the case in spite of the fact that the orbits of the deterministic Lotka–Volterra model lack the property of asymptotic stability!

E.   Concluding Remarks

Fluctuations are an essential aspect of the evolution of macroscopic systems undergoing nonlinear kinetics away from equilibrium. This is particularly true of systems in the vicinity of a point lacking the property of asymptotic stability, where the evolution acquires an essentially statistical character. Eventually the decisive macroscopic fluctuation needed by the system is produced, this chaotic stage of evolution ends, and a new stable macroscopic state is attained.

References

1.   I. Prigogine, *Etude Thermodynamique des Processus Irréversibles*, Desoer, Liège, 1947.
2.   P. Glansdorff and I. Prigogine, *Thermodynamic Theory of Structure, Stability, and Fluctuations*, Wiley–Interscience, New York, 1971.
3.   G. Nicolis and I. Prigogine, *Self-Organization in Nonequilibrium Systems*, Wiley, New York, 1977.
4.   R. Thom, *Stabilité Structurelle et Morphogénèse*, Benjamin, Reading, Mass., 1972.
5.   D. Sattinger, *Topics in Stability and Bifurcation Theory*, Lectures Notes in Mathematics, Vol. 309, Springer-Verlag, New York, 1973.
6.   B. J. Matkowsky and E. L. Reiss, *SIAM J. Appl. Math.*, **33,** 230 (1977).
7.   M. Herschkowitz-Kaufman, *Bull. Math. Biol.*, **37,** 589 (1975).
8.   P. Hanusse, *C.R. Acad. Sci. (Paris)*, **274C,** 1245 (1972).
9.   J. Tyson and J. C. Light, *J. Chem. Phys.*, **59,** 4164 (1973).
10.   T. Erneux and M. Herschkowitz-Kaufman, *Biophys. Chem.*, **3,** 345 (1975).
11.   J. F. G. Auchmuty and G. Nicolis, *Bull. Math. Biol.*, **38,** 325 (1976).
12.   T. J. Mahar and B. J. Matkowsky, *SIAM J. Appl. Math.*, **32,** 394 (1977).
13.   J. P. Keener, *Stud. Appl. Math.*, **55,** 187 (1976).
14.   T. Erneux and M. Herschkowitz-Kaufman, 1977, in preparation. T. Erneux and M. Herschkowitz-Kaufman, *Bull. Math. Biol.*, 1978, in press.
15.   L. Cesari, *Asymptotic Behavior and Stability Problems in Ordinary Differential Equations*, Academic, New York, 1963.
16.   G. Nicolis and R. Lefever, Eds., *Membranes, Dissipative Structures, and Evolution*, Wiley, New York, 1975.

17. P. Fife, Bulletin AMS (1978), to appear. Proc. of NSF-CBMS Regional Conference on Nonlinear Diffusion, Research Notes in Math. **14,** Pitman, London.
18. R. Lefever, M. Herschkowitz-Kaufman, and J. W. Turner, *Phys. Lett.*, **60A,** 389 (1977).
19. T. Y. Li and J. A. Jorke, *Am. Math. Mon.*, **82,** 985 (1975).
20. E. N. Lorenz, *J. Atmos. Sci.*, **20,** 130 (1963).
21. D. Ruelle and F. Takens, *Commun. Math. Phys.*, **20,** 167 (1971).
22. R. M. May, *Nature*, **261,** 459 (1976).
23. O. E. Rössler, *Z. Naturforsch.*, **31a,** 259 (1976).
24. D. McQuarrie, in *Suppl. Rev. Ser. Appl. Prob.*, Methuen, London, 1967.
25. S. Ma, *Modern Theory of Critical Phenomena*, Benjamin, Reading, Mass., 1976.
26. F. Schlögl, *Z. Phys.*, **248,** 446 (1971).
27. H. Stanley, *Introduction to Phase Transitions and Critical Phenomena*, Clarendon, Oxford, 1971.
28. E. Ince, *Ordinary Differential Equations*, Dover, New York, 1956, pp. 184, 201, 202, 468.
29. G. Nicolis and J. W. Turner, *Physica*, **89A,** 326 (1977).
30. H. Janssen, *Z. Phys.* **270,** 67 (1974).
31. M. Lavrentiev and B. Chabat, *Méthodes de la Théorie des Fonctions d'une Variable complexe*, Mir, Moscow, 1972.
32. J. Cole, *Perturbation Methods in Applied Mathematics*, Blaisdell, Waltham, Mass., 1968.
33. C. Mou, G. Nicolis and R. Mazo, *J. Stat. Phys.*, 1977, in press.
34. P. Hanusse, Thèse de Doctorat ès Sciences, University of Bordeaux, 1976; preprint, M.I.T., Cambridge, Mass., 1977.
35. G. Nicolis, M. Malek-Mansour, K. Kitahara, and A. Van Nypelseer, *Phys. Lett.*, **48A,** 217 (1974); *J. Stat. Phys.*, **14,** 414 (1976).
36. W. Horsthemke and M. Malek-Mansour, *Z. Phys.*, **B24,** 307 (1976).
37. L. Arnold, *Stochastische Differential-gleichungen*, Oldenbourg, Munich.
38. W. Horsthemke and J. Stucki, 1977, to be published.
39. G. Dewel, P. Borckmans and D. Walgraef, *Z. Physik*, **B28,** 235 (1977).
40. L. Arnold, W. Horsthemke, and R. Lefever, *Z. Physik*, **B29,** 367 (1978). W. Horsthemke and R. Lefever, *Phys. Lett.*, **64A,** 19 (1977).

INSTABILITY AND FAR-FROM-EQUILIBRIUM STATES OF CHEMICALLY REACTING SYSTEMS

P. HANUSSE and JOHN ROSS

Department of Chemistry
Massachusetts Institute of Technology
Cambridge, Massachusetts

P. ORTOLEVA

Department of Chemistry
Indiana University
Bloomington, Indiana

CONTENTS

I. INTRODUCTION

In this article we discuss some aspects of nonlinear chemically reactive systems under far-from-equilibrium conditions. This field, sometimes referred to as chemical instabilities, has progressed considerably in the last 10 years. Professor I. Prigogine and his co-workers in Brussels have made major contributions to the field, and we welcome the opportunity to dedicate this article to Professor Prigogine.

The purpose of this article is an introduction to the field of chemical instabilities, in particular the variety of interesting phenomena such as multiple stationary states, oscillations, formation of macroscopic spatial structures, and chemical waves. Theoretical approaches will be discussed qualitatively and in some cases compared with experiment. For related and other topics in this field see the article by G. Nicolis in this volume. No

317

attempt is made at a comprehensive survey of the field or the literature; several reviews and monographs have been published in the last few years.[1-15]

A few words about nomenclature: We use the words homogeneous system to refer to a one-phase system. A heterogeneous system contains more than one phase. In a uniform system the state variables, such as concentration or temperature, do not have any spatial dependence, whereas in a nonuniform system they do. Thus homogeneous systems may be uniform or nonuniform.

In Section II we discuss uniform nonlinear systems under far-from-equilibrium conditions. We treat stationary states, their stability, transitions between stationary states, and periodic and chaotic evolution of state variables. In Section III we turn to nonuniform systems and consider the formation of spatial structures and chemical waves. Finally, in Section IV we discuss heterogeneous systems: single sites and arrays of sites on which catalysis may occur. The arrays of sites show interesting cooperative instability behavior.

II. UNIFORM SYSTEMS

A. Kinds of States in Uniform Systems

The concepts of system, state, state variable, and constraints have been defined and used for a long time in classical equilibrium thermodynamics.[16] Since we are reviewing some aspects of phenomena far from equilibrium, it is useful to restate these definitions and extend them on an operational basis to experimental situations relevant to our purpose.[10]

1. Stationary States.   Among all the possible states of a system, one of great importance is the stationary state, in which none of the thermodynamic properties of the system is changing with time. The properties may vary from point to point in space and the intensive properties of the system may be discontinuous at the boundary. Thus in general, exchanges of matter and energy between the system and the surroundings may take place at the boundary. When the system is in a stationary state the corresponding fluxes of mass and energy are constant in time. The system is said to be under constraints, since several parameters, particularly those characterizing the state of the surroundings (such as temperature, pressure, chemical potentials), are kept constant or at least are not influenced by the state of the system. In fact, the distinction between system and surroundings requires the assumption that the latter influence the former but not the opposite. We maintain here the usual meaning of constraint in equilibrium thermodynamics,[16] and in particular we do not restrict it to mean "driving force," even though we shall be mainly interested in this aspect of the constraints.

2. Equilibrium State. Equilibrium is defined as a stationary state in which the intensive properties of the system are continuous across the boundary. In other words, the fluxes of mass and/or energy are zero at the boundary.

Let us now consider the various possible constraints acting on a chemical system that are of interest to us.

3. Constraints and Boundary Conditions. The first kind of constraint concerns exchanges of matter between the system and the surroundings. In the simplest situation, we can distinguish between two kinds of chemical species in the system—intermediary species that are not directly controlled and constraint species the concentration of which is supposed to be constant and usually uniform. This kind of constraint is commonly used in the study of theoretical models,[17] although it presents problems experimentally. It has even been claimed that in some cases no physically reasonable controlled fluxes between the system and the surroundings could maintain constant concentration of some species.[18] In fact, the best experimental approximation to this kind of constraint can be found in closed systems. Of course no real stationary state could be attained in such systems, but in many cases (such as the Belousov-Zhabotinsky reaction, which can oscillate hundreds of times before finally reaching equilibrium[4]) the net consumption of the reactants that act as the constraints is much slower than the phenomena under study. For the same reason that constraints may be taken to be constant in time, they can be considered to remain uniform in space, since gradients in other species, if any, are not expected to induce large gradients of constraint species.

The condition of uniformity of the constraints can be removed if we fix the concentration of the constraint species at the boundary only and allow diffusion through the system. A concentration gradient of these species would then generally appear, which modifies the behavior obtained with uniform concentration[19] or may even give rise to new effects.[20] This kind of boundary condition has been used in theoretical models,[19-21] and can be achieved experimentally by using a membrane between the system and the surroundings.[22] It should be pointed out that if the thickness of the membrane (scaled by the membrane transport coefficients) is large relative to the size of the system or to the length scale of the phenomena occurring in the system (scaled by the system transport coefficients), then the concentration is not fixed at the inner boundary, but at the outer one, which may lead to different behavior.

Instead of controlling concentrations we can control fluxes. As far as theoretical models are concerned, this leads to an increase of complexity, since for a given system the number of independent variables is increased in going from a situation where the concentration of a chemical species is kept constant to a situation where, instead, the input and/or output flux of that species is controlled. Furthermore, it can be shown, either by studying simple models[18,23] or experimentally,[24] that the behavior of the system may depend substantially on whether the concentrations or the fluxes have been fixed. A large number of experimental studies have been performed in steady-flow reactors or continuous-stirred tank reactors (CSTR) in which flux constraints can be achieved experimentally.[9] In this case the input fluxes of constraints are controlled, whereas the output fluxes, strictly speaking, are not, although the operation of the system is well defined. Obviously, the behavior of a reaction like the Belousov-Zhabotinsky reaction is somewhat different when studied in a closed system or in a CSTR. Some of the features may be simply altered, such as the frequency of oscillations, or the position of the transition point between oscillating and non-oscillating states; moreover, new features may appear when going from a closed system to an open one [for instance, new stationary states (see Section II.B.3)].

Another kind of constraint concerns exchanges of energy as a driving force. The case of illuminated systems has been studied theoretically[25] as well as experimentally[26] (see Section III.B).

All the constraints referred to previously may be held constant in time, the usual case; may be known functions of time, particularly periodic functions; or may be considered as randomly fluctuating. The relative value of the time scale of these disturbances and the time scales of the system is of course of importance. Slow variations of the constraints will maintain a particular state and may be used to record a hysteresis loop, for instance.[26] Perturbations of the time scale of the system may be used to synchronize a spontaneous oscillation with a periodic constraint[27] or to measure relaxation times.[26] Finally, fast random variations have been considered.[28,63] We also have to consider the boundary conditions that apply to the intermediary species not directly controlled. In theoretical studies three kinds of boundary conditions have been considered. We may have a "zero-flux" condition at the boundary; that is, the chemical potential gradients vanish at the boundary, which is impermeable to the corresponding species.

Second, we can consider the concentration to be fixed at the boundary at its homogeneous stationary value. (For a typical phenomenon with such boundary conditions, see Ref. 29.) There is then no major difference between such an "intermediary" species and a constraint species fixed at the boundary, except that its value is determined by the steady-state condition. Third, we can also consider a "periodic" boundary condition, in which case a one-dimensional system, for instance, has the geometry of a ring.[30,73]

Experiments provide examples of some of these boundary conditions. In closed systems "zero-flux" conditions apply, whereas in experiments on membranes, concentrations may be fixed at the boundary.[22] The question does not arise for experiments achieved in a CSTR where uniformity is imposed.

Given the constraints and boundary conditions, the system reaches a stationary state or, more generally, a stationary regime (such as a limit cycle) unless chaotic behavior is observed. Periodic or fluctuation constraints can usually be considered as a perturbation added to the constant constraint that keeps the system far from equilibrium. Therefore let us first consider the most usual case with constant constraints and the simplest behavior, the stationary state.

Let ψ be the vector of the independent variables of the system. The components of this vector are typically the concentrations of the various species, but can include temperature or other state variables. Let Λ be the set of constraints. The equation of motion of such a system can typically be written for electrically neutral species as

$$\frac{\partial}{\partial t}\psi = R(\psi, \Lambda) + D\nabla^2\psi \qquad (\text{II.1})$$

where R accounts for all the "reactive processes," such as chemical reactions, and the second term represents the transport of the ψ variables with a transport (diffusion, thermal conduction) matrix D, which is usually taken as isotropic in the ψ-space but may include crosseffects.

The stationary state ψ_s is then simply defined by

$$R(\psi_s, \Lambda) + D\nabla^2\psi_s = 0 \qquad (\text{II.2})$$

for the general case, and

$$R(\psi_s, \Lambda) = 0 \qquad (\text{II.3})$$

for the uniform solution, which is usually sought first and which is often used as a reference state. Equation II.3 may have several solutions, in which case the system has multiple stationary states.

For "weak" constraints (i.e., near equilibrium), the stationary state is stable,[1] but far from equilibrium it may become unstable. This is the regime of interest here.

4. Stability of Stationary States. In the question of the stability of a state, we have to include the description of the perturbation to be considered. We may wish to study infinitesimal perturbations or finite perturbations, and uniform (space-independent) or nonuniform perturbations. Most commonly one considers infinitesimal uniform or nonuniform perturbations. Other and more general directions regarding stability have been investigated by Feinberg and Horn, who established the so-called zero-deficiency theorem;[31a] Clark, who developed a method based on graph theory;[31b] and Franck, who studied various forms of feedback in reaction schemes.[32] A technique using diagrams for feedback in multiple time scale systems has also been introduced.[33]

To investigate the stability of a uniform stationary state with respect to infinitesimal perturbations, the equation of motion (II.1) is simply linearized about this state ψ_s and we obtain the linearized equation of motion

$$\frac{\partial}{\partial t} \delta\psi = \{\Omega^R(\Lambda) + D\nabla^2\} \delta\psi \tag{II.4}$$

where the matrix Ω^R is given by

$$\Omega_{ij}^R(\Lambda) = \left(\frac{\partial R_i}{\partial \psi_j}\right)_{\psi_j = \psi_{js}(\Lambda)}$$

with $\delta\psi_i = \psi_i - \psi_{is}$, $i = 1, \ldots, n$.

We consider the perturbations $\delta\psi$ of the form

$$\delta\psi = \delta\psi^0(t) e^{i\mathbf{k}\cdot\mathbf{r}} \tag{II.5}$$

where $\mathbf{k}$ is the wave vector

$$\frac{\partial}{\partial t} \delta\psi(\mathbf{r}, t) = \Omega(\Lambda, \mathbf{k}) \delta\psi(\mathbf{r}, t) \tag{II.6}$$

$$\Omega = \Omega^R - D\mathbf{k}^2 \tag{II.7}$$

The stationary state is asymptotically stable if all the eigenvalues of the matrix Ω have *negative real parts*.[34] These eigenvalues, corresponding to "normal modes" of the system for the stationary state under consideration, depend on the wave vector $\mathbf{k}$. The eigenvalues are not required, only the signs of their real parts. Several criteria can provide this information. For example, one may use the coefficients of the characteristic equation

$$f(\omega) = \det[\Omega - \omega I] \tag{II.8}$$

where I is the unit matrix. The eigenvalues ω are the solutions of this polynomial of degree n (number of components of ψ). The Hurwitz criterion,[35] for instance, can tell us whether there is one eigenvalue with positive real part, in principle for any value of n, but this criterion is tractable only for small numbers of variables. Recently, very simple criteria have been obtained[36] that give the number of unstable eigenvalues for systems involving up to five independent variables.

In general, the stability of the stationary state is expressed in terms of some parameter that could be one of the Λ_i or a function of them, called a bifurcation parameter. Let us call it B. No matter what criterion has been used, there exists a relation, involving B and the wave number k, that corresponds to marginal stability points (e.g., the existence of one or several eigenvalues with zero real part). Such a relation, written as $f(B, k) = 0$, defines a curve in the plane (B, k); on one side of this curve the stationary state is unstable, because of the existence of at least one eigenvalue with positive real part, and on the other side of the line the stationary state is stable to perturbations of the wave vector $\mathbf{k}$ (Fig. 1). Several curves may exist in this plane, corresponding to various properties of the eigenvalues (number of unstable eigenvalues, real or complex), and these curves determine marginal stability conditions for various modes. When B is increased from low values there exists in general a critical value B_c which gives rise to an instability of the stationary state with respect to fluctuations with the wave number k_c. The study of what happens beyond this point is the object of bifurcation theory (see Section III.A.5 and the article by Nicolis et al. in this volume).

The structure of the stability diagram depends on the specific system under consideration. However, some general results independent of the details of the reaction scheme have been obtained.[37] Some of the properties of the equations of motion of type (II.1) result from the form of the rate laws for the chemical reactions. For instance, when only mono- and bimolecular reactions are involved and no stationary-state assumption is made for some species (e.g., when the rate equations are strictly bilinear),

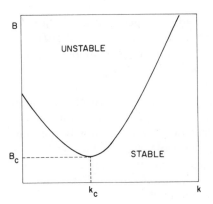

Fig. 1. Bifurcation parameter B versus wave number k at marginal stability. When B reaches B_c from below, the homogeneous stationary state becomes unstable with respect to perturbations of wavelength $2\pi/k_c$.

the diagonal elements of the matrix Ω^R (II.4) are always negative or zero, for any reaction scheme and for any number of variables.[37c] Consequently, the analysis of the stability shows that certain kinds of instability cannot appear in two-variable systems. For example, in such systems there cannot occur complex conjugate unstable modes, which may lead to a limit cycle type oscillation. Nor can there appear instability in a finite domain of wave vector (not including $k = 0$), which leads to a system unstable with respect to inhomogeneous perturbations only; such a condition is favorable for the formation of spatial structures.

The stability analysis described above gives no information about what happens beyond a marginal stability point. It does not describe the new stable state or regime reached subsequently by the system. For slightly supercritical conditions bifurcation analysis can provide a quantitative description of these new states.[38] (See the article by Nicolis et al. in this volume.)

5. Periodic Oscillations. So far we have been dealing with states independent of time, namely, equilibrium and stationary states. We extend here the concept of a stability regime to phenomena that are periodic in time. A stable limit cycle[34] can be a solution of (II.1). In this case, the point describing the state of the system in the space of the state variables follows a trajectory that converges on a closed path independent of the initial state of the system, at least in a finite domain of attraction. Limit cycle behavior may appear when there exists an unstable stationary state with two complex conjugate stability eigenvalues with positive real part (this, however, is not a sufficient condition).

Another kind of oscillation is the Lotka–Volterra type.[39] Such oscillations correspond to a continuum of concentric cycles, each of which is characterized by an amplitude parameter. These oscillations are not asymptotically stable and as such are not good candidates for the description of observed stable oscillations.

There exists much experimental evidence of periodic oscillations in various types of systems: isothermal homogeneous chemical reactions,[40] biological or biochemical systems,[2,12] heterogeneous systems involving several phases,[8,9] and chemical reactions involving temperature or other state variables in the main feedback loop.[9,41]

In all the instances it is easy, particularly in steady-flow reactors, to show that the observed periodic variations are limit cycle type oscillations; after the imposition of a perturbation the oscillation is recovered with the same frequency and amplitude[24,42] (but with a possible change of phase).

More details on the properties of these oscillations, and experimental examples, are given later. Some results on the stability of oscillatory systems to infinitesimal perturbations have been obtained.[73]

6. Chaotic Evolution. In addition to regular periodic oscillations, it is very interesting that deterministic rate equations of uniform systems may lead to temporal aperiodic, chaotic variations of state variables. Consider the reaction mechanism[43]

$$A \xrightarrow{\ k_1\ } 2A \quad \text{(a)}$$

$$A \xrightarrow{\ k_2, K, (B)\ } \quad \text{(b)}$$

$$A \xrightarrow{\ k_3\ } B \quad \text{(c)} \qquad \text{(II.9)}$$

$$B \xrightarrow{\ k_4\ } \quad \text{(d)}$$

$$\xrightarrow{\ k_5\ } A \quad \text{(e)}$$

involving two species A and B and where the baths of constant concentration (sources and sinks) have been omitted. These reactions are taking place in a two-compartment system in which the concentrations are A, B and A', B', respectively; the compartments are connected by the permeation reaction

$$B \underset{D}{\overset{D}{\rightleftarrows}} B' \quad \text{(f)} \qquad \text{(II.9)}$$

All the preceding reaction steps are to be considered as elementary reactions, except step (b), which stands for a Michaelis–Menten reaction with constant K and involves B as a catalyst; thus the deterministic equations read

$$\frac{dA}{dt} = (k_1 - k_3)A - \frac{k_2 BA}{A + K} + k_5$$

$$\frac{dB}{dt} = k_3 A - k_4 B + D(B' - B)$$

$$\frac{dA'}{dt} = (k_1 - k_3)A' - \frac{k_2 B'A'}{A' + K} + k_5 \qquad \text{(II.10)}$$

$$\frac{dB'}{dt} = k_3 A' - k_4 B' + D(B - B')$$

Rössler has shown[43c] by computational methods of solution that the concentration follows the temporal variation shown in Fig. 2. Spatiotem-

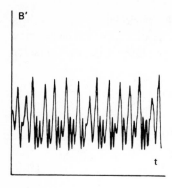

Fig. 2. Time behavior of species B' of mechanism (II.9) from $t = 0$ to 64.4 units; initial conditions: $A(0) = 0.75$, $B(0) = B'(0) = 0.6$, $A'(0) = 0.005$; $k_1 = 10.8$, $k_2 = 6$, $k_3 = 6$, $k_4 = 3$, $k_5 = 1$, $K = 0.03$, $D = 12$. (After O. E. Rössler, Ref. 46b.)

poral chaos in chemical reactions has also been demonstrated[71] and, earlier, examples have been suggested by Ruelle and Takens[45] and Lorenz[44] with possible application to turbulence.

7. Pseudostationary States. We define a pseudostationary state as one in which the temporal evolution is very slow compared to other characteristic macroscopic time scales of the system. Formally, we may say that (II.1) defines a velocity vector field in the phase space, the space of ψ variables. As we have seen before, a stationary state is located at the point where the amplitude of the velocity is zero. Similarly, a pseudostationary state corresponds to a nonzero minimum of the velocity amplitude. If in this region the rate of change of the concentrations is slow compared to other regions, then, on a short time scale, the system appears to be stationary. This concept turns out to be useful in the presentation of a coherent description of several phenomena, along with the other "states" described earlier. The existence of several time scales in the evolution of chemical systems is a typical feature of nonlinear dynamics. We shall see later how it allows useful approximations to the description of complex phenomena (Section III) and, in particular, the invocation of catastrophe theory.[82]

8. Continua of Stationary States. Systems with a limit cycle or certain degeneracies may have a continuum of stationary states. For example, the frequency of a limit cycle depends on the conditions or constraints under which the system is maintained. If this frequency passes through zero, then the limit cycle may represent a closed-orbit continuum of stationary states in the phase space.[14b] This phenomenon also occurs when multiple feedback mechanisms are present and the conditions are such that two species play an essentially identical role at a given constraint.[15] Alternatively (and more typically) it represents a homoclinic orbit.[34]

B. Transitions Between States

In this section we describe several features of nonequilibrium phenomena in terms of transitions, forced or spontaneous, between the types of "states" discussed: stationary states, periodic oscillations, pseudo-stationary states, or the other phenomena considered in Section A.

1. Transitions Due to a Change in Constraints

State Diagram. As the system is driven farther and farther from equilibrium by a change in imposed constraints, the stationary state, which is initially stable, may become unstable in the constraint space. Thereupon the system evolves to a new steady or other strictly far-from-equilibrium state. It is therefore possible to determine the "state diagram" or stability diagram that maps out the domains of existence of the various states, in the space of the constraints, exactly in the same way as an equilibrium state diagram describes the state of the system in the (P, V, T) space. In some regions of the diagram the stable state or regime may, for instance, be oscillating; in others it is nonoscillating. These various regions are separated by transition surfaces that indicate first or second order transition as in the equilibrium case.

Stability diagrams have been determined for the Belousov-Zhabotinsky reaction[47] and for one similar to the Bray reaction.[48] Besides detailing the nature of the steady states these diagrams give information about the influence of the various constraints on the characteristics of these steady states or regimes (position, shape, and frequency of oscillations). Interesting features of these state diagrams are the transition regions. From an experimental standpoint one can define two types of transitions: those at which a given property of the system changes discontinuously (called hard transitions) and those at which the transition occurs smoothly (called soft transitions). Furthermore, as we shall see, a transition may be "reversible," that is, the position of the transition point does not depend on the way the constraint is changed, or it may be "nonreversible." At a transition there is always a discontinuity of some property of the system (or its derivative) that allows distinction between the two states. Therefore to determine the character of the transition, hard or soft, we always refer to a hierarchy of properties.

Soft Transitions. Soft transitions observed experimentally usually involve a stationary state (SS) and a limit cycle oscillation (LC), but soft transitions between two stationary states are conceivable.[15,49,50] For the case of a transition to a homogeneous or spatially patterned state it has been shown that "critical" conditions exist for the transition to be soft. As the transition is approached from a LC region, the amplitude of the limit cycle may decrease continuously and vanish at the transition. The amplitude of the oscillation can be used as an "order parameter" to describe the transition. This kind of transition bears a close analogy to a second-order phase transition.[14,15,49,50] Bifurcation theory can be used to calculate the behavior of the system as the bifurcation parameter is changed.[38,46] At the transition the sign of the real part of two complex conjugate stability eigenvalues changes (see Section II.A.4). This accounts for the emergence of the limit cycle but not for the "soft" feature of the transition; the final limit cycle attained depends, of course, on the nature of the nonlinearities.

Hard Transitions.   Essentially three kinds of hard transitions may be defined: those in which the system goes from one stationary state to another ($SS_1 \leftrightarrow SS_2$); those between a stationary state and a time-dependent state, such as a limit cycle ($SS \leftrightarrow LC$); and finally, those between two time-dependent states ($LC_1 \leftrightarrow LC_2$ or chaos or waves transitions).

The transitions of the first type are nonreversible except when the system dynamics contain a degeneracy leading to a continuum of states at a given value of the constraints.[15]

The second type of transition ($SS \leftrightarrow LC$ or waves) may be accounted for with simple arguments based on the topology of the trajectories in the phase plane.[52] Consider a two-variable system described by the equations

$$\frac{dX}{dt} = F(X, Y) \tag{II.11}$$

$$\frac{dY}{dt} = G(X, Y, C) \tag{II.12}$$

where the parametric dependence on the constraint C has been made explicit. We suppose that the rate of X does not depend on this constraint.

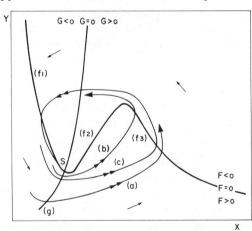

Fig. 3. Stable stationary state S near a hard transition to a limit cycle. All trajectories (a, b, c) converge to S, which is excitable (cf. Sec. II.B.3.b). Trajectories are horizontal when crossing curve (g), vertical when crossing curve (f).

In Fig. 3 the curves (f) $(F = 0)$ and (g) $(G = 0)$ have been plotted for a given value of C, along with a few trajectories (a, b, c). Note the particular shape of curve (f) with three branches (f_1, f_2, f_3). There is one stable stationary state (S). In Fig. 4 we have the same system, but C has been slightly changed so that the curve (g) is shifted: we now have a limit cycle. At the transition between these two regimes the limit cycle appears with a finite amplitude and zero frequency. Such a transition is reversible and has been observed in several experimental circumstances.[47,48]

Linear stability analysis does not distinguish a hard transition, between a stationary state and a limit cycle, from a soft transition. In fact, the features of the trajectories near the stationary state may be the same in both cases. The difference comes from the phase space characteristics far from the stationary state, that is the presence of the branches f_1 and f_2 of curve f (Fig. 4), which is responsible for the shape of trajectories such as b and c. We discuss in more detail the properties of such a topology of the trajectories. Hard transitions between a stationary state and an oscillating one can also result from the coalescence of a stable and an unstable limit cycle.[9]

For the third kind of hard transition there seems to be some experimental evidence for the $LC_1 \rightarrow LC_2$ case.[53]

Hysteresis and Bistability.   As previously mentioned, the transition may not be reversible. We already met an example of a transition between two stationary states. There is ample experimental evidence of such transitions[22c,26,47b,54,9] as well as several theoretical models.[51,55] The nonreversible feature of the transition results in hysteresis associated with the existence of two different stationary states in the same region of the state

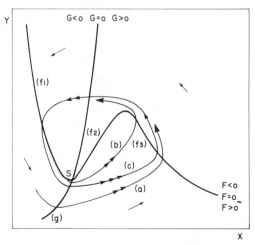

Fig. 4. Stable limit cycle (c) near a hard transition to a stable stationary state. Here S is unstable. All trajectories (a, b) converge to (c).

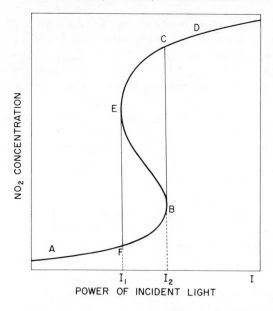

Fig. 5. Bistability and hysteresis (schematically). Stationary concentration of NO_2 in the illuminated system $2NO_2 \rightleftharpoons N_2O_4$ versus light intensity. For I lower then I_1 or greater than I_2 there is only one stationary state, A and D respectively. Between I_1 and I_2 there are three stationary states, two stable (branches FB and EC), one unstable (branch EB). When light intensity is increased starting from zero, transition occurs at point B. When decreased, transition occurs at point E. (After Creel and Ross, Ref. 26.)

diagram (Fig. 5). For such systems the macroscopic equations show that there are two stable stationary states separated by an unstable one (Fig. 5). But a very interesting question arises when fluctuations of concentrations are taken into account, that is, when one does a stochastic treatment of such a system. Hysteresis may then disappear (Fig. 6): For a given value of the constraints one of the two states becomes metastable (in other words, has a finite lifetime), whereas the other remains strictly stable. The transitions between branches of stable states likely occur inhomogeneously so that a nucleation process may be predicted[56] (see also the paper by Nicolis et al. in this volume) and observed in stochastic simulation studies.[57a] Other statistical properties have also been studied by nonequilibrium computer molecular dynamics.[57b]

As far as phenomenological behavior is concerned, the concept of bistability can be extended to states other than stationary ones. Some experiments[24] and model systems[9,58,82] provide examples of bistability involving a stationary state and a limit cycle (Fig. 7). Also, we shall see that a system may have two pseudostationary states separated by an unstable one (Section II.B.3). Furthermore, systems may be able to propagate

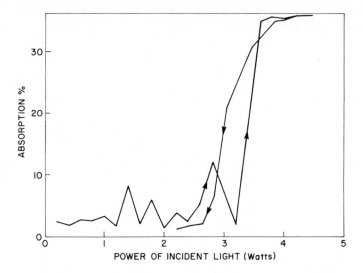

Fig. 6. Effect of fluctuations on a bistable system. Same system as in Fig. 5. Light intensity has been swept up and down but now fluctuations are present: when averaging over many loops the hysteresis disappears (obtained by integration of macroscopic equations to which was added a Gaussian noise, cf. Ref. 26).

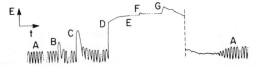

Fig. 7. Example of bistability involving a stationary state (E) and a limit cycle (A). Electrochemical potential as a function of time. System: KIO_3, 0.024 M; Malonic acid, 0.056 M; H_2O_2, 1.2 M; $HClO_4$, 0.058 M; $MnSO_4$, 0.004 M; 25°C, in a CSTR. At points B and C the limit cycle is perturbed by fast injection of I^- ions; the system returned to the limit cycle. At point D the perturbation was sufficient for the system to reach a new stationary state E. At points F and G malonic acid is injected. The last perturbation is sufficient to make the system go back to limit cycle A. (After Pacault, de Kepper, and Hanusse, Ref. 24.)

multiple, qualitatively different types of waves at a given value of constraint parameters (see Ref. 78, Appendix and Ref. 15).

2. Transitions at Constant Constraint. So far we have considered transitions from one state to another one resulting from a change in the constraints. When bistability is observed, it is possible to induce a transition from one state to the other by perturbing the system, at constant constraint. If the perturbation is applied for a sufficiently short time compared to the relaxation time of the system, then one can determine the surface or part of the surface separating the attraction domains of each stationary state in the descriptive-variable (ψ) space (Fig. 8). In some cases it is possible to stabilize the unstable state located between the two stable

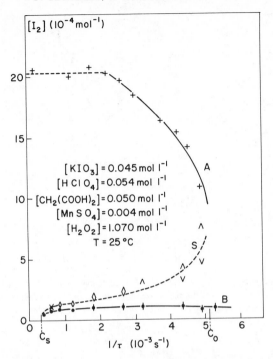

Fig. 8. Bistability and hysteresis—experiment. System as indicated. Stable stationary state (branch A) and stable oscillation (branch B, average concentration) as a function of the constraint (inverse residence time in the CSTR). Branch S is the locus of threshold of perturbations for transitions from branch B to branch A. (After de Kepper, Pacault, and Rossi, Ref. 48b.)

states by using an external feedback and so determine the position of this unstable state. This is easily done when temperature is one of the pertinent variables of the system.[9]

3. Transitions Involving Pseudostationary States

Relaxation Oscillation. Consider the limit cycle of Fig. 4. We said that the peculiar shape of branch (f_3) was responsible for the hard transition from a stationary state. In fact, this feature is most easily understood in the presence of two time scales in the evolution of the system. Consider the trajectory (c) in Fig. 4, starting right below the stationary state S. Since the slope of the trajectory is given by

$$\frac{dY/dt}{dX/dt}$$

the trajectory moves away from S because the rate of change of X is much higher than that of Y. As a result, the representative point of the system in the X-Y phase plane jumps toward branch (f_3), where the rate of change of X is slow again. Branch (f_3) acts as an attractor and, indeed, corresponds to a pseudostationary state, but this is not sufficient. In order to be reached, this branch must lie not too far from the stationary state S, or equivalently, the rate of change from the neighborhood of S to this branch must be high. If we were to scale the X and Y variables by the rate of change in the X and Y directions, respectively, so that equal distances in the plane (X, Y) correspond to equal intervals in time, then the pseudostationary state would appear very close to S. On the other hand, if such a difference of time scale did not exist, then the pseudostationary state would be so "far" from S that it would have no influence on the trajectories in the neighborhood of S. Similarly, a rapid transition occurs later between branch (f_3) and branch (f_1).

Oscillations that involve two well-separated time scales have been called relaxation oscillations. Most of the experimentally observed oscillations in chemical systems are of this type.

From what precedes one can say that a relaxation oscillation results from the interaction in the phase plane of two pseudostationary states, here the region around branch (f_3) and the region around branch (f_1), and stationary state S.

Suppose now that species X is an input or external parameter for another part of a mechanism that is oscillating with a high frequency when X is near branch (f_3) and nonoscillating otherwise. Then the system switches periodically from a high-frequency oscillating mode to a nonoscillating mode; in other words, we observe a two-frequency oscillation or composed oscillation. This sort of argument has been used to construct models of composed oscillations,[23] and there exist several experimental examples of this phenomenon[48,53,59] (Fig. 9). Using the previous language, we can say that in such a situation a pseudostationary state interacts with a pseudo-oscillating state, but in fact there is no simple way to depict such a situation in a two-dimensional phase space.

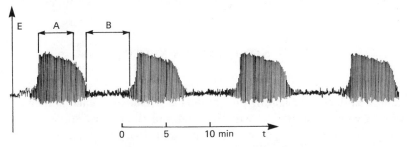

Fig. 9. Composed oscillations. (A) Pseudo-oscillating regime; (B) pseudostationary regime. Same reaction as Fig. 7, Ref. 48b.

From a mechanistic viewpoint, the occurrence of such oscillations can be related to the nature of various feedback loops present in the chain of processes taking place in the system.[1,33] For instance, in a CSTR where a single-step reaction is occurring, the coupling between temperature control and chemical reaction may give rise to periodic oscillations and many other peculiar phenomena.[9] Feedback loops may also exist in the structure of complex reaction mechanisms that, because of the nonlinearity of reaction rates, may give rise to isothermal oscillations, even in closed systems, like the Belousov-Zhabotinsky reaction. (Of course, strictly isothermal oscillations are not possible since, in general, chemical reaction is not an isothermal process; and indeed temperature oscillations have been observed.[60] Nevertheless, when studying such a reaction in a CSTR one may have to distinguish between phenomena that result from the features of the reaction mechanism itself and those resulting from the CSTR mode of operation.)

Excitability. Let us return now to Fig. 3, where the stationary state is again stable. We see immediately that, because of the features discussed earlier, if a perturbation lowers slightly the concentration of species Y, then the system moves along a large loop like (b) or (c) in Fig. 3 before going back to the stationary state. On the other hand, very small perturbations decay with traversal of regions of phase space restricted to the neighborhood of the stationary state. The relation between the loop followed by the system and the limit cycle that will appear beyond the transition point already discussed is clear. Such a relation has been demonstrated experimentally.[61]

The phenomenon of threshold excitation has been observed in many types of systems,[62] particularly in the so-called excitable membranes. It is related to an important type of chemical wave, as is discussed later. Finally, when such a feature is associated with fluctuations of external parameters or with internal fluctuations, then spontaneous excitation may occur leading to quasiperiodic behavior and renormalization (shifting) of the point of onset of oscillation.[63]

III. NONUNIFORM SYSTEMS

A. Chemical Waves

The wealth of propagating phenomena in chemically reacting media has been well demonstrated in experimental and theoretical studies. In this section we outline first some of the theoretical advances on the study of waves in unbounded media (i.e., systems so large in the direction of propagation that for all times of interest the boundary conditions at the ends of the system do not affect the results). For a discussion of waves and structures in bounded systems see the article by G. Nicolis et al. in this volume. Since the theory for one spatial dimension has received the most complete treatment, we concentrate on it as our model problem but also briefly review wave phenomena of other geometrical configurations. Then we discuss the experimental situation; the gap between theory and experiment is substantial, but some useful correlations have been made.

Work in closely related fields such as nerve axon propagation and biomorphogenesis has stimulated many of the ideas that have led to the development of this field. Relevant references to this literature may be found in many of the references to work on chemical waves cited here (see Section III.A.9).

1. Three Fundamental Propagating Phenomena in One Dimension.   A variety of propagating phenomena occur in a reacting system and it is useful to start our discussion by defining some basic concepts. A front is a propagating disturbance in the state of the system (i.e., concentration, pressure, fluid velocity, and temperature) that leaves the system in its wake in a different state than that in the medium in advance of its arrival. In contrast to this, a pulse (or finite train of pulses) leaves the system in its original state after its passage at a given point of observation. For some systems the propagating disturbance is an infinite wave train that may consist of a periodic or aperiodic sequence of disturbances. All these basic phenomena may propagate unattenuated in reaction-diffusion systems maintained far from chemical equilibrium.

2. The Macroscopic Theory of Chemical Waves.   The starting point of the theory is the phenomenological continuity equations of a reacting hydrodynamic continuum.[64] In our discussions here we limit ourselves to the case of isothermal, isobaric systems without center of mass flow, and hence the local descriptive variables for the system are the chemical concentrations $\psi_1, \psi_2, \ldots, \psi_N$ for an N-species system. For compactness of notation we denote the local state by a column vector ψ such that $\psi = \text{col}[\psi_1, \psi_2, \ldots, \psi_N]$. If we introduce the column vector $\mathbf{J}$ of fluxes (moles/area-time) the continuity equation for the balance of the rate of influx of material, $-\nabla \cdot \mathbf{J}$, and local production of molecules $\mathscr{R}$, with the local rate of change of state, $\partial \psi / \partial t$, is

$$\frac{\partial \psi}{\partial t} = -\nabla \cdot \mathbf{J} + \mathscr{R} \qquad \text{(III.1)}$$

It is instructive to examine this equation even at this general level to determine what propagating phenomena may occur. Clearly, the disturbance can involve composition variations. But this is not the entire story, since the chemical species involved may be ionic. Because of the possibility of widely different ionic mobilities and the tendency toward local charge neutrality, it has been shown that a liquid-junction-type potential may arise in conjunction with the wave and may propagate along with it.[65]

In addition to these possibilities for variation in the descriptive variable, the propagation may occur in a bulk medium or along a membrane or catalytic wall. At such an interface (solution-membrane or solution-wall) the rate of efflux to the solution must be just equal to the rate of interfacial reaction and chemidesorption. Let n be the unit normal (to the surface separating the two phases) pointing into the solution. Then this balance is written as[66]

$$n \cdot \mathbf{J} = G \qquad (III.2)$$

where G_i is the interfacial production rate of species i and $G = \mathrm{col}\,\{G_1, G_2, \ldots, G_N\}$. This balance condition serves as the boundary condition for the problem of propagation along an active interface.

Since under the far-from-equilibrium conditions of interest at least one of the rates $\mathscr{R}$ and G must be a nonlinear functional of ψ, our problem is strictly a nonlinear one. Nevertheless, as we shall see, a variety of nonlinear perturbation techniques have been developed that enable one to understand many of the qualitative features of a variety of propagating phenomena. With judicious choice of zeroth-order solutions and smallness parameters, these perturbation schemes have been developed to provide quantitative descriptions.

Chemical waves can occur in systems with various geometries. If the reacting medium is a thin layer of solution, then the waves may be two-dimensional with circular and spiral geometries.[67] In three dimensions complex scroll patterns may also develop.[67] Waves may propagate along a catalytic membrane or wall, being attenuated in the direction perpendicular to the active surface.[66] Finally, waves may propagate in rings[68] and within thick membranes.[69] As a prototype for the study of the previously mentioned phenomena we focus on the propagation of planar or one-dimensional waves in an infinite homogeneous medium, but comment on the more complex structures at appropriate points in the development. Furthermore, we limit ourselves to nonionic media and assume that Fick's law holds for diffusion,

$$\mathbf{J} = -\mathscr{D}\boldsymbol{\nabla}\psi \qquad (III.3)$$

where $\mathscr{D}$ is a matrix of diffusion coefficients, which, for simplicity of presentation, we assume to be constant. We take our one-dimensional medium to be in the x direction and with this the continuity equation

(III.1) with Fick diffusion (III.3) becomes

$$\frac{\partial \psi}{\partial t} = \mathscr{D}\frac{\partial^2 \psi}{\partial x^2} + \mathscr{R} \tag{III.4}$$

It may seem at first surprising that this diffusion equation has solutions corresponding to disturbances that can propagate without attenuation. This possibility is inherently a consequence of the nonlinearity of $\mathscr{R}$ and the fact that the reactions are maintained sufficiently far from equilibrium by constraints imposed by baths or light fields[70,25] for disturbances from equilibrium (III.4) shows only attenuation with distance.

General theorems defining the limits on the possible type of propagating disturbances, even for the one-dimensional system (III.4), are not available. Thus the disturbance may advance with constant or time-varying velocity; it may have a profile that eventually settles down to a constant shape or that is oscillatory or even chaotic.[71] The case that has received most attention is the simple one of waves that move with a constant velocity and profile, and we now consider that case in some detail.

It is convenient to transform coordinates to a reference frame moving with the velocity v of the wave. The relative coordinate ϕ is given by

$$\phi = x - vt \tag{III.5}$$

Let the wave profile be denoted by $\chi(\phi)$ and assume the wave has attained this constant form; thus $\psi(x, t)$ approaches $\chi(\phi)$ and from (III.4) we obtain the wave equation

$$\mathscr{D}\frac{d^2\chi}{d\phi^2} + v\frac{d\chi}{d\phi} + \mathscr{R} = 0 \tag{III.6}$$

An essential part of the theory of chemical waves has been to find known elementary solutions of this equation and to define smallness parameters by which the full wave solutions may be derived by appropriate perturbation schemes. The elementary solutions include steady or almost steady states, homogeneous oscillations, and static spatial structures. The smallness parameters include the wave amplitude, vector, velocity, and the time and length scale ratios. We now examine these various possibilities. Roughly speaking, the reference states either are attractors (i.e., a given class of initial conditions eventually leads to the reference states) or are weakly unstable to a given class of perturbations.

3. Limit Cycle Systems. In this section we encounter our first example of how one attractor or stable state of the nonlinear reaction-diffusion equations can lead to a family of propagating solutions. The case at hand is the homogeneous oscillation. Let us assume that the spatially uniform system has a limit cycle solution[34,35] ψ_c with frequency ω_c for the temporal evolution,

$$\frac{d\psi_c}{dt} = \mathcal{R}[\psi_c] \tag{III.7}$$

The existence of this state of oscillation, stable to uniform perturbations, suggests that there may be a family of closely related solutions to the partial differential equation that are always near some phase of the unperturbed cycle at each point in space. This idea has been developed in a variety of directions, including transients and generation of waves,[72,73] plane waves,[74,75] stability of plane waves,[74] waves in rings,[73] and circular waves.[75]

Let us consider in some detail the study of plane waves as long-wavelength extensions from the homogeneous limit cycle. To extract the correct dimensionless parameter for the system we introduce scaled variables into the wave equation (III.6). Denoting the wave vector for the infinite wave train by $\mathbf{k}$, we define a dimensionless coordinate ρ

$$\rho = k\phi \tag{III.8}$$

and thus χ is a 2π-periodic function of ρ depending parametrically on $\mathbf{k}$. The kinematic relation between angular frequency $\omega(\mathbf{k})$, wave vector, and wave velocity $v(\mathbf{k})$ must hold,

$$\omega(\mathbf{k}) = kv(\mathbf{k}) \tag{III.9}$$

for the propagation of an infinite wave train of any physical system. As $k \to 0$, $kv(\mathbf{k}) \to \omega_c$ and hence $\rho \to -\omega_c t$. Thus we expect that for long wavelengths there is a family of solutions for some reaction diffusion systems such that

$$\chi \underset{k \to 0}{\sim} \psi_c\left(\frac{-\rho}{\omega_c}\right) \tag{III.10}$$

for sufficiently small k. It is convenient to introduce dimensionless units into the plane wave equation by scaling the chemical rate with the limit cycle frequency ω_c, and diffusion with a typical diffusion coefficient $\bar{D}$,

$$\mathcal{R} = \omega_c R \tag{III.11}$$

$$\mathcal{D} = \bar{D}D \tag{III.12}$$

With this (III.5) becomes

$$\alpha D \frac{d^2\chi}{d\rho^2} + \beta \frac{d\chi}{d\rho} + R = 0 \tag{III.13}$$

where

$$\beta = \frac{\omega(\mathbf{k})}{\omega_c} \qquad \text{(III.14)}$$

$$\alpha = (\mathbf{k}l_c)^2 \qquad \text{(III.15)}$$

The frequency ratio or "frequency renormalization" β expresses the dispersion or variation of the frequency with wave vector. The cycle diffusion length $l_c \equiv (\tilde{D}/\omega_c)^{1/2}$ is a typical distance over which material diffuses in one cycle. Thus we see that the natural dimensionless smallness parameter for the system, α, is the square of the ratio of the cycle diffusion length to the wavelength. The theory proceeds[75] by expanding the profile (a 2π-periodic function of ρ) and the renormalization factor β in powers of α, inserting the expansions in (III.13), and collecting terms to various orders.

The plane wave equation (III.13) parametrized in terms of α is a problem in singular perturbation theory, since as $\alpha \to 0$ the order of the equation changes from second to first order. The problem may be cast in the form of a set of $2N$ first-order equations for the variables $\{\chi_1, \ldots, \chi_N, d\chi_1/d\rho, \ldots, d\chi_N/d\rho\}$.[75] In this form one may use a theorem from singular perturbation theory derived by Wasow.[76] Application of Wasow's theorem to the present problem ensures that long-wavelength waves exist as extensions from a limit cycle, which is stable to homogeneous perturbations, if the matrix D has no pure imaginary eigenvalues. This condition is assured if we assume that the system is not so far from equilibrium that Onsager's relations[64] for diffusion are no longer valid. Clearly, however, unless the system is sufficiently far from chemical equilibrium so that Onsager's relations for R linearized about a steady state become invalid, no limit cycle can exist.

Finally, we note that an iterative procedure for calculating oscillatorlike wave solutions may be obtained by transforming (III.13) into an integral equation.[74] This method provides an alternative procedure of proving existence of the solutions and appears to be a rapidly convergent method for calculating wave solutions.

4. Catastrophe and Propagation. In a reaction-diffusion system processes may occur over a range of time and length scales.[77] Chemical rate coefficients may differ by several orders of magnitude and diffusion coefficients may also vary over one or two factors of 10 (i.e., from the large diffusion coefficient for H^+ to the much smaller ones for large protein molecules). Thus the ratios of these coefficients present themselves as possible expansion parameters, and a number of approaches to the theory of plane waves have used this concept.[77–83] In this section we focus on the separation of time and length scales, its relation to catastrophe theory, and the resulting propagating phenomena.

The particular way in which the time-scale separation enters the problem profoundly affects the nature of the resulting behavior. Consider, for example, the case for which the rate contribution $\mathscr{R}$ from (III.5) may be written[82]

$$\mathscr{R} = \epsilon^{-H} F \qquad (III.16)$$

where H is a diagonal matrix whose elements are either one or zero; for simplicity we order the subscripts so that for the first f elements $H_{ii} = 1$, $(1 \le i \le f)$ and for the remaining $H_{ii} = 0$ ($f < i \le N$, where N is the number of variables). The parameter ϵ is taken to be a measure of the time scale ratio and as $\epsilon \to 0$ the F_i $(1 \le i \le N)$ are taken to be finite. We pose the question, "What is the behavior of the system (III.4) as $\epsilon \to 0$?"

We name the variables for which $H_{ii} = 1$ the fast variables and those for which $H_{ii} = 0$ the slow or control variables. Then it is clear that in order for χ to be a solution of (III.4) either $F_{i \le f}$ must be of order ϵ or the derivative terms in (III.4) must diverge as $\epsilon \to 0$. Thus the wave profile may be qualitatively divided into smoothly varying regions where

$$F_i[\chi] = 0, \qquad i = \text{fast variable} \qquad (III.17)$$

and these are connected by regions in which the fast variables change rapidly so that there are quasidiscontinuous jumps as $\epsilon \to 0$. It is at this point that the close relationship between the geometry of the behavior surfaces (III.17) and multiple scaling becomes important.

The condition $F_i = 0$ for each fast variable determines a hypersurface in the N-dimensional phase space of the descriptive-variable (concentration) $\{\psi_1, \psi_2, \ldots, \psi_N\}$. Catastrophe theory[84] has been used to show that in some cases one may specify the number and types of the most general topological features available to the intersection of these surfaces, the "behavior surface," for a system of a given value of N and f. For example, it was shown that for one fast variable and four slow variables the most general topological features or elementary catastrophes available to the system are the cuspoids—the fold, cusp, swallow's tail, and butterfly. Furthermore, the appearance of elementary catastrophes is a strictly far-from-equilibrium phenomenon, and if all the processes on the fast time scale are reversible, then no catastrophes can occur on the behavior surface.[82]

A simple example shows how the occurrence of folds and other catastrophes can lead to propagating disturbances. Consider a two-variable system $\{\psi_1, \psi_2\}$ that obeys the homogeneous kinetic equations[78]

$$\frac{d\psi_1}{dt} = \frac{1}{\epsilon} F_1[\mathbf{\psi}] \qquad (III.18)$$

$$\frac{d\psi_2}{dt} = F_2[\boldsymbol{\psi}] \tag{III.19}$$

In Fig. 10 we see a behavior line $F_1 = 0$ plotted in the $\{\psi_1, \psi_2\}$ plane. Also shown is the line $F_2 = 0$ whose intersections 1, 2, 3 with the behavior line give the stationary states 1, 2, and 3 of the system. From the signs of F_1 and F_2 indicated it is seen that states 1 and 3 are stable while state 2 is unstable. Furthermore, evolution of a phase point in the vicinity of $F_1 = 0$ to the left of point l or to the right of point r tends toward these stable or "attractor" branches of the behavior lines, whereas the segment between l and r is unstable. Suppose now we consider the initial value problem where the one-dimensional system is started off in state 1 to the left of the origin of the spatial coordinate system and in state 3 to the right of the origin. What is the subsequent evolution of the system? A matched asymptotic expansion technique has been used to show that for this case, Fig. 10, a front develops that takes the system either from state 1 to state 3 or the reverse, depending, qualitatively speaking, on the relative proximity of 1 and 3 to l and r, respectively.[78] In Fig. 11 we see a schematic profile of a front sweeping through the system such that a $1 \to 3$ transition occurs with an overshoot. The profile is seen to be in qualitative agreement with the phase plane trajectory $1 \to Q \to 3$ shown in Fig. 10.

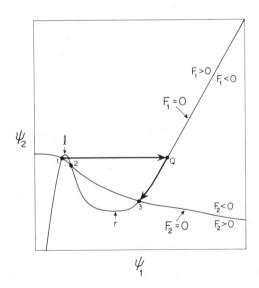

Fig. 10. Behavior of lines $F_1 = 0$ and $F_2 = 0$ in the (ψ_1, ψ_2) space corresponding to (III.18) and (III.19) when there are two stable stationary states (1 and 3) and an unstable one (2), and trajectory of transition between states 1 and 3 $(1 \to Q \to 3)$ for widely separated time scales.

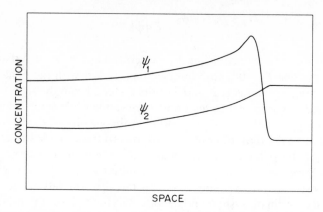

Fig. 11. Schematic profile of a front sweeping through the system so that transition $1 \to 3$ occurs with an overshoot as qualitatively described by trajectory $1 \to Q \to 3$ in Fig. 10.

We may expect the transition to point Q from steady state 1 since the branch on which Q resides is an attractor and as $\epsilon \to 0$ the slow variable ψ_2 changes on a length or time scale much longer than ψ_1 and is thus essentially constant during the abrupt change in the fast variable ψ_1. The proximity of state 1 to l allows the system to be pulled over the repelling barrier of the segment l to r (of the behavior line) via the force of diffusion. Thus propagation is seen to result from two opposing tendencies present in a system with multiple time scales: on the one hand, for the system to reside on stable branches of the behavior surface; on the other, for it to respond to the uniformization of diffusion.

All these qualitative arguments based on the geometry of the stationary-state regions in the phase plane and the attractor and repellor tendency of the branches of the behavior surface have been put on firm theoretical ground by the application of matched asymptotic expansion techniques.[78-83] Without going into the details, let us consider some of the physical motivation for the multiple-scaling theories. By the preceding geometrical arguments it is clear that the plane wave profile $\chi(\phi)$ must have at least two qualitatively different regions of behavior characterized by at least two length scales. The regions of rapid transition are dominated by the opposing tendency of diffusion and the fast time scale τ_f associated with the evolution to the attractor branches. If $\bar{D}$ is a characteristic diffusion coefficient of the fast species then the natural length that arises is $(\bar{D}\tau_f)^{1/2}$. On the other hand, the wave passes through such a region in a time τ_f so that the velocity is expected to be of order $(\bar{D}/\tau_f)^{1/2}$. After the quasidiscontinuous transition the slow processes, on a characteristic time scale τ_s, take place and hence following the branch jumping there is a profile on the length scale of τ_s times the velocity, $(\bar{D}/\tau_f)^{1/2}\tau_s = (\bar{D}\tau_s)^{1/2}/\epsilon^{1/2}$, where ϵ is the time-scale ratio, $\epsilon = \tau_f/\tau_s$. Thus if we define an intermediate length scale $L = (\bar{D}\tau_s)^{1/2}$, there are three possible length scales in the wave profile: the short scale $L\epsilon^{1/2}$, the intermediate scale L, and the long-length scale $L\epsilon^{-1/2}$. Depending on the placement of the stationary states (1 and 3 for the preceding two-variable example) two or three of these scales are important. In addition, if diffusion coefficients are widely separated in value, additional length scales arise. For the case of Fig. 10, with roughly equal diffusion coefficients, only the two extreme scales have been shown to be important (except, roughly speaking, when the two states 1 and 3 are equally close to l and r).[77] The matched asymptotic technique[85] consists of introducing a variety of new scaled space variables such as $\epsilon^{-1/2}\phi$, ϕ and $\epsilon^{1/2}\phi$, and writing the wave profile χ as a sum of terms that depend on the various scales.[78-83] These functions and the wave velocity are then expanded in powers of ϵ and the former is appropriately matched on either side of the discontinuity in the fast variables $\chi_1, \chi_2, \ldots, \chi_f$.

This matched asymptotic procedure has been applied to the analysis of propagation associated with catastrophes of various types.[82] An exciting aspect of this approach is the emphasis on the use of the geometric properties of catastrophe theory to unfold the great variety of propagating phenomena that can occur in reaction diffusion systems and provides a technique for quantitatively evaluating their velocity and wave profile. It has been found possible to study analytically the encroachment of a wave train on a steady state; the propagation of a pulse with a single jump; complex multispecies fronts; the propagation of finite trains of pulses; and propagating dephasing zones in an otherwise homogeneous oscillation. In another study[78] it was shown that, for certain relatively simple systems, multiple qualitatively different stable modes of wave propagation may occur for given conditions, depending on the initial concentration distribution.

5. Bifurcation Theory. In the previous two sections propagation has been seen in connection with the existence of attractors (limit cycles and stable branches of the behavior surface) that are in the associated ordinary differential equation for the chemical kinetics in the uniform system. In this section we turn to the study of the bifurcation of chemical waves, that is, the existence of propagating phenomena that occur because of the loss of stability of a reference state to patterned perturbations.[30,86,1] This concept has been applied to the study of planar chemical waves.[74,75,87] An important aspect of a bifurcation development is the identification of a parameter, the bifurcation parameter, which passes through a critical value at which the reference state becomes marginally stable.[38a,88] In the study of plane waves one parameter that has been used is the wave vector. It emphasizes the fact that in an infinite system there exist a family of wave solutions parametrized by the wave vector.[75]

The reference state to which the theory is most easily applied is the homogeneous stationary state ψ_h, which is independent of space and time. From (III.1) we obtain

$$\mathcal{R}[\psi_h] = 0$$

To examine the stability of this state we linearize the continuity equation (III.1) or in its simpler form (III.4) about the homogeneous stationary state ψ_h. For the case of a one-dimensional system with Fickian diffusion we get

$$\frac{\partial \delta \psi}{\partial t} = \left[D \frac{\partial^2}{\partial x^2} + \Omega_h \right] \delta \psi \tag{III.20}$$

where $\delta \psi = \psi - \psi_h$ and Ω, defined by

$$\Omega = (\partial \mathcal{R} / \partial \psi)_{\psi = \psi_h} \tag{III.21}$$

represents the matrix of linear coefficients derived from the expansion of $\mathcal{R}$ about ψ_h. Taking the spatial Fourier transform $(x \to \mathbf{k})$ of (III.20), we obtain

$$\hat{\psi}(\mathbf{k}, t) = e^{+[\Omega_h - k^2 \mathscr{D}]t} \psi(\mathbf{k}, 0) \tag{III.22}$$

and hence if the matrix $\Omega_h - \mathbf{k}^2 \mathscr{D}$ has one or more eigenvalues with positive real part the perturbations of the wave vector $\mathbf{k}$ grow. If the eigenvalues are complex then the perturbation has an oscillatory character. For some systems the linear growth of these perturbations will eventually be balanced off by the terms nonlinear in the deviation from ψ_h to produce propagating nonlinear waves or static spatial patterns of finite amplitude. Bifurcation theory provides a technique to carry out an analysis of such possibilities.

Let us assume that in the vicinity of $k = k_c$ there is a complex conjugate pair of eigenvalues of $\Omega_h - \mathbf{k}^2 \mathscr{D}$ whose real part passes through zero at k_c and, for concreteness, is positive for $k > k_c$. Then for $\mathbf{k}$ near k_c we expect that wave solutions should have small amplitude. One of the tasks of a bifurcation theory is to determine the analytical behavior of the amplitude of the waves for k near k_c.

It is convenient to carry out the calculations in the dimensionless variable $\rho = k\phi$ introduced earlier. Using $\omega = kv$ we may thus put the planar wave equation (III.6) into the form

$$k^2 \mathscr{D} \frac{d^2 \chi}{d\rho^2} + \omega \frac{d\chi}{d\rho} + \mathcal{R} = 0 \tag{III.23}$$

The trivial solution is $\chi = \psi_h$ corresponding to the homogeneous stationary state. To study the bifurcation of solutions from this state we expand χ, k^2, and ω in a dimensionless amplitude parameter A, which is a measure of the deviation of χ from the trivial solution. From the linear stability analysis we expect that $A \to 0$ as $k^2 \to k_c^2$. There is some arbitrariness in the definition of A as introduced to this point. This is easily removed, for example, by fixing the arbitrary first coefficient χ_1 in the expansion of χ ($\chi = \psi_h + \chi_1 A + \chi_2 A^2 + \cdots$); that can always be done, since χ_1 obeys a linear homogeneous equation (i.e., $[k_c^2\, d^2/d\rho^2 + \omega_c\, d/d\rho + \Omega_h]\chi_1 = 0$).

When the expansion of k^2 in terms of A is inverted we determine the dependence of the wave amplitude on wave vector. Under the assumption that there exists a complex conjugate pair of eigenvalues of $(\Omega_h - k^2\mathscr{D})$ whose real part passes through zero linearly at k_c^2, we find that the emergent family of waves (parametrized by k) has an amplitude that has the analytical form

$$A \propto \left(\frac{k^2 - k_c^2}{k_2^2} \right)^{1/2}$$

and it is seen that the branch of waves near k_c^2 exists for $(k^2 - k_c^2)/k_2^2 > 0$, where k_2^2 is the coefficient of A^2 in the expansion of k^2 in terms of A. The branch of waves may bifurcate toward the domain of either stable or unstable perturbations (i.e., the region of k^2 values around k_c^2 where the real part of the eigenvalues $z_\pm(k^2)$ of $(\Omega_h - k^2\mathscr{D})$ is negative or positive, respectively). The situation which arises depends on the nonlinear terms of the calculation of k_2^2. Finally the dispersion of waves, $\omega(k^2)$, is also determined by this procedure.

Thus far we have discussed the bifurcation of spatiotemporal patterns from homogeneous stationary states. Bifurcation theory has also been applied to the study of the emergence of patterned phenomena due to the instability of the state of homogeneous evolution to patterned perturbations.[75] It was shown that systems may show periodic, multiply periodic, and aperiodic phenomena.

6. Autonomous Centers.

The theory of leading centers in one-dimensional, circular, spiral, and other wave geometries has not been developed as far as that for plane waves. Schemes closely related to the oscillator perturbation theory of subsection 3 have been used to treat periodic circular and spiral waves.[75] Particularly in the latter case the theory breaks down near the center of the wave where, because of strong gradients, the solution is fundamentally different from the oscillator and thus cannot be expressed as a small deviation from some phase of the oscillation. However the theory is believed to be valid for circular and spiral waves away from the center, and for certain cases it is valid everywhere for circular waves. The result of the theory is a nonlinear equation for the phase of the oscillation $\alpha(r)$ involving both $\nabla^2\alpha$ and $|\nabla\alpha|^2$. A closely related equation has been derived for the time-dependent local phase of oscillation.[89] This latter work treats two-variable systems in the phase plane and has been used to show the possibility of spiral waves in systems without a homogeneous cycle. As in the more general treatments of limit cycle systems the theory breaks down near the center of the spiral wave. It is claimed that spiral waves are demonstrated in this work[89] even though the ultimate resolution of the center divergence is resolved by solving the full coupled polar concentration phase plane coordinate equations numerically. Since these equations are equivalent to the original coupled partial differential equations for the concentrations, no analytical resolution of the problem is actually obtained. Thus for all limit cycle perturbation-type theories a breakdown at the center is found. However, both bifurcation and limit cycle perturbation theories predict that away from the center the concentrations are in the form of the involute of a circle as found experimentally.[67]

Core properties of circular and spiral waves were studied by means of an expansion in the distance r from the center of the phenomena.[75] It was shown that in the core region the isoconcentration lines (isocons) could not be the involute of a circle.[67] However, both bifurcation and limit cycle perturbation theories show that far from the center the isocons are indeed involutes of a circle (i.e., if r is the distance from the center and θ is the polar angle, then at a fixed time the isocons satisfy the equation $r = \lambda\theta/2\pi + $ constant where λ is the wavelength far from the center as found experimentally).[67] In all treatments (bifurcation, limit cycle perturbation, and core expansions) the possibility of multiply armed spiral waves is predicted, although these have not been produced experimentally.

Multiple scaling methods would appear to be a very important method for attacking these problems and have been used to study leading centers in one dimension.[90] Bifurcation expansions have been used to study standing and rotating circular waves.[75] However, the method breaks down for radially propagating circular or spiral waves because of a nonuniform convergence of the theory near the wave center.[75]

Padé approximants have been used to obtain periodic and aperiodic center wave solutions valid for all space.[71b]

Circular and spiral waves have been studied for a model of the diffusion process that replaces the effect of diffusion by an averaging operator that tends, in analogy to diffusion, to make concentrations more uniform in space.[91]

7. Transient Behavior and Wave Stability.   Thus far we have discussed solutions for the wave profile without inquiring about the attainment of that profile from an arbitrary initial distribution. A second, closely related topic is the investigation of the stability of wave solutions to small perturbations. The rigorous determination of wave stability within the framework of a given approximation scheme such as those studied in subsections 3 to 5 is rather difficult in regard to proving stability, since it can always be argued that in principle there may exist perturbations that grow but that are outside the realm of a given expansion scheme like that used for limit cycle or multiple-scale theories. The question of wave instability is somewhat easier, since if one can find a perturbation that grows within the framework of the given expansion procedure, then one has proved instability.

The limit cycle perturbation theory of Section III.A.3 has been generalized to include transient situations.[14b,73] It has been found that phase gradients and frequency shift may arise both autonomously and in the presence of heterogeneities or imposed gradients.[15,72,73] This approach results in a linear,[72,73] and more recently a nonlinear,[15] phase diffusion theory and has been used to study the pseudo or kinematic waves that one finds in the early stages of systems with effective gradients of frequency.[92] The phase diffusion theory showed that because of diffusion, not considered in the kinematic picture, a system with a stable limit cycle oscillation ultimately settles down to a state of uniform (renormalized) frequency and a static gradient of phase.[73,75] The boundary conditions are accounted for in the phase diffusion theory and it is found that the waves that on a short time scale are purely kinematic are true reaction diffusion waves but of long wavelength. The phase diffusion theory was also used to show that the long-wavelength extension of the homogeneous limit cycle, discussed in Section III.A.3, are stable to small perturbations.[73]

There are a number of studies on existence and wave stabilities that have been carried out on reaction-diffusion and closely related equations and a bibliography is given in Ref. 93 and in Refs. 79, 94, 95. (Ref. 95 gives a nonexistence result for the Lotka–Volterra mechanism.)

8. Experiments on Chemical Waves.   A number of experiments have been reported on wave phenomena in chemical systems such as the Belousov-Zhabotinsky reaction, in nerve conduction,[100] in plants,[101] and in chemotaxis.[102] We comment here briefly only on chemical systems. The variety of predicted wave propagation and the experimental difficulties associated with control of the variables of concentration and temperature have made interpretations of experiments (i.e., correlation of experiment with theory) difficult. However, measurements have been made and interpreted to provide evidence for kinematic waves and trigger waves.[67]

As we said before, kinematic waves occur in oscillatory systems for a limited time after an initial condition has been set up for which the frequency of oscillation varies sufficiently weakly in space. This can be and has been achieved in the Belousov-Zhabotinsky reaction by establishing a temperature or concentration gradient[103,53,92] in a cylindrical tube containing the reaction system (effectively a one-dimensional geometry). Because of the frequency gradient waves appear that are *not* blocked by a solid barrier; that is, material diffusion does not play a significant role in a limited time period after initiation of the oscillations.

Trigger waves, on the other hand, depend on both reaction and diffusion. They are emitted by source points, or leading, autonomous centers (see Section III.A.6) in a thin layer of reagent. Both single pulses and wave trains have been observed in the Belousov-Zhabotinsky reaction.[104] They propagate with constant velocity of the order of 5 mm/min, have a wavelength of the order of 3 mm, and may form circular or scroll waves. Autonomous centers appear generally randomly, apparently because of dust particles or scratches on the dish, but can also be triggered deliberately.[105] Several centers with different frequency usually appear during the same experiment. Very little is known about events near and inside the center. Attempts have been made to describe the features of the propagating front,[106,52d] but until now no quantitative measurements of concentration profiles have been done.

9. Chemiacoustic Instabilities. Gas phase chemical reactions cause local changes in temperature and pressure because of enthalpy changes and changes in number of moles as the reaction occurs. These changes allow for the coupling of chemical reactions and acoustic modes and hence for the possibility of instabilities involving coupled acoustic and reactive degrees of freedom.

The macroscopic theory of these phenomena starts with the continuity equations for a reacting hydrodynamic continuum (i.e., the equations for mass, momentum, and energy conservation). The theory thus far has been limited to the study of the instability of acoustic-reactive modes as small deviations from uniform states that are transient[96-98] or from stationary states that are maintained out of equilibrium by either illumination[25] or buffering reactions.[99]

Typically, the acoustic period is short relative to the time scales of reaction and thermal conduction. Thus much of the work on stability has been based on an asymptotic expansion in the acoustic frequency.[98] The result of this linear stability analysis is a new term in the acoustic eigenvalues that can overcome the viscous and thermal conductive damping of sound so that sound amplification, driven by the chemical reaction, can occur. A closely related phenomenon in insect flight has also been studied.[15]

B. Time-Independent Spatial Structures

In the discussion of the stability of a steady state we were concerned with both uniform and nonuniform (spatially dependent) perturbations. Let us consider the general linearized reaction-diffusion equation (III.6) and assume a simple perturbation of the concentration of a species X of the form

$$\delta X \sim e^{i\mathbf{k}\cdot\mathbf{r}} e^{\omega t} \qquad (III.24)$$

which is a Fourier component of a general nonuniform perturbation (see Section II.A.4). Substitution of this form of the perturbation into the linearized reaction-diffusion equation shows (III.7) that because of diffusion alone we have $\omega \sim -k^2 D_X$. This of course is expected. For positive diffusion coefficients the dissipative process of diffusion is stable for any $\mathbf{k}$ (since $\omega < 0$). (With electrolyte solutions subjected to an applied electric field it seems that effective negative diffusion or migration coefficients might be possible.[107]) Since the wave vector $\mathbf{k}$ is simply related to the wavelength λ, $k = 2\pi\lambda^{-1}$, we see that the larger k, the smaller λ, the more swiftly decays the perturbation: Larger gradients disappear more quickly than smaller ones.

A combination of reaction and diffusion, however, may act to produce a positive eigenvalue ω, an instability, for a certain range of k values.[30,108] For this to occur the reaction terms must have appropriate feedback loops. Let us consider a simple example.[25] Take two gases, A and B, and shine on this system light of a frequency absorbed only by A; after absorption the photon energy is turned into heat. In addition we require the pair A, B to be chosen such that in a separate thermal diffusion experiment A is enriched in the hotter region. Now as light of constant intensity I shines on the A, B mixture at uniform temperature, imagine a temperature fluctuation in a given small spatial region of the system. If the temperature in that region is increased then, for appropriate time scales of diffusion and thermal diffusion, the region becomes more concentrated in A. The enhancement of the concentration of A, because of absorption of light, increases the heat input into that region and the temperature increases still further above that of the remainder of the system. Thus the presence of a positive feedback loop may bring about the formation of macroscopic spatial structures (i.e., variations of state variables in space). A linear stability analysis of this system for a nonreactive pair of gases shows that a minimum light intensity is necessary to obtain a spatial structure; the wavelength of that structure is determined by the size of the system. We refer to such cases as extrinsic structures. For more complicated mechanisms (feedback loops) it is possible to obtain intrinsic structures for which the characteristic dimensions are determined by the rate and transport coefficients of the system and not by its size (provided of course that the size considerably exceeds the characteristic dimensions of the structure).

It has been proposed[109] that the periodic precipitation phenomenon known as Liesegang rings is an example of a time-independent structure formed by a chemical instability. When two solutions, say one of $Pb(NO_3)_2$ and one of KI, are placed next to each other (in an effectively one-dimensional geometry as in a long test tube), the yellow precipitate PbI_2 appears after some time (hours if the solutions are in gel). At first the precipitate is seen at the junction of the solutions; later one would expect a spatially monotonic process of precipitation as diffusion takes place between the two solutions. However, the observed precipitation is discontinuous, with rings of PbI_2 perpendicular to the axis of the test tube appearing at regular but unequal spacings.

We note that in the experiment just described gradients of concentrations (Pb^{++}, I^-) are present and a variety of previous theories presented to explain this phenomenon require such gradients. Following a suggestion of P. Ortoleva, Michele Flicker[109] showed that discontinuous precipitation of PbI_2 occurs even if the initial condition—in this case uniform, nearly colorless sol of PbI_2 in gel (small colloidal particles)—has no gradients in concentration, temperature, or external potentials (gravity). The spatial structures developed are of the intrinsic type.

IV. LOCALIZED CHEMICAL INSTABILITIES

A. Introduction

In this section we investigate effects associated with the occurrence of reactions with potentially unstable mechanisms on local sites such as may occur in membranes or heterogeneous catalysis. The local sites may be a catalytic particle (point or a given volume), an electrode (a line or surface), a membrane or solid (a surface). We consider the cases of single and multiple sites. The sites are immersed in a bulk medium that either is stable or has potentially unstable reactions. For the various cases we find a number of interesting phenomena, including local stable undulatory spatial structures around a heterogeneous catalytic site; dependence of the frequency of potentially oscillatory reaction mechanisms occurring on a local site on both rate and transport coefficients; the possibility of waves and static concentration patterns on catalytic surfaces; and cooperative effects in regular and random arrays of localized catalytic sites dependent on site density, such as numbers of available stationary states and conditions necessary to attain instability.

B. Heterogeneous Catalysis[110]

Let us first analyze a reaction mechanism in a homogeneous bulk medium enhanced by a heterogeneous catalyst located at a (y, z) plane at $x = 0$. We first consider only variations in the bulk that are uniform in the directions parallel to the plane $x = 0$. For all the pertinent species we write a concentration vector $\psi(x, t)$ and diffusion matrix D so that the reaction diffusion equations in one dimension are

$$\frac{\partial \psi}{\partial t} = D\frac{\partial^2 \psi}{\partial x^2} + F[\psi] + \gamma G[\psi]\delta(x) \tag{IV.1}$$

Here $F[\psi]$ denotes the homogeneous reactions and $G[\psi]$ the heterogeneous reactions; γ is a catalytic strength parameter. We seek perturbation solutions for stationary states inhomogeneous near the catalytic site at $x = 0$ and homogeneous in the asymptotic limits $|x| \to \infty$. For that purpose we expand the concentration vector in a power series in γ

$$\psi(x, t) = \sum_n \psi(x, t|n)\gamma^n \tag{IV.2}$$

and find to first order in γ

$$D\frac{\partial^2 \psi}{\partial x^2}(x|1) + \Omega\psi(x|1) + G[\psi(\ |0)]\delta(x) = 0 \tag{IV.3}$$

for the stationary state, $(\partial \psi/\partial t) = 0$, and for a homogeneous zeroth-order state, $\psi(x|0) = \psi(\ |0)$. The symbol Ω denotes $(\partial F/\partial \psi)_{\psi = \psi(x, t|0)}$. Integrating (IV.3) over a small region around $x = 0$, we have the conservation equation at the heterogeneity

$$D\left[\frac{\partial \psi}{\partial x}(0^+|1) - \frac{\partial \psi}{\partial x}(0^-|1)\right] + G[\psi(10)] = 0 \tag{IV.4}$$

The solution of (IV.3) for $x \neq 0$ may be written in terms of the eigenvectors of $D^{-1}\Omega$.

$$\psi(x|1) = \sum_{n=1}^{\infty} \psi_n \chi_n e^{-\kappa_n|x|} \tag{IV.5}$$

with κ_n being the solution of the secular equation

$$|\kappa^2 D + \Omega| = 0 \tag{IV.6}$$

and the eigenfunctions χ_n the solution of the eigenvalue problem

$$D^{-1}\Omega\chi_n = -\kappa_n^2\chi_n \tag{IV.7}$$

Let us consider the Prigogine–Lefever reaction to take place homogeneously

$$A \to X$$

$$B + X \to Y + B$$

$$2X + Y \to 3X \qquad \text{(IV.8)}$$

$$X \to E$$

and let a catalyst C at $x = 0$ enhance the autocatalytic step $2X + Y + C \to 3X + C$. We take the special case such that all rate coefficients are unity, $A = 1$; $D_x = D_y = D$. We obtain for the "wave vector" κ_m the result:

$$\kappa_\pm = 2D^{-1/2}[(2 - B) \pm i(4B - B^2)^{1/2}]^{1/2} \qquad \text{(IV.9)}$$

Under the stated conditions this reaction mechanism in a homogeneous bulk system becomes unstable at $B = 2$. For $B < 2$ we see that $\kappa_\pm$ is complex with the condition Re $\kappa_\pm > 0$. Therefore from (IV.9) we see that an undulatory static spatial structure exists in the vicinity of the heterogeneous catalyst at $x = 0$ of extent given by $(\text{Re } \kappa_n)^{-1}$ and wavelength $(I_m \kappa_n)^{-1}$. As B approaches the value of 2 the extent of the spatial structure approaches infinity; that is, the entire system, originally homogeneous, becomes unstable and a global dissipative structure is formed.

For the Lotka–Volterra reaction mechanism in the bulk

$$Q + X \to 2X$$

$$X + Y \to 2Y \qquad \text{(IV.10)}$$

$$Y \to P$$

with a heterogeneous catalyst at $x = 0$ enhancing the reaction $X + Y + C \to 2Y + C$ we find the "wave vector" κ to be

$$\kappa_\pm = (1 \pm i)\left(\frac{\alpha Q}{4D^2}\right)^{1/4} \qquad \text{(IV.11)}$$

We have again set all rate coefficients to unity and define $D_x = D$, $D_y = D\alpha^{-1}$. There exists a localized undulatory spatial structure near the heterogeneity of wavelength $2\pi(4D^2/\alpha Q)^{1/4}$ and of similar extent.

Thus we note the important feature that although a global dissipative structure, which would require Re $\kappa_1 \geq 0$, may not be possible for certain systems, nonetheless localized structures can be found.

C. Chemical Instabilities at Localized Sites[66]

Consider a homogeneous bulk system in which only stable reaction mechanisms take place. Add to such a system a reaction site (heterogeneous catalyst) localized to a plane at $x = 0$ and let reactions occur on the catalyst, which either by themselves or in conjunction with reactions in the bulk are potentially unstable. The total system, bulk and reaction site, obeys the reaction diffusion equations

$$\frac{\partial \psi}{\partial t} = D\nabla^2 \psi + F[\psi] + G[\psi]\, \delta(x) \tag{IV.12}$$

where ψ is a concentration vector, D a diffusion matrix, $F[\psi]$ describes the bulk kinetics, and $G[\psi]$ the kinetics at the localized site. In the absence of the site the bulk system is in a stable stationary state, $\psi_0(\mathbf{r})$, which obeys the equation

$$D\nabla^2 \psi_0(\mathbf{r}) + F[\psi_0(\mathbf{r})] = 0 \tag{IV.13}$$

Let the localized site now introduce a change in $\psi(\mathbf{r})$, which we write

$$\psi(\mathbf{r}, t) = \psi_0(\mathbf{r}) + \Delta\psi(\mathbf{r}, t) \tag{IV.14}$$

Since $\psi(\mathbf{r}, t)$ and $\psi_0(\mathbf{r})$ obey the same boundary conditions, we must have $\Delta\psi(\mathbf{r}, t)$ zero on the boundaries if concentrations are fixed there or the diffusion flux associated with $\Delta\psi$ normal to the boundary surface must vanish if this quantity is specified there. Next we linearize the bulk reactions about the stable stationary state but retain the full nonlinearity of the local site reactions, and obtain for the deviation $\Delta\psi(\mathbf{r}, t)$

$$\frac{\partial \Delta\psi}{\partial t} = \{D\nabla^2 + \Omega(\mathbf{r})\}\,\Delta\psi + G[\psi_0(\mathbf{r}) + \Delta\psi(\mathbf{r}, t)]\, \delta(x) \tag{IV.15}$$

where, as before, $\Omega = (\partial F/\partial \psi)_{\psi_0}$. With the definition of the matrix propagator $\Xi(\mathbf{r}, \mathbf{r}'; t)$ for the bulk reactions, which satisfies the equations

$$\frac{\partial \Xi}{\partial t} = \{D\nabla^2 + \Omega(\mathbf{r})\}\Xi \tag{IV.16a}$$

$$\Xi(\mathbf{r}, \mathbf{r}'; t = 0) = \delta(\mathbf{r} - \mathbf{r}')I \tag{IV.16b}$$

the solution to (IV.15) is

$$\Delta\psi(\mathbf{r}, t) = \int d\mathbf{r}' \Xi(\mathbf{r}, \mathbf{r}'; t)\, \Delta\psi(\mathbf{r}', 0)$$
$$+ \int d^2\mathbf{r}'_\parallel \int_0^t dt'\, \Xi[\mathbf{r}, (\mathbf{r}'_\parallel, 0); t - t']\, G[\psi_0(\mathbf{r}'_\parallel, 0) + \Delta\psi((\mathbf{r}'_\parallel, 0); t')] \tag{IV.17}$$

where $\mathbf{r}_\parallel$ is a two-dimensional vector parallel to the plane at $x = 0$.

In obtaining the result (IV.17) we have achieved a reduction of the problem originally described by partial differential equations to one described by much more easily solvable nonlinear integral equations. The first term in (IV.17) describes the development of the bulk kinetics, that is, the propagation of a perturbation in the absence of local site reactions. Since the bulk reaction is assumed to be stable this first term is always transient. The second term on the rhs of (IV.17) describes the perturbing effect of the local site reactions, which are in turn affected by bulk reactions and transport. The result (IV.17) has been used for the analysis of stationary states and their stability, oscillations, static spatial structure, and wave propagation localized to the site at $x = 0$.

As an illustration let us consider again the Prigogine–Lefever reaction mechanism. This mechanism in a homogeneous phase, with the concentrations of species A, B, D, and E kept constant, gives a limit cycle oscillation sufficiently far from equilibrium (sufficient concentration of B, for instance). Now let this reaction mechanism occur on a plane located at $x = 0$, let there be no bulk reactions ($F = 0$), and neglect variations in the $\mathbf{r}_\parallel$ direction. Then with the simplifications of all rate coefficients set to a constant k; the concentrations A, B, D, and E kept constant; $A = 1$; and $D_x = D_y$, we find from a linear stability analysis that the critical value of B necessary for the onset of oscillation is $B_c = 2 + \sqrt{2}$, as compared to 2 for the homogeneous case, and the corresponding critical frequency of oscillation

$$\omega_c = \frac{k}{4D} \tag{IV.18}$$

as compared to k for the homogeneous case. It is to be expected that diffusion of the species X and Y to and from the localized site of reaction influences the frequency of oscillation and that the presence of diffusion necessitates a different value of B_c. For the case studied, $D_x = D_y = D$, B_c is increased (although it is in principle possible that unequal diffusion could actually serve to lower B_c).

Multiple stationary states may also exist at a localized reaction site for some reaction mechanisms. The details depend not only on the local reaction parameters but also on the transport and rate coefficients for the bulk dynamics.

D. Cooperative Phenomena in Arrays of Catalytic Sites

We now generalize the discussion of the last section to arrays of many reaction sites[112-114] and consider first a one-dimensional regular array. Let N reaction sites be located at x_m; the reaction diffusion equations are

$$\frac{\partial \psi}{\partial t} = D \nabla^2 \psi + F[\psi] + \sum_m G_m[\psi(x_m, t)] \, \delta(x - x_m) \tag{IV.19}$$

where the symbols are the same as introduced earlier except that the subscript m on G allows for the possibility of different types of local sites. The system, bulk, and local reaction sites are taken to be isothermal with no convective transport. Again we linearize the bulk kinetics about a homogeneous stationary state but retain all nonlinearities of the kinetics at the local sites. The partial differential reaction-diffusion equations can thereby be reduced to a set of ordinary nonlinear integral equations by setting $x = x_n$ in the following:

$$\psi(x;t) = \psi_h + \int dx' \Xi(x, x'; t)[\psi(x', t=0) - \psi_h]$$

$$+ \sum_m \int_0^t dt' \Xi(x, x_m; t-t') G_m[\psi(x_m; t')] \qquad \text{(IV.20)}$$

Note that the concentrations ψ depend parametrically on the $\{x_m\}$; ψ_h are the homogeneous concentrations in the absence of the local sites. The propagator obeys (IV.16).

The integral equation can be reduced further for a regular array of given lattice spacing a. Analytic results can be obtained for some typical cases. Consider a system in which the bulk reactions are

$$X \to D$$

$$B \to Y \to C \qquad \text{(IV.21)}$$

At the localized sites there occur the reactions

$$X + E'' \rightleftarrows E'$$

$$X + E' \rightleftarrows E \qquad \text{(IV.22)}$$

$$Y + E \to E + X$$

The first two reactions of (IV.22) are assumed to equilibrate on a time scale much shorter than typical for all other reactions. The mechanism is product-enhanced enzymatic catalysis.

In Fig. 12 we graph the quantity B_c, the critical concentration of B necessary to attain the condition of multiple stationary states for a case where $D_Y \gg D_X$, and show the variation of B_c with the logarithm of the localized reaction site density (the density is inversely proportional to the intersite distance a). In the region $B > B_c$ three stationary states are available to the system.

For a fixed value of B there exists the possibility of altering the number of available stationary states by changing the reaction site density. Thus along the line 1–2 the system has three stationary states except in the interval 1'–2', where there is only one stationary state. We thus have an example of the interesting cooperative phenomena of variable numbers of

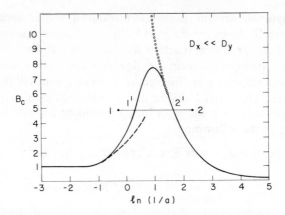

Fig. 12. Critical concentration B_c, necessary for multiple stationary states plotted (full line) against the natural logarithm of the density of sites for model system (IV.21) for the case $D_x \ll D_y$. Dashed and dotted lines are for the continuum limit and isolated site limit respectively. Values of parameters used are $D_x = 0.01$, $D_y = 100$, $k_1 = 4$. (After Bimpong-Bota, Nitzan, Ortoleva, and Ross, Ref. 111.)

stationary states depending on site density. The cooperative effects come about in the following way: As the intersite distance a decreases from infinity there begins an interaction in that the autocatalytic process at one site is augmented by the increase in the concentration of X produced at neighboring sites. Thus as a decreases we should expect B_c to decrease. Whether it does so depends on the diffusion coefficients D_X and D_Y. If we have $D_Y \gg D_X$, then the depletion of Y near one site by diffusion to and reaction at a neighboring site takes precedence and B_c at first increases as the density increases. These two opposing effects bring about the maximum in Fig. 12. Similar cooperative processes occur in reaction mechanisms that may lead to oscillation.

Cooperative effects are also evident in the case of two localized reaction sites, for instance with mechanisms that allow multiple stationary states. For certain ranges of the intersite distance both symmetric and asymmetric stationary states are possible for the two-site system.[112]

The results discussed thus far have been for one-dimensional arrays or single planar sites. The theory for two- and three-dimensional sites like catalyst wires and particles is a much more difficult problem. Catalyst wires and beds with flow have been discussed in the chemical engineering literature.[113-116] Suspensions of catalyst particles are discussed in Ref. 116. Arrays of sites with flow in chemical reaction beds are studied in Refs. 113 and 114.

E. Waves at Local Sites

At a catalytic wall or at the interface between a membrane and electrolyte solution localized surface reactions may take place. In keeping with our focus on the propagation of disturbances in infinite media we limit our discussion of these heterogeneous systems to geometrical situations where the system is infinite in one direction, the x direction, and discuss the propagation of chemical disturbances in this direction. Examples of such situations are large planar membranes, catalytic surfaces, catalyst wires, parallel arrays of wires, and coaxial membrane configurations (as in the nerve axon). The formulation of these problems varies slightly from geometry to geometry, but most of the basic concepts are essentially contained in the prototype problem of propagation along a planar catalyst wall, to which we now limit our discussion.

The basic equations are as outlined in Section IV.B with special emphasis, however, put on the boundary condition (IV.2), which introduces the influence of the wall on the bulk medium. It is found most convenient to convert the partial differential equation (III.1) with nonlinear boundary condition (III.2) to an integral equation.[66] The theory is particularly simple when the bulk medium has only linear kinetics or when the variations of interest are sufficiently small so that the bulk reaction nonlinearities are unimportant (linearizable), even though the surface nonlinearities are strong. In such cases of linear bulk kinetics the Green's function for the bulk dynamics is a complete description and hence the problem can be reduced to solving a closed integral equation for the bulk concentrations at the surface.[66]

Using the integral equation for the surface concentrations, a wave profile integral equation analogous to (III.6) has been derived and solved analytically for a model system that shows infinite periodic trains of surface disturbances.[66] Oscillator and bifurcation schemes analogous to those of subsections C and E may also be applied to surface waves.[117]

Acknowledgments

This work was supported in part by the National Science Foundation, the Air Force Office of Scientific Research, and the National Aeronautics and Space Administration.

References

1. P. Glansdorff and I. Prigogine, *Thermodynamic Theory of Structure, Stability, and Fluctuations*, Wiley, New York, 1971.
2. B. Hess and A. Boiteux, *Ann. Rev. Biochem.*, **40**, 237 (1971).
3. G. Nicolis and J. Portnow, *Chem. Rev.*, **73**, 365 (1973).
4. B. Chance, E. K. Pye, A. K. Ghosh, and B. Hess, Eds., *Biological and Biochemical Oscillations*, Academic, New York, 1973.

5. *Physical Chemistry of Oscillatory Phenomena*, Faraday Symposium No. 9, London, 1974.

6. R. M. Noyes and R. J. Field, *Ann. Rev. Phys. Chem.*, **25**, 95 (1974).

7. G. Nicolis and R. Lefever, *Membranes, Dissipative Structures, and Evolution*, Advances in Chemical Physics, Vol. 29, Wiley, New York, 1975.

8. R. Aris, *The Mathematical Theory of Diffusion and Reaction in Permeable Catalysts*, Vol. I, Oxford University Press, Oxford, 1975.

9. R. Schmitz, in *Chemical Reaction Engineering Reviews*, Advances in Chemistry Series 148, American Chemical Society, Washington, D.C., 1975, p. 156.

10. A. Pacault, P. Hanusse, P. de Kepper, C. Vidal, and J. Boissonade, *Acc. Chem. Res.*, **9**, 438 (1976).

11. J. Ross, *Ber. Bunsenges. Phys. Chem.*, **80**, 1113 (1976).

12. A. Goldbeter and S. R. Kaplan, *Ann. Rev. Biophys. Bioeng.*, **5**, 449 (1976).

13. (a) J. J. Tyson, *The Belousov-Zhabotinsky Reaction*, Lecture Notes in Biomathematics, Vol. 10, Springer-Verlag, 1976; (b) J. D. Murray, "Biological Oscillations: Spatial Structure and Nonlinear Wave Propagation in Reacting Systems, preprint.

14. H. Eyring, Ed., *Periodicities in Chemistry and Biology*, Advances in Theoretical Chemistry, Academic, New York, 1978.

15. P. Ortoleva, in Ref. 14.

16. H. Reiss, *Methods of Thermodynamics*, Blaisdell, New York, 1965.

17. I. Prigogine and R. Lefever, *J. Chem. Phys.*, **48**, 1695 (1968).

18. B. F. Gray and L. J. Aarons, in Ref. 5, p. 129.

19. M. Herschkowitz-Kaufman and G. Nicolis, *J. Chem. Phys.*, **56**, 1890 (1972).

20. C. Vidal, *C.R. Acad. Sc. Paris*, **274C**, 1713 (1972); **275C**, 523 (1972).

21. P. Ortoleva and J. Ross, *Biophys. Chem.*, **1**, 87 (1973).

22. (a) Y. Kobatake, in Ref. 7, p. 319; (b) G. Brown, D. Thomas, and E. Selegny, *J. Membrane Biol.*, **8**, 313 (1972); (c) A. Naparstek, J. L. Romette, J. P. Kervenez, and D. Thomas, *Nature*, **249**, 496 (1974).

23. J. Boissonade, *J. Chem. Phys.*, **73**, 540 (1976).

24. A. Pacault, P. de Kepper, and P. Hanusse, *C.R. Acad. Sc. Paris*, **280B**, 157 (1975).

25. A. Nitzan, P. Ortoleva, and J. Ross, *J. Chem. Phys.*, **60**, 3134 (1974).

26. C. L. Creel and J. Ross, *J. Chem. Phys.*, **65**, 3779 (1976).

27. (a) A. N. Zaikin and A. M. Zhabotinsky, in Ref. 4, p. 81; (b) H. G. Busse, in Ref. 4, p. 63; (c) M. Marek and I. Stuchl, *Biophys. Chem.*, **3**, 291 (1975).

28. W. Horsthemke and M. Malek-Mansour, *Z. Phys.*, **B24**, 1307 (1976).

29. P. Hanusse and A. Pacault, in P. Barret, Ed., *Reaction Kinetics in Heterogeneous Chemical Systems*, Elsevier, Amsterdam, 1975, p. 68.

30. A. M. Türing, *Phil. Trans. Roy. Soc. (London)*, **B237**, 37 (1952).

31a. H. Feinberg and F. J. Horn, *Chem. Eng. Sci.*, **29**, 775 (1974).

31b. B. Clark, *J. Chem. Phys.*, **62**, 3726 (1975).

32. U. F. Franck, in Ref. 4, p. 7, and in Ref. 5, p. 137.

33. H.-S. Hahn, P. Ortoleva, and J. Ross, *J. Theoret. Biol.* **41**, 503 (1977).

34. N. Minorsky, *Nonlinear Oscillations*, Van Nostrand, Princeton, N.J., 1962.

35. L. Cesari, *Asymptotic Behavior and Stability Problems in Ordinary Differential Equations*, Academic, New York, 1963.

36. P. Hanusse, Ph.D. Dissertation, University of Bordeaux I, France, 1976; also *C.R. Acad. Sc. Paris*, **277C**, 263 (1973).

37. (a) P. Hanusse, *C.R. Acad. Sc. Paris*, **274C**, 1245 (1972); (b) J. J. Tyson, *J. Chem. Phys.*, **58**, 3919 (1973); (c) P. Hanusse, *C.R. Acad. Sc. Paris*, **277C**, 263 (1973); also Ref. 32 and references therein.

38. (a) D. Sattinger, *Topics in Stability and Bifurcation Theory*, Springer-Verlag, New York, 1975; (b) J. F. G. Auchmuty and G. Nicolis, *Bull. Math. Biol.*, **37**, 322 (1975).

39. (a) A. J. Lotka, *J. Phys. Chem.*, **14**, 271 (1910); (b) *J. Am. Chem. Soc.*, **42**, 1595 (1920); (c) V. Volterra, *Leçons sur la théorie mathématique de la lutte pour la vie*, Gauthier-Villars, Paris, 1931.

40. The two most famous reactions are the Belousov-Zhabotinsky reaction (see Refs. 4, 9, 13), and the Bray reaction (W. C. Bray, *J. Am. Chem. Soc.*, **43**, 1262 (1921) and Ref. 5).

41. See for instance, P. Gray, J. F. Griffiths, and R. J. Moule, in Ref. 5, p. 103; C. H. Yang, in Ref. 5, p. 114.

42. H. G. Busse, in Ref. 4, p. 63.

43. (a) O. E. Rössler and F. F. Seelig, *Z. Naturforsch.*, **27b**, 1444 (1972); (b) O. E. Rössler, *Z. Naturforsch.*, **31a**, 259 (1976); (c) ibid., **31a**, 1168 (1976); (d) ibid., **31a**, 1664 (1976); O. Rössler and P. Ortoleva, Lecture Notes in Biomathematics, Vol. 21, p. 51.

44. E. N. Lorenz, *J. Atmos. Sci.*, **20**, 130 (1963).

45. D. Ruelle and F. Takens, *Commun. Math. Phys.*, **20**, 167 (1971).

46. M. DelleDonne and P. Ortoleva, *J. Chem. Phys.*, **67**, 1861 (1977).

47. (a) Y. A. Vavilin, A. N. Zhabotinsky, and A. N. Zaikin, in Ref. 54, p. 71; (b) P. de Kepper, A. Rossi, and A. Pacault, *C.R. Acad. Sc. Paris*, **283C**, 371 (1976); (c) K. R. Graziani, J. L. Hudson, and R. A. Schmitz, *Chem. Eng. J.*, **12**, 9 (1976).

48. (a) A. Pacault, P. de Kepper, P. Hanusse, and A. Rossi, *C.R. Acad. Sc. Paris*, **281C**, 215 (1975); (b) P. de Kepper, A. Pacault, and A. Rossi, *C.R. Acad. Sc. Paris*, **282C**, 199 (1976).

49. K. J. McNeil and D. F. Walls, *J. Statist. Phys.*, **10**, 439 (1974).

50a. A. Nitzan, P. Ortoleva, J. Deutch, and J. Ross, *J. Chem. Phys.*, **61**, 1056 (1974).

50b. A. Nitzan, *Phys. Rev.* **A17**, 1513 (1978).

50c. A. Nitzan and P. Ortoleva (submitted for publication).

51. I. Matheson, D. F. Walls, and C. W. Gardiner, *J. Statist. Phys.*, **12**, 21 (1975).

52. (a) R. FitzHugh, *J. Gen. Physiol.*, **43**, 867 (1960); (b) P. C. Fife, *Singular Perturbation by a Quasilinear Operator*, Lecture Notes in Mathematics, No. 322, Springer, Berlin, 1972; (c) U. F. Franck, in Ref. 4, p. 7, and in Ref. 5, p. 137; (d) J. J. Tyson, *J. Chem. Phys.*, **66**, 905 (1977).

53. M. Marek and E. Svoboda, *Biophys. Chem.*, **3**, 263 (1975).

54. Y. Kobatake, *Physica*, **48**, 301 (1970).

55. (a) R. A. Sprangler and F. M. Snell, *J. Theor. Biol.*, **16**, 381 (1967); (b) R. Blumenthal, J. P. Changeux, and R. Lefever, *J. Membrane Biol.*, **2**, 351 (1970); (c) B. B. Edelstein, *J. Theor. Biol.*, **29**, 57 (1970); (d) F. Schlögl, *Z. Phys.*, **253**, 147 (1972).

56. (a) G. Nicolis, M. Malek-Mansour, K. Kitahara, and A. Van Nypelseer, *Phys. Lett.*, **48A**, 217 (1974); (b) A. Nitzan, P. Ortoleva, and J. Ross, in Ref. 5, p. 241.

57. (a) P. Hanusse, *J. Chem. Phys.*, **67**, 1282 (1977), and Ref. 36; (b) P. Ortoleva and S. Yip, *J. Chem. Phys.*, **65**, 2046 (1976).

58. L. K. Kaczmarek, *Biophys. Chem.*, **4**, 249 (1976).

59. P. G. Sørensen, in Ref. 5, p. 88.

60. U. Franck and W. Geiseler, *Naturwiss.*, **58**, 52 (1971) and Ref. 24b.

61. P. de Kepper, *C.R. Acad. Sc. Paris*, **283C**, 25 (1976).

62. See, for instance, papers in Ref. 7 and references therein: D. Thomas, p. 113; Y. Kobatake, p. 319; R. Lefever and J. L. Denenbourg, p. 349.

63. H. S. Hahn, A. Nitzan, P. Ortoleva, and J. Ross, *Proc. Nat. Acad. Sci.* (USA), **71**, 4067 (1974).

64. D. D. Fitts, *Nonequilibrium Thermodynamics*, McGraw-Hill, New York, 1962.

65. S. Schmidt and P. Ortoleva, *J. Chem. Phys.*, **67**, 3771 (1977).

66. K. Bimpong-Bota, P. Ortoleva, and J. Ross, *J. Chem. Phys.*, **60**, 3124 (1974).

67. A. T. Winfree, in Ref. 5, p. 38.

360   P. HANUSSE, J. ROSS, P. ORTOLEVA

68.   A. T. Winfree, in H. Mel, Ed., *Aharon Katchalsky Memorial Symposium, Science and Humanism: Partners in Human Progress,* University of California Press, Berkeley and Los Angeles, in press.
69.   (a) A. Naparstek, D. Thomas, and S. R. Caplan, *Biochem. Biophys. Acta,* **232,** 643 (1973); (b) D. A. Larsen, "On Models for Two Dimensionally Structured Chemo-Diffusional Propagation," preprint.
70.   A. Nitzan and J. Ross, *J. Chem. Phys.,* **59,** 291 (1973).
71a. M. DelleDonne and P. Ortoleva, *Z. Naturförsch.* **339,** 558 (1978).
71b. P. Ortoleva, *J. Chem. Phys.* July (1978).
72.   P. Ortoleva and J. Ross, *J. Chem. Phys.,* **58,** 5673 (1973).
73.   P. Ortoleva, *J. Chem. Phys.,* **64,** 1395 (1976).
74.   N. Kopell and L. Howard, *Stud. Appl. Math.,* **52,** 291 (1973).
75.   P. Ortoleva and J. Ross, *J. Chem. Phys.,* **60,** 5090 (1974).
76.   W. Wasow, *Asymptotic Expansions for Ordinary Differential Equations,* Wiley, New York, 1965.
77.   B. Lavenda, G. Nicolis, and M. Herschkowitz-Kaufman, *J. Theor. Biol.,* **32,** 283 (1971).
78.   P. Ortoleva and J. Ross, *J. Chem. Phys.,* **63,** 3398 (1975).
79.   (a) P. C. Fife, *Proc. AMS-SIAM Symp. on Asymptotic Methods and Singular Perturbations,* New York, 1976; *J. Chem. Phys.,* **64,** 554 (1976); (b) P. C. Fife and J. B. McLeod, *Arch. Ration. Mech. Anal.,* **65,** 335 (1977).
80.   J. Stanshine, Ph.D. Thesis, Massachusetts Institute of Technology, Cambridge, 1975.
81.   P. C. Fife, "Asymptotic Analysis of Reaction Diffusion Wave Fronts," preprint.
82.   D. Feinn and P. Ortoleva, *J. Chem. Phys.,* **67,** 2119 (1977).
83.   S. Schmidt, A. T. Winfree, and P. Ortoleva, "Chemical Waves in Ring-shaped Reactors," unpublished.
84.   R. Thom, *Stability, Structure, and Morphogenesis,* Benjamin, New York, 1972.
85.   A. H. Nayfeh, *Perturbation Methods,* Wiley, New York, 1973.
86.   J. I. Gmitro and L. E. Scriven, in J. B. Warren, Ed., *Intracellular Transport,* Academic, New York, 1966, p. 221.
87.   J. A. Boa, Ph.D. Thesis, California Institute of Technology, Pasadena, 1974.
88.   I. Statgold, *SIAM Rev.,* **13,** 289 (1971).
89.   T. Yamada and Y. Kuramoto, *Prog. Theor. Phys.,* **55,** 2035 (1976).
90.   V. G. Yakhno, *Biofizika,* **20,** 669 (1975).
91.   T. Pavlidis, *J. Chem. Phys.,* **63,** 5269 (1975).
92.   N. Kopell and L. Howard, *Science,* **180,** 1171 (1973).
93.   D. G. Aronson and H. F. Weinberger, *Nonlinear Diffusion in Population Genetics, Combustion, and Nerve Propagation,* Proc. of Tulane Program in Partial Differential Equations and Related Topics, Springer-Verlag, New York, 1975.
94.   D. A. Larsen and J. D. Murray, "Finite Amplitude Traveling Solitary Waves in a Model for the Belousov-Zaikin-Zhabotinskii Reaction," preprint.
95.   J. D. Murray, *J. Theor. Biol.,* **52,** 495 (1975).
96.   T. Y. Toong, *Combust. Flame,* **18,** 207 (1972).
97.   R. Gilbert, H. S. Hahn, P. Ortoleva, and J. Ross, *J. Chem. Phys.,* **57,** 2672 (1972).
98.   R. Gilbert, P. Ortoleva, and J. Ross, *J. Chem. Phys.,* **58,** 3625 (1973).
99.   P. Ortoleva and J. Ross, *J. Chem. Phys.,* **55,** 4378 (1971).
100. (a) A. L. Hodgkin and A. F. Huxley, *J. Physiol.,* **117,** 500 (1952); (b) *The Conduction of Nervous Impulse,* Liverpool University Press, 1967; (c) E. C. Zeeman, in C. H. Waddington, Ed. *Towards a Theoretical Biology,* Vol. 4, Aldine-Atherton, Chicago, 1972, p. 8.

101. B. Novak, in Ref. 5, p. 281, and references therein.

102. J. P. Boon, in Ref. 5, p. 169, and references therein.

103. (a) H. G. Büsse, *J. Phys. Chem.*, **73,** 750 (1969); (b) D. F. Tatterson and L. N. Howard, *Science*, **180,** 1177 (1973).

104. (a) M. Herschkowitz-Kaufman, *C.R. Acad. Sc. Paris*, **270C,** 1049 (1976); (b) A. T. Winfree, *Science*, **175,** 634 (1972); (c) A. T. Winfree, *Sci. Am.*, **230,** no. 6, 82 (1974).

105. R. M. Noyes, *J. Am. Chem. Soc.*, **98,** 3730 (1976).

106. (a) R. J. Field and R. M. Noyes, *J. Am. Chem. Soc.*, **96,** 2001 (1974); (b) J. D. Murray, *J. Theor. Biol.*, **56,** 329 (1976).

107. J. Jorné, *J. Theor. Biol.*, **43,** 375 (1974).

108. (a) H. Poincaré, *Acta Math.*, **7,** 259 (1885); (b) H. Rashevsky, *Bull. Math. Biophys.*, **2,** 15, 65, 109 (1940).

109. M. Flicker and J. Ross, *J. Chem. Phys.*, **60,** 3458 (1974).

110. P. Ortoleva and J. Ross, *J. Chem. Phys.*, **56,** 4397 (1972).

111. K. Bimpong-Bota, A. Nitzan, P. Ortoleva, and J. Ross, *J. Chem. Phys.*, **66,** 3650 (1977).

112. R. M. Shymko and L. Glass, *J. Chem. Phys.*, **60,** 835 (1974).

113. H.-K. Rhee and N. Amundsen, *Chem. Eng. Sci.*, **28,** 55 (1973).

114. R. M. Turian, *Chem. Eng. Sci.*, **28,** 2021 (1973).

115. H. D. Thames and A. D. Elster, *J. Theor. Biol.*, **59,** 415 (1976).

116. P. Ortoleva, "Dressed Reaction and Transport in Catalytic Particle Suspensions," submitted for publication.

117. P. Ortoleva, *Limit Cycle Perturbation and Bifurcation Theory for Surface Waves*, unpublished notes.

TEMPORAL, SPATIAL, AND FUNCTIONAL ORDER IN REGULATED BIOCHEMICAL AND CELLULAR SYSTEMS

B. HESS

Max Planck-Institut für Ernährungsphysiologie
Dortmund, FRG

A. GOLDBETER
R. LEFEVER

Faculté des Sciences, Université Libre de Bruxelles, Brussels, Belgium

CONTENTS

I. INTRODUCTION

A. Historical Context

Over the last 50 years the extension of thermodynamics to open and far-from-equilibrium conditions has uncovered the macroscopic mechanisms of self-organization. In a first stage the work of Onsager,[1] Meixner,[2] and Prigogine[3] has built up a consistent phenomenological theory of irreversible processes. In this theory the entropy balance equation plays a central role. It expresses that, under nonequilibrium conditions, the entropy changes in time of some volume element involve two factors: first, an entropy production due to the irreversible processes, which always tend to increase entropy; second, an entropy flow that is exchanged with the surroundings and that may be either positive or negative. A negative entropy flow overwhelming the effect of entropy production is the fundamental thermodynamic prerequisite for self-organization.

Prigogine also demonstrated that under conditions assuming the validity of Onsager's reciprocity relations, nonequilibrium steady states are asymptotically stable and correspond to the minimum level of entropy production compatible with the nonequilibrium constraints imposed on the system.[3,4] Steady states then belong to what he later called the thermodynamic branch[5] as it contains the equilibrium state as a particular case.

From the beginning these developments of nonequilibrium thermodynamics have brought with them a discussion of the microscopic foundation of irreversibility (see the paper by Balescu and Résibois in this volume) and found applications in many domains of physics, chemistry,[6,7] and biology.[8,9] However, at that time nonequilibrium thermodynamics was dealing exclusively with linear phenomena, that is, phenomena in which thermodynamic forces and fluxes are linearly related. This rather strongly limited the applications in chemistry or biochemistry and was a strong motivation to extend thermodynamics beyond the linear domain.[10] Such an extension has been accomplished during the last 20 years. It led to a systematic investigation of the thermodynamic properties of macroscopic systems under far-from-equilibrium conditions and to a new breakthrough in the understanding of the mechanisms of self-organization (for a detailed discussion see Ref. 11). The central result is the existence, outside the linear domain, of a thermodynamic threshold beyond which the steady states of the thermodynamic branch may become unstable and may be replaced by new classes of regimes having completely different properties. To describe these regimes the concept of *dissipative structure* was introduced,[5,10] and it was readily used in the study of complex and highly organized systems, such as those occurring in chemistry (see the papers by Nicolis et al. and Hanusse et al. in this volume) and biology. The latter field, being the very domain of nonequilibrium periodic and spatial organization, lends itself as a beautiful object for application, and indeed now in biology it is fitting to say, *"erst die Theorie entscheidet darüber was man beobachten kann"* (Einstein). In addition, the new concept clearly defines the limits of a strictly reductionistic description of macroscopic systems.

The development of the thermodynamic theory of dissipative structures was accompanied by a rediscovery of the importance of autocatalysis—originally established by Lotka 70 years ago[12]—in the general function of biological systems. Simple autocatalysis and other types of control mechanisms occur in the form of feedback and feed-forward phenomena in biochemical and biological processes. It is furthermore interesting to note that the mechanisms of feedback regulation described in the theory of allosteric enzymes[13,14] as cooperative ligand-protein interactions have been discovered almost simultaneously with the formulation of the theory of dissipative structures. These developments strongly stimulated the investigations on the newly discovered oscillating reactions in biochemical and biological systems. Oscillatory enzyme reactions brought the first direct experimental evidence that biological systems can indeed function beyond some threshold of instability; therefore these reactions are a true example of dissipative structures. It is expected that a thorough consideration of such nonequilibrium dynamic phenomena will greatly add to our understanding of functions and structures in living systems.

B. Nonequilibrium Instabilities in Biochemical Systems

When investigating biochemical systems, one soon realizes that the following properties that are characteristic of living systems precisely favor the occurrence of nonequilibrium instabilities.

A continuous energy or matter supply as the necessary condition for self-organization in an open system is indeed satisfied; cells can live only if they are fed with various chemical substrates that enter the cellular compartment via the plasma membrane by diffusion, or by facilitated or active transport. Several organelles such as mitochondria or chloroplasts need oxygen, carbon dioxide, or light for their functioning.

Biochemical kinetics are the prototype of nonlinear kinetics in chemistry. There are three causes for this: Allosteric enzymes that control cell metabolism[15] respond in a *cooperative* manner to small changes in metabolite concentrations; multiple types of positive or negative feedback, often exerted simultaneously by a large number of interacting and recycling metabolites, ensure the control of biochemical pathways; finally, the fact that the concentration of the enzymes is often smaller than that of the metabolites allows simplification of the kinetic description of biochemical systems by making possible a quasi-steady-state hypothesis for the enzymatic forms.[16] This not only has the effect of substantially diminishing the number of variables to be considered in the mathematical analysis, but also renders the remaining equations highly nonlinear. In addition, many enzymes are bound to membranes, such as the mitochondrial or plasma membranes. The activity of membrane-bound enzymes is controlled not only by the phospholipid components of the membrane but by the nature of the membrane-protein-directed diffusion processes, yielding highly complex and nonlinear behavior.

Controlled biochemical reactions usually function far from equilibrium;[17-19] moreover, cells and organelles are subjected to various kinds of concentration gradients across membranes, which all displace related metabolic pathways from the state of equilibrium.

These remarks illustrate the privileged status of biochemical systems with regard to the occurrence of instabilities far from thermodynamic equilibrium. They explain why dissipative structures are more frequent in biochemistry than in other areas of chemistry,* at least judging by the numerous examples of periodicities observed in the synthesis or in the activity of enzymes.[24,25] It is thus legitimate to conclude that dissipative structures are preferentially associated with the type of kinetics that prevails in the living cell. The oscillatory enzyme reactions have provided the first evidence that essential metabolic pathways function, in certain definite conditions that may be physiological, beyond the point of instability of the thermodynamic branch. The clearest evidences in this respect are found with enzymes regulated by positive feedback; the phosphofructokinase (PFK) reaction, which we discuss below, is representative of this class of enzymes. The possibility of periodic behaviors in enzymatic chains of reactions regulated by end product inhibition has been established theoretically by several authors[26–28], but is not yet supported by experimental evidence.

The instabilities we analyze in this article arise in the regulation of enzyme activity and cellular communication processes. We consider examples of the classes of so-called soluble as well as membrane-bound enzyme systems. We do not discuss the effect of genetic or posttranscriptional regulation that may give rise to "epigenetic" oscillations in the synthesis of various proteins.[24,25] Thus the enzymatic systems we consider in Sections II to III exhibit time periodicities that participate either in the temporal organization of cells (as in the phosphofructokinase reaction) or in the mechanism of intercellular communication between identical cells (as in the slime mold *Dictyostelium discoideum*). In Section IV we discuss some aspects of cellular competition between antagonist cells, in relation with the problem of immune surveillance against cancer. We focus our attention on examples for which experimental data support the existence of dissipative structures and allow a detailed analysis of the mechanisms of instability. Our purpose here is not to present an exhaustive account of dissipative structures in biochemistry, since recent reviews have been devoted to the subject.[24,25]

The first system, glycolysis, represents the prototype of sustained periodicity in a metabolic pathway. Because the mechanism of oscillating glycolysis is well understood, it is possible to discuss the onset and the characteristics of periodic behavior in great detail. The second system relates to the problem of intercellular communication by chemical signals.

---

* A notable exception is the Belousov-Zhabotinsky reaction in organic chemistry. This ensemble of some 14 elementary reactions undergoes oscillatory kinetics and wave propagation.[20,21] Though it represents an excellent model system, it remains an isolated case except for a few other examples (see Refs. 22 and 23 for reviews of chemical oscillations).

Here identical cells communicate by means of cyclic AMP pulses. The latter signals are generated by adenylate cyclase; they not only govern the chemotactic movement of the amoebae in the course of aggregation, but also control the process of cell differentiation. Besides giving rise to autonomous periodic signals, the adenylate cyclase system is also able to produce single pulses of cyclic AMP as a response to minute signals of the same chemical. The system then functions as a biochemical amplifier of external signals, and a compartmental model is needed to investigate this kind of behavior in order to link the intracellular response to the concentration of the chemicals in the extracellular medium. Besides the enzymes catalyzing the synthesis and the breakdown of cyclic AMP, the underlying biochemical mechanism involves other membrane-linked proteins responsible for the transduction of the chemical signal through the membrane and for the signal reception.

Finally, in the last type of system considered we show that the interaction between cancer cells and antagonist cytotoxic cells can appropriately be described within the framework of a Michaelian type of kinetics. The similarities of this cytotoxic reaction with the regulatory properties of simple enzymes leads to results suggesting that the phenomenon of cancerization is under some condition analogous to a first-order type of nonequilibrium transition. We discuss the role of sensibilization and cellular motion in this transition.

II. GLYCOLYTIC OSCILLATIONS

A. Experimental Observations

Some 15 years ago glycolytic oscillations in a suspension of intact yeast cells and later in cell-free extracts of yeast and muscle were described. These observations were followed by a series of studies carried out in a number of laboratories, which led to a general view of the mechanism of glycolytic oscillations (see, for example, Refs. 24, 25, and 29). Indeed, the studies had shown that all glycolytic intermediates oscillate under proper conditions, as a result of the periodic activity change of the glycolytic enzymes. The oscillatory range of the actual rates of the enzymic reactions were determined by variation of the input rate of glycolytic metabolites.

A typical experiment indicating the dependency of the oscillation on the rate of substrate input is shown in Fig. 1. The addition of fructose with an injection rate of 40 mM/h drives the system into an oscillatory state with periods of 3.5 min and a nonsinusoidal waveform. A decrease in the injection rate to 20 mM/hr leads without a significant phase shift to a double periodicity, which slowly merges into the previous waveform. Finally, the increased injection rate of 80 mM/hr leads almost immediately

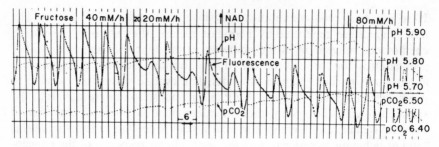

Fig. 1. Recording of NADH by fluorimetry in an extract of *S. carlsbergensis* (48). NADH scale in arbitrary units. pCO_2 scale corresponds to pH-meter readings. The arrows indicate changes in input rate.

to a quasisinusoidal waveform with a period of 2.8 min. This experiment demonstrates a type of double periodicity usually observed at low input rates. Experiments performed at the upper margin of the oscillatory domain show that for input rates larger than 160 mM/hr, oscillations are damped within several cycles, and the system settles on a stable state corresponding to a maximum level of NAD.[30]

Recently the conditions for entrainment of glycolytic oscillations and the effect of random noise on these periodicities have been determined experimentally.[31] Not only entrainment by a periodic source of substrate was found, but also subharmonic entrainment for $\frac{1}{2}$-harmonic and $\frac{1}{3}$-harmonic of a quasisinusoidal input of substrate. The discovery of the subharmonic entrainment indicated the nonlinear nature of the glycolytic oscillator (see below). It was also found that stochastic variation of the substrate input led to sustained oscillations of a stable period but of somewhat irregular waveform.

The quantitative analysis of glycolytic fluxes, of concentration changes, of phase relationships of the concentrations during oscillations as well as of the phase sensitivity of the glycolytic oscillation to the addition of glycolytic intermediates allow us to identify the primary source of the oscillations as being the enzyme phosphofructokinase (for review of these studies see Refs. 24, 32). The activity of this enzyme was found to vary under oscillating conditions between a maximum activity on the order of 70% of V_{max} and a minimum activity on the order of 1% (see Fig. 2). A change of activity of this enzyme is controlled by the rate of generation and consumption of the second substrate of the enzyme, ATP, and its allosteric ligand, AMP. The two adenine nucleotides are coupled via the reaction product ADP with the other phosphotransferases of the glycolytic pathway. They control the propagation of the pulse production of intermediates along this enzymic chain. The function of phosphofructokinase in controlling glycolytic oscillations is consistent with the generally accepted view of its function in the control of glycolysis in many biological systems.

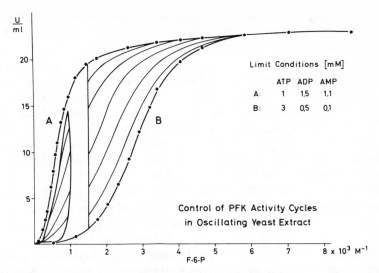

Fig. 2.   Activity of phosphofructokinase as a function of the concentration of fructose-6-phosphate in the presence of ATP, ADP, and AMP (36). The cycle indicates the periodic change in PFK activity during glycolytic oscillations.

It is obvious that two types of enzymes in the glycolytic pathway must be considered: enzymes such as phosphofructokinase, with a potential for autonomous and stable oscillations and, on the other hand, enzymes with kinetic properties insufficient to maintain an autonomous oscillating state. In glycolysis it can be concluded that most of the enzymes are driven into an oscillatory state by induction through the autonomous activity change of phosphofructokinase. However, some enzymes such as pyruvate kinase might follow their own autonomous periodicity for a given concentration range of phosphoenolpyruvate,[33,34] but settle on the same frequency as phosphofructokinase, being entrained by pulses of ADP. Since ATP pulses produced by pyruvate kinase feed back to phosphofructokinase, a synchronization of the whole pathway is achieved.

The mechanism of glycolytic oscillations in yeast and other tissues focuses our attention on the molecular properties of phosphofructokinase. Its molecular weight was found to be 720.000, composed of at least four protomers, dissociating into eight subunits of two types with molecular weights of approximately 86.000 and 94.000 daltons. The kinetics of the enzyme follow an allosteric model with the homotropic effectors fructose-6-phosphate and ATP as substrates and the strong heterotropic activator AMP, indicating a highly cooperative response of the enzyme function.[35] The kinetics of the enzyme were analyzed in the yeast extract under the conditions of glycolytic oscillations; pH-dependent Hill coefficients up to 4.9 were observed.[36] The pH dependency of the allosteric property of this enzyme coincided with the pH dependency of the glycolytic oscillation.[37] The detailed structure analysis as well as the kinetic mechanism of this highly complex enzyme are still a matter of investigation.

So far the study of glycolytic oscillations was restricted to conditions of spatial homogeneity. However, the question was raised whether glycolytic oscillations might generate spatial inhomogeneities and finally form a regular structural pattern. Prigogine et al.[38] analyzed the model of Sel'kov[39] and predicted inhomogeneities with critical wavelengths within 10^{-4} to 10^{-2} cm when diffusion is considered. Since the intracellular molarity of glycolytic enzymes is high, with a mean molecular distance between enzymes of only 40 to 50 Å, a transit time for diffusion of the metabolites in the range of 1 μsec can be computed. Thus the distance and time are small compared to the predicted range of spatial propagation. We do not know whether spatial inhomogeneities of low concentration of molecular intermediates within the cellular compartment of yeast can occur. This might well be the case if some of the enzymes are partially immobilized within the cellular compartment and diffusion is controlled by reduction of dimensionality. In principle, inhomogeneities might occur on a different scale of spatial dimensions, and, indeed, inhomogeneities of the phosphofructokinase model were computed (see below and Ref. 40). The theoretical analysis was recently supported by experiments in cell-free extracts of yeast, where periodic structure formation synchronized with the glycolytic oscillation could be recorded (A. Boiteux and B. Hess, unpublished observations).

B. Phenomenological Models for Glycolytic Oscillations

As soon as the role of phosphofructokinase in the mechanism of glycolytic oscillations was demonstrated, a model for these metabolic periodicities was proposed by Higgins, who considered an enzyme activated by its reaction product.[41] Higgins showed, by analogue computer simulations of his model, the possibility of sustained biochemical oscillations. Sel'kov carried this analysis a step further by obtaining, for a similar system, a condition for instability as a function of a critical substrate injection rate below which limit cycle oscillations would occur in the phosphofructokinase reaction.[39]

In the original models of Higgins and Sel'kov, as well as in their subsequent extensions,[42,43] the activation of PFK by a reaction product represents the destabilizing process required for evolving toward a dissipative structure in the form of sustained temporal oscillations. The autocatalytic step in these models is always represented empirically. This qualitative procedure allows an analytical treatment that yields simple stability criteria.[39]

When a quantitative approach of the phenomenon is sought, however, the preceding treatment presents certain drawbacks. The main one concerns the autocatalytic step in the phosphofructokinase reaction. This step plays an essential role in the instability mechanism; hence it is important to represent the nonlinearity introduced by autocatalysis as plausibly as possible.

C. Allosteric Model for the Oscillatory Phosphofructokinase Reaction

Several models account for the cooperative behavior of allosteric proteins. These models fall into two main classes based, respectively, on a concerted[13] or sequential[14] conformational change in the protein subunits, on binding of the ligand(s). These models have been developed mainly in order to account for equilibrium saturation curves representing the binding of a ligand to a multisubunit protein, or for steady-state kinetic data.

It is possible to analyze either one of these models under nonequilibrium conditions to determine the dynamic behavior of an open system such as that of the phosphofructokinase reaction. The interest of such a study is to treat a biochemical mechanism giving rise to a nonequilibrium instability in the frame of a plausible theory for allosteric enzymes. Then it becomes possible to represent the binding of the substrate and feedback processes by a sequence of bimolecular reactions between the metabolites and the enzymatic forms.

Such a model has been developed for the PFK reaction[44,45] in the frame of the concerted transition theory of Monod, Wyman, and Changeux.[13] A similar study, leading to comparable results, could be conducted with a sequential theory of the kind described by Koshland, Némethy, and Filmer.[14] The parameters of the concerted model have been determined for the phosphofructokinase from *Escherischia coli*,[46] and similar values have been found for the yeast enzyme.[35] These values were used in the numerical analysis of the allosteric model proposed for the oscillatory PFK reaction.

Phosphofructokinase from a variety of sources is an allosteric protein that generally consists of four subunits.[47] Regulation of the enzyme can be achieved through the binding of a large number of positive or negative effectors; association-dissociation phenomena and covalent modification of the protein are other ways in which the enzyme is controlled.[47] As in our earlier publications, we restrict ourselves to the simple situation of a monosubstrate dimer enzyme activated by the reaction product. The influence of the number of protomers is discussed below. We thus consider a single molecular enzyme species existing under two conformations, R and T. Regulation of the enzyme is achieved through binding of the reaction product to a regulatory site on the R conformation. A single substrate-product couple is considered, namely, that of ATP-ADP, and the couple F6P-FDP was neglected.[48] Though these simplifications are drastic, the resulting model satisfactorily accounts for a large number of experimental observations on glycolytic oscillations in yeast and muscle. Under these conditions we may infer that the essential variables and the suitable control mechanism have been well identified in oscillating glycolysis.

The model is represented in Fig. 3; this figure, although given in several previous publications, is shown here for clarity. When the effect of diffusion is considered, spatiotemporal structures arise in the PFK model, in the form of propagating or standing concentration waves, for definite values of the boundary conditions and of the size of the system.[40,48] We treat here the case of a homogeneous system. This situation, which corresponds to the conditions of experiments in continuously stirred extracts of yeast or muscle, allows us to describe the time evolution of metabolite and enzyme concentrations by a set of ordinary differential equations.

Let us denote by R_{ij} the concentration of the enzyme species in the R state with i molecules of substrate and j molecules of product bound, by T_i the concentration of the enzyme species in the T state with i molecules of substrate bound, and by S and P the concentrations of substrate and product. Furthermore, we denote (see Fig. 3) by k and k' the catalytic constants of the R and T states; a and a_2 are the rate constants for association of substrate and product to the R state, whereas d and d_2 are

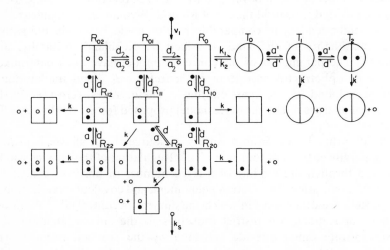

Fig. 3.   Concerted model of a dimer allosteric enzyme activated by the reaction product (44, 45). The substrate (●) enters at a constant rate v_1; the product (○) leaves the system at a rate proportional to its concentration (see details in text).

the rate constants for the reverse processes; a' and d' are the rate constants for association and dissociation of substrate to the T state; k_1 and k_2 are rate constants for interconversion of the forms R_0 into T_0; k_s is the rate constant for the assumed linear output of the product. We then obtain kinetic equations of the form

$$\frac{dR_0}{dt} = -k_1 R_0 + k_2 T_0 - 2a_2 PR_0 + d_2 R_{01} - 2aSR_0 + (d+k)R_{10}$$

$$\cdots$$

$$\frac{dT_0}{dt} = k_1 R_0 - k_2 T_0 - 2a'ST_0 + (d'+k')T_1$$

$$\cdots$$

$$\frac{dS}{dt} = v_1 - 2aS\Sigma_1 + d\Sigma_2 - aS\Sigma_2 + 2d\Sigma_3 - 2a'ST_0 + d'T_1 - a'ST_1 + 2d'T_2$$

$$\frac{dP}{dt} = -2a_2 PR_0 + d_2 R_{01} - a_2 PR_{01} + 2d_2 R_{02} + k\Sigma_2 + 2k\Sigma_3$$

$$+ k'T_1 + 2k'T_2 - k_2 P \tag{II.1}$$

Similar equations are obtained for other enzymatic forms in the T and R conformations. A total of 13 equations is thus obtained, together with the conservation relation for the total enzyme quantity: $D_0 = \Sigma R + \Sigma T$, with $\Sigma R = \Sigma_1 + \Sigma_2 + \Sigma_3$, and $\Sigma_1 = (R_0 + R_{01} + R_{02})$, $\Sigma_2 = (R_{10} + R_{11} + R_{12})$, $\Sigma_3 = (R_{20} + R_{21} + R_{22})$.

When a quasi-steady-state assumption is made for the enzyme in the T and R states, system (II.1) reduces to the two equations

$$\frac{d\alpha}{dt} = \sigma_1 - \sigma_M \phi$$

$$\frac{d\gamma}{dt} = q\sigma_M \phi - k_s \gamma \tag{II.2}$$

in which

$$\phi = \frac{\alpha e(1+\alpha e)(1+\gamma)^2 + L\theta\alpha ce'(1+\alpha ce')}{L(1+\alpha ce')^2 + (1+\alpha e)^2(1+\gamma)^2} \tag{II.3}$$

The variables and parameters appearing in (II.2) to (II.3) are defined as $\gamma = P/K_P$, $\alpha = S/K_{R(S)}$, $c = K_{R(S)}/K_{T(S)}$, $K_P = d_2/a_2$, $K_{R(S)} = d/a$, $K_{T(S)} = d'/a'$, $\theta = k'/k$, $q = K_{R(S)}/K_P$, $L = k_1/k_2$, $\sigma_1 = v_1/K_{R(S)}$, $\sigma_M = (2kD_0)/K_{R(S)}$, $\epsilon = k/d$, $\epsilon' = k'/d'$, $e = (1+\epsilon)^{-1}$, $e' = (1+\epsilon')^{-1}$.

Function ϕ is the rate function (v/V_M) for the PFK reaction;[45] it includes the term $(1+\gamma)^2$, which expresses the activation of the enzyme by its product. Without this term, (II.2) to (II.3) would still be nonlinear, but the system would not admit any nonequilibrium instability; instead, it would evolve toward a unique stable steady state or substrate accumulation would take place (this happens when $\sigma_1 > \sigma_M$).

We show below that a large number of experimental observations on the oscillatory dynamics of the glycolytic system are accounted for by the set of equations (II.2). Computer simulations of the complete glycolytic pathway have also been performed.[49]

1. Stability Analysis and Limit Cycle Behavior.   The system governed by (II.2) to (II.3) admits a single steady state in which the product concentration is

$$\gamma_0 = \frac{q\sigma_1}{k_s} \tag{II.4}$$

The steady-state concentration of the substrate α_0 is solution of the second-degree equation

$$\alpha_0^2[(\sigma_1 - \sigma_M\theta)L(ce')^2 + (\sigma_1 - \sigma_M)(1+\gamma_0)^2 e^2]$$
$$+ \alpha_0[(2\sigma_1 - \sigma_M)e(1+\gamma_0)^2 + (2\sigma_1 - \sigma_M\theta)Lce'] + \sigma_1[L + (1+\gamma_0)^2] = 0 \tag{II.5}$$

The stability properties of the steady state (α_0, γ_0) can be investigated by normal mode analysis. When the condition $k_s < \sigma_M[q(\partial\phi/\partial\gamma)_0 - (\partial\phi/\partial\alpha)_0]$ is satisfied, the steady state is an unstable node or focus enclosed by a limit cycle in the phase plane (α, γ).

The limit cycle oscillations in the allosteric model for PFK have been analyzed in detail and compared with relevant experimental findings on the following points:[44,45] variation of period and amplitude with the substrate input, waveform, periodic variation in enzyme activity, phase shift of the oscillations on addition of the reaction product ADP. The results of this comparison are summarized in Table I for the case of a constant input of substrate. The case of a periodic or stochastic substrate input has also been studied, both theoretically, and experimentally in yeast extracts (see Fig. 4).[31] Computer simulations of the allosteric model for PFK agree with these experiments as to the domains of entrainment and subharmonic entrainment of the glycolytic oscillator by a sinusoidal source of substrate (Fig. 5), and as to the stability of periodic behavior in the presence of a randomly varying substrate source.

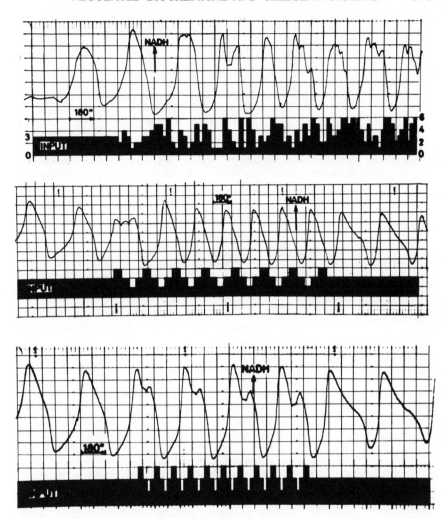

Fig. 4. *From top to bottom*: Effect of a stochastic variation of the substrate injection rate, and entrainment of glycolytic oscillations by the fundamental frequency and by the 1/3 harmonic of a periodic substrate input (from Ref. 31). The upper part of each figure shows the fluorescence of NADH; the lower part (black trace) gives the variation of the glucose input. Experiments were performed in yeast extracts.

When $q \gg 1$, that is, when the affinity of the product for the enzyme is much larger than that of the substrate, sawtooth waveforms for the substrate oscillations are obtained; furthermore, these periodicities correspond to pulses in the product concentration (see Fig. 6). A similar waveform has been observed in a reconstituted glycolytic system,[48] in

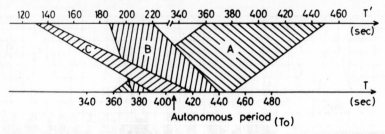

Fig. 5. Entrainment (domain A) and subharmonic entrainment (domains B and C) in the PFK model for glycolytic oscillations (from Ref. 31). Periods T and T' refer, respectively, to the period of the sinusoidal substrate input and to the period of the enzyme after entrainment.

TABLE I

Comparison of Glycolytic Oscillations in Model and Experiment, for a Constant Source of Substrate[a]

Sustained oscillations	Model	Experiment
Oscillatory range of substrate injection rate (ν_1)	19–246 mM/hr	20–160 mM/hr
Period	Of the order of min; decreases by a factor ≤ 10 as ν_1 increases	Of the order of min; decreases by a factor ≤ 10 as ν_1 increases
Amplitude	In the range $10^{-5} - 10^{-3}$ M; passes through a maximum as ν_1 increases	In the range $10^{-5} - 10^{-3}$ M; passes through a maximum as ν_1 increases
Periodic change in PFK activity (in % V_M)	Minimum: 0.95; maximum: 73; mean:17.5; activation factor: 77	Minimum: 1; maximum: 80; mean: 16; activation factor : 80
Phase shift by ADP	Delay of 1–3 min upon addition of 0.7 mM ADP (14 units of γ) around the minimum of ADP oscillations of 5 min period; small phase advance when the addition precedes ADP maximum	Delay of 1.5 min upon addition of 0.7 mM ADP at the minimum of ADP oscillations of 5 min period; small phase advance when the addition precedes ADP maximum

[a] From Ref. 31.

which glycolytic enzymes and the corresponding metabolites have been brought together and supplied with a continuous input of substrate. The difference in waveform observed in the reconstituted system and in the yeast extract could thus be due to a difference in the affinity of the substrate and/or of the product for PFK in the two experimental systems. The variety of waveforms exhibited by the glycolytic oscillator has been repeatedly emphasized, both in experimental[50,51] and theoretical[52] studies.

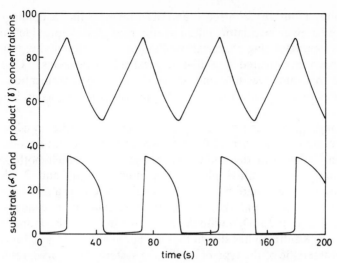

Fig. 6. Sustained oscillations obtained by integration of (II.2) for $\sigma_1 = 1$, $4\,s^{-1}$, $\sigma_M = 4\,s^{-1}$, $k_s = 10\,s^{-1}$, $q = 100$, $L = 10^6$, $\epsilon = \epsilon' = 0.1$, $c = 10^{-5}$, $\theta = 1$. Initial conditions: $\alpha = 40$, $\gamma = 5$.

Whereas in the case $q = 1$ the amplitude of the oscillations passes smoothly through a maximum as the input rate σ_1 increases, the amplitude of the limit cycle periodicities remains practically independent of this parameter when $q \gg 1$. At the same time the half-width of the product pulses increases with σ_1. The period of the oscillations still decreases as the substrate input increases, except near the upper boundary of the oscillatory domain of σ_1 values.

The possibility of observing an excitable behavior in the PFK reaction has recently been predicted by analysis of the allosteric model.[139] Such behavior consists in the pulsatory amplification of ADP pulses beyond a threshold, for substrate injection rates close to those that produce glycolytic oscillations. The phenomenon of excitability is further discussed in Section III.

2. Arc Discontinuity of the PFK Oscillator. Chemical perturbations provide a means of analyzing the mechanism of oscillations in metabolic pathways, as well as in other oscillatory systems.[53] In glycolysis, addition of adenine nucleotides to oscillating yeast extracts was used to demonstrate the control of periodicities by the couple ATP/ADP.[48,54,55] Administration of oxygen pulses in suspensions of intact yeast cells, that presumably perturb the ATP/ADP ratio, have also shed light on the dynamics of oscillating glycolysis.[56]

As pointed out by Winfree[57] and Kauffman,[58] the coupling of two samples of a given oscillator, taken at different phases, provides another means of characterizing chemical oscillations. Experimentally, this question has been investigated in oscillating yeast cell suspensions.[59] One of the results of this study was to demonstrate the rapid synchronization of two populations of yeast cells, the nature of the synchronizing agent being still unknown.

The coupling between two samples of a given chemical oscillator can well be studied on the basis of the arc discontinuity concept introduced by Kauffman.[58] One can determine the phase ϕ_s on which both samples synchronize, for an initial phase of samples 1 (ϕ_1) and 2 (ϕ_2). By examination of the variation of ϕ_s as a function of ϕ_2 for a given ϕ_1, it can be seen that ϕ_s will undergo one or more discontinuous changes as ϕ_2 varies from zero to 2π. Denoting the phase at which the discontinuity takes place by ϕ^*, Kauffman has shown that the curve giving ϕ^* as a function of ϕ_1 is characteristic of the type of oscillating system. This curve, referred to as the *arc discontinuity*,[58] reflects the degree of nonlinearity of the oscillations.

An interesting application of this method has been made on the mitotic cycle of the slime mold *Physarum polycephalum*. Kauffman and Wille proposed that mitosis in *Physarum* is controlled by a nonlinear biochemical oscillator of a quasirelaxation nature. From experiments with mixes of two *Physarum* populations they concluded indeed that the arc discontinuity observed is characteristic of a quasirelaxation oscillator.[60,140]

By using the procedure described by Kauffman et al.,[58,61] we have obtained the arc discontinuity characteristic of the glycolytic oscillator. We have considered two samples of the allosteric model for the PFK reaction (II.2), coupled by diffusion of the reactant species (simulations were performed for $q = 1$):

$$\frac{d\alpha_1}{dt} = \sigma_1 - \sigma_M \phi(\alpha_1, \gamma_1) + D_\alpha(\alpha_2 - \alpha_1)$$

$$\frac{d\gamma_1}{dt} = \sigma_M \phi(\alpha_1, \gamma_1) - k_s \gamma_1 + D_\gamma(\gamma_2 - \gamma_1)$$

$$\frac{d\alpha_2}{dt} = \sigma_1 - \sigma_M \phi(\alpha_2, \gamma_2) + D_\alpha(\alpha_1 - \alpha_2)$$

$$\frac{d\gamma_2}{dt} = \sigma_M \phi(\alpha_2, \gamma_2) - k_s \gamma_2 + D_\gamma(\gamma_1 - \gamma_2)$$

(II.6)

Subscripts 1 and 2 refer to the metabolite concentrations in the two samples of the oscillatory system. Parameters D_α and D_γ are permeability coefficients controlling the exchange of metabolites between the two samples. The synchronization of two yeast cell populations undergoing glycolytic oscillations takes only a few seconds.[59] Here the coefficients D_α and D_γ were chosen equal to 0.01 sec^{-1}. For this value the two samples of the PFK system were fully synchronized after some 20 sec, which time is small in comparison with the period of oscillation considered, that is, 311 sec.

By determining, for a given initial phase ϕ_1, the phases ϕ_s of synchronization of the two oscillators obtained for different initial phases ϕ_2, one obtains the arc discontinuity of the PFK model, that is, the set of critical phases ϕ^* for which ϕ_s undergoes a discontinuous change as ϕ_2 is varied, for all possible values of ϕ_1; this curve is shown in Fig. 7. It is characteristic of a quasirelaxation oscillator; moreover, it exhibits a region

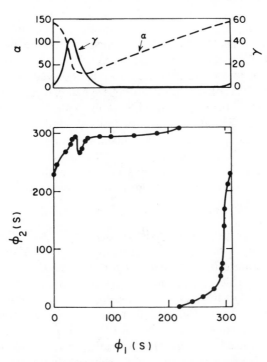

Fig. 7. Arc discontinuity for the allosteric model of glycolytic oscillations (lower diagram). The curve is constructed as indicated in text, by integration of (II.6) for $\sigma_1 = 0.5\ s^{-1}$, $\sigma_M = 8\ s^{-1}$, $k_s = 0.1\ s^{-1}$, $q = 1$, $L = 5 \times 10^6$, $\epsilon = \epsilon' = 10^{-3}$, $c = 10^{-5}$, $\theta = 1$, $D_\alpha = D_\gamma = 10^{-2}\ s^{-1}$; these parameter values yield a limit cycle with a period of 311 s. Near $\phi_1 = 45\ s$ and $\phi_1 = 300\ s$, the system admits a triple arc discontinuity; only one discontinuity is represented for these phases. In the upper part of the diagram, the oscillations in product (γ) and substrate (α) concentrations are represented; phase $\phi_1 = 0$ is taken arbitrarily as the phase corresponding to the maximum in α.

of negative slope that corresponds to a triple discontinuity. With respect to the arc discontinuity, the PFK model thus shares two properties with the putative mitotic oscillator of *Physarum*, namely, the existence of a triple arc discontinuity for some phases, in association with a region of negative slope.[60]

The preceding results of simulations on the mixing of two samples of the glycolytic oscillator can also be compared with the experiments on the synchronization of two yeast cell populations undergoing glycolytic oscillations. In the model all mixes with the first half of the first quadrant ($\phi_1 = 0$ to 40 sec) synchronize with the latter phase that corresponds here to the region of increase in ADP and of decrease in ATP (see the upper part of Fig. 7). All other mixes synchronize at a phase intermediary between the initial phases of the two mixed oscillators.

In the experiments of Ghosh, Chance, and Pye,[59] all mixes with the second half of the second quadrant or with the third quadrant synchronize with the latter phases, which then correspond to the beginning of the increase in NADH; the latter metabolite oscillates in phase with ADP.[24] Synchronization to an intermediate phase is observed in other situations. Though the patterns of synchronization between two yeast cell populations depend on pH,[59] they compare favorably with the theoretical results obtained on the PFK model. In both cases a small portion of the initial phases behaves as an attractor when mixed with an oscillator whose phase is outside this domain. Moreover, these attractor phases correspond to the increase in ADP concentration; this increase in the product of PFK is autocatalytic and therefore corresponds to the region of fastest movement on the limit cycle.

3. Role of Enzyme Cooperativity in Glycolytic Oscillations.   We have already emphasized in the preceding sections that the allosteric nature of phosphofructokinase is of primary importance in the onset of glycolytic oscillations. The role of enzyme cooperativity in the onset of metabolic oscillations in glycolysis can be investigated more quantitatively, by determination of the Hill coefficient related to the substrate, at the stationary state, in the allosteric model for PFK.[45,62] The Hill coefficient is the most commonly used measure of cooperative interactions in multisubunit enzymes. Positive cooperativity, negative cooperativity, and the absence of cooperative interactions correspond, respectively, to a Hill coefficient larger or smaller than unity, or equal to unity.[63] Furthermore, the analytical expression of the Hill coefficient in the concerted allosteric model has been determined both for binding[64] and for kinetic data.[65,66]

Previous studies of the concerted model[67] have shown that under equilibrium conditions, cooperativity is maximum when the nonexclusive binding coefficient of the substrate c goes to zero. This observation remains valid far from equilibrium.[66] As to sustained oscillations, they occur in the PFK model only when the coefficient c is smaller than a critical value.[45] When the two states of the enzyme have an equal catalytic activity ($\theta = 1$), that is, in perfect K systems, values of c that give rise to oscillations correspond to values of the Hill coefficient close to the number of protomers, which is two in the case considered. In K-V systems, when the catalytic activity differs in the T and R states ($\theta < 1$), smaller values of the Hill coefficient at the unstable steady state are compatible with oscillations. Indeed, limit cycle behavior can then be obtained for larger values of c, of the order of 10^{-2}, since binding of the substrate to the less active T state enhances the effect of autocatalysis by the reaction product.[45] However, substrate inhibition is not a prerequisite for periodic behavior, since oscillations also occur when the substrate binds exclusively to the R state ($c = 0$).

The dependence of the Hill coefficient at the steady state on the allosteric constant L is shown in Fig. 8 for different values of the nonexclusive binding coefficient c. These curves are obtained for $q = 1$, that is, for equal dissociation constants $K_{R(S)}$ and K_P, which is approximately the case for ATP and ADP in the PFK reaction.[46] The dashed line indicates the oscillatory domain on each curve. As is the case for the maximum Hill coefficient under equilibrium conditions,[67] bell-shaped curves are obtained for the dependence of the Hill coefficient at the stationary state on the allosteric constant. When the coefficient c is larger than zero, the oscillatory domain extends over a finite domain of L values; moreover, no oscillations are observed below a critical value of L for a given value of c.

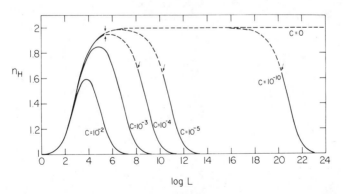

Fig. 8. Hill coefficient at the steady state and oscillatory domain in the dimer PFK model, as a function of the allosteric constant L and of the nonexclusive binding coefficient c (62). The data are obtained for a perfect K system ($\theta = 1$), with $q = 1$. The dashed line on each curve represents the domain of sustained oscillations.

The most significant results of the analysis of the role of cooperativity in the onset of oscillations are twofold. First, the nonequilibrium instability and the limit cycle behavior are associated with large values of the Hill coefficient at the stationary state. As shown in Fig. 9, the Hill coefficient remains close to its maximum value in the course of sustained oscillations. In the case $q = 1$, a necessary (but not sufficient) condition for instability is a Hill coefficient larger than 1.6 at the steady state.[45,66] Smaller values of the Hill coefficient at the steady state are compatible with periodic behavior when $q \gg 1$. The most general result regarding this point is that positive cooperativity is a necessary prerequisite for periodic behavior in the PFK model. This result compares well with the fact that glycolytic oscillations correspond to the region of maximum cooperativity in the saturation curve of PFK by its substrate F6P (see Fig. 2).

Second, the domain of sustained oscillations corresponds to a finite domain of L values when $c > 0$; this result can be related to several experimental observations. Positive effectors of PFK, such as ammonium ions,[48] and negative effectors, such as citrate,[68] inhibit glycolytic oscillations. The theoretical results of Fig. 8 suggest that this phenomenon can be attributed to a decrease in PFK cooperativity in the presence of positive or negative effectors. Indeed, the Hill coefficient at the steady state can decrease below some critical value on addition of an activator of PFK that shifts the curve n_H versus L to smaller values of L, or upon addition of an inhibitor that shifts the curve to larger values of the apparent allosteric constant. In the case of exclusive binding of the substrate to the R state

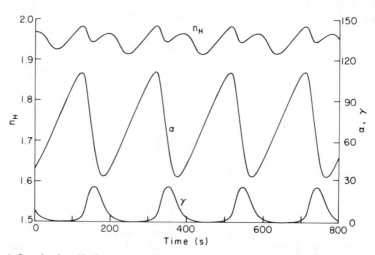

Fig. 9. Sustained oscillations of the Hill coefficient and of the metabolite concentrations in the dimer allosteric model for the PFK reaction (62). The conditions are as in Fig. 6.

$(c = 0)$, the analysis predicts that the periodicities will be suppressed by addition of a positive effector only; addition of an inhibitor will only cause an increase in the period of the phenomenon.[45]

The calculations on the Hill coefficient have been carried out on a dimer enzyme model. Dimers are of special interest for the determination of the role of cooperativity in the mechanisms of sustained oscillations, since they represent the minimal situation in which cooperative interactions can take place in multisubunit enzymes. The results on the role of positive cooperativity in sustained oscillations have been extended to larger numbers of protomers.[141] A Hill coefficient larger than unity is still a prerequisite for oscillatory behavior; the value of the ratio (n_H/n) at steady state required for instability diminishes, however, as the number of protomers increases. The analysis of the PFK model has also been extended to a study of a two-substrate model and of a three-enzyme model with five metabolic intermediates approximating the function of the glycolytic pathway. The analysis yields the boundaries of the oscillatory domain and the waveform as a function of the dynamic state. In addition, a mathematical proof for the existence of a stable limit cycle for enzyme catalyzed reactions with positive feedback was obtained, verifying the numerical simulations and linear stability analysis of the allosteric model of glycolytic oscillations.[34,142,143] Another study has recently shown[144] that a slow interconversion between the T and R conformations of PFK tends to reduce the domain of sustained oscillations.

The experimental and theoretical results obtained for the PFK reaction lead to the conclusion that enzyme cooperativity, in addition to regulatory feedback, plays an essential role in the onset of glycolytic oscillations. This conclusion likely extends to other metabolic oscillations,[26,69] as well as to periodic phenomena that result from genetic regulation[26,70] and from cooperative transport processes in membranes.[71] It thus appears that cooperative allosteric transitions at the genetic, enzyme, or membrane level might well represent the common molecular basis for many periodic phenomena in biology.[24]

III. THE CYCLIC-AMP SIGNALING SYSTEM IN *DICTYOSTELIUM DISCOIDEUM*

A. Experimental Results

The cellular slime mold *Dictyostelium discoideum* represents a major model in the study of differentiation and intercellular communication.[72] When deprived of nutrients, amoebae of this species pass through an interphase of several hours, aggregate around centers by a chemotactic response to cyclic AMP,[73] and further develop into multicellular fruiting bodies.

The process of aggregation has a periodicity of several minutes: waves of inward amoeboid movement appear to propagate outward from the center toward the periphery of the aggregation field.[74] This phenomenon has long suggested that aggregation centers periodically release the chemotactic factor; other cells move toward the source of attractant and relay the chemotactic signal.[75] In other species of cellular slime molds, such as *Dictyostelium minutum*, the process of aggregation is nonperiodic; a comparative study with *D. discoideum* suggests that aggregation centers in the latter species control larger aggregation territories through a mechanism for the periodic generation and relay of chemotactic signals.[74]

In addition to their role as chemotactic signals, periodic cAMP pulses in *D. discoideum* promote cell differentiation. Indeed, aggregateless mutants subjected to periodic cAMP pulses are able to complete their development; these mutants do not respond to a continuous supply of cAMP.[76] Moreover, cAMP oscillations accelerate differentiation in the wild type, by inducing the synthesis of specific enzymes such as phosphodiesterase.[77] A similar role of periodic cAMP pulses in the control of differentiation has been demonstrated in *D. discoideum* cells at the late stationary phase, which are unable to develop unless subjected to periodic cAMP signals.[78] The latter signals induce the appearance of cAMP receptors and of specific contact sites on the cell surface that are characteristic of aggregation-competent cells.

The cAMP-signaling system in *D. discoideum* thus represents a particularly efficient mode of intercellular communication. By emitting periodic cAMP pulses that propagate by diffusion before being relayed by cells in their vicinity, aggregation centers are able to synchronize the development of amoebae in an aggregation territory and to control the chemotactic process that leads to the formation of a multicellular system.

Recent experiments in suspensions of *D. discoideum* amoebae support the existence of the hypothesized mechanism of oscillations and relay of the chemotactic attractant. The first observation[36,79] indicated the existence of spontaneous oscillations, reflected by changes in the light scattering, in a population of interphase amoebae (Fig. 10). These periodicities were shown

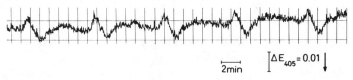

$$\overline{2\text{min}} \qquad \left\rceil \Delta E_{405} = 0.01 \right\rfloor$$

Fig. 10. Spontaneous oscillations of light scattering in cell suspensions of *D. discoideum* (79).

to be under control of cAMP; indeed, pulses of the latter chemical shift the phase of the oscillations, as shown in Fig. 11. The existence of intracellular and extracellular oscillations of cAMP (Fig. 12) corresponding to the periodic changes in light scattering was later demonstrated.[80] The next step demonstrated that tiny pulses of extracellular cAMP can elicit the synthesis of a large intracellular pulse of cAMP,[81,82] proving that cells are able to relay cAMP signals.

The phenomena of relay and oscillation of cAMP both involve the regulation of adenylate cyclase, which is the membrane-bound enzyme that transforms ATP into cAMP. Whereas the oscillations can be explained by models based on intracellular or extracellular regulation of the cyclase, the mechanism of relay involves activation of the enzyme via the binding of cAMP to a specific cell surface receptor.[83] The detailed mechanism by which the receptor-mediated response leads to intracellular cAMP synthesis is still under investigation.

We next examine the nonequilibrium behavior of models for the adenylate cyclase reaction that account for some observed properties of the cAMP-signaling system in interphase amoebae. The models suggest an explanation for the appearance of center-founding cells in aggregation territories, and for a sequence of developmental events observed during interphase. We consider models based on both the intracellular and extracellular regulations of adenylate cyclase.

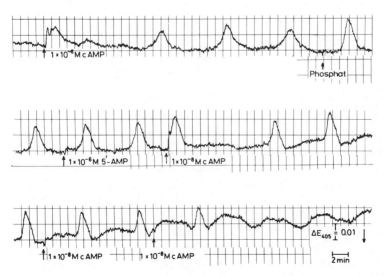

Fig. 11. Phase-shift of oscillations by cAMP pulses in *D. discoideum* suspensions (79).

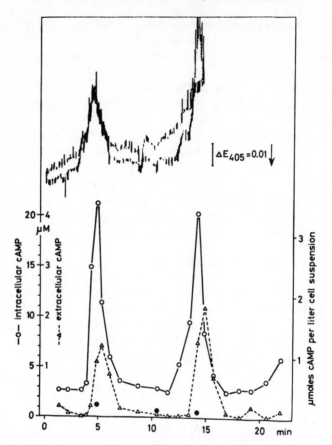

Fig. 12. Sustained oscillations in intracellular and extracellular cAMP concentrations, in suspensions of *D. discoideum* cells (80).

B.   Intracellular Mechanism for the Periodic Synthesis of cAMP in *D. discoideum*

The first study of adenylate cyclase in *D. discoideum* is due to Rossomando and Sussman.[84] Their experiments indicate that cAMP synthesis is under the control of two positive feedback loops, exerted on two enzymes that take a direct part in cAMP metabolism. These enzymes, adenylate cyclase and ATP pyrophosphohydrolase, transform ATP into cAMP and 5'AMP, respectively. In addition, cAMP is transformed into 5'AMP via the phosphodiesterase reaction. The main point regarding oscillations is the finding that 5'AMP activates adenylate cyclase, whereas cAMP activates ATP pyrophosphohydrolase. Rossomando and Sussman suggested that this cross-catalytic control may give rise to periodicities in cAMP synthesis. The analysis of a model based on these positive regulatory interactions confirms that sustained oscillations of cAMP synthesis can take place around a nonequilibrium unstable steady state.[69]

The model describes the intracellular synthesis of cAMP; the diffusion of the metabolites inside the cell or in the extracellular space is not considered (see Section III.C). The three normalized concentration variables are those of ATP (α), cAMP (β), and 5'AMP (γ). The time variation of these concentrations is governed by the following differential equations:[69]

$$\frac{d\alpha}{dt} = v - \epsilon_1 \phi_1 - \epsilon_2 \phi_2$$

$$\frac{d\beta}{dt} = \epsilon_2 \phi_2 - k\beta \tag{III.1}$$

$$\frac{d\gamma}{dt} = k\beta + \epsilon_1 \phi_1 - k\gamma$$

where

$$\phi_1 = \frac{\alpha(1+\alpha)(1+\gamma)^2}{[L_1 + (1+\alpha)^2(1+\gamma)^2]}$$

$$\phi_2 = \frac{\alpha(1+\alpha)(1+\beta)^2}{[L_2 + (1+\alpha)^2(1+\beta)^2]} \tag{III.2}$$

Parameter v denotes a constant ATP input; ϕ_1 and ϕ_2 are the rate functions for ATP pyrophosphohydrolase and adenylate cyclase; L_1, L_2 and ϵ_1, ϵ_2 are the allosteric constants and the maximum activities of these enzymes; the rate constants k_1 and k_2 relate to the phosphodiesterase and 5'-nucleotidase reactions. Equations (III.1) are obtained assuming that the two enzymes are allosteric dimers obeying the concerted model,[13] with exclusive binding to the R state. Furthermore, a quasisteady state for the enzymatic forms is considered.

System (III.1) admits a single stationary state solution, whose stability properties have been determined by normal mode analysis as a function of the main parameters of the model.[69] Such an analysis indicates (Fig. 13) the existence, in the parameter space formed by adenylate cyclase concentration and ATP input rate, of a closed domain in which the steady state is unstable; sustained oscillations in the metabolite concentrations develop in these conditions. As shown in Fig. 14, the waveform of cAMP oscillations is highly pulsatile, in accordance with the measurements of the periodic variation of intracellular cAMP in *D. discoideum* suspensions (see Fig. 12 and Ref. 80).

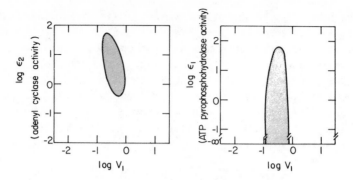

Fig. 13. Stability diagrams for the intracellular mechanism of adenylate cyclase regulation (69). In the dashed area, the steady state is unstable and sustained oscillations occur. Parameter v_1 denotes the rate of ATP input.

System (III.1) still admits oscillations in the absence of ATP pyrophosphohydrolase.[69] This is due to the presence of phosphodiesterase, which transforms the activation of adenylate cyclase by 5'AMP into an autocatalytic control. The detailed mechanism of this regulation might involve the reversal or prevention by 5'AMP of an ATP inactivation of the enzyme.[85]

In the absence of ATP pyrophosphohydrolase, (III.1) resembles the equations that govern the phosphofructokinase reaction [see (II.2) in

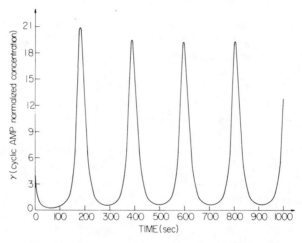

Fig. 14. Sustained oscillations of cAMP produced by the intracellular mechanism of adenylate cyclase regulation (69). The curve is obtained by integration of (III.1) for $\epsilon_1 = 0.4\ s^{-1}$, $\epsilon_2 = 10\ s^{-1}$, $k_1 = k_2 = 0.1\ s^{-1}$, $v_1 = 0.5\ s^{-1}$, $L_1 = L_2 = 10^6$.

Section II.C]. The equations would become identical were adenylate cyclase directly activated by cAMP, rather than by 5′AMP. This stresses the similarity of oscillatory mechanism in yeast and muscle glycolysis, and in the synthesis of cAMP in *D. discoideum*. In support of this conclusion is the observation that both phenomena exhibit similar temperature dependences.[86–88]

C. Compartmental Model for Relay and Oscillation of cAMP

The model of Section III.B accounts for the existence of intracellular oscillations of cAMP. It cannot account, however, for the control of these oscillations by extracellular cAMP,[79] nor for the relay of extracellular cAMP signals,[81,82] which plays an important role in the course of aggregation.

For these reasons Goldbeter and Segel have extended the model of Section III.B by analyzing a compartmental model for the extracellular regulation of adenylate cyclase.[89] This model can give rise to sustained oscillations of cAMP and to the relay of cAMP pulses; the two kinds of behavior correspond to two different parameter regimes.

The model is based on the observation by Roos and Gerisch[83] that extracellular cAMP activates adenyl cyclase, via the binding to a cAMP receptor that might be specific for relay.[90] The activation of adenylate cyclase by extracellular cAMP could be mediated by an intracellular effector such as cyclic GMP[145] possibly acting on a protein kinase.[146] Given that the catalytic site of adenylate cyclase faces the interior of the cell,[91] we make the simplest assumption that the cAMP receptor on the cell surface behaves as a regulatory subunit of adenylate cyclase.

Two compartments are considered, namely, the intracellular and extracellular media, separated by the cell membrane. As for the intracellular model, binding of the substrate and regulation by the product are considered as being cooperative. For definiteness, the receptor-cyclase

TABLE II
Comparison of Relay in Model and Experiment[a]

Relay	Relative amplitude	Time for maximal relay (s)	Half-width (sec)	Delay between extra- and intracellular cAMP (sec)
Experiment	10–25	100–120	60	30–40
Model	20	113	53	3

[a] See Ref. 89. The experimental data have been obtained by Gerisch et al. in *D. discoideum* suspensions[92]; the theoretical data are obtained for the parameter values of Fig. 13.

complex is assumed to possess two catalytic sites for ATP as well as two regulatory (receptor) sites for external cAMP. The three variables are now the normalized concentrations of intracellular ATP (α), intracellular cAMP (β), and extracellular cAMP (γ). The time variation of these metabolite concentrations is given, in homogeneous conditions that correspond to the experiments in *D. discoideum* stirred suspensions, by the following kinetic equations:[89]

$$\frac{d\alpha}{dt} = v - \sigma\phi$$

$$\frac{d\beta}{dt} = q\sigma\phi - k_t\beta \tag{III.3}$$

$$\frac{d\gamma}{dt} = \left(\frac{k_t\beta}{h}\right) - k\gamma$$

where

$$\phi = \frac{\alpha(1+\alpha)(1+\gamma)^2}{[L+(1+\alpha)^2(1+\gamma)^2]} \tag{III.4}$$

Here v denotes a constant ATP input, divided by the Michaelis constant K_S of adenylate cyclase for ATP; σ and L denote the maximum activity, divided by K_S, and the allosteric constant of adenylate cyclase; $q = K_S/K_P$, where K_P is the dissociation constant for the cAMP receptor; k_t and k relate, respectively, to the assumed linear rates of the cAMP transport across the membrane and of the phosphodiesterase reaction; h denotes the dilution factor, that is, the ratio of extracellular to intracellular fluid volume. Actual metabolite concentrations are obtained by multiplying α, β, and γ by K_S, K_P, and K_P, respectively.

It should be noted that (III.3) and (III.4) formally reduce to (III.1) and (III.2) in the absence of ATP pyrophosphohydrolase, when $h = 1$. The intracellular and extracellular mechanisms of adenylate cyclase regulation are thus likely to give similar types of dynamic behavior. Equations (III.3) admit a single steady-state solution (α_0, β_0, γ_0). As for (III.1), it is possible to determine by normal mode analysis a domain of the parameter space in which equations (III.3) produce sustained oscillations in the metabolite concentrations. This domain generally extends over several orders of magnitude of the adenylate cyclase maximum activity σ, and over one order of magnitude of the substrate input v and of the phosphodiesterase catalytic constant k. As found for (III.1), the waveform of intracellular cAMP oscillations is highly pulsatile, in agreement with the experimental observations.

The next question was to determine whether equations (III.3) are capable of relaying extracellular cAMP signals, that is, of synthesizing a large pulse of intracellular cAMP upon stimulation by a tiny external pulse of this metabolite. To test this, equations (III.3) were integrated numerically for different parameter triplets (σ, k, v) outside the oscillatory domain. Intracellular ATP and cAMP were taken initially at their steady-state values, whereas extracellular cAMP was slightly risen above the steady state. Many attempts to obtain relay failed; the system responded by showing at most a tiny increase in intracellular cAMP. A successful attempt is shown in curve a of Fig. 15. Here the parameters correspond to a state of the system in the immediate vicinity of the oscillatory domain. A pulse of intracellular cAMP is obtained whose relative amplitude is of the order of 20. The relative amplitude is defined as the ratio of the maximum of the intracellular cAMP pulse divided by the basal (steady-state) level of intracellular cAMP before stimulation by the external signal.

The relay response occurs only when the external stimulation exceeds a threshold, which corresponds, in the case of Fig. 15, to an extracellular cAMP concentration close to 0.2 μM when the value of parameter K_P is taken as 0.1 μM.[89] Below this threshold no relay occurs, as shown in curve b of Fig. 15. Above the threshold, the amplitude of relay remains practically independent of the external stimulation, whereas the time for maximal relay drops by an order of magnitude (from 100 to 10 sec) as the amplitude of the external signal increase (Fig. 16).[89] The model thus accounts for the existence of a threshold for relay, which has been postulated in a theoretical analysis of slime mold aggregation.[93]

The simulations on relay can be compared with the experiments of Gerisch et al.[92] on cAMP relay in suspensions of *D. discoideum* cells (see

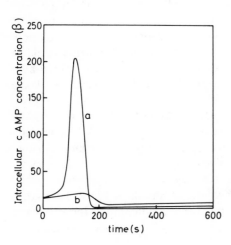

Fig. 15. Relay of cAMP signals. Curve a shows a successful relay with a relative amplitude of 20 (redrawn from Ref. 89). Curve b shows the response to a subthreshold excitation. The initial concentration of extracellular cAMP is $\gamma = 2$ in a, and $\gamma = 1.843$ in b. The curves are obtained by integration of (III.3) for $v = 0.04\ s^{-1}$, $\sigma = 1.2\ s^{-1}$, $k = k_t = 0.4\ s^{-1}$, $L = 10^6$, $q = 100$, $h = 10$.

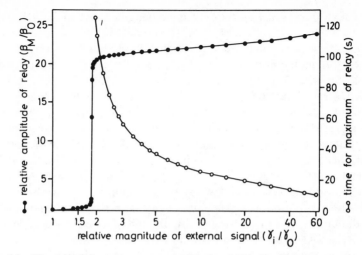

Fig. 16. Theoretical dose-response curves for the cAMP signalling system. The relative amplitude of relay is given as the ratio of the maximum of the intracellular cAMP peak, β_M, divided by the steady-state level β_0. The magnitude of the extracellular signal is given as the ratio of the initial cAMP concentration, γ_i, divided by the steady-state level γ_0.

Table II). Satisfactory agreement is reached as to the relative amplitude, half-width, and time for maximal relay, as well as to the values of metabolite concentrations at the steady state and in the course of relay.[89] The agreement is less satisfactory for the delay that separates the peak in extracellular cAMP from that in intracellular cAMP. This observation suggests that the assumed linear transport of cAMP across the membrane is an oversimplification. Nevertheless, the simulations indicate that major components of the signaling system may well have been identified. The ideas underlying the model doubtlessly will have to be modified to some extent as more is learned about transport and adenylate cyclase regulation. Other effectors of the cyclase, such as Ca^{++} ions, could, for instance, play a role in the signal mechanism.[94] Results analogous to those presented here have been obtained in a qualitative study of the signalling system based on a putative catabolite regulation of cAMP synthesis.[147]

A phase plane analysis of equations (III.3) shows that the relay of cAMP pulses represents an excitable response of adenylate cyclase.[148] Excitability is the capability of a system, initially at a stable steady state, to amplify small perturbations in a pulsatory manner. The association of relay capability with oscillation demonstrated in the adenylate cyclase system is a widespread phenomenon. A similar association between excitability and oscillations has been found in models for the nerve membrane,[149] in models for enzymatic and non-enzymatic[95,150] chemical systems, as well as in experiments on the latter type of reactions.[96,97] The enzymatic systems analyzed for excitability are the PFK reaction[139] and a reaction exhibiting pH-dependent auto-catalytic kinetics.[98] In view of their relative simplicity, the adenylate cyclase and phosphofructokinase systems offer suitable models for the analysis of excitability in biochemical reactions.

As to the differentiation of *D. discoideum*, the model of the cAMP signalling system suggests an explanation for a sequence of developmental events observed in interphase amoebae.[89] In a few hours following star-vation, amoebae that were first inert with respect to the generation of cAMP pulses become capable of relaying cAMP signals, and later begin to generate periodic autonomous signals of cAMP.[99] Such a sequence can be explained by the adenylate cyclase model, assuming that in the beginning of inter-phase, cells are in a region located far from the oscillatory domain in the parameter space. Then, because of slow parameter changes, cells pass through a region near the oscillatory domain, in a location where relay is possible, before entering the domain of sustained oscillations. According to this view, cells that are the first to reach the domain of instability of the cyclase reaction become the aggregation centers. It is thus possible to explain the observed sequence of developmental events by assuming a slow shift in the values of some key parameters of the model, such as phos-phodiesterase and adenylate cyclase activities.

IV. CELL-MEDIATED IMMUNE SURVEILLANCE AGAINST CANCER

The immune response and its control mechanisms become the subject of many theoretical studies. (For a recent review of the field, see Ref. 100.) The spirit of these studies is phenomenological, the goal being to formulate mathematically the overall kinetic behavior of cellular populations during the immune response. As in the biochemical systems discussed in Sections II to III, the stability problems arising at this cellular level are often reducible to a small number of parameters in which the details of molecular architecture or interactions need not appear explicitly. Models of this sort have been set up, on the basis of the clonal hypothesis[101–103] or along the lines of Jerne's network theory,[104–105] to account for the selection and synthesis of specific antibodies in response to varying doses of antigens. More recently also several studies have been devoted to various aspects of the immune response against cancer.[106–108] At the cellular level, cell repli-cation is an obvious autocatalytic process; in this section we discuss the conditions under which the cellular replication of cancer cells leads to instabilities that break up the pattern of interactions between cells charac-terizing normal tissues. We base our discussion on a model proposed by

Lefever and Garay[108-111] and show that the predator-prey type of control that the immune system exerts on cancer cells may lead to nonequilibrium phenomena bearing similarities to first-order phase transitions. We also analyze the properties of the diffusion of cytotoxic cells and particularly its influence on the threshold of cancerization.

Three classes of phenomena are involved in the cancerization of a tissue: (1) the transformation of a normal cell into a cancerous one; (2) the replication of the cancer cells; (3) the interactions of the transformed cells with the host, in particular with its immune system. Each of these phenomena is ruled by complex molecular mechanisms and by numerous systemic and environmental factors. The coupling of these mechanisms and factors determines whether or not at the macroscopic level of tissues the conditions are favorable for cancerization. These conditions, as we indicated earlier, can be formulated phenomenologically in terms of a small number of dimensionless control parameters. In order to introduce these parameters and the model, let us first comment briefly on some general questions underlying points (1) and (3) (for a more detailed discussion, see Ref. 108).

A.   Nature of the Cancerous Transformation and the Immune Surveillance Against Cancer

1. The Origin of Cancer Cells.   It is generally admitted that the accumulation of an effective amount of mutations in the genetic apparatus of single normal cells can transform them into malignant cells. At the time Burnet formulated his theory of immune surveillance against cancer, the high incidence of point mutations, which is of the order of 10^6 to 10^7 mutations per gene $\times$ day in the human being,[112] led him to think that spontaneous somatic mutations were the natural source of malignant clones. In the course of evolution this process would be the selective pressure maintaining cell-mediated immunity.[113] The experimental evidence that has accumulated in the last years is against the existence of such a relation between somatic mutations and carcinogenesis. The experiments with the athymic mouse[114] have shown that under germ-free conditions not a single spontaneous malignant tumor appears for a total of 15,700 such mice studied for a total of approximately 5600 years of mouse life. On the contrary, the athymic mice have been found to be much more sensitive to some oncogenic viruses.

Thus the point of view generally prevailing now is that the cancerous transformations require the presence of environmental physical, chemical, or biological agents inducing some sort of alterations at the level of single cells. The exact nature of these alterations is unknown. For the purpose of the approach considered here the molecular mechanism of carcinogenesis need not be formulated explicitly. The essential fact to be retained is that there might exist in tissues a continuous occurrence of cancer cells. This process will be characterized kinetically by the value of a phenomenological constant.

2. Antigenic Properties of Cancer Cells.   The idea that most cancer cells possess antigenic qualities different from those of the original cells from which they are derived is at the basis of the theory of immunological surveillance against cancer. The antigenic properties of induced tumors, by chemicals[115,116] or by viruses,[117,118] were clearly established with respect to the immune system of the host. They consist of the loss of some normal antigens and/or the appearance of new antigens on the tumor cell surface.[113,119,120] It was found later that the tumor cell surface is also immunogenic to the primary host.[121] There is evidence that the immune response against these antigens is mediated by the thymus-dependent system. A cytotoxic activity of T-lymphocytes (thymus lymphocytes) has been demonstrated against a great variety of cancer cells either in vitro or in vivo.[122,123] In vivo, however, despite a high cytotoxic activity, this immune response is frequently unable to induce tumor rejection in the primary host[121] or in syngeneic animals.[125] This failure may be associated with various kinds of immunosuppressive agents operating in vivo.[124,125]

Besides T-lymphocytes, it has been established in vivo and in vitro that macrophages have a cytotoxic activity against cancer cells.[126] Their population in developing tumors may in some cases be up to 56% of the total number of cells, the percentage of infiltration being proportional to the tumor resistance in syngeneically transplanted rats.[124] The efficiency of the cytotoxic reaction is enhanced by unspecific factors like BCG (Bacillus Calmette Guérin) and phytohemaglutinin or by other factors like lymphokines, which are specific and released by the T-lymphocytes in contact with cancerous cells.

In the following model the exact nature of the cytotoxic cells involved is not defined.* It is simply assumed that in normal or tumoral tissues there exists a population of cytotoxic cells that can interact with and destroy cancer cells. The properties of this cytotoxic reaction, as suggested by the data available for macrophages and T-lymphocytes, can be summarized as follows:

3. Properties of the Cytotoxic Reaction.   The cytotoxic reaction of T-lymphocytes and macrophages with cancer cells involves two steps: first, a cellular contact is formed between a free cytotoxic cell and a cancer

---

* Some authors[127] consider the possibility that the immune response against tumors functions as a thymus-independent nonadaptative phenomenon. The predictions of the model here do not depend crucially on these controversial questions.

cell;[128,129] second, the bound cancer cell is killed. In addition, it has been established that in contact with cancer cells T-lymphocytes emit chemical signals, usually called lymphokines,[130] that enhance the cytotoxic activity by the cooperation of several effects: (1) the lymphokines may modify the motion of cytotoxic cells and induce their accumulation in the cancerous regions; (2) the kinetic constants of the cytotoxic reaction are enhanced; (3) the replication of cytotoxic cells is stimulated. The best-known lymphokine, usually called MIF (Migration Inhibition Factor), mediates at least the first of these effects.[131–134] This is well established for macrophages, and some recent data also indicate that it has an effect on the migration of other cells.

B.   A Predator-Prey Model for the Immune Response Against Cancer

The basic features of phenomena (1) to (3) above may be schematized in the following way:[108–111]

(1) Cellular transformation: normal cells $\xrightarrow{A} X$ (IV.1)

(2) Replication: $X \xrightarrow{\lambda} 2X$ (IV.2)

(3) Cellular interactions $X + E_0 \xrightarrow{k_1} E_1 \xrightarrow{k_2} E_0 + P$ (IV.3)

Here A is the phenomenological constant whose value determines the rate of transformation of normal cells into the cancerous ones X. This constant is difficult to evaluate but certainly is very low. For example, Burnet reported that the probability for a plasmoblast to be transformed and to escape the immune surveillance is of the order of 10^{-16} to 10^{-17}. An order of magnitude of 10^{-17} for the probability of carcinogenesis during replication has also been given by Folkman[135] based on cancer statistics in the United States. Using these values, for an ideal tissue of 10^6 cells/mm^3, with a generation time of 10 days, the rate of carcinogenesis A is found to be of the order of 10^{-11} to 10^{-12} cancer cell/day $\times$ mm^3. The replication constant λ can easily be estimated from the data on tumor growth. For most cancer cells it lies in the range: $0.2 \lesssim \lambda \lesssim 1.4$ day^{-1}. E_0 and E_1 represent the free cytotoxic cells and the cytotoxic cells having bound one cancer cell; P corresponds to the killed cancer cells; k_1 and k_2 are the kinetic constants associated with the binding and destruction processes of the cancer cells by the cytotoxic cells. In general, k_1, k_2, and $E_t = E_0 + E_1$ may vary in time in the course of the process of sensibilization. In the case of macrophages, for example, E_t may vary from a few percents up to 50% of the total number of cells present in the tissue. We do not consider the kinetics of sensibilization here. We consider states of regime where these quantities are likely to be constant. This is the case of the normal state, when no cancer cells are present, or of the tumoral state, after sensibilization has produced its effects. It is also generally the case in in vitro experiments. The kinetic equations describing phenomena (1) to (3) may be written as follows[108,109] (we consider later the effect of cellular motion):

$$\frac{dx}{dt'} = \alpha + (1 - \theta x)x - \beta e_0 x \qquad (IV.4)$$

$$\frac{de_0}{dt''} = -(1 + x)e_0 + 1 \qquad (IV.5)$$

with

$$x = \left(\frac{k_1 N}{k_2}\right)\frac{X}{N}; \qquad e_0 = \frac{E_0}{E_t}; \qquad t' = (\lambda - A)t; \qquad t'' = k_2 t$$

$$\alpha = \frac{Ak_1 N}{k_2(\lambda - A)}; \qquad \beta = \frac{k_1 E_t}{\lambda - A}; \qquad \theta = \frac{\lambda k_2}{k_1 N(\lambda - A)}; \qquad E_t = E_0 + E_1$$

(IV.6)

where N is the total number of cells (normal and cancerous) that form the volume element of tissue ΔV under consideration; α, β, θ are the dimensionless parameters associated with the rate of carcinogenesis, the activity of the cytotoxic reaction, and the maximum number of cells in ΔV. The values of k_1 and k_2 have been determined from the in vitro data of various authors, assuming that the cytotoxic reaction in first approximation obeys a Michaelian type of kinetics. Accordingly, the initial growth of the X cell population is given by

$$\frac{dx}{dt'} = x - \frac{\beta x}{1 + x} \qquad (IV.7)$$

In the in vitro experiments analyzed on the basis of (IV.7), β is a constant experimentally controlled and α may be taken as zero. As an example, in Fig. 17 we report the fitting by (IV.7) of the experimental data (circles) of Evans and Alexander.[128] Satisfactory results are also obtained with other systems.[111] The values of k_1 and k_2 obtained by this method for various systems are reported in Table III.

C. Microcancer States and the Threshold for Cancerization

The orders of magnitude of α, β, θ are as follows:

$$10^{-10} \sim 10^{-11} \simeq \alpha \ll 1; \qquad 10^{-2} \lesssim \beta \lesssim 10^2; \qquad 10^{-1} \lesssim \theta \lesssim 50$$

(IV.8)

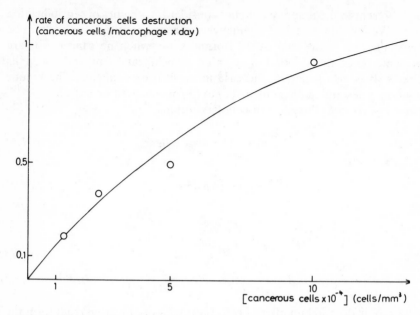

Fig. 17. Cytotoxic effect of immune macrophage on syngeneic lymphoma L5178Y cells. This curve was obtained from the in vitro growth curves of cancerous cell populations in the presence of immune macrophages (Ref. 128). The initial rate of cytotoxic activity was obtained by subtracting the growth rate of lymphoma cells in the presence of macrophages from the control curves. The experiment was made in a monolayer; then the surface concentrations were transformed in volume concentrations, assuming $N = 10^6$ cells/mm^3. The circles correspond to the values calculated from the experimental curves.

TABLE III

Effector cell	k_1N (mm^3) day^{-1}	k_2 day^{-1}	Target cell
Macrophage nonactivated	0.09	0.045	neoplasia mouse embryo[136]
Macrophage activated with BCG	0.36	0.18	neoplasia mouse embryo[136]
Macrophage activated with lymphokine	1.56	0.22	lymphoma[128]
Natural killer cells from CBA mouse spleen	0.07	2.8	Moloney lymphoma of mouse[137]
Sensitized peritoneal lympho-cyte T from BALB/c	6	18	EL 4 leukemia C 57 BL/6 mouse[138]

Equations (IV.4) and (IV.5) admit three steady-state solutions when β lies in the interval

$$\beta_1 = (1 + \sqrt{\alpha})^2 \leq \beta \leq \tfrac{1}{2}\left(1 + \frac{1}{2\theta} + \frac{\theta}{2}\right) = \beta_2 \qquad (IV.9)$$

provided the inequalities

$$\alpha < \frac{1}{27\theta^2} \qquad (IV.10a)$$

$$\theta < 1 \qquad (IV.10b)$$

be satisfied. We see that two situations arise in the case of Table III. With natural killer cells or sensitized peritoneal T-lymphocytes, inequality (IV.10b) is not fulfilled. As a result there is a continuous variation of the steady-state value of X as a function of β. This means that when the value of β increases from zero up to values greater than one (see Fig. 18), one passes *continuously* from tissues in which the proliferating cancer population is large to tissues in which it is vanishingly small (zero in the limit $\alpha \rightarrow 0$). Thus near $\beta = 0$ the steady states should be considered as

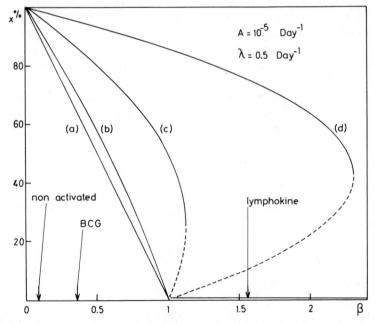

Fig. 18. Percentage of cancerous proliferating cells at the steady state as a function of β. The curves (a), (b), (c), (d) correspond respectively (see Table III) to Natural Killer Cells, lymphocyte T, macrophages (nonactivated or activated with BCG), macrophages activated with lymphokine.

cancerous, whereas on the contrary for $\beta > 1$ we have normal states. The latter because of their small cancer population are referred to as *micro-cancer states*. Interestingly, if we take $E_t = N/2$ (a value quite frequent in tumors) we see that in the case of sensitized peritoneal T-lymphocytes from BALB/C $\beta \simeq 6$ (with $\lambda = 0.5$) and thus the allograph tumoral cells of EL4 leukemia C57BL/6 mouse would be rejected (cf. Table III). This prediction of the model is in agreement with the experimental observations: Allograph transplantations of cancer cells are indeed commonly rejected.

Let us now consider the somewhat more complex situation that arises when $\theta < 1$, as in the macrophage systems of Table III. When θ decreases, as we have already emphasized,[108,109] the steady-state curves behave in a way that presents similarities with the pressure-volume isotherms of the Van der Waals equation, the role of temperature being played by θ. We now analyze this fact in relation to the data of Table III.

When three distinct steady-state solutions for X exist for each β in the interval (IV.9), two are stable (continuous line in Fig. 18) and one (dashed line) is unstable and plays the role of threshold. The steady states of the lower branch (x_1) correspond to microcancer states; those of the upper branch (x_3) correspond to cancerous states. As in equilibrium first-order phase transitions, there exists at β equal to some critical value β_c, a thermodynamic transition point between the microcancer and cancer states. When β is varied and passes through β_c the statistical mean of the cellular populations (in the limit $N \to \infty$) jumps from values that lie on the upper branch (x_3) to values that lie on the lower branch (x_1) (or inversely). Exactly at $\beta = \beta_c$ the probabilities $P(X_3)$, $P(X_1)$ of finding the system in state X_2 or in state X_1 are equal. These probabilities may be evaluated by a stochastic master equation representation of processes (IV.1) to (IV.3). At $\beta = \beta_c$ this yields

$$\frac{P(X_3)}{P(X_1)} = \exp\left[N(\phi(X_3) - \phi(X_1))\right] = 1 \qquad \text{(IV.11)}$$

The term $\phi(X)$ corresponds to free energy in equilibrium phase transitions and one has here

$$\phi(x) = \int_0^X \ln\left[\frac{(\alpha + (1 - \theta X')X')(1 + X')}{\beta_c X'}\right] dX' \qquad \text{(IV.12)}$$

The stochastic mean $\langle X \rangle$:

$$\langle X \rangle = \sum_{X:0}^{N} X P(X) \qquad \text{(IV.13)}$$

has been plotted in Fig. 19 as a function of β. It can indeed be seen that in the limit $N \to \infty$ an abrupt jump takes place between microcancer and

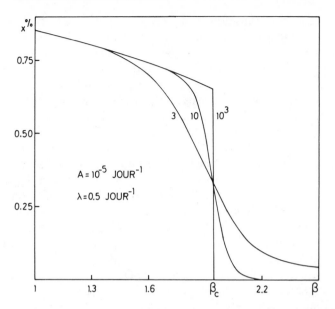

Fig. 19. Mean percentage of cancer cells as obtained on the basis of equations (IV.12 and IV.13) with three different values of N.

cancer states in $\beta = \beta_c$. In agreement with the classical situation it can be shown[108] that in the interval $\beta_1 < \beta < \beta_c$ the x_1 branch of steady states is metastable whereas in the interval $\beta_c < \beta < \beta_2$ it is the x_3 branch that is metastable.

In the light of these results let us consider again the numerical values for macrophages in Table III and Fig. 18. The arrows indicate the values of β corresponding to the various states of activation of the macrophages, the ratio E_t/N being taken equal to one-half. (For nonactivated macrophages this is a value that is rather high; such favorable values do not seem to occur in normal tissues.) One sees that without activation or with BCG activation, the values of β are smaller than β_1 ($\beta_1 \simeq 1$). When the macrophages are activated by lymphokines we see two effects:* (1) A higher value of β expresses a higher efficiency of the cytotoxic reaction; this value of β is sufficiently high to enter the metastable domain of X_1. (2) The sigmoïd character of the steady-state curves increases, the value of β_2 increases, and the cancer steady-state branch becomes bigger.

---

* In this comparison the target cells are different. It has not been possible to find data for the three kinds of macrophage reacting with the same type of cancer cell.

Consequently, the transition from X_3 back to X_1 for a system that has reached the cancer branch requires larger values of β after activation by lymphokines. This example also shows clearly that an increase of β due to activation has an optimal effect if it originates from an increase of E_t rather than from a variation of the ratio k_1/k_2. The latter is inversely proportional to θ; consequently, if k_1 increases relatively to k_2 the sigmoïd character of the curves increases, which means that the threshold of cancerization for a given value of β decreases.

This indicates that under these conditions the immune response of macrophages is likely to be beneficial only if the process of sensitization by lymphokines is very precocious. Indeed, if we admit that during sensitization β passes from values smaller than β_1 to values larger than β_1, this means that there is a time interval during which the proliferation of cancer cells takes place freely. When β becomes larger than β_1 the tumoral nucleus that has formed will be rejected or not, depending on whether it exceeds some critical size.

The possibility that the threshold decreases during activation indicates that under some conditions a tumor would develop *more easily after sensitization* than before. Suppose, for example, a system whose values of β without activation would lie in the interval $1 \lesssim \beta \lesssim 1.1$ (curve c), whereas after activation they would lie on the interval $1 < \beta < 1.9$. If we inject increasing doses of cancer cells, the threshold to exceed to initiate the development of a tumor would be lower after activation.

Let us now investigate the behavior of the threshold of cancerization, when it exists, as a function of the properties of the cellular motion of cytotoxic cells.

D. The Threshold of Cancerization as a Function of the Diffusion of Cytotoxic Cells

The properties of the diffusion of cytotoxic cells in tissues may vary considerably depending on the cells present: (1) If no cancer cells are present, the cytotoxic cells have a seemingly random motion and spatial distribution tends toward homogeneity; in first approximation this motion seems to obey a Fickian type of diffusion law. (2) In the presence of cancer cells the homogeneous spatial distribution of cytotoxic cells is generally altered and accumulations of cytotoxic cells are found in the cancerous regions. The cellular fluxes depend then not only on the cytotoxic cells gradients but also on the cancerous population.

We formulate these properties mathematically on the basis of the data concerning the diffusion of macrophages (cf. Ref. 110). The motion of these cells is particularly well described experimentally and its properties can be summarized as follows:

1. The cancerous cells are recognized by T-lymphocytes, which at their contact liberate in the extracellular medium chemical substances or lymphokines.
2. It has been established at least for one of the these lymphokines, the MIF, that it slows down the cellular motion of macrophages and as a result decreases their rate of diffusion. The MIF concentration being high in the neighborhood of cancer cells, the macrophages tend to accumulate in these regions.
3. The MIF is eliminated from tissues by diffusion or by metabolic degradation processes.

These results suggest for the diffusion flux J of macrophages a law of the form:

$$J = D \frac{d}{dr}\left(\frac{E}{1+KX}\right) \qquad (IV.14)$$

Here x is the spatial coordinate; E and X, the macrophage and cancer populations, respectively; D is Fick's diffusion coefficient; K is a positive constant whose magnitude depends on the intensity of the migration inhibition effect.

Taking (IV.14) into account, the spatiotemporal behavior of the cellular population is now given by the set of equations:

$$\frac{\partial x}{\partial t'} = \alpha + (1 - \theta x)x - \beta e_0 x \qquad (IV.15)$$

$$\frac{\partial e_0}{dt''} = -e_0 x + e_1 + \frac{\partial^2}{\partial r'^2}\left(\frac{e_0}{1+\gamma x}\right) \qquad (IV.16)$$

$$\frac{\partial e_1}{\partial t''} = e_0 x - e_1 + \mu \frac{\partial^2}{\partial r'^2}\left(\frac{e_1}{1+\gamma x}\right) \qquad (IV.17)$$

with

$$\mu = \frac{D_0}{D_1}, \qquad r' = \frac{k_2 r}{D_0^{1/2}}, \qquad \gamma = \left(\frac{k_2}{k_1 N}\right)K \qquad (IV.18)$$

The diffusion coefficient of macrophages can be estimated from in vitro migration experiments. The order of magnitude is

$$D_0 \simeq 10^{-8} \text{ mm}^2 \text{ sec}^{-1}$$

We now study the dependency of the threshold of cancerization on diffusion and more precisely on the size of the tissue volume element ΔV

considered. We suppose first that ΔV is part of a tumor of volume V and look for conditions such that inside ΔV the cancer cell population may be low or may regress toward the microcancer states x_1. For simplicity we put $\alpha = 0$ and assume that the medium is unidimensional and of length l. The cancerous population of $V - \Delta V$ being (x_3) we take as boundary conditions

$$x(0) = x(l) = x_3; \qquad e_0(0) = e_0(l) = \xi = \frac{1}{1+x_3}; \qquad e_1(0) = e_1(l) = \frac{x_3}{1+x_3}$$

$$(\text{IV.19})$$

The values (IV.19) are also a homogeneous steady-state solution of ΔV. Besides this homogeneous solution, (IV.15) to (IV.17) may admit different inhomogeneous steady-state solution that correspond either to micro-cancer states or to the threshold. Let us first consider the case of Fickian diffusion ($\gamma = 0$). The inhomogeneous steady-state solutions corresponding to the threshold can then be calculated exactly. Their free cytotoxic population is then in terms of the new variable

$$u(r') = e_0(r') - \xi \qquad (\text{IV.20})$$

given by the solutions of

$$\left(\frac{d^2 u}{dr'}\right)^2 = K - F(u) \qquad (\text{IV.21})$$

with

$$F(u) = (2\beta\xi - 1 - \mu\theta)\frac{u^2}{\theta} + \frac{2\beta}{3\theta}u^3 \qquad (\text{IV.22})$$

where $\pm K^{1/2}$ is the gradient of $u(r')$ in $r' = 0$ and $r' = l$. In Fig. 20 we have plotted $F(u)$ for different values of β. The value K must necessarily be positive. Furthermore, it is bounded above by the value $K_{max} = F(1/\beta - \xi)$ beyond which the solutions corresponding to $x(r')$ become negative over some regions in space. The parts of the curves that yield these nonphysical solutions are in interrupted line. According to the value of β two situations may then arise:

1. If

$$\beta < \beta_1' = \frac{[4(1 + \theta + 3\mu\theta) - (3\mu\theta - 1)^2]}{16\theta} \qquad (\text{IV.23})$$

the only acceptable steady-state solution, whatever the length l, is the homogeneous cancerous solution (IV.19). One can see that β_1' is significantly bigger than β_1. In the domain of inequality (IV.23), although

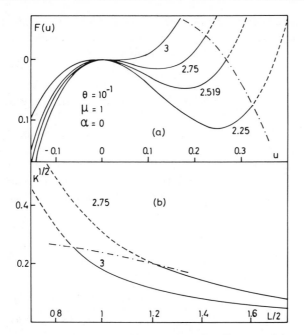

Fig. 20. (a) Plot of $F(u)$ versus u for different values of β. The parts of the curves that are in dashed line correspond to physically unacceptable solutions. The boundary value ξ corresponds to $u = 0$. (b) Relation between $K^{1/2}$ and the size of the threshold steady-state solution; the threshold and the homogeneous steady-state solution (IV.19) have their point of bifurcation at l infinite. In a system of finite length the transition between cancer and microcancer solution require perturbations of finite amplitude the greater that the size is the smaller.

(IV.4) to (IV.5) of the problem without diffusion may admit a microcancer solution, it is impossible that within ΔV one approaches it locally. The cancerous state is asymptotically stable in ΔV whatever the size or amplitude of fluctuations to which it is submitted. The behavior of ΔV is entirely dominated by the value of its boundary conditions.

2. If

$$\beta'_2 = \frac{[(1+\theta)^2 - (1-\mu)^2\theta^2]}{4\theta} > \beta > \beta'_1 \tag{IV.24}$$

(IV.21) admits besides the solution (IV.19), one inhomogeneous unstable solution (threshold) provided l be equal to

$$L(K) = 2 \int_0^{u_K} \frac{du'}{[K - F(u')]^{1/2}} \tag{IV.25}$$

where $0 < K < F(1/\beta - \xi)$ and where in u_K the elliptic integral (IV.25) becomes singular. In $r' = l/2$, the deviation of the threshold with respect to the reference state $e_0(r') = \xi$ is maximum and we have $u(l/2) = u_K$. For any $K > 0$ the integrand of (IV.25) behaves as $[u_K - u']^{1/2}$ in the neighborhood of u_K and $L(K)$ is finite. When $K = 0$, the integrand of (IV.25) diverges as $[u_K = u']^{-1}$ and $L(K)$ is infinite. The amplitude of the threshold is thus a decreasing function of the size of ΔV (see Fig. 20b). In other words, if the size of ΔV diminishes it is necessary in order to reach the threshold to apply larger and larger deviations with respect to the cancerous state maintained at the boundaries. Below some critical size (given by the interrupted line in Fig. 20b), the transition toward x_1 becomes impossible as it would require that $x(r')$ could become negative. The solutions of (IV.25) are given by

$$L(K) = F(\psi(90° - \alpha°)) \qquad (IV.26)$$

where $F(\psi(90° - \alpha°))$ is the incomplete elliptic integral of the first kind whose modular angle and amplitude are

$$\sin^2 \alpha = \frac{1}{2} \frac{F''(u_K)}{8[F'(u_K)]^{1/2}}; \qquad \cos \psi = \frac{[F'(u_K)]^{1/2} - u_K}{[F'(u_K)]^{1/2} + u_K} \qquad (IV.27)$$

These solutions are represented in Fig. 20b. When in this figure $L(K) = 1$, the corresponding length in real units is approximately 10^{-4} cm. However this evaluation must be considered with caution. In reality indeed, the ratio $\mu = D_0/D_1$ could be much larger than one as probably the bound cytotoxic cells lose much of their mobility. In that case, the lengths calculated here could increase by several orders of magnitude.

When $\gamma \neq 0$, (IV.15) to (IV.17) are no longer soluble exactly. Figure 21 illustrates qualitatively the effects of MIF in that case, for a discrete model where ΔV is supposed homogeneous and of constant length $\Delta r' = l = 1$. At the steady state, the total cytotoxic populations $\sigma(\Delta V)$ is

$$\sigma(\Delta V) = \frac{E_t[1 + \gamma x(\Delta V)]}{1 + \gamma x(V - \Delta V)} \qquad (IV.28)$$

where E_t is the total constant cytotoxic population of $V - \Delta V$. The curve in a dashed line represents the cancerous population of ΔV when it is equal to the boundary value (IV.19). The curves in a dotted line represent for two values of γ, the threshold with the same boundary conditions. When $\gamma = 0$, as in the continuous case, β_1' is shifted with respect to β_1; the length l being now fixed, the gap between x_3, x_1 remains finite over the entire domain of β. If $\gamma > 0$, one observes an increasing *stabilization* of the tumoral state by the immune response.

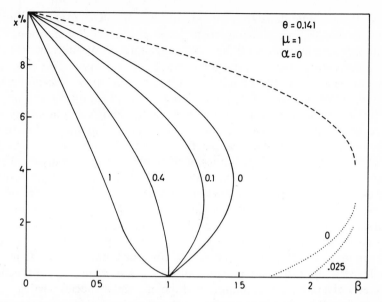

Fig. 21. Cancerous population in % of N as a function of β and for different values of γ. The curves in continuous line are obtained with (IV.29) as boundary conditions; those in dashed or dotted line correspond to (IV.19) as boundary condition.

The curves in a continuous line are obtained for the boundary conditions of a normal tissue:

$$x(0) = x(l) = 0, \qquad e_0(0) = e_0(l) = 1, \qquad e_1(0) = e_1(l) = 0 \quad \text{(IV.29)}$$

The immune response is then stabilizing the microcancer state. When γ increases β_2 decreases and the sigmoïd character of the steady-state curves diminishes; simultaneously the threshold increases. When $\gamma > 0.4$, only the microcancer state $(x(\Delta V) = 0)$ exists for $\beta > 1$.

In conclusion, these results show that the diffusion of cytotoxic cells may lead to contradictory effects depending on the nature and size of the tissue considered. The effects are favorable, that is, susceptible to stop cancerous proliferation when the cancer cell population is small. In that case the nucleus of cancer cells is strongly constrained to the boundaries of a normal tissue, which tend to restore an x_1 type of steady state.

V. CONCLUDING REMARKS

We would like to comment particularly on three remarkable aspects of the nonequilibrium phenomena that we have analyzed.

First, these phenomena are cooperative. They involve a great number of molecules or cells, over a characteristic time-space scale of dimensions that is macroscopic compared with the short-range interactions determining the values of kinetic constants, diffusion coefficients, and so on. This cooperativity thus amounts to the establishment of a precise *supramolecular* coordination or "regulation" of the functioning of metabolic reaction chains or of cellular communication processes. Specific to biological systems are the nature and ingeniosity of the molecular events responsible for these phenomena. In a usual chemical reaction the equilibrium constant and the rate of reaction are simply functions of the molecular species involved and of their relative abundancies. In reactions involving such complex molecules as enzymes and proteins, the kinetics becomes critically dependent on such factors as the conformational state of these molecules or the mechanisms regulating conformational transitions. By comparison with the properties of nonbiological systems exhibiting dissipative structures (see, for example, the Belousov-Zhabotinsky reaction), one can see that this complexity in molecular architecture allows for a simplification of reaction schemes and a reduction at minimum of the molecular species participating in the nonequilibrium phenomena.

Second, this type of order is intimately linked with the notion of irreversibility. The nonequilibrium constraints that maintain a negative entropy flow also orient time evolution unidirectionally. If the constraints are reversed the phenomena disappear. Furthermore, although in some chemical clocks there exists a conserved quantity that plays the role of energy in mechanics and permits formally a Hamiltonian type of description of these systems, it must be stressed that this analogy is only fortuitous. For example, the sense of rotation of an oscillating chemical system in its appropriate phase space is unique and intrinsic to the reaction chain considered. In mechanics because of the time reversibility of equations this sense of rotation would simply be a function of the initial condition. On the other hand, because nonequilibrium order is the result of instabilities, as any type of critical phenomena, its description to be complete needs to incorporate the effect of fluctuations. This aspect was not the subject of this paper (for a detailed treatment, see Ref. 11) and we would like simply to stress the limitations on experimental predictions and observations due to the importance of fluctuations in the occurrence of dissipative structures. The general situation once the thermodynamic threshold of instability has been passed is that there exists a multiplicity of organized states whose occurrence critically depends on the properties of fluctuations. The evolution from the thermodynamic branch to some particular states always involves some stochastic element. Inversely, questions concerning the onset of the phenomena, which in a strictly deterministic description could be straightforwardly answered, escape to this framework. For example, it would be meaningless to trace back the moment of onset of some tumor by reversing time in the equations describing its growth. As a result of the existence of threshold phenomena, the time that elapses between the formation of the first cancer cell and the escape of a critical tumoral nucleus from the immunological surveillance is indeterminate.

Third, dissipative structures indicate limits of reductionism in the description of biological order. Far from equilibrium, the relation between chemical kinetics and the space-time structure of reacting systems cannot be described in strictly local terms. The interactions that determine the value of kinetic constants and transport coefficients (valency forces, hydrogen bounds, Van der Waals forces) result from short-range effects. However, the solutions of kinetic equations depend in addition on such global features as the system size or geometry. This dependence, which near equilibrium, on the thermodynamic branch, is rather trivial, becomes decisive in chemical systems working under far-from-equilibrium conditions.

Acknowledgments

One of the authors (A.G.) is Chargé de Recherches du Fonds National Belge de la Recherche Scientifique. This work has been partially supported by the Fonds Cancérologique de la Caisse Générale d'Epargne et de Retraite.

References

1. L. Onsager, *Phys. Rev.*, **37**, 405 (1931); **38**, 2265 (1931).
2. J. Meixner, *Ann. Phys.*, **39**, 333 (1941); **41**, 409 (1942); **43**, 244 (1943).
3. I. Prigogine, *Etude Thermodynamique des Processus Irréversibles*, Desoër, Liège, 1947.
4. I. Prigogine, *Bull. Acad. Roy. Belg.*, **31**, 600 (1945).
5. I. Prigogine, in M. Marois, Ed., *Theoretical Physics and Biology*, North-Holland, Amsterdam, 1969, p. 23.
6. S. R. de Groot and P. Mazur, *Nonequilibrium Thermodynamics*, North-Holland, Amsterdam, 1962.
7. I. Prigogine, *Introduction to Thermodynamics of Irreversible Processes*, 3rd ed., Wiley, New York, 1967.
8. I. Prigogine and J. M. Wiame, *Experientia*, **2**, 451 (1946).
9. A. Katchalsky and P. F. Curran, *Nonequilibrium Thermodynamics in Biology*, Harvard University Press, Cambridge, 1965.
10. P. Glansdorff and I. Prigogine, *Thermodynamic Theory of Structure, Stability, and Fluctuations*, Wiley, New York, 1971.
11. G. Nicolis and I. Prigogine, *Self-Organization in Nonequilibrium Systems*, Wiley, New York, 1977.
12. A. J. Lotka, *J. Phys. Chem.*, **14**, 271 (1910).
13. J. Monod, J. Wyman, and J. P. Changeux, *J. Mol. Biol.*, **12**, 88 (1965).
14. D. E. Koshland, G. Némethy, and D. Filmer, *Biochemistry*, **5**, 365 (1966).
15. E. A. Newsholme and C. Start, *Regulation in Metabolism*, Wiley, New York (1973).

16. F. G. Heineken, H. M. Tsuchiya, and R. Aris, *Math. Biosci.*, **1**, 95 (1967).
17. B. Hess, in B. Wright, Ed., *Control Mechanisms in Respiration and Fermentation*, Ronald, New York, 1963, p. 333.
18. D. Atkinson, *Science*, **150**, 851 (1965).
19. B. Hess, *Ciba Found. Symp.*, **31**, 369 (1975).
20. A. N. Zaikin and A. M. Zhabotinsky, *Nature*, **225**, 535 (1970).
21. *The Physical Chemistry of Oscillatory Phenomena, Faraday Soc. Symp.*, **9** (1975).
22. G. Nicolis and J. Portnow, *Chem. Rev.*, **73**, 365 (1973).
23. R. M. Noyes and R. J. Field, *Ann. Rev. Phys. Chem.*, **25**, 95 (1974).
24. B. Hess and A. Boiteux, *Ann. Rev. Biochem.*, **40**, 237 (1971).
25. A. Goldbeter and S. R. Caplan, *Ann. Rev. Biophys. Bioeng.*, **5**, 449 (1976).
26. C. Walter, *J. Theor. Biol.*, **27**, 259 (1970).
27. A. Hunding, *Biophys. Struct. Mech.*, **1**, 47 (1974).
28. J. J. Tyson, *J. Math. Biol.*, **1**, 311 (1975).
29. B. Chance, E. K. Pye, A. K. Ghosh, and B. Hess, Eds., *Biological and Biochemical Oscillators*, Academic, New York, 1973.
30. B. Hess, A. Boiteux, and J. Krüger, *Adv. Enzyme Regul.*, **7**, 149 (1969).
31. A. Boiteux, A. Goldbeter, and B. Hess, *Proc. Nat. Acad. Sci. (USA)*, **72**, 3829 (1975).
32. A. Boiteux and B. Hess, *Faraday Soc. Symp.*, **9**, 202 (1975).
33. V. V. Dynnik and E. E. Sel'kov, *FEBS Lett.*, **37**, 342 (1973).
34. T. Plesser, in *Proc. 7th Internat. Conf. Nonlinear Oscillations*, Akademie der Wissenschaften, Berlin, DDR, 1977, Vol. II, p. 273.
35. N. Tamaki and B. Hess, *Hoppe-Seyler's Z. Physiol. Chem.*, **356**, 399 (1975).
36. B. Hess, A. Boiteux, H. G. Busse, and G. Gerisch, in G. Nicolis and R. Lefever, Eds., *Membranes, Dissipative Structures, and Evolution*, Wiley, New York, 1975, p. 137.
37. B. Hess and A. Boiteux, *Hoppe-Seyler's Z. Physiol. Chem.*, **349**, 1567 (1968).
38. I. Prigogine, R. Lefever, A. Goldbeter, and M. Herschkowitz-Kaufman, *Nature*, **223**, 913 (1969).
39. E. E. Sel'kov, *Eur. J. Biochem.*, **4**, 79 (1968).
40. A. Goldbeter, *Proc. Nat. Acad. Sci. (USA)*, **70**, 3255 (1973).
41. J. Higgins, *Proc. Nat. Acad. Sci. (USA)*, **51**, 989 (1964).
42. J. Higgins, *Ind. Eng. Chem.*, **59**, 18 (1967).
43. E. E. Sel'kov, *Mol. Biol.*, **2**, 208 (1968).
44. A. Goldbeter and R. Lefever, *Biophys. J.*, **12**, 1302 (1972).
45. A. Goldbeter and G. Nicolis, *Progr. Theor. Biol.*, **4**, 65 (1976).
46. D. Blangy, H. Buc, and J. Monod, *J. Mol. Biol.*, **31**, 13 (1968).
47. T. E. Mansour, *Curr. Top. Cell. Regul.*, **5**, 1 (1972).
48. B. Hess and A. Boiteux, in J. Järnefelt, Ed., *Regulatory Functions of Biological Membranes*, Elsevier, Amsterdam, 1968, p. 148.
49. B. Hess, *Nova Acta Leopoldina*, **33**, 195 (1968).
50. B. Chance, E. K. Pye, and J. Higgins, *IEEE Spectrum*, **4**, 79 (1967).
51. B. Hess and A. Boiteux, in B. Chance, E. K. Pye, A. K. Ghosh, and B. Hess, Eds., *Biological and Biochemical Oscillations*, Academic, New York, 1973, p. 229.
52. J. Higgins, R. Frenkel, E. Hulme, A. Lucas, and G. Rangazas, *ibid.*, p. 137.
53. A. Winfree, *Nature*, **253**, 315 (1975).
54. B. Chance, B. Schoener, and S. Elsaesser, *Proc. Nat. Acad. Sci. USA*, **52**, 337 (1964).
55. E. K. Pye, *Can. J. Bot.*, **47**, 271 (1969).
56. A. T. Winfree, *Arch. Biochem. Biophys.*, **149**, 388 (1972).
57. A. T. Winfree, *J. Math. Biol.*, **1**, 73 (1974).
58. S. Kauffman, *Bull. Math. Biol.*, **36**, 171 (1974).

59. A. K. Ghosh, B. Chance, and E. K. Pye, *Arch. Biochem. Biophys.*, **145**, 319 (1971).
60. S. A. Kauffman and J. Wille, *J. Theor. Biol.*, **55**, 47 (1975).
61. J. Tyson and S. Kauffman, *J. Math. Biol.*, **1**, 289 (1975).
62. A. Goldbeter, *Biophys. Chem.*, **6**, 95 (1976).
63. D. E. Koshland, in P. D. Boyer, Ed., *The Enzymes*, Vol. 1, Academic, New York, 1970, p. 341.
64. R. J. Watts-Tobin, *J. Mol. Biol.*, **23**, 305 (1967).
65. H. Paulus and J. K. DeRiel, *J. Mol. Biol.*, **97**, 667 (1975).
66. A. Goldbeter, *Biophys. Chem.*, **4**, 159 (1976).
67. M. M. Rubin and J. P. Changeux, *J. Mol. Biol.*, **21**, 265 (1966).
68. R. Frenkel, *Arch. Biochem. Biophys.*, **125**, 151 (1968).
69. A. Goldbeter, *Nature*, **253**, 540 (1975).
70. J. S. Griffith, *J. Theor. Biol.*, **20**, 202 (1968).
71. R. Blumenthal, *J. Theor. Biol.*, **49**, 219 (1975).
72. J. T. Bonner, *The Cellular Slime Molds*, 2nd ed., Princeton University Press, Princeton, N.J., 1967.
73. T. M. Konijn, J. G. C. van de Meene, J. T. Bonner, and D. S. Barkley, *Proc. Nat. Acad. Sci. (USA)*, **58**, 1152 (1967).
74. G. Gerisch, *Curr. Top. Dev. Biol.*, **3**, 157 (1968).
75. B. M. Shaffer, *Adv. Morphog.*, **2**, 109 (1962).
76. M. Darmon, P. Brachet, and L. H. Pereira da Silva, *Proc. Nat. Acad. Sci. USA*, **72**, 3163 (1975).
77. C. Klein and M. Darmon, *Biochem. Biophys. Res. Commun.*, **67**, 440 (1975).
78. G. Gerisch, H. Fromm, A. Huesgen, and U. Wick, *Nature*, **225**, 547 (1975).
79. G. Gerisch and B. Hess, *Proc. Nat. Acad. Sci. (USA)*, **71**, 2118 (1974).
80. G. Gerisch and U. Wick, *Biochem. Biophys. Res. Commun.*, **65**, 364 (1975).
81. W. Roos, V. Nanjundiah, D. Malchow, and G. Gerisch, *FEBS Lett.*, **53**, 139 (1975).
82. B. M. Shaffer, *Nature*, **255**, 549 (1975).
83. W. Roos and G. Gerisch, *FEBS Lett.*, **68**, 170 (1976).
84. E. F. Rossomando and M. Sussman, *Proc. Nat. Acad. Sci. (USA)*, **70**, 1254 (1973).
85. E. F. Rossomando and M. A. Hesla, *J. Biol. Chem.*, **251**, 6568 (1976).
86. B. Hess, K. Brand, and E. K. Pye, *Biochem. Biophys. Res. Commun.*, **23**, 102 (1966).
87. B. Wurster, *Nature*, **260**, 703 (1976).
88. V. Nanjundiah, K. Hara, and T. M. Konijn, *Nature*, **260**, 705 (1976).
89. A. Goldbeter and L. A. Segel, *Proc. Nat. Acad. Sci. (USA)*, **74**, 1543 (1977).
90. A. A. Green and P. C. Newell, *Cell*, **6**, 129 (1975).
91. L. S. Cutler and E. F. Rossomando, *Exp. Cell Res.*, **95**, 79 (1975).
92. G. Gerisch, D. Malchow, W. Roos, U. Wick, and B. Wurster, in L. Saxen and L. Weiss, Eds., *Cell Interactions in Differentiation*, Academic, New York, 1978, in press.
93. M. H. Cohen and A. Robertson, *J. Theor. Biol.*, **31**, 101 (1971).
94. P. E. Rapp and M. J. Berridge, *J. Theor. Biol.*, **66**, 497 (1977).
95. R. J. Field and R. M. Noyes, *Faraday Soc. Symp.*, **9**, 21 (1975).
96. A. T. Winfree, *Science*, **175**, 634 (1972).
97. P. De Kepper, *C.R. Hebd. Acad. Sci. Ser. C*, **283**, 25 (1976).
98. H. S. Hahn, A. Nitzan, P. Ortoleva, and J. Ross, *Proc. Nat. Acad. Sci., USA*, **71**, 4067 (1974).
99. A. Robertson, D. J. Drage, and M. H. Cohen, *Science*, **175**, 33 (1972).
100. G. Bell, A. Perelson, and G. Pimbley, Eds., *Theoretical Immunology*, Marcel Dekker, New York, 1978.
101. G. I. Bell, *J. Theor. Biol.*, **29**, 191 (1970); **33**, 339 (1971); **33**, 379 (1971).

412   B. HESS, A. GOLDBETER, R. LEFEVER

102. G. H. Pimbley, Jr., *Math. Biosci.*, **20,** 27 (1974); **21,** 251 (1974); *Arch. Ration. Mech. Anal.*, **55,** 43 (1974).
103. C. Bruni, M. A. Giovenco, G. Koch, and R. Strom, *Math. Biosci.*, **27,** 191 (1975).
104. P. H. Richter, *Eur. J. Immunol.*, **5,** 350 (1975).
105. G. W. Hoffmann, *Eur. J. Immunol.*, **5,** 638 (1975).
106. C. DeLisi and A. Rescigno, *Bull. Math. Biol.*, **39,** 201 (1977).
107. A. Rescigno and C. DeLisi, *Bull. Math. Biol.*, **39,** 487 (1977).
108. R. Lefever and R. Garay, in G. Bell, A. Perelson, and G. Pimbley, Eds., *Theoretical Immunology,* Marcel Dekker, New York, 1978, p. 481.
109. R. Garay and R. Lefever, *C.R. Hebd. Acad. Sci. Ser. D,* **285,** 741 (1977).
110. R. Lefever and R. Garay, *C.R. Hebd. Acad. Sci. Ser. D,* **285,** 845 (1977).
111. R. Garay and R. Lefever, *J. Theor. Biol.,* in press (1978).
112. F. M. Burnet, *Lancet,* **1,** 1171 (1967).
113. F. M. Burnet, *Transplant. Rev.,* **7,** 3 (1971).
114. J. Rygaard and C. O. Poulsen, *Transplant. Rev.,* **28,** 43 (1976).
115. E. J. Foley, *Cancer Res.,* **13,** 835 (1953).
116. R. T. Prehn and J. M. Main, *J. Nat. Cancer Inst.,* **18,** 769 (1957).
117. K. Habel, *Proc. Soc. Exp. Biol. (N.Y.),* **106,** 722 (1961).
118. H. O. Sjögren, I. Hellström, and G. Klein, *Exp. Cell Res.,* **23,** 204 (1961).
119. L. A. Zilber, *Adv. Cancer Res.,* **5,** 291 (1958).
120. R. W. Baldwin and M. Moore, *Nature,* **220,** 287 (1968).
121. G. Haughton and A. C. Whitmore, *Transplant. Rev.,* **28,** 76 (1976).
122. K. T. Brunner, J. Mavel, J. C. Cerottini, and B. Chapuis, *Immunology,* **14,** 181 (1968).
123. I. Hellström and K. E. Hellström, in B. R. Bloom and P. R. Glade, Eds., *In Vitro Methods in Cell-Mediated Immunity,* Academic, New York (1971); *Fed. Proc., Fed. Amer. Soc. Exp. Biol.,* **32,** 156 (1974); *Adv. Immunol.,* **18,** 209 (1974).
124. R. W. Baldwin, *Transplant. Rev.,* **28,** 62 (1976).
125. O. J. Plescia, A. H. Smith, and K. Grenwich, *Proc. Nat. Acad. Sci. USA,* **72,** 1848 (1975).
126. R. Evans, *Transplantation,* **14,** 468 (1972).
127. A. H. Greenberg and M. Greene, *Nature,* **264,** 356 (1976).
128. R. Evans and P. Alexander, *Nature,* **228,** 620 (1970).
129. M. L. Lohmann-Matthes and H. Fischer, *Transplant. Rev.,* **17,** 150 (1973).
130. D. C. Dumonde, R. A. Wolstencroft, G. S. Panyi, M. Matthew, J. Morley, and W. T. Howson, *Nature,* **224,** 38 (1969).
131. B. R. Bloom and B. Bennett, *Science,* **153,** 80 (1966).
132. J. Svejcar, J. Johanovsky, and J. Pekarek, *Z. Immun. Forsch.,* **133,** 259 (1967).
133. I. D. Bernstein, D. E. Thor, B. Zbar, and H. J. Rapp, *Science,* **173,** 729 (1971).
134. B. Hochova, J. Krejci, J. Pekarek, J. Svejcar, and J. Johanovsky, *Immunology,* **29,** 231 (1975).
135. J. Folkman, *Sci. Am.,* **234,** 58 (1976).
136. M. S. Meltzer, R. W. Tucker, K. K. Sanford, and E. J. Leonard, *J. Nat. Cancer Inst.,* **54,** 1177 (1975).
137. R. Kiesling, E. Klein, and H. Wigzell, *Eur. J. Immunol.,* **5,** 112 (1975).
138. G. Berke and B. Amos, *Transplant. Rev.,* **17,** 71 (1973).
139. A. Goldbeter and T. Erneux, *C.R. Hebd. Acad. Sci. Ser C,* **286,** 63 (1978).
140. J. J. Wille, C. Sheffey, and S. A. Kauffman, *J. Cell. Sci.,* **27,** 91 (1977).
141. T. Erneux, D. Venieratos, and A. Goldbeter, manuscript in preparation.
142. B. Hess and T. Plesser, *Ann. NY Acad. Sci.,* in press (1978).
143. T. Erle, K. H. Meyer, and T. Plesser, submitted for publication to *J. Math. Biol.* (1978).

144.   J. Demongeot, Thèse de Doctorat d'Etat, Université de Grenoble (1978).
145.   J. M. Mato, P. J. M. Van Haastert, F. A. Krens, E. H. Rhijnsburger, F. C. P. M. Dobbe, and T. M. Konijn, *FEBS Lett.*, **79,** 331 (1977).
146.   G. Gerisch, Y. Maeda, D. Malchow, W. Roos, U. Wick, and B. Wurster, in P. Cappuccinelli and J. M. Ashworth, Eds., *Development and Differentiation in the Cellular Slime Moulds*, Elsevier/North Holland Biomed. Press, 1977, p. 105.
147.   M. S. Cohen, *J. Theor. Biol.*, **69,** 57 (1977).
148.   A. Goldbeter, T. Erneux, and L. A. Segel, *FEBS Lett.*, **80** (1978).
149.   R. Fitzhugh, *Biophys. J.*, **1,** 445 (1961).
150.   J. J. Tyson, *J. Chem. Phys.*, **66,** 905 (1977).

ELECTRIC FIELDS AND SELF-COHERENT PATTERNS AND STRUCTURES IN CHEMICAL SYSTEMS: LARGE-SCALE EFFECTS AND BIOLOGICAL IMPLICATIONS

A. D. NAZAREA*

*Center for Statistical Mechanics and Thermodynamics
University of Texas at Austin
Austin, Texas*

CONTENTS

I. INTRODUCTION

In the complicated reaction systems encountered in molecular biology, the overwhelming majority of the reactants carry charge in solution either as simple ions or as polyvalent macroions (polyelectrolytes). Since biological order at the molecular level is invariably manifested as self-coherent nonequilibrium structures and patterns in the number density distribution (spatial as well as temporal) of biological molecules, an external electric

* Also: Solvay Institutes (*Instituts Internationaux de Physique et de Chimie*), Brussels.

415

field can be expected to exert *large-scale* effects for which there does not exist at present an encompassing theory. I have made an attempt to present such a theory here, comprehensive enough to predict unambiguously what the systematic (i.e., large-scale) effects will be. In view of the implications that it holds for biology, attention is devoted to the case where the reaction system is in solution with charged reactants, either simple ions or poly-electrolytes.

In multicomponent nonlinear chemical reaction systems that are main-tained closed with respect to some of the reactants and buffered (open) with respect to the rest, there can occur, beyond some critical value of an external parameter, instabilities that give rise to a number density dis-tribution (of the reactants) that repeats regularly in space and, depending on the type of instability, in time as well. Such dissipative structures, that is, self-coherent spatiotemporal patterns in the number density distribution, are formed by the interplay between the nonlinear kinetics of the chemical transformations, on the one hand, and particle transport, on the other. These parametric instabilities giving rise to dissipative structures must, of necessity, be modulated by the presence of an external electric field whenever at least one of the reactants with respect to which the system is closed carries net charge in solution.

An external electric field modifies the transport properties in solution caused by finite ionic drift and anisotropic translational diffusion and—taking both reactive and charge fluctuations explicitly into account—it is shown in this work, for the first time, that the distinct large-scale effects of an imposed uniform electric field on symmetry-lowering instabilities in multicomponent chemical reaction systems are *threefold*: (1) a field-dependent frequency shift in the chemical oscillations, (2) a field-depen-dent rearrangement of the characteristic size and shape of the repeating spatial pattern, and (3) an inherent asymmetry in the frequency shift of the chemical oscillations with respect to polarity reversal of the field.

The theoretical setting of the problem is made precise in Section II for an n-component nonlinear chemical reaction system having buffered and unbuffered reactants subjected to a uniform electric field. Because of a failure to equipartition the translational energy gained from the field, anisotropic translational diffusion occurs. So does field-induced ionic drift. From the Poisson equation defining the dynamics of the internal field due to the ions, one can derive the divergence of the fluctuation in the particle density current caused by ionic drift alone. This, in conjunction with the equation that defines the divergence of the fluctuation of that part of the particle density current resulting from anisotropic translational diffusion, allows us to determine, to first order in fluctuations/perturbations, the dynamics of the Fourier components of the fluctuations—parametrized, for a given wave vector, by the intensity of the imposed external electric field, the number densities of the buffered reactants, the transverse and longitudinal diffusion matrices, the mobility matrix, and the average dielectric constant of the solution. The set of free parameters (those that can be externally varied) are the first two mentioned above. The dynamics of the Fourier components of the fluctuations is expressed in terms of a certain complex linear operator—whose spectrum determines the possible instabilities—and in this operator the external electric field enters holomorphically as a parameter, allowing us to derive several conclusions using some simple consequences of the theory of analytic perturbation of operators. The Poisson equation guarantees that the fluctuation dynamics are obtained self-consistently.

In Section III, two main types of symmetry-lowering instabilities are considered, namely, those giving rise to symmetry lowering in space alone and those giving rise to symmetry lowering in both space and time. In the former, self-coherence is manifested as a regularly repeating structure in space (in the number density distribution) whereas in the latter, self-coherence is manifested, additionally, as temporal oscillations in each spatially repeating unit. For symmetry-lowering instabilities of both types, infinitesimal perturbations elicit a finite response from the reacting system. As such, these phenomena occur on a large (i.e., macroscopic) scale, and in the present instance the size of the system must exceed the relevant Debye–Hückel length for the solution as well as another critical length defined by the transport properties and the kinetics of the chemical transformations.

The detailed analysis of the modulation by the electric field of the two types of instabilities considered is made both for a weak external field and for moderate ones. This analysis leads to precisely the conclusions given in the third paragraph of the present section. The restriction to moderate field strengths allows the neglect of both the effect of Joule heating (and the attendant convection instabilities) and of the second Wien effect on weakly ionized reactants. In Appendix 2, experimental protocol is suggested for the minimization of the effects of Joule heating as well as for the provision of electrodes best suited for biological experimentation.

Section IV and Appendix 1 are devoted to a discussion of how the three distinct macroscopic effects, deduced and predicted in this paper for the first time, can be verified experimentally, and their implications for the study of biochemical oscillators, the spatiotemporal patterns in the activity of the central nervous system and in embryogenesis: in particular, how, among other things, these predicted effects can help in the understanding of field-induced inversion of a morphogenetic pattern, of the polarity asymmetry in cortical polarization experiments (anodal and cathodal polarization), of the predicted shift in circadian rhythm caused by hypo-thalamic polarization, and of field-induced frequency modulation of intrinsic biological periodicities and rhythms at the level of the single eukaryotic cell. Appendix 5 rounds out and updates the discussion.

II. FIRST-ORDER DYNAMICS OF COMBINED REACTIVE AND CHARGE FLUCTUATIONS WHEN NONLINEAR CHEMICAL REACTIONS COUPLE TO IONIC DRIFT AND ANISOTROPIC DIFFUSION

We consider the case in which the uniform electric field $\vec{E}_{ext}$ is in the longitudinal direction, namely, $\vec{E}_{ext} = (0, 0, E_{\parallel})$; a direction that can be defined arbitrarily whenever the reaction system has no inherent aniso-tropy with respect to translational diffusion transport in the *absence* of the field. Note that the latter condition—for biological reaction systems, where commonly the reactants have hydrodynamically anisometric conformations (e.g., α-helixes or double-stranded helixes behaving as rigid rods in solu-tion)—implies a certain degree of diluteness, since solutions of anisometrically conformed chain molecules exhibit a measurable aniso-tropy with respect to translational diffusion in sufficiently concentrated solutions even in the absence of an external field. (This in fact is the likely explanation of why the spectrum of light scattered from, say, moderately concentrated solutions of DNA has a non-Lorentzian shape.[1])

The presence of an externally imposed electric field induces a systematic anisotropy in the translational diffusion of the components, and hence the diffusion of each of them acquires a tensor character in which the trans-lational diffusion constant transverse to the electric field differs from the diffusion constant parallel to it. Thus, in general, one has to take into account two distinct symmetric translational diffusion matrices $\mathbf{D}_{\parallel}$ and $\mathbf{D}_{\perp}$ (with elements $\delta_{\parallel}^{jk}$ and $\delta_{\perp}^{jk}$), which tend to become identical only in the limit of a vanishingly weak field. Stated in the simplest terms, the reason why the ratios $(\delta_{\parallel}^{jk}/\delta_{\perp}^{jk})$—for both diagonal and (nonzero) off-diagonal (cross-diffusive) elements of the diffusion matrix—differ from unity in the presence of the field is that the translational energy gained by ionic reac-tant molecules is not equipartitioned between the longitudinal and trans-verse components of the (translational) diffusive motion of the ions. Uncharged reactants will also tend to have differing transverse and longi-tudinal diffusion coefficients owing to the viscous drag in the direction of ionic drift.

Note that, historically, this failure to equipartition the translational energy gained from the field by charged particles, leading to the anisotropy of translational diffusion, was first predicted for the case of electrons and ions in a neutral gas subjected to a static electric field.[2] The phenomenon was detected first for electrons diffusing in several different types of neutral atomic and molecular gases,[3] using time-of-flight techniques; and subsequently for positive ions. Under the influence of not too weak fields, anisotropic translational diffusion tends to occur in liquid solutions as well.

The purely field-induced ionic drift is parametrized by the diagonal matrix $\mathbf{M}_\parallel$ of specific ionic mobilities whose elements $\bar{m}_{jj}$ have the dimension $cm^2/volt\text{-}sec$ if the field intensity is expressed in volts per centimeter. The quantity $\bar{m}_{jj}E_\parallel$ equals the drift velocity of the jth component in a field of intensity $E_\parallel$. Such quantities are signed since there are two directions in which ionic drift can occur. (In the case of simple spherical ions, from Debye–Hückel theory the absolute value of the specific mobility is determined by its radius, the thickness of the counterion layer (which in turn depends on the dielectric constant, the temperature, and the total ionic strength of the solution), its zeta potential, and the viscosity of the solution. For flexible polyelectrolytes, making use of the porous-sphere model,[4,5] it is determined by the total charge on the polyelectrolyte, its translational frictional coefficient, the hydrodynamic shielding depth (a function of the viscosity, the volume of a monomer unit, and, again, the translational frictional coefficients of the monomer units), and the thickness of the counterion layer.

For generality, let there be present a total of $(n+m)$ distinct reactant types in solution, m of which are maintained buffered by means of external particle transfer, work being continually done on the system to accomplish such buffering—*at the expense of a finite rate of energy dissipation*. The nonequilibrium system is otherwise closed for the n remaining reactants. Denote by $\{\beta_\mu^0\}$, $\mu = 1, \ldots, m$, the mean number density (externally buffered values) of those reactants for which the system is maintained open. In addition, assume present in solution s inert (nonreactive) components, including the solvent, and let $\{\gamma_\sigma^0\}$, $\sigma = 1, \ldots, s$, denote their mean number density. Of the m buffered reactants we let, say, m' $(\leq m)$ of them carry charge in solution and, likewise, let $s'(\leq s)$ of the inert components carry charge. Finally, denoting by $\hat{\nu}$ the number density of the n reactants for which the reaction system is closed, let us write the nonlinear kinetics of the purely chemical transformations as $[\partial_t \hat{\nu}]_{chem} = \hat{\Xi}(\hat{\nu}; \{\beta_\mu^0\})$. Of these n components, say n' $(\leq n)$ carry charge in solution.

The local (total) electric field in the solution is composed of two parts,

$$\vec{E}(\vec{r}, t) = \vec{E}_{ext} + \vec{E}_{ion}(\vec{r}, t)$$

where the ionic part $\vec{E}_{ion}(\vec{r}, t)$ satisfies (denoting the average dielectric constant of the solution as ε) the Poisson equation

$$\Sigma_\alpha \vec{\partial}_\alpha E_{ion}^{(\alpha)}(\vec{r}, t) = \left(\frac{4\pi}{\varepsilon}\right)[\Sigma_i z_i \hat{\nu}_i(\vec{r}, t) + \Sigma_\mu^{\neq} z_\mu \beta_\mu(\vec{r}, t) + \Sigma_\sigma^{\neq} z_\sigma \gamma_\sigma(\vec{r}, t)] \quad (1)$$

Here, $\alpha = 1, 2, 3$ denotes the component index of spatial vectors, and the superscript ($\neq$) over the last two summations indicates that these summations are over charged components only. The z_i's, z_μ's and z_σ's denote the charge valences (ionic numbers) of the components. (In the first summation only n' of the z_i's are, of course, nonzero.) Taking a reference steady state $[\hat{\boldsymbol{\nu}}^0, \{\beta_\mu^0\}, \{\gamma_\sigma^0\}]$, then if overall electroneutrality prevails in the open system, that is, if

$$\Sigma_i z_i \nu_i^0 + \Sigma_\mu^{\neq} z_\mu \beta_\mu^0 + \Sigma_\sigma^{\neq} z_\sigma \gamma_\sigma^0 = 0$$

then, it is easily shown to first order in perturbations or fluctuations, the Poisson equation takes the form

$$\Sigma_\alpha \vec{\partial}_\alpha E_{\text{ion}}^{(\alpha)}(\vec{r}, t) = \left(\frac{4\pi}{\varepsilon}\right)[\Sigma_i z_i \, \delta\hat{\nu}_i(\vec{r}, t) + \Sigma_\mu^{\neq} z_\mu \, \delta\beta_\mu(\vec{r}, t) + \Sigma_\sigma^{\neq} z_\sigma \, \delta\gamma_\sigma(\vec{r}, t)] \quad (2)$$

We need to evaluate, to first order in small perturbations/fluctuations about the reference steady state, the divergence of the fluctuation of that part of the particle density current attributable to ionic drift alone of all reactants for which the system is closed, as well as of all those buffered reactants and inert components that carry charge in solution. There are exactly a total of $(n + m' + s')$ of these, and we denote their number density vector as $\boldsymbol{\nu}$. The abovementioned divergence is given, taking note of the two parts of the local (total) electric field and keeping in mind that $\vec{E}_{\text{ext}}$ has only a longitudinal component, by

$$\Sigma_\alpha \vec{\partial}_\alpha \delta J_{\text{drift}}^{(\alpha)}(\vec{r}, t) = E_\parallel \mathbf{M}_\parallel \partial_\parallel \, \delta\boldsymbol{\nu}(\vec{r}, t) + \mathbf{M}_\parallel \boldsymbol{\nu}^0 \Sigma_\alpha \vec{\partial}_\alpha E_{\text{ion}}^{(\alpha)}(\vec{r}, t) \quad (3)$$

The preceding equation can be arrived at by noting, for instance, that a divergence term of the form $\mathbf{M}_\parallel \Sigma_\alpha \vec{\partial}_\alpha \, \delta\nu E_{\text{ion}}^{(\alpha)}(\vec{r}, t)$ is *approximately* $O(\delta^2 \boldsymbol{\nu}(\vec{r}, t))$, owing to (2), and can therefore be neglected. $\mathbf{M}_\parallel$ is, of course, an $(n + m' + s')$-dimensional diagonal matrix among whose first n (diagonal) elements there are exactly $(n - n')$ zeros.

From the continuity equation defining the coupling of the nonlinear kinetics of the chemical transformations, the finite ionic drift, and of anisotropic translational diffusion, one then obtains, to first order in perturbations/fluctuations, the space-time evolution of these fluctuations:

$$\partial_t \, \delta\boldsymbol{\nu}(\vec{r}, t) + \Sigma_\alpha \vec{\partial}_\alpha (\delta J_{\text{drift}}^{(\alpha)}(\vec{r}, t) + \delta J_{\text{diff}}^{(\alpha)}(\vec{r}, t)) = \mathbf{F}_\Xi(\boldsymbol{\nu}^0; \{\beta_\mu^0\}) \quad (4)$$

In the preceding expression, $\mathbf{F}_\Xi(\boldsymbol{\nu}^0; \{\beta_\mu^0\})$ is the Jacobian (Fréchet derivative) of the purely reactive flux $\hat{\boldsymbol{\Xi}}(\hat{\boldsymbol{\nu}}; \{\beta_\mu^0\})$, for the unbuffered reactants, *augmented* with $(m' + s')$ zeros—that is, equal to the number of buffered and inert components carrying charge in solution. Thus, whereas $\hat{\boldsymbol{\Xi}}$ is an

n-dimensional vector, Ξ is $(n + m' + s')$ dimensional. (The preceding Jacobian is, of course, evaluated at $\boldsymbol{\nu}^0$.) Using the usual hydrodynamic equation—parametrized by the transverse and longitudinal translational diffusion matrices—that expresses, to first order in perturbations/fluctuations, the divergence of the fluctuation of that part of the particle density current attributable to anisotropic translational diffusion alone, namely, of $\Sigma_\alpha \vec{\partial}_\alpha \delta \mathbf{J}_{\text{diff}}^{(\alpha)}(\vec{r}, t)$, then, from (2), (3), and (4), the time evaluation of the spatially Fourier-transformed perturbations/fluctuations can be written (denoting *spatial Fourier transformation* with an asterisk) as

$$\partial_t \, \delta^* \boldsymbol{\nu}(\vec{k}, t) = \boldsymbol{\Omega}(\vec{k}; \varepsilon, \{\beta_\mu^0\}, E_\parallel) \, \delta^* \boldsymbol{\nu}(\vec{k}, t) \tag{5}$$

where the wave vector has components $\vec{k} = (k_\perp(1), k_\perp(2), k_\parallel(3))$ and the complex linear operator $\boldsymbol{\Omega}$ is given explicitly by

$$\boldsymbol{\Omega} = [\mathbf{F}_\Xi(\boldsymbol{\nu}^0; \{\beta_\mu^0\}) - k_\perp^2 \mathbf{D}_\perp - (k_\parallel(3))^2 \mathbf{D}_\parallel + \left(\frac{4\pi}{\varepsilon}\right) \dot{\mathbf{M}}_\parallel - ik_\parallel(3) E_\parallel \mathbf{M}_\parallel] \tag{6}$$

In the preceding expression $k_\perp^2$ denotes the sum of squares of the two transverse wave numbers and the matrix $\dot{\mathbf{M}}_\parallel$ is $(n + m' + s')$ dimensional with the jkth term giving by $(\bar{m}_{jj} \nu_j^0 z_k)$. Note that the charge valences appearing in the Poisson equation must now be subscripted in a single sequence. Observe also that the Poisson equation guarantees that (5) is self-consistently obtained.

Clearly, there are at least $(m + 1)$ *free* parameters in $\boldsymbol{\Omega}$ that can be *externally* varied—namely, the externally buffered values $\{\beta_\mu^0\}$ of the n reactants for which the reaction system is maintained open and the intensity $E_\parallel$ of the external electric field. In particular cases (e.g., in enzyme-catalyzed reactions or in reactions where photochemical steps are of importance), $\{\beta_\mu^0\}$ could be augmented with other kinetic parameters that could be directly or indirectly varied by external means.

As regards the electric field intensity itself as an external parameter, we note, first, certain conclusions that can readily be deduced from an application of the theory of analytic functions of matrices.[6] Whenever sp $(\boldsymbol{\Omega})$, the spectrum of the complex operator $\boldsymbol{\Omega}$, is *distinct* (all eigenvalues having algebraic multiplicity 1), then every eigenvalue is a *holomorphic* function of the elements of $\boldsymbol{\Omega}$, and of $i\bar{m}_{jj} k_\parallel(3) E_\parallel$ in particular (and therefore of the electric field intensity), and thus can be expanded, by definition of holomorphicity, in a convergent power series. The second point that must be noted concerns certain consequences that follow if the $\boldsymbol{\Omega}$-operator is *globally* diagonalizable.[7] Consider a decomposition

$$\boldsymbol{\Omega}_\lambda = \boldsymbol{\Omega}_0 + \lambda \boldsymbol{\Omega}_1$$

where

$$\Omega_1 = -k_{\parallel}(3)\mathbf{M}_{\parallel}$$

and

$$\Omega_0 = [\mathbf{F}_{\Xi}(\boldsymbol{\nu}^0; \{\beta_{\mu}^0\}) - k_{\perp}^2\mathbf{D}_{\perp} - (k_{\parallel}(3))^2\mathbf{D}_{\parallel} + (4\pi/\varepsilon)\dot{\mathbf{M}}_{\parallel}]$$

If Ω_{λ} is diagonalizable for any choice of complex number λ, the spectrum of the Ω-operator can be shown to be representable, by a suitable ordering of the eigenvalues, in the form

$$\text{sp}(\Omega) = \text{sp}(\Omega_0) + iE_{\parallel}\,\text{sp}(\Omega_1) \tag{7}$$

and all eigenvalues are therefore complex, provided $(n - n') = 0$. A consequence of this result for the parametrization of the characteristic frequency, at threshold, of field-induced hard-mode instabilities is given in the next section. For now we merely remark that since sp (Ω_{λ}) is discrete, the distinctness of this spectrum (for all complex λ) is a necessary and sufficient condition for the global diagonalizability of Ω_{λ} and hence for (7) to hold. Furthermore, since Ω_1 is itself already diagonal, the global diagonalizability of Ω_{λ} implies that Ω_0 *commutes* with Ω_1 whenever no two (or more) values of the ionic mobilities $\bar{m}_{jj}$, $\bar{m}_{kk}$ $(j \neq k)$ are identical and $(n - n') = 0$.

III. ELECTRIC FIELD EFFECTS ON SOFT MODE AND HARD MODE INSTABILITIES DUE TO COMBINED REACTIVE AND CHARGE FLUCTUATIONS

It is clear from (5) that whether instabilities can arise owing to small perturbations or fluctuations depends on sp (Ω). A reference stationary state $\boldsymbol{\nu}^0$ remains stable to small perturbations/fluctuations whenever the spectrum of Ω lies entirely to the left of the imaginary axis (considered as the axis of ordinates) on the complex plane. To see this we merely have to note that in terms of the resolvent

$$\mathbf{R}_{\lambda}(\Omega(\vec{\mathbf{k}})) = [\text{diag}(\lambda) - \Omega(\vec{\mathbf{k}}; \varepsilon, \{\beta_{\mu}^0\}, E_{\parallel})]^{-1}$$

of the complex Ω-operator, the time evolution of the Fourier components of fluctuations with wave vector $\vec{\mathbf{k}}$ can be expressed as

$$\delta^*\boldsymbol{\nu}(\vec{\mathbf{k}}, t) = \left[\left(\frac{1}{2\pi i}\right)\oint_{c(\Omega(\vec{\mathbf{k}}))} d\lambda\,\mathbf{R}_{\lambda}(\Omega(\vec{\mathbf{k}}))\exp(\lambda t)\right]\delta^*\boldsymbol{\nu}(\vec{\mathbf{k}}, 0)$$

where $c(\Omega(\vec{k}))$ is a contour (Jordan curve) on the complex λ-plane enclosing sp $(\Omega(\vec{k}))$. Taking the spectral norm of both sides results in

$$\|\delta^*\nu(\vec{k}, t)\| \le \left(\frac{1}{2\pi}\right)\|\delta^*\nu(\vec{k}, 0)\|\oint_{c(\Omega(\vec{k}))} |d\lambda| \, \|\mathbf{R}_\lambda(\Omega(\vec{k}))\| \, |\exp(\lambda t)|$$

If sp $(\Omega(\vec{k}))$ lies entirely to the left of the half-plane Re $\lambda < \rho_\Omega$, then

$$\|\delta^*\nu(\vec{k}, t)\| \le \|\delta^*\nu(\vec{k}, 0)\|\left(\frac{\Gamma_\Omega}{2\pi}\right)\left(\max_{c(\Omega(\vec{k}))} \|\mathbf{R}_\lambda(\Omega(\vec{k}))\|\right) \exp(\rho_\Omega t)$$

where Γ_Ω is the length of the contour $c(\Omega(\vec{k}))$. Hence infinitesimal perturbations/fluctuations cannot grow whenever $\rho_\Omega < 0$. In this case $1/|\rho_\Omega|$ forms an *upper bound* on the *relaxation times* of all distinct eigenmodes. Of course, ρ_Ω is a function of the relevant wave vector, the number densities of the buffered reactions, the intensity of the external electric field, and the average dielectric constant of the solution.

A *soft mode* instability occurs whenever a real eigenvalue in sp (Ω) crosses the imaginary axis as an external free parameter is made to exceed some critical value, the rest of the spectrum remaining to the left of the imaginary axis. Such an instability corresponds to a zero frequency mode becoming critical for a particular set of *compatible* nonzero longitudinal and transverse wave numbers. In this case the reaction system, beyond criticality, breaks up spontaneously into a *time-independent* regularly repeating structure in space in the number densities of the reactants. The particular set of wave numbers dictates the characteristic size of the repeating spatial pattern; compatibility requires, among other things, that the pattern that emerges must repeat an integral number of times subject to the configuration of the boundary. Thus a soft mode instability induces a symmetry lowering in space alone in the number density distribution.

On the other hand, a *hard mode* instability occurs whenever either a complex eigenvalue or a complex conjugate pair (in both cases of algebraic multiplicity 1) in sp (Ω) crosses the imaginary axis at a finite rate as an external free parameter is made to exceed some critical value. Such an instability corresponds to a mode with finite frequency becoming critical for a particular set of compatible nonzero longitudinal and transverse wave numbers. In this case the reaction system, beyond criticality, breaks up spontaneously into a repeating pattern in space in the number densities of the n reactants—but for which, in each and every repeating unit, the time evolution in the number densities of these reactants is *periodic* with threshold frequency given by $\omega_{crit} = |\text{Im } \xi_{crit}(\Omega)|/2\pi$, where the numerator denotes the absolute value of the imaginary part of the complex eigenvalue(s) in sp (Ω) that become critical. The particular set of compatible longitudinal and transverse wave numbers would then again determine the characteristic size of the repeating spatial structure. In the case of a hard mode instability therefore, self-coherence is manifested as a *cellularization* of space (in terms of the number density distribution of the reactants), accompanied by temporal oscillations taking place in each "unit cell"—and thus exhibits symmetry lowering in *both* space and time in the number density distribution. *There is an important principle that dictates our interest in instabilities of the type described in the preceding paragraphs:* Whenever they occur a *lowering* of the spatiotemporal symmetry entails a *gain* in structure.

Thus, beyond criticality, for the two types of spatiotemporal transitions described above, infinitesimal perturbations or fluctuations elicit a *finite* response in the reacting system.[8,9] Such transitions are strongly self-coherent phenomena and occur on a macroscopic scale. Indeed, in the present case the characteristic (minimum) dimension of the *entire* reaction system has to exceed, first, the Debye–Hückel length (which, of course, is determined by the total ionic strength, the dielectric constant, and the temperature of the solution), so that it acts as a true electrolyte solution and not merely as a collection of free charges; and, second, the characteristic dimension of the whole reaction system has to exceed as well the intrinsic critical length below which translational diffusion alone would be capable of wiping out any spatiotemporal patterns that could arise. In dilute solutions this intrinsic critical length can be parametrized by the frequency of the chemical oscillations, the equivalent Stokes radius of the slowest diffusing reactant, and the viscosity and temperature of the solution.[10,11]

We note that in the *initial* approach to the threshold of instability, one can infer from classical theory[12] that the Fourier components of the fluctuations will have relaxation times diverging as the reciprocal of the real part of the eigenvalue(s) in sp $(\Omega(\vec{k}))$ that become critical. The two-time correlation function of the Fourier components of fluctuations with wave vector $\vec{k}$ is given by (again in terms of Dunford integrals)

$$\langle \delta^* \boldsymbol{\nu}(\vec{k}, t)\, \delta^* \boldsymbol{\nu}^\dagger(\mathbf{k}, 0) \rangle =$$

$$\left[\left(\frac{1}{2\pi i} \right) \int_{c(\Omega(\vec{k}))} d\lambda\ \mathbf{R}_\lambda(\Omega(\vec{k})) \exp(\lambda t) \right] \langle \delta^* \boldsymbol{\nu}(\vec{k}, 0)\, \delta^* \boldsymbol{\nu}^\dagger(\vec{k}, 0) \rangle \qquad (8)$$

And from the known[13] spectral dependence of the elements of the static correlation function in the preceding expression, it straightforwardly follows that the initial amplitude of the Fourier components of the growing fluctuations will go as the reciprocal of the real part of the critical eigenvalue; or, if there are more than one of them, of the product of their real parts. Defining $\Lambda(\vec{k}; \vec{k}') = \langle \delta^* \nu(\vec{k}, 0) \, \delta^* \nu^\dagger(\vec{k}', 0) \rangle$, the general correlation function for the Fourier components of the fluctuations can be written as

$$\langle \delta^* \nu(\vec{k}, t) \, \delta^* \nu^\dagger(\vec{k}', t') \rangle = \left[\left(\frac{1}{2\pi i} \right) \oint_{c(\Omega(\vec{k}))} d\lambda \, \mathbf{R}_\lambda (\Omega(\vec{k})) \exp(\lambda t) \right] \Lambda(\vec{k}; \vec{k}')$$

$$\times \left[\left(\frac{1}{2\pi i} \right) \oint_{c(\Omega(\vec{k}'))} d\lambda \, \mathbf{R}_\lambda (\Omega(\vec{k}')) \exp(\lambda t') \right]^\dagger \qquad (9)$$

If one had full knowledge of Ω—and of its spectral properties—for a particular system, it would in principle be possible to study at the threshold of instability the initial behavior of the critical fluctuations. Of course, neither the divergence of the critical relaxation times nor that of the critical amplitudes of the growing Fourier components of the fluctuations can become infinite as the critical eigenvalue(s) cross the imaginary axis. This is owing to the nonlinearity of the system.

In the full (three-dimensional) $\vec{k}$-space, given ε, ν^0, $\{z_k\}$, $\mathbf{M}_{\|}$, $\mathbf{D}_{\|}$ and $\mathbf{D}_{\perp}$, a specific set of values of the external parameters $\{\beta_\mu^0\}$ should prescribe, through the equation $\det \mathbf{R}_\lambda^{-1}(\Omega(\vec{k})) = 0$ nested surfaces parametrized by $E_{\|}$. Denote the (spatial) dimensions of the reaction system as (L_1, L_2, L_3). Each surface of constant $E_{\|}$ has isolated point(s) of intersection with the set of wave vectors $\vec{k}_L = (2\pi N_r/L_1, 2\pi N_r/L_2, 2\pi N_r/L_3)$ for all $N_r = \pm 1, \pm 2, \ldots$, up to an *upper cutoff* dictated by the relevant Debye–Hückel length for the solution. A set of such points, denoted by $\{\vec{k}_c(E_{\|})\}$, and parametrized by a particular $E_{\|}$, forms the set of what was spoken of in earlier paragraphs as *compatible* wave vectors. The collection of such points, for all values of $E_{\|}$ in the interval $0 < |E_{\|}| \le |E_{\|}^{\max}|$ is denoted in the sequel as $\{\vec{k}_c(E_{\|})\}_N$. Note that in this work we deliberately restrict our attention to instabilities with *real* wave vectors. This restriction will be relaxed in a subsequent paper.

We are now in a position to proceed to study the effects of a static electric field on soft mode and hard mode instabilities due to combined reactive and charge fluctuations. The real part of sp (Ω) can be bounded as

$$\xi_s^0 \le \text{Re} \{\text{sp}(\Omega)\} \le \xi_s^{00} \qquad (10)$$

where ξ_s^0 and ξ_s^{00} are, respectively, the lower and upper spectral bounds of the real symmetric operator $\Omega_s = \frac{1}{2}(\Omega + \Omega^\dagger)$,

$$\Omega_s = \frac{1}{2}[\mathbf{F}_{\equiv}(\nu^0; \{\beta_\mu^0\}) + \mathbf{F}_{\equiv}^\dagger(\nu^0; \{\beta_\mu^0\}) - 2k_\perp^2 \mathbf{D}_\perp$$

$$- 2(k_{\|}(3))^2 \mathbf{D}_{\|} + \left(\frac{4\pi}{\varepsilon}\right) (\dot{\mathbf{M}}_{\|} + \dot{\mathbf{M}}_{\|}^\dagger)] \qquad (11)$$

From the earlier paragraphs of this section it is clear that given a particular set of parameters $\{\beta_\mu^0\}$, and $\mathbf{D}_\perp$ and $\mathbf{D}_\parallel$, no soft mode or hard mode instability corresponding to a particular spatial pattern (parametrized by a set of compatible nonzero wave numbers) can arise if ξ_s^{00} is not greater than zero.

Several conclusions follow directly from (10) and (11). In the case of a very weak imposed external field, such that the anisotropy of translational diffusion can be neglected, $(\mathbf{D}_\perp \simeq \mathbf{D}_\parallel)$, the inception from a given reference stationary state of either a soft mode or hard mode instability should be *relatively* insensitive to the field and should depend much more on the external "chemical" parameters $\{\beta_\mu^0\}$, the kinetics of the chemical transformations (through the Jacobian), and the ionic and transport properties of the solution. This is so since the real part of sp $(\mathbf{\Omega})$ is bounded above by ξ_s^{00}, which is the upper spectral bound of an effectively *field-independent** real symmetric operator. Notice though that the *type* of instability that would first arise, as an external parameter is made to exceed a critical value, could very well—and very likely would—be changed by the presence of the field, even a weak one. For it is known from the theory of analytic perturbations of linear operators that the eigenvalues in the spectrum of an operator such as $\mathbf{\Omega} = \mathbf{\Omega}_0 + iE_\parallel \mathbf{\Omega}_1$ are (in general, multiple-valued) branches of analytic functions of $iE_\parallel$. As the electric field is switched on, the eigenvalues of $\mathbf{\Omega}_0$ (the zero-field operator) can split and coalesce in a very complicated manner; and, in fact, if the algebraic multiplicities are not taken into account, the number of eigenvalues change as $E_\parallel$ is varied; and so would the number of distinct (nondegenerate) eigenmodes. In any case it is shown in Appendix 3 that it is possible to *calculate how weak field intensities have to be in order to preserve the sum of spectral multiplicities within a minimal contour on the complex λ-plane defined by the spectrum of the zero-field operator.*

If the external field is not too weak, $\mathbf{\Omega}_s$ has an explicit field dependence since now $\mathbf{D}_\perp \neq \mathbf{D}_\parallel$, and the divergence of $(\delta_\parallel^{jk}/\delta_\perp^{jk})$ from unity varies with external field strength. If this field dependence is known for a particular chemical reaction system, the influence of the operator $\mathbf{D}(\vec{k})$, $\mathbf{D}(\vec{k}) = -2[k_\perp^2 \mathbf{D}_\perp + (k_\parallel(3))^2 \mathbf{D}_\parallel]$, on the spectral properties of $\mathbf{\Omega}_s$, as a function of field strength, can be calculated for this particular reaction system. In the general case, however, we are interested here only in the question of whether a sufficiently strong field can induce a soft mode or hard mode instability to occur. Recall that from physical considerations,[14] translational diffusion matrices can be expected to be positive definite (this

---

* This effective field-independence obtains whenever the *sum* $[k_\perp^2 + (k_\parallel(3))^2]$ is only weakly dependent on $|E_\parallel|$.

being immediately obvious for dilute solutions, for which the off-diagonal (cross-diffusive) terms are negligibly small) and strongly dominated by their diagonal elements, which are positive. Owing to this fact, the contribution of $\mathbf{D}(\vec{\mathbf{k}})$ to $\mathbf{\Omega}_s$ is inherently "stabilizing" in the sense that given any set $\{\beta_\mu^0\}$ and replacing $\mathbf{D}(\vec{\mathbf{k}})$ by $\theta\mathbf{D}(\vec{\mathbf{k}})$ in (11) where θ is any positive constant, a decrease in θ will always result in an increase in the upper spectral bound ξ_s^{00}, and vice versa. It is therefore entirely possible that given a set of chemical parameters for which ξ_s^{00} is already near criticality, the imposition of a static electric field of sufficient strength could cause $\mathbf{D}(\vec{\mathbf{k}})$ to become less stabilizing, and sufficiently so that ξ_s^{00} becomes positive. This can be proved rigorously.

Turning to the imaginary part of sp $(\mathbf{\Omega})$, one can again deduce that

$$\xi_h^0 \le \mathrm{Im}\,\{\mathrm{sp}\,(\mathbf{\Omega})\} \le \xi_h^{00} \tag{12}$$

where ξ_h^0 and ξ_h^{00} are the lower and upper spectral bounds, respectively, of the hermitian operator $\mathbf{\Omega}_h = (1/2i)(\mathbf{\Omega} - \mathbf{\Omega}^\dagger)$,

$$\mathbf{\Omega}_h = \left(\frac{1}{2i}\right)[\mathbf{F}_\Xi(\mathbf{\nu}^0; \{\beta_\mu^0\}) - \mathbf{F}_\Xi^\dagger(\mathbf{\nu}^0; \{\beta_\mu^0\}) + \left(\frac{4\pi}{\varepsilon}\right)(\dot{\mathbf{M}}_\| - \dot{\mathbf{M}}_\|^\dagger) - 2ik_\|(3)E_\|\mathbf{M}_\|] \tag{13}$$

From this one sees that if a hard mode instability is induced in a reaction system either through the use of the parameters $\{\beta_\mu^0\}$ or through the imposition of a sufficiently strong field, or both together, the resulting frequency of oscillations in each spatially repeating unit is *field dependent*. This is clear since ξ_h^0 and ξ_h^{00} are the spectral bounds of an *explicitly field-dependent* hermitian operator. Writing $\xi_h^{**} = \min\,(|\xi_h^0|, |\xi_h^{00}|)$, the frequency of oscillations admits at threshold the upper bound $\bar{\omega}_{\mathrm{crit}} = \xi_h^{**}/2\pi$. This frequency is, of course, determined by the usual chemical, ionic, and transport properties in addition to the electric field intensity. Notice that the operator $\mathbf{\Omega}_h$ is independent of the transverse wave numbers. A further analysis of the induction of instabilities is given in Appendix 4.

If the $\mathbf{\Omega}_\lambda$-operator defined in the last section is globally diagonalizable for all complex λ, then the representation (7) exhibits most transparently how the electric field intensity would parametrize—keeping in mind that sp $(\mathbf{\Omega}_1)$ is entirely real—the threshold frequency of oscillations of field-induced hard mode instabilities. It should, of course, be clear that whenever the $\mathbf{\Omega}_\lambda$-operator is globally diagonalizable, soft mode instabilities can *never* be field induced if $(n - n') = 0$.

Some further comments: (*1*) The question of the detailed stability properties of the structures that emerge beyond any of the instabilities whose existence we have established would be quite difficult to incorporate into a global theory like the one developed here. Resolution of such questions could best be accomplished on an individual basis in the context of a given particular reaction system, and taking into account the boundary configuration. (2) The theory as developed here should be most appropriate in the case where moderate field strengths (up to around the order of a couple of hundred volts per centimeter) are involved. This is the reason why the field dependence of the ionic mobilities as well as the effect of Joule heating are both neglected. In the same spirit we have, for instance, not taken into account the second Wien effect—on the assumption that the field strengths of interest lie much below the lower limit (in the order of kilovolts per centimeter) where field dissociation of any weakly ionized components can no longer be neglected. (3) An approximation to the present analysis—appropriate for dealing with very complicated multicomponent solutions (with šufficiently high ionic strength) in which the type of *inert* ionic components cannot be specified or characterized completely (except for the one fact that they are simple ions) but are nevertheless known to make the principal contribution to the total ionic strength—is to drop the *"dielectric" contribution* $[(4\pi/\varepsilon)\dot{\mathbf{M}}_\|]$ when analyzing the complex Ω-operator and to collapse the dimension of the problem to n. This approximation has, for instance, been employed in the theoretical analysis of field-induced effects in the cell-free glycolytic system.[15]

IV. BIOPHYSICAL IMPLICATIONS OF THE THEORY

From the preceding theoretical results, it is clear that as a consequence of the switching on of a longitudinal electric field, the effect of combined reactive and charge fluctuations on the *induction* of instabilities can be quite radically different and it should be possible to field-induce certain hard mode spatiotemporal structures that can never be made to emerge in the absence of the field. (Such has been shown to be the case for glycolytic instabilities.[15]) The underlying reason is, of course, the transition of the Ω-operator from real to complex as the field is switched on (Section III).

In the case where an external field is applied to a reaction system where a coherent spatiotemporal pattern is already present even before the field is switched on, the result could be any one of the following. (1) The switching on shifts the spectrum of the resultant (complex) Ω-operator to the left of the imaginary axis and there is no coherent pattern for the given field strength. This can happen only fortuitously. (2) The "same" critical mode (as the one for $E_\| = 0$) remains the critical mode under conditions of finite $E_\|$. This would be an unlikely occurrence but is entirely possible if the eigenvalue belonging to the zero-field critical mode does not split off as the field is switched on. Note that the modulus $|\xi_i(0) - \xi_i(iE_\|)|$ of an eigenvalue that does not become degenerate as the field is switched on has value $\sim 0(|iE_\parallel|) = 0(|E_\parallel|)$. (3) The original critical mode is replaced by a different eigenmode, so that a completely new spatiotemporal pattern is established for the given nonzero $E_\parallel$. This can be expected to be the most common outcome. (In all these, keep in mind that, as mentioned in Section III, the eigenvalues of the Ω-operator are multiple-valued branches of analytic functions of $iE_\parallel$.)

In all cases, whether an existing coherent spatiotemporal pattern in the number density distribution for nonzero $E_\parallel$, the analysis in the last two sections allows us to infer *three principal large-scale effects*, three effects *whose existence is directly due to the presence of the field*: First, there is a shift in the frequency of oscillations, a frequency shift that can be modulated by varying the field strength. Second, the characteristic size of any spatially repeating pattern can be modulated by varying the field strength. Third, a distinct effect is manifested by reversal of the polarity of the field—for on reversal, the operator $[-2ik_\parallel(3)E_\parallel \mathbf{M}_\parallel]$ suffers a change in sign as a consequence of the reversal, and this operator then does not give the same spectral contribution to Ω_h. Thus, in general, and using an obvious notation, sp $(\Omega_h(\uparrow)) \neq$ sp $(\Omega_h(\downarrow))$—so that the frequency shift effect of a ($\uparrow$)-field can be expected to differ from that of a ($\downarrow$)-field.

There are several different levels at which the three effects predicted above can be experimentally verified. Only a brief account is given here of some possible experimental approaches, but more detailed suggestions will be published elsewhere. First of all, verification can be carried out at the fundamental level of fairly well-characterized chemical reaction systems capable of sustaining dissipative structures in aqueous solution. The likeliest ones that come immediately to mind are the Belousov-Zhabotinskii reaction[16] and the system of cell-free glycolytic reactions.[17,18] In both these systems one or more of the reactants carry net charge in solution either as simple ions or as polyelectrolytes. A first step has been taken in Ref. 15 in an attempt to predict theoretically, using an allosteric model for the phosphofructokinase reaction,[19,20] what the detailed experimental findings will be in the cell free glycolytic system. This is an essential first step, since it leads us naturally to the next level of experimental verification of the three effects predicted in this paper, namely, that of biological periodicities (rhythms).

The view is held by many—see Ref. 21 for an introduction and survey—that at the heart of higher-frequency as well as circadian oscillations are biochemical oscillators (at least for nonphotosynthetic cells). Among the ones that have been studied to date, the set of reactions constituting the glycolytic pathway still remains one of the likelier candidates for the role of the principal oscillator. Theoretically, it has been shown that under certain specific conditions the period of oscillation of the cell-free glycolytic system can approach circadian values.[20,22] But even if it turns out that it is not one oscillator but a larger, interconnected network of biochemical reactions that is involved, having some ionic or polyelectrolyte reactants, then the analysis in this paper rigorously shows that at the molecular level a frequency shift effect in biological rhythms can be induced by and modulated with an external static electric field. Furthermore, because of the asymmetry implied in the third predicted effect, an *alternating* homogeneous electric field would have a more pronounced effect on biological periodicities and rhythms.

It must be pointed out that model systems for circadian rhythms, at the level of the single eukaryotic cell, other than those depending exclusively on biochemical reactions, have been proposed—most notably, compartmentalized membrane models.[23,24] It cannot, and should not, be asserted without further calculations that these other models will, when placed in an electric field, exhibit the analogous effects predicted in this paper. On the other hand, this circumstance may turn out to be all for the better, since the effects predicted here—most importantly the third one, involving the asymmetry in the frequency shift of the oscillations, and its corollary that an alternating-polarity field will have a more drastic effect (even not taking into account the displacement current) than a static field of equal intensity—can then serve as *diagnostic* tools to help discriminate between different proposed model systems. Some further remarks on this point will be published elsewhere. But, in the last few paragraphs of this section, an additional account is given of an example of circadian rhythms mediated by the central nervous system—and of how these rhythms, when one knows that the relevant neurotransmitter molecules are charged, furnish another mechanism for field-induced shifts in circadian periodicity, but at a higher level of cellular organization than that discussed so far.

The fullest interplay between the first and second effects should in principle be observed in experiments dealing with embryogenesis under the influence of external electric fields. This interplay is due to the fact that here we have processes occurring in which mitotic activity (with all the inherent periodicities in the biochemical events that this term implies) is intimately linked with morphogenesis. In Appendix 1 we propose two concrete experimental model systems involving the field-induced modulation of the developmental events in echinoderms and dipterans. If, on the other hand, one is interested in the effect of electric fields on mitotic activity alone, it seems one could best experiment with synchronously growing cell cultures—normal or malignant—to see whether shifts occur in the frequencies of the chemical events that together constitute the cell cycle, including polarity-dependent shifts.

An intriguing idea is to find a way to use the third effect predicted here to completely *invert* a given morphogenetic axis (e.g., the dorsoventral) in developing organisms, in much the same way that this has actually been accomplished in amphibian embryos by *selective* restriction of the oxygen supply.[25] Should this type of field-induced inversion be proved experimentally feasible, it could furnish a versatile probe for the study of coherent spatiotemporal patterns in developmental biology. (Dr. J. Hiernaux did kindly draw my attention subsequently to the fact that electric field-induced inversion of the longitudinal morphogenetic pattern in *Acetabularia* has indeed been unambiguously observed in recent experiments.[26]) Appendix 5 should be consulted for more detail.

The results derived in this paper could provide a mechanism as to how neurophysiological interactions would be modulated in the large by an external electric field. The level of activity of neural masses can be viewed as being parametrized, first, by the *spatiotemporal synthesis and destruction* of those excitatory and inhibitory neurotransmitter molecules (e.g., acetylcholine, γ-aminobutyric acid, dopamine, adrenaline, noradrenaline, 5-hydroxytryptamine, and others) that biochemically mediate neuronal (synaptic) interactions in a particular *type* of neural mass; and, second, by the known concomitant role of Ca^{+2} ions.[27] (By the term *type* is simply meant here whether, for instance, the neural mass of interest is composed of cholinergic neurons and/or adrenergic neurons, and so on.) Some of the known neurotransmitter molecules (not to mention Ca^{+2}) carry net charge in solution, as do the several enzymes associated with acetylcholine and monoamine metabolism (e.g., acetylcholinesterase, cholinacetylase, cholinephosphokinase, monamine oxidase, catechol-O-methyl transferase, and others), as well as other metabolites such as the brain peptides that are now known to act on a fairly large scale as neurohormones.[28] Thus it would seem that dc-polarization of, say, the cerebral cortex must have among its many interrelated and overlapping effects one that is due to the frequency shift mechanism discussed here.

For discussions of—and references to the literature on—frequency shifts in cortical polarization experiments due to static as well as modulated fields, see the contributions of Morrell and of Adey in Ref. 29. (It is worth noting that in experiments with *M. nemestrina*, where the field is applied through an intervening dielectric—air—it has been found that the reproducible effect of a field of intensity as small as 0.1 volt/cm modulated at 7 Hz was to *alter* the subjective *time scale* with which the animal measures the *passage of time*.[30] This can certainly be interpreted as a shift in the frequency of a fundamental oscillator. The fact that such a frequency shift can be effected with an electric field argues strongly for a biochemical type of oscillator with charged reactants.)

Considering the third predicted effect once more by itself, one finds that in actual cortical polarization experiments the frequency modulation is in fact *asymmetric* (see the relevant references in Ref. 29), that is, the observed frequency shifts are *not* identical when the electrode polarities are reversed; thus the names *anodal* and *cathodal* polarization are often found in the literature on cortical polarization experiments. However, a further subtlety is involved because the fields used are usually inhomogeneous fields. It can be expected that the asymmetric effect (on the frequency shifts) due to polarity reversal of nonuniform electric fields would be much more pronounced, since, as can be shown, in the case where not all the reactants are charged, relation A.13 below fails to be satisfied. Hence an *additional* intrinsic asymmetry inherent in inhomogeneous fields is manifested by polarity reversal. A later paper will clarify this point.

We now turn finally to what the present theory would lead us to conclude regarding the effect of *hypothalamic polarization*. In the intact higher organism there are in fact many functional variables that oscillate with a circadian period—the number being of the order of 50^{31} in man. A most important endogenous circadian rhythm,[32] connected with the diurnal cycle of waking and sleeping, is in fact found in the neurally mediated release of CRF (corticotrophin-releasing factor), and therefore of ACTH (adrenocorticotropic hormone) and thence of various corticosteroids.[33] Since ACTH release is neurally mediated through neurotransmitter regulation, it should not be surprising that the (localized) concentrations of various neurotransmitter molecules in higher animals should have *circadian* periodicity. This has actually been found to be the case for acetylcholine,[34] noradrenaline,[35] and 5-hydroxytryptamine.[36] Conversely, it has been possible to show that changing the concentration or blocking neurotransmitter action can abolish circadian oscillation in plasma corticosteroid content. Specifically, in the case of 17-hydroxycorticosteroids, this fact has already been verified for acetylcholine[37] and 5-hydroxytryptamine.[38]

Thus it strongly appears that circadian oscillations in the concentrations of corticosteroids are due to the regulatory role of the central nervous system in the periodic synthesis and release of CRF and ACTH. Since all the neurotransmitter molecules enumerated above carry charge, the same conclusions follow—marshalling the same arguments are those used for the cortical polarization experiments discussed before—as ·to the effects of static electric fields acting on the hypothalamus and areas adjacent to it. One can expect that hypothalamic polarization can induce changes in the periodicities of the corticosteroid cycle, and hence of the circadian rhythms and functions connected with and controlled by it (previous paragraph), in higher animals. Moreover, the present theory predicts that a reversal of the polarity of the field in the hypothalamic-limbic area of the brain (detailed note forthcoming) will induce an unequal change in the frequency modulation of the corticosteroid cycle for an identical absolute value of the field intensity. Note that in the same way as in cortical polarization experiments, one cannot hope to have the polarizing field homogeneous; therefore the remarks made earlier concerning the additional intrinsic asymmetry inherent in inhomogeneous fields apply here as well. (Consult Appendix 5.)

In a later work the analysis begun here will be completed by treating several aspects of field-induced chemical wave propagation in complex multicomponent reacting systems.

Acknowledgments

For the opportunity to contribute to this volume in honor of the sixtieth birthday of Professor Ilya Prigogine, it is a pleasure for me to thank Professor Stuart A. Rice. I am also grateful to Martina Grafmüller for her help in the preparation of the manuscript. This study was undertaken with the support of the Robert A. Welch Foundation and of the *Instituts Internationaux de Physique et de Chimie*.

APPENDIX 1. EXPERIMENTAL MODEL SYSTEMS

As was stated in the text, the fullest interplay between the first and second effects predicted in this paper should in principle be observed in experiments dealing with embryogenesis under the influence of external electric fields. Here we have processes occurring in which *mitotic activity* (with all the inherent periodicities in the biochemical events that this term implies) is inextricably linked with *morphogenesis*. Two experimental model systems are suggested here: (1) fertilized or artificially activated eggs of echinoderms, of sea urchins in particular, which are well characterized[39] and for which the biochemical changes accompanying development are quite well known;[40,41] and (2) eggs of dipterans (e.g., of *Protophormia* or *Drosophila*) for which the normal spatiotemporal patterns and structures arising in early embryonic development have already been studied.[42]

The role of simple ions, in particular K^+, in the developmental process of sea urchins is well documented[41,43] and there is no doubt that other charge-carrying reactants (ions as well as polyelectrolytes) must play significant roles in the overall process of embryogenesis in dipterans as well. (On the other hand, in the development of *Fucus* in seawater, the flux of Ca^{+2} is known to crucially determine the polarity of the developing organism.) There is one reason why it can be argued that the use of dipteran eggs might be a somewhat better-posed experiment. In embryogenesis, dipteran eggs undergo *superficial cleavage* so that spatiotemporal patterns and structures associated with development essentially arise even at the unicellular level—so that the first two effects predicted here would manifest themselves at a correspondingly early stage.

For the first model system proposed in this Appendix, confinement in the same type of seawater-filled cuvette (or capillary) capped at both ends with BSA-coated electrodes (Appendix 2) would be most appropriate; whereas for the second model system, use of capillary electrodes filled with conducting agar gel—of the type often used in cortical polarization experiments—might prove better.

The distance between the cap electrodes (in the first model system) and between the gel electrodes (in the second system) is dictated, of course, by the diameter of the egg. In most cases the effective intensity of the resultant electric field inside the egg is not completely determinable because of the presence of the outer membrane. But the applied potential difference between the electrodes should be very low, *taking into account both the outer membrane surface conduction and the electrode overpotential.*

APPENDIX 2.   EXPERIMENTAL PROTOCOL

A.   Appropriate Electrodes

The type of electrodes employed is crucial in the implementation of experiments designed specifically to study the effects for nonequilibrium systems predicted in this paper. It is quite clear that platinized platinum electrodes (i.e., black platinum electrodes) would be most suitable (see, however, Appendix 1) because of their inert surface, coupled with their inherently low electrode polarization and large surface capacitance. In experiments dealing with biological systems, it would be imperative to avoid the occurrence of redox or other types of surface reactions, whether gaseous or not.

Black platinum electrodes are excellent in the sense that gaseous surface reactions are eliminated, at least up to the moderate field strengths of interest to us here ($E_{\parallel}^{max} \sim 100$ volts/cm). Based on recent work,[44,45] it is possible to eliminate other surface reactions as well, reactions that would be intolerable in experiments with biological model systems. The method requires the coating of the black platinum surface with a porous monolayer coating of BSA (bovine serum albumin[44]). This is accomplished by simple dipping of the electrodes in a 10% solution of BSA for several minutes and washing with distilled water. The electrodes so treated can be stored in distilled water as well, and the integrity of the BSA monolayer can be maintained by simple redipping in BSA solution. It has been found experimentally[45] in laser Doppler spectroscopy work that such BSA-coated black platinum electrodes completely eliminate the temporal decay of the Doppler spectrum attributable to the nongaseous electrode surface reactions.

B.   Length of the Experimental System and Joule Heating

The electric field intensity $E_\parallel$ in the solution is, of course, given by the difference between the total potential (applied to the electrodes) and the electrode overpotential, the difference being divided by the gap distance between the electrodes. For a given strength of the electric field $E_\parallel$ in the solution, the rise in temperature at the midpoint between the electrodes is directly proportional to the distance between the electrodes and is known[46] to go as $\sim d_e^2 E_\parallel^2 / r_e K$, where d_e is the distance between the electrodes, r_e the electrical resistivity and K the thermal conductivity of the solution.

It is therefore appropriate to use the shortest possible distance between the electrodes consistent with the scale of the structures and patterns being studied, so as to reduce Joule heating to a minimum. This rule would, at the same time, reduce to a minimum the possibility of the occurrence of the attendant convection instabilities. In biological work it would in fact be imperative to have some type of thermostatic control of the experimental cuvette even at low field strengths.

APPENDIX 3.   CALCULATION OF FIELD STRENGTHS PRESERVING THE SUM OF SPECTRAL MULTIPLICITIES OF THE ZERO-FIELD OPERATOR

Consider the smallest possible contour (closed Jordan curve) on the complex λ-plane that is capable of enclosing the entire spectrum of the zero-field operator $[\Omega_0(\vec{k})]$ defined in Section III, for all possible variations in the wave vector in some neighborhood of the set of points $\{\vec{k}_c(E_\parallel)\}_N$ constituting the set of compatible wave vectors in $\vec{k}$-space (Section III). Recall that in defining $\{\vec{k}_c(E_\parallel)\}_N$, we have let $|E_\parallel|$ range from zero to the maximum field strength of interest to us. As discussed in Section III, the set of compatible wave vectors are determined not only by $E_\parallel$ but also, among other parameters, by the finite dimensions of the system as well as by the upper cutoff imposed by the relevant Debye–Hückel length. Here we denote the minimal contour by C_0^* and its length by Γ_0^*.

Since this minimal contour is chosen to contain the entire spectrum of the bare zero-field operator for all $\vec{k} \in \{\vec{k}_c(E_\parallel)\}_N$, the sum of spectral multiplicities within it for $[\Omega_0(\vec{k})]$ is exactly equal to $(n + m' + s')$—which is also equal to the "total" number of eigenmodes, provided degenerate eigenmodes are counted repeatedly according to their degeneracies. A field of strength $E_\parallel$ is considered weak, *in the spectral sense*, if upon switching it on the resulting *full* Ω-operator has an identical sum of spectral multiplicities *within* the minimal contour defined earlier for the zero-field operator. When switched on therefore, a weak field has the property of keeping invariant the sum of spectral multiplicities contained within the specified contour C_0^*. It necessarily follows that for such fields the *entire* spectrum of the full $\boldsymbol{\Omega}$-operator lies within C_0^*. We wish to investigate in this Appendix what condition (on the magnitude of the intensity) is necessary so that the switching on of an electric field still allows the preservation of the sum of spectral multiplicities within the contour C_0^*, in the exact sense defined above.

We take the space of Fourier components of the reactive and charge fluctuations as a complex Banach space $\mathscr{F}$ in which the norm employed is the *uniform norm* for operators in the space. The sum of spectral multiplicities within C_0^* for the bare zero-field operator is given by

$$n^*(\boldsymbol{\Omega}_0(\vec{\mathbf{k}})) = (n + m' + s') = \dim \mathscr{P}_{\Omega_0/C_0^*} \mathscr{F} \tag{A.1}$$

where $\mathscr{P}_{\Omega_0/C_0^*}$ is the eigenprojection operator commuting with $\boldsymbol{\Omega}_0(\mathbf{k})$:

$$\mathscr{P}_{\Omega_0/C_0^*} = (\tfrac{1}{2}\pi i) \oint_{C_0^*} d\lambda \ \mathbf{R}_\lambda(\boldsymbol{\Omega}_0(\vec{\mathbf{k}})) \tag{A.2}$$

One can define similarly the eigenprojection operator commuting with the full (i.e., $E_\| \neq 0$) $\boldsymbol{\Omega}$-operator as

$$\mathscr{P}_{\Omega/C_0^*} = (\tfrac{1}{2}\pi i) \oint_{C_0^*} d\lambda \ \mathbf{R}_\lambda(\boldsymbol{\Omega}(\vec{\mathbf{k}})) \tag{A.3}$$

Furthermore, define by ρ_0^* the quantity

$$\rho_0^* = \max_{\lambda \in C_0^*} \|\mathbf{R}_\lambda(\boldsymbol{\Omega}_0(\vec{\mathbf{k}})\|, \qquad \text{for all } \vec{\mathbf{k}} \in \{\vec{\mathbf{k}}_C(E_\|)\}_N \tag{A.4}$$

We shall need to evaluate the norm of the difference of the two eigenprojection operators (A.2) and (A.3), which can be shown to be expressible as

$$\|\mathscr{P}_{\Omega_0/C_0^*} - \mathscr{P}_{\Omega/C_0^*}\| = (\tfrac{1}{2}\pi) \left\| \oint_{C_0^*} d\lambda \ \mathbf{R}_\lambda(\boldsymbol{\Omega}(\vec{\mathbf{k}})) \sum_{\alpha'=1}^\infty \{\|ik_\|(3)E_\|\mathbf{M}_\|\|\mathbf{R}_\lambda(\boldsymbol{\Omega}_0(\vec{\mathbf{k}}))\}^{\alpha'} \right\| \tag{A.5}$$

with $\vec{\mathbf{k}} \in \{\vec{\mathbf{k}}_c(E_\|)\}_N$. This gives

$$\|\mathscr{P}_{\Omega_0/C_0^*} - \mathscr{P}_{\Omega/C_0^*}\| \leq \left(\frac{\Gamma_0^*}{2\pi}\right) \max_{\substack{\lambda \in C_0^* \\ \vec{\mathbf{k}} \in \{\vec{\mathbf{k}}_C(E_\|)\}_N}} \left\{ \frac{\|ik_\|(3)E_\|\mathbf{M}_\|\| \|\mathbf{R}_\lambda(\boldsymbol{\Omega}_0(\vec{\mathbf{k}}))\|^2}{1 - \|ik_\|(3)E_\|\mathbf{M}_\|\| \|\mathbf{R}_\lambda(\boldsymbol{\Omega}_0(\vec{\mathbf{k}}))\|^2} \right\} \tag{A.6}$$

For the electric field when switched on to keep invariant the sum of spectral multiplicities within C_0^*, it would be necessary to have

$$\dim \mathscr{P}_{\Omega/C_0^*} \mathscr{F} = (n + m' + s') \tag{A.7}$$

It is known from the classic theorem due to Sz.-Nagy[47] that, for the Banach space $\mathscr{F}$, a necessary and sufficient condition for (A.7) to be satisfied—taking note of (A.1)—is that the norm of the difference between the two eigenprojection operators $\mathscr{P}_{\Omega_0/C_0^*}$ and $\mathscr{P}_{\Omega/C_0^*}$ be less than unity. Making use of (A.6), this requirement leads directly to the following computationally accessible condition:

$$\|ik_\|(3)E_\|M_\|\| \leq 2\pi[\rho_0^*(2\pi + \rho_0^*\Gamma_0^*)]^{-1} \tag{A.8}$$

The preceding inequality constitutes a sufficient condition under which, despite the fact that the electric field is switched on, the entire spectrum of the full Ω-operator would nonetheless lie within the same minimal contour C_0^* defined by the zero-field operator. In particular, if the minimal contour lies entirely to the left of the imaginary axis, then fields that do not violate the inequality (A.8) are incapable of inducing a symmetry-lowering instability (neither soft mode nor hard mode). It will not escape the reader's attention that the condition (A.8) is determined by the fullest interplay between the ionic, kinetic, and transport properties of the reacting system. Since only weak fields can preserve spectral multiplicities in the manner described here, account has not been taken in (A.8) of the field dependence of the ratios $(\delta_\|^{jk}/\delta_\perp^{jk})$ of the elements of the longitudinal and transverse diffusion matrices.

APPENDIX 4. FURTHER REMARKS ON THE FIELD INDUCTION OF INSTABILITIES

We take in this appendix a different point of view—different from the predominantly *spectral* viewpoint that has been adopted so far—in order to demonstrate most clearly and concisely how the switching on of a sufficiently strong electric field can induce either a soft mode or hard mode instability in an otherwise stable reacting system.

It can be shown quite directly that for sp (Ω) to lie entirely to the left of the imaginary axis we must have

$$\text{Re}\,(\Omega^{jj}) < -\sum_{\substack{k=1 \\ k \neq j}}^{n+m'+s'} |\Omega^{jk}|, \qquad \text{for each } j \tag{A.9}$$

$j = 1, 2, \ldots, (n + m' + s')$. Since the mobility matrix does not have off-diagonal elements, this condition reduces, for each j, to

$$\left\{ \phi_{\underline{\underline{=}}}^{jj} - k_\perp^2 \delta_\perp^{jj} - (k_\|(3))^2 \delta_\|^{jj} + \left(\frac{4\pi}{\varepsilon}\right) \bar{m}_{jj} \nu_j^0 z_j \right\}$$

$$< -\sum_{\substack{k=1 \\ k \neq j}}^{n+m'+s'} \left\{ \left| \phi_{\underline{\underline{=}}}^{jk} - k_\perp^2 \delta_\perp^{jk} - (k_\|(3))^2 \delta_\|^{jk} + \left(\frac{4\pi}{\varepsilon}\right) \bar{m}_{jj} \nu_j^0 z_k \right| \right\} \tag{A.10}$$

in which expressions the field dependence is implicit, as might be expected from the discussion in Section III. In the above, of course, the term $\phi_{\equiv}^{jk}(\nu^0; \{\beta_\mu^0\})$ denotes the jkth element of the Jacobian $\mathbf{F}_{\equiv}(\nu^0; \{\beta_\mu^0\})$ of the purely reactive flux. The field dependence of (A.10) stems, obviously, from the facts that (a) the accessible wavenumbers $k_\perp(1)$, $k_\perp(2)$ and $k_\parallel(3)$ must belong to $\{\vec{\mathbf{k}}_c(E_\parallel)\}$ and (b) the ratios $(\delta_\parallel^{jk}/\delta_\perp^{jk})$ are dependent on $E_\parallel$. The use of the above "nonspectral" condition for the induction of instabilities (soft mode as well as hard mode) is best illustrated by means of an example. Let us then take the case, say, of $n = 5$, $m' = 3$ and $s' = 2$; with the signs of the charge valences being taken as

$$\{\operatorname{sgn} z_k\} = (0, +, +, -, 0), \qquad k = 1, \ldots, 5$$

$$\{\operatorname{sgn} z_k\} = (-, -, +), \qquad k = 6, \ldots, 8$$

$$\{\operatorname{sgn} z_k\} = (+, -), \qquad k = 9, 10$$

Denote by $\mathcal{M}_{\text{diag}}^j(\vec{\mathbf{k}}; \varepsilon, z_j, E_\parallel)$ the quantity

$$
\begin{aligned}
&\mathcal{M}_{\text{diag}}^j(\vec{\mathbf{k}}; \varepsilon, z_j, E_\parallel) \\
&= \max_{\vec{\mathbf{k}} \in \{\vec{\mathbf{k}}_c(E_\parallel)\}} \left[\phi_{\equiv}^{jj} - k_\perp^2 \delta_\perp^{jj} - (k_\parallel(3))^2 \delta_\parallel^{jj} + \left(\frac{4\pi}{\varepsilon}\right) \bar{m}_{jj} \nu_j^0 z_j \right] \quad \text{(A.11a)}
\end{aligned}
$$

A necessary condition for stability is that

$$\mathcal{M}_{\text{diag}}^j(\vec{\mathbf{k}}; \varepsilon, z_j, E_\parallel) \le 0, \qquad j = 1, \ldots, 10 \quad \text{(A.11b)}$$

subject to the reference state electroneutrality condition

$$\sum_{i=1}^{10} \nu_j^0 z_j = 0 \quad \text{(A.11c)}$$

A more than sufficient condition for an instability is easily inferred, namely that

$$\mathcal{M}_{\text{diag}}^j(\vec{\mathbf{k}}; \varepsilon, z_j, E_\parallel) > 0$$

for any j, subject to (A.11c). Assume, merely for the sake of illustration, that we can neglect the off-diagonal elements of the diffusion matrices $\mathbf{D}_\perp$ and $\mathbf{D}_\parallel$. Define by $\mathcal{M}_{\text{off}}^j(\{z_k\}, \varepsilon)$ the quantity

$$\mathcal{M}_{\text{off}}^j(\{z_k\}, \varepsilon) = -\sum_{\substack{k=1 \\ k \ne j}}^{10} \left| \phi_{\equiv}^{jk} + \left(\frac{4\pi}{\varepsilon}\right) \bar{m}_{jj} \nu_j^0 z_k \right| \quad \text{(A.12a)}$$

$k = 1, \ldots, 10$. Then a sufficient set of conditions for stability consists of (A.11c) together with

$$\mathcal{M}^j_{\text{diag}}(\vec{\mathbf{k}}; \varepsilon, z_j, E_\parallel) < \mathcal{M}^j_{\text{off}}(\{z_k\}, \varepsilon) \qquad \text{(A.12b)}$$

with (A.11b) of course being automatically satisfied.

We take note of two further observations concerning this example. The most obvious is that $\mathcal{M}^j_{\text{diag}}(\vec{\mathbf{k}}; \varepsilon, z_j, E_\parallel)$ is independent of the charge valences z_j for $j = 1$ and $j = 5$, and only the interplay between translational diffusion and the chemical reactions therefore, determine the sign of this quantity for the subset of indices $\{1, 5\}$. Second, we note that $\mathcal{M}^j_{\text{off}}(\{z_k\}, \varepsilon)$ is independent of $\phi^{jk}_{\equiv}$, the jkth element of the Jacobian of the purely reactive flux, for $j = 6, \ldots, 10$, since for this particular subset of indices we have $\phi^{jk}_{\equiv}$ vanishing identically. Thus, owing to the additional fact that for the present example z_6, z_7 and z_{10} are negative, one finds that (A.11b) is always satisfied for this subset of indices. Further analysis of the example is left to the interested reader.

APPENDIX 5. PARADIGMS AND PERSPECTIVES

The composition of the foregoing was completed in the early spring of 1977, although earlier preprints of the main results and experimental predictions were continuously available and were circulated since the summer of the previous year. In this appendix I wish to narrow as much as possible whatever distance may still remain between the analytical predictions deduced and presented for the first time above and the experimental paradigms which can substantiate them. Some of the comparisons discussed in this appendix between theoretical results and recent experimental findings will be pursued much further and in greater detail in later papers. I have found it a great help that in the meantime the survey by Jaffe and Nuccitelli has appeared[48] which must surely figure as the most successful effort to date in gathering together and imposing order on the main—and largely heterogeneous—corpus of experimental results and problems in the area of electrical controls in developmental biology.

I should like to start by presenting a detailed recapitulation of the analytical results of the present paper. As mentioned earlier, it can be deduced *self-consistently* that as a consequence of the switching on of a longitudinal electric field the effect of combined reactive and charge fluctuations on the induction of symmetry-lowering instabilities can be quite radically different and, as a consequence, it should be possible to field-induce certain soft (s-mode) and hard (h-mode) spatiotemporal patterns that can otherwise not be made to emerge in the absence of the field. The underlying reason is the transition of the $\boldsymbol{\Omega}$-operator from real to complex as the field is switched on. For it is known from the theory of analytic perturbations of operators that the eigenvalues in the spectrum of an operator such as $\Omega = \Omega_0 + iE_\parallel \Omega_1$ are, in general, multiple-valued branches of analytical functions of $iE_\parallel$. As the electric field is switched on, the eigenvalues of Ω_0, the zero-field operator, can split and coalesce in a very complicated manner; and, in fact, if the algebraic multiplicities are not taken into account, the number of eigenvalues change as $E_\parallel$ is varied—and so would the number of *distinct* (nondegenerate) eigenmodes.

In the case where an external field is applied to a reacting system wherein a self-coherent spatiotemporal pattern is already present even before the field is switched on, the result could be any of the following:

1. The switching on shifts the spectrum of the resultant (complex) Ω-operator to the left of the imaginary axis, and there is no self-coherent large-scale pattern for the given field strength. This can happen only fortuitously.
2. The "same" critical mode (as the one for $E_\parallel = 0$) remains the critical mode under conditions of finite $E_\parallel$ and the basic spatiotemporal pattern remains the same. This would be an unlikely occurrence but cannot be entirely ruled out. And indeed it becomes possible if the corresponding eigenvalue belonging to the zero-field critical mode does not split off on switching on the field. We take note again that the modulus $|\xi_i(0) - \xi_i(iE_\parallel)|$ of an eigenvalue that does not become degenerate as the field is switched on has the value $|\xi_i(0) - \xi_i(iE_\parallel)| \sim 0(|iE_\parallel|) = 0(|E_\parallel|)$. This can be shown to follow from the analytic perturbation theory of finite dimensional linear operators.
3. The original critical mode is replaced by a different eigenmode, so that a completely new self-coherent pattern is established for the given finite $E_\parallel$.

Thus in brief: upon switching on an external field, any new structure or pattern that emerges—s-mode or h-mode—can either be one which was essentially preexisting but now field modulated (corresponding to Case 2), which we shall more conveniently refer to as of Type I; or, there may emerge a self-coherent pattern which is completely field induced (corresponding to Case 3), which we shall conveniently refer to as of Type II. Type II coherent patterns would be *much likelier* to emerge in actual chemical systems because their occurrence requires no stringent precondition, like the one which requires the zero-field critical eigenvalue to remain *both* critical and nondegenerate on switching on the field—the latter condition being necessary for Type I patterns to persist under conditions of external polarization.

The bounds on the imaginary part of the spectrum of the Ω-operator are given by the lower and upper spectral bounds $\xi_h^0(E_\parallel; \{\beta_\mu^0\})$ and

$\xi_h^{00}(E_\parallel; \{\beta_\mu^0\})$ of the hermitian operator Im $(\mathbf{\Omega}) = (1/2i)(\mathbf{\Omega} - \mathbf{\Omega}^\dagger)$ as given before. If ξ_h^0 and ξ_h^{00} both have the same sign (either both positive or both negative), then only h-mode patterns are in fact possible under polarization. We note again that if an h-mode pattern is induced through the imposition of an external field (with or without a concomitant external modulation of $\{\beta_\mu^0\}$), the resulting frequency of oscillations in each spatially repeating unit is *field dependent*. This is clear since ξ_h^0 and ξ_h^{00} are spectral bounds of an explicitly field-dependent hermitian operator. This frequency is of course determined by the full interplay of chemical, ionic, and transport properties, in addition to the electric field intensity. It has already been indicated, merely in the way of a simple-minded illustration, that in the chance event that the $\mathbf{\Omega}_\lambda$-operator defined in Section II is globally diagonalizable for all complex λ, then the representation (7) exhibits most transparently how the electric field intensity would parametrize—keeping in mind that sp $(\mathbf{\Omega}_1)$ is entirely real—the threshold frequency of oscillations of either Type I or Type II h-mode patterns. (It was also noted before that if, fortuitously, the operator should indeed be globally diagonalizable, s-mode patterns cannot arise when the field is switched on if the condition $n = n'$ is met.)

In both cases therefore—whether a self-coherent pattern is preexistent (Type I) or completely field induced (Type II)—the present analysis allows one to infer *three principal large-scale effects* of the imposition of an external field:

E1.　In the case of an h-mode pattern, a shift in the frequency of oscillations can be made to occur—a frequency shift that can be continuously modulated by varying the field strength. This assertion follows from the form of the operator Im $(\mathbf{\Omega})$.

E2.　The characteristic size itself of any spatially repeating pattern can be modulated—but, in general, not continuously—by varying the field strength. This assertion follows from the discreteness of the set $\{\vec{\mathbf{k}}_c(E_\parallel)\}$ of the compatible wave vectors.

E3.　In the case of a self-coherent pattern that has spatio-*temporal* dependence, a distinct asymmetric effect will be exhibited by simple *polarity reversal* of the field. This will take the form of an asymmetric frequency shift of the oscillations (for anodal and cathodal polarizations) for the same absolute value of the field strength—unless, through a most singular happenstance, the consequent shift in the compatible wave vectors $\{\vec{\mathbf{k}}_c(E_\parallel)\} \to \{\vec{\mathbf{k}}_c(-E_\parallel)\}$ dictates a parallel shift in the sign of the longitudinal wavenumber $k_\parallel(3)$ *without* a change in its absolute value. Anodal and cathodal frequency shifts are therefore direct consequences of the asymmetric spectral shifts in the operator Im $(\mathbf{\Omega})$ upon polarity reversal. These asymmetric shifts will cellular entity) under consideration—and depends on the very complicated interplay of several factors: (a) The signs of the charge valences $\{\text{sgn } z_k\}$, $k = 1, \ldots, (n + m' + s')$, and their absolute magnitudes relative to each other. (b) The types of chemical reactions occurring and the consequent nonlinearities of the purely chemical flux, and hence of the elements of the Fréchet derivative $\mathbf{F}_{\equiv}(\nu^0; \{\beta_\mu^0\})$. ($c$) The diagonal and off-diagonal diffusion coefficients of the components and their respective magnitudes (transverse and longitudinal) relative to each other. (d) The levels at which the number densities (concentrations) of the externally buffered reactants $\{\beta_\mu^0\}$, $\mu = 1, \ldots, m$, are maintained constant. (e) Any restrictions (not necessarily anisotropic) on the electrophoretic mobilities $\bar{m}_{jj}$, $j = 1, \ldots, (n + m' + s')$ and their magnitudes relative to each other. (f) The overall size (L_1, L_2, L_3) of the reaction system, which determines, along with the magnitude an direction of the electric field, what wavenumbers $k_\perp(1)$, $k_\perp(2)$, $k_\parallel(3)$ will be accessible to and compatible with the reacting system under polarization. (g) The magnitude and direction of the electric field or, more accurately, of the complex quantity $iE_\parallel$—the eigenvalues of the Ω-operator being multiple-valued branches of analytic functions of this quantity. (h) The presence of structural anisotropies (as well as the configuration and geometry of the polarizing electrodes) which cause a permanent warping of the field lines of the external electric field thus inducing a response to polarity reversal that is markedly more asymmetric.

Thus it has been observed in the same series of experiments cited earlier[62] that certain batches of *Pelvetia* eggs grew "anomalously" toward the cathodal rather than the (expected) anodal side of an imposed electric field. I wish to suggest that this particular observation has to be taken in the light of the arguments of the previous paragraph. On the other hand, in the growth of cultured embryonic neural cells under the influence of an external field, the "preferred" growth direction, insofar as is now experimentally known,[63,48] is cathodal. Studies involving explants of embryonic chick medulla and dorsal root ganglia, respectively, have been reported in the last cited references. In the former paper, for field strengths above 0.5 V/cm, nerve outgrowths tending in the direction of the anode were inhibited in their growth while the outgrowths tending toward the cathode proliferated with uncommon vigor; and in fact curved growth toward the cathode was even observed for initially opposite-facing oblique growth. In the latter and more recent work, the threshold value of 0.5 V/cm of a longitudinal field was essentially verified—together with the fact that nerve outgrowths from the explant growing toward the anode were entirely suppressed, or even caused to retract, while the opposite-facing ones were greatly favored, and flourished and spread longitudinally parallel to the field. But a crucially important experimental observation[48] stands out—

diencephalon and rostral midbrain elicits, and, in particular, that of the hypothalamic area itself. Such behavioral patterns have been particularly well studied in the cat, where, in accordance with by now classic experiments,[69] the stimulation of the lateral and posterior hypothalamus gives rise to a rage pattern of response possessing both motor and visceral components. In fact, a directed attack pattern is usually the final outcome,[70,71] a pattern of behavior possessing, in addition, certain motivational properties.[72] A little reflection will show however that these facts will not prove to be experimental hindrances at all and might well prove themselves to be blessings in disguise. For since the paradigm is proposed mainly to show the *asymmetrical* effect of hypothalamic polarization, it would in fact be equally valid and perhaps simpler and more convenient to study the asymmetric effects of polarity reversal on the components of the rage pattern of response themselves. This would then reduce the observational discrimination between the effects of anodal and cathodal hypothalamic polarization to the same plane (i.e., the behavioral one) on which the discrimination between the effects of anodal and cathodal cortical polarization is customarily made. Further details will be given in a forthcoming paper.

I indulge in a final comment. The reader may perhaps have noted that in the formalism developed here I have neglected to speak of the *indirect* field effects on the equilibration (i.e., on the K_{eq}) of the chemical transformations constituting the reactive process. This is of course because in the *presence* of the field one cannot strictly speak of thermodynamic equilibration of those chemical reactions involving charged reactants and/or products—notwithstanding the fact that the notion of a (pulsed) *shift* in K_{eq} has been found useful in the study of the second Wien effect by Onsager. But as will be discussed, in somewhat more detail, in a later paper where I deal with anisotropically warped fields, this neglect in considering expressly the shift in K_{eq} for the moderate field strengths (which are of sole interest to us in this paper) is fully justified, whether charged or merely polarizable reactants (and products) are involved. In the later paper too will be discussed polyelectrolyte effects on K_{eq} shifts.

References

1. S. B. Dubin, J. H. Lunacek, and G. B. Benedek, *Proc. Natl. Acad. Sci. (USA)*, **57**, 164 (1967).
2. G. H. Wannier, *Bell Syst. Tech. J.*, **32**, 170 (1953).
3. E. B. Wagner, F. J. Davis, and G. S. Hurst, *J. Chem. Phys.*, **47**, 3138 (1967).
4. P. Debye, and A. M. Bueche, *J. Chem. Phys.*, **16**, 573 (1948).
5. J. J. Hermans, *J. Polymer Sci.*, **18**, 527 (1955).
6. W. O. Portmann, *Proc. Am. Math. Soc.*, **11**, 97 (1960).

7. T. S. Motzkin and O. Taussky, *Trans. Am. Math. Soc.*, **80,** 387 (1955); see also the earlier article in the series.

8. P. Glansdorff and I. Prigogine, *Thermodynamic Theory of Structure, Stability, and Fluctuations*, Wiley–Interscience, New York, 1971.

9. G. Nicolis and I. Prigogine, *Self-Organization in Nonequilibrium Systems*, Wiley, New York, 1977.

10. A. D. Nazarea, *Proc. Natl. Acad. Sci. (USA)*, **71,** 3751 (1974).

11. A. D. Nazarea, *J. Theor. Biol.*, **64,** 311 (1977).

12. M. Lax, *Rev. Mod. Phys.*, **32,** 25 (1960).

13. P. Mazo, *J. Chem. Phys.*, **52,** 3306 (1970).

14. J. S. Kirkaldy, D. Wiechert, and Zia-Ul-Haq, *Can. J. Phys.*, **41,** 2166 (1963).

15. D. K. Kondepudi and A. D. Nazarea, *Biophys. Chem.*, **8,** 71 (1978).

16. R. J. Field, E. Körös, and R. M. Noyes, *J. Am. Chem. Soc.*, **94,** 8649 (1972).

17. B. Hess and A. Boiteux, *Annu. Rev. Biochem.*, **40,** 237 (1971).

18. A. Goldbeter and S. R. Caplan, *Annu. Rev. Biophys. Bioeng.*, **5,** 449 (1976).

19. A. Goldbeter and R. Lefever, *Biophys. J.*, **12,** 1302 (1972).

20. A. Goldbeter and G. Nicolis, *Prog. Theor. Biol.*, **4,** 65 (1976).

21. E. Bünning, *The Physiological Clock*, Springer-Verlag, New York, 1973; M. Menaker, Ed., *Biochronometry*, National Academy of Sciences, Washington, 1971; J. Aschoff, Ed., *Circadian Clocks*, North-Holland, Amsterdam, 1965; A. Chovnick, Ed., *Biological Clocks (Cold Spring Harbor Symposium, Vol. XXV)*, Long Island Biological Association, New York, 1961.

22. A. Boiteux, A. Goldbeter, and B. Hess, *Proc. Natl. Acad. Sci. (USA)*, **72,** 3829 (1975).

23. B. M. Sweeney, *Int. J. Chronobiol.* **2,** 25 (1974).

24. D. Njus, F. Sulzman, and J. W. Hastings, *Nature*, **248,** 116 (1974).

25. S. Løvtrup and A. Pigon, *J. Embryol. Exp. Morphol.*, **6,** 486 (1958).

26. B. Novak and C. Sironval, *Plt. Sci. Lett.*, **5,** 183 (1975).

27. J. I. Hubbard, *Progr. Biophys. Molec. Biol.*, **21,** 33 (1970).

28. J. L. Barker, *Physiol. Rev.*, **56,** 435 (1976).

29. V. Rowland and R. Blumenthal, Eds., "Dynamic Patterns of Brain Cell Assemblies (Report on a Neurosciences Research Program Work Session)", *Neurosci. Res. Progr. Bull.*, **12,** 132 (1974).

30. R. J. Gavalas, D. O. Walter, J. Hamer, and W. R. Adey, *Brain Res.*, **18,** 491 (1970).

31. J. Aschoff, *Science*, **148,** 1427 (1965).

32. F. Halberg, *Annu. Rev. Physiol.*, **31,** 675 (1969).

33. D. T. Krieger, in L. W. Hedlund, J. M. Franz, and A. D. Kenny, Eds., *Biological Rhythms and Endocrine Function*, Plenum, New York, 1975, p. 169.

34. I. Hanin, R. Massarelli, and E. Costa, *Science*, **170,** 341 (1970).

35. D. J. Reis, M. Weinbren, and A. Corvelli, *J. Pharmacol. Exp. Ther.*, **164,** 135 (1968).

36. D. J. Reis, A. Corvelli, and J. Conners, *J. Pharmacol. Exp. Ther.*, **167,** 328 (1969).

37. D. T. Krieger, A. I. Silberberg, F. Rizzo, and H. P. Krieger, *Am. J. Physiol.*, **215,** 959 (1968).

38. D. T. Krieger and F. Rizzo, *Am. J. Physiol.*, **217,** 1703 (1969).

39. D. Mazia, in J. Brachet and A. E. Mirsky, Eds., *The Cell*, Vol. VIII, Academic, New York, 1961, p. 80.

40. T. Gustafson, in A. A. Moscona and A. Monroy, Eds., *The Biochemistry of Animal Development*, Vol. II, Academic, New York, 1965, p. 1.

41. A. Monroy, in M. Florkin and E. H. Stotz, Eds., *Comprehensive Biochemistry*, Vol. 28, Elsevier, Amsterdam, 1967, p. 1.

42. W. Herth and K. Sander, *Wilhelm Roux' Arch. Entwicklungsmech. Org.*, **172,** 1 (1973).

43. R. Hori, *Embryologia*, **9,** 34 (1965).
44. I. Giaver, *J. Immunol.*, **110,** 1424 (1973).
45. E. E. Uzgiris and J. H. Kaplan, *Rev. Sci. Instrum.*, **45,** 120 (1974).
46. F. Kohlrausch, *Ann. Phys.*, **1,** 132 (1900).
47. B. Sz.-Nagy, *Commun. Math. Helv.*, **19,** 347 (1947); *Acta Sci. Math.*, **14,** 125 (1951).
48. L. F. Jaffe and R. Nuccitelli, *Ann. Rev. Biophys. Bioeng.*, **6,** 445 (1977).
49. K. R. Robinson and L. F. Jaffe, *Science*, **187,** 70 (1975).
50. L. F. Jaffe, K. R. Robinson, and B. F. Picologlu, *J. Theor. Biol.*, **45,** 593 (1974).
51. L. F. Jaffe, K. R. Robinson, and R. Nuccitelli, *Ann. NY Acad. Sci.*, **238,** 372 (1974).
52. P. R. Baker, *Progr. Biophys. Mol. Biol.*, **24,** 185 (1972).
53. M. J. Kushmerick and R. J. Podolsky, *Science*, **166,** 1297 (1969).
54. P. Baker and A. C. Crawford, *J. Physiol.*, **227,** 855 (1972).
55. D. J. Woodward, *J. Gen. Physiol.*, **52,** 509 (1968).
56. S. W. De Laat, W. Wouters, M. M. Marquez da Silva Pimenta Guarda, and M. A. da Silva Guarda, *Exp. Cell. Res.*, **91,** 15 (1975).
57. G. B. Moment, *J. Exp. Zool.*, **112,** 1 (1949).
58. I. Kurtz and A. R. Schrank, *Physiol. Zool.*, **28,** 322 (1955).
59. R. B. Borgens, J. W. Vanable, Jr., and L. F. Jaffe, *Proc. Natl. Acad. Sci. (USA)*, **74,** 4528 (1977).
60. R. B. Borgens, J. W. Vanable, Jr. and L. F. Jaffe, *J. Exp. Zool.*, **200,** 403 (1977).
61. B. Novak and F. W. Bentrup, *Planta*, **108,** 227 (1972).
62. H. B. Peng and L. F. Jaffe, *Dev. Biol.*, **53,** 227 (1976).
63. G. Marsh and H. W. Beams, *J. Cell. Comp. Physiol.*, **27,** 139 (1946).
64. L. F. Jaffe and R. Nuccitelli, *J. Cell Biol.*, **63,** 614 (1974).
65. R. Nuccitelli, M.-m. Poo, and L. F. Jaffe, *J. Gen. Physiol.*, **69,** 743 (1977).
66. H. H. Ussing, *Harvey Lectures*, **59,** 1 (1964).
67. R. O. Becker, *Nature*, **235,** 109 (1972).
68. S. D. Smith, *Ann. NY Acad. Sci.*, **238,** 500 (1974).
69. W. R. Hess and M. Brügger, *Helv. Physiol. Pharmacol. Acta*, **1,** 33 (1943).
70. R. W. Hunsperger, *Helv. Physiol. Pharmacol. Acta*, **14,** 70 (1956).
71. M. Wasman and J. P. Flynn, *Arch. Neurol.*, **6,** 220 (1962).
72. W. W. Roberts and H. O. Kiess, *J. Comp. Physiol. Psychol.*, **58,** 187 (1964).

AUTHOR INDEX

Numbers in parentheses are reference numbers and indicate that the author's work is referred to although his name is not mentioned in the text. Numbers in italics show the pages on which the complete references are listed.

451

SUBJECT INDEX